Bioenergy

Biomass to Biofuels

Bioenergy
Biomass to Biofuels

Edited By

Anju Dahiya

AMSTERDAM • BOSTON • HEIDELBERG • LONDON • NEW YORK • OXFORD
PARIS • SAN DIEGO • SAN FRANCISCO • SINGAPORE • SYDNEY • TOKYO

Academic Press is an imprint of Elsevier

ACADEMIC
PRESS

Academic Press is an imprint of Elsevier
32 Jamestown Road, London NW1 7BY, UK
525 B Street, Suite 1800, San Diego, CA 92101-4495, USA
225 Wyman Street, Waltham, MA 02451, USA
The Boulevard, Langford Lane, Kidlington, Oxford OX5 1GB, UK

Notices
Knowledge and best practice in this field are constantly changing. As new research and experience broaden our
understanding, changes in research methods, professional practices, or medical treatment may become
necessary.

Practitioners and researchers must always rely on their own experience and knowledge in evaluating and using
any information, methods, compounds, or experiments described herein. In using such information or methods
they should be mindful of their own safety and the safety of others, including parties for whom they have
a professional responsibility.

To the fullest extent of the law, neither the Publisher nor the authors, contributors, or editors, assume any
liability for any injury and/or damage to persons or property as a matter of products liability, negligence or
otherwise, or from any use or operation of any methods, products, instructions, or ideas contained in the
material herein.

ISBN: 978-0-12-407909-0

Library of Congress Cataloging-in-Publication Data
Dahiya, Anju, author.
    Bioenergy: biomass to biofuels / Anju Dahiya, Plant & Soil Science, Jeffords Hall, Burlington, VT.
        pages cm
    ISBN 978-0-12-407909-0 (hardback)
    1. Biomass energy. I. Title.
TP339.D34 2015
662'.88–dc23

                                    2014035313

British Library Cataloguing in Publication Data
A catalogue record for this book is available from the British Library

For information on all Academic Press publications
visit our web site at http://store.elsevier.com

Working together
to grow libraries in
developing countries

www.elsevier.com • www.bookaid.org

Dedicated to the man who was ahead of his time and lived
with a very big heart, who approached life with strength, dignity, compassion,
and above all with selfless love.
Who changed many lives for the better!

To My dear father

Jaikishan Dahiya

Papa, here's a secret note to you (☺):

2011 Father's Day was on June 19th around when the first time I shared
my intension of putting together this book. The following year, father's day
was so empty for me
I did not know how to wish you—with tears or gratitude. I somehow managed to
meditate on this book and formalized the agreement. It is June 15th today,
which is father's day—I've finally completed the final touches on this book
and it is ready to fly.

I can only imagine how proud you would have felt holding this book in your hands.
(I know in spirit you will!) And after looking through it you would have said,
"So nice!" (and after a pause) "What's next?"

Hmm... My answer would be, "how's the technology of growing fuel sounds?"
You would give a thumbs up!

Cheers Pa!
And yes, Happy Father's Day 2014!
- Anju

Contents

PART I BIOENERGY—BIOMASS TO BIOFUELS: AN OVERVIEW
Part Introduction by Anju Dahiya

PART II WOOD AND GRASS BIOMASS AS BIOFUELS
Part Introduction by Anju Dahiya

PART III BIOMASS TO LIQUID BIOFUELS
Part Introduction by Anju Dahiya

PART IV GASEOUS FUELS AND BIOELECTRICITY
Part Introduction by Anju Dahiya

PART V CONVERSION PATHWAYS FOR COST-EFFECTIVE BIOFUEL PRODUCTION

Part Introduction by Anju Dahiya

PART VI BIOFUELS ECONOMICS, SUSTAINABILITY, ENVIRONMENTAL AND POLICY

Part Introduction by Anju Dahiya

PART VII QUIZZES

List of Contributors

Chapter Authors
Bioenergy – Biomass to Biofuels Program Student Contributors
Contributing Organizations
Editor and Author

CHAPTER AUTHORS

Part I Bioenergy - Biomass to Biofuels: an Overview

Dr. Carol L. Williams
Research Scientist
Wisconsin Energy Institute
University of Wisconsin – Madison

Pam Porter
Midwest Office Director
Environmental Resource Center,
University of Wisconsin – Madison

Dr. Anju Dahiya

Part II Wood and Grass Biomass as Biofuels

Dr. William G. Hubbard
Southern Regional Extension Forester
Southern Regional Extension Forestry, The University of Georgia

Dr. Sidney C. Bosworth
Professor, Agronomist
Plant and Soil Science Department, UVM Extension, University of Vermont

Part III Biomass to Liquid Biofuels

Dennis Pennington
Senior Bioenergy Educator, Michigan State University Extension

Dr. Heather Darby
Associate Professor, University of Vermont Extension

Christopher W. Callahan
Engineer / Extension Assistant Professor of Agricultural Engineering
University of Vermont Extension

Dr. Eric L. Garza
Lecturer
University of Vermont

Dr. Mark Hanna
Extension Agricultural Engineer
Ag & Biosystems Engineering, Iowa State University

Scott Sanford
Sr. Outreach Specialist
Rural Energy Program
Biological Systems Engineering, University of Wisconsin-Madison

Dr. Nathan S. Mosier
Associate Professor Agricultural & Biological Engineering/Lorr
Department of Agricultural and Biological Engineering, Purdue University

Robert G. Hedden
NORA Director of Education, National Oilheat Research Alliance (NORA)
Efficient Heating Consultant and Educator

Dr. Anju Dahiya

Part IV Gaseous Fuels and Bioelectricity

Dr. Robert G. Jenkins
Professor Emeritus CEng,
The School of Engineering, University of Vermont

Dr. Klein E. Ileleji
Associate Professor of Agricultural & Biological Engineering
Associate Professor of Mechanical Engineering (by Courtesy)
Extension Engineer in Agricultural & Biological Engineering
Department of Agricultural & Biological Engineering, Purdue University

Chad Martin
Renewable Energy Extension Specialist
Department of Agricultural & Biological Engineering
Purdue University

Dr. Don Jones
Professor and Extension Agricultural Engineer
Department of Agricultural and Biological Engineering
Purdue University

M. Charles Gould
Extension Educator-Agricultural Bioenergy and Energy Conservation
Agriculture and Agribusiness Institute, Michigan State University

Part V Conversion Pathways for Cost-Effective Biofuel Production

Rudy Pruszko
Biofuels Consultant,
Dubuque, Iowa

Sean M. McCarthy
Department of Chemistry, University of Vermont

Jonathan H. Melman
Department of Chemistry, University of Vermont

Omar K. Reffell
Department of Chemistry, University of Vermont

Dr. Scott W. Gordon-Wylie
Former Chemistry Professor, Department of Chemistry, University of Vermont
Founder & CEO, Green Technologies Inc.

Dr. Anju Dahiya

Dr. Lieve M.L. Laurens
Senior research scientist,
Bioprocess Research and Development group.
National Renewable Energy Laboratory, Department of Energy

Nathan S. Mosier
Associate Professor
Department of Agricultural and Biological Engineering
Purdue University

Dr. Klein E. Ileleji

Dr. Jonathan R. Mielenz
Former Senior Staff Scientist & Group Leader
Bioconversion Research Group, Bioscience Division,
Oak Ridge National Laboratory (ORNL)
President and Chief Scientific Officer, White Cliff Biosystems Co.
Director at the Society for Industrial Microbiology and Biotechnology

Brennan Pecha
Doctoral Candidate
Biological Systems Engineering, Washington State University

Dr. Manuel Garcia-Perez
Assistant Professor/Scientist, Adjunct Faculty
Biological Systems Engineering, Washington State University

Richard Altman
Executive Director Emeritus
Commercial Aviation Alternative Fuel Association

Dr. Anju Dahiya

Part VI Biofuels Economics, Sustainability, Environmental and Policy

Dr. Wallace E. Tyner
James and Lois Ackerman Professor
Department of Agricultural Economics, Purdue University

Dr. Bob Parsons
Professor - Extension Ag Economist
Department of Community Development and Applied Economics
University of Vermont Extension

Matt Cota
Executive Director
Vermont Fuel Dealers Association

Daniel Ciolkosz
Penn State Biomass Energy Center and Department of Agricultural and
Biological Engineering,
The Pennsylvania State University

F. John Hay
Extension Educator – Energy
Department of Biosystems Engineering,
University of Nebraska Lincoln Extension

Samuel Gorton
Process Engineer, R3 Fusion Inc.
Intern Engineer, New Cycle Bioneering
Doctoral Student, Gund Institute for Ecological Economics, University of Vermont

Jason McCune-Sanders
Environmental Engineer,
Principal, Forest Enterprises LLC
Doctoral Candidate, Department of Engineering, University of Vermont

Dr. Anju Dahiya

BIOENERGY – BIOMASS TO BIOFUELS PROGRAM STUDENT CONTRIBUTORS

Wood and Grass Energy Related Service Learning Projects & Case Studies:

Tom Tailer
Executive Director
Vermont Sustainable Heating Initiative

Ron McGarvey
Former Director, Residential Energy Services
Efficiency Vermont

Heather M. Snow
Founder, Addie's Acres, Vermont

Liquid Biofuels Related Service Learning Projects & Case Studies:

Chuck Custeau
Loan Officer
Yankee Farm Credit, Vermont

Ethan Bellavance
Energy Consultant, Engineering
Efficiency Vermont

Tracey McCowen
Doctoral candidate, University of Vermont

Gaseous Fuels Related Service Learning Projects & Case Studies:

Samantha Csapilla
Technician, Avatar Energy LLC

Grant Troester
Student, University of Vermont

Adam Riggen
Student, University of Vermont

Ariadne Brancato
Student, University of Vermont

Deandra Perruccio
Graduate student,
Community Development and Applied Economics, University of Vermont

Fuel Conversion Related Service Learning Projects & Case Studies:

Richard O. Barwin
Animal Science teacher, High School, Vermont

William R. Riggs
Environmental Studies Student, University of Vermont

John C. O'Shea
Environmental Studies Student, University of Vermont

Thomas G Joslin
Environmental Engineer, Vermont Department of Environmental Conservation

Vanessia B. Lam
Student, University of Vermont

Richard P. Smith
Engineering student
University of Vermont

CONTRIBUTING ORGANIZATIONS

National Biodiesel Board
National Renewable Energy Laboratory (NREL). US Department of Energy
Office of Energy Efficiency and Renewable Energy, U.S. Department of Energy

EDITOR AND AUTHOR

Dr. Anju Dahiya
Bioenergy Instructor, University of Vermont
President and Chief Science Officer, GSR Solutions LLC

Foreword

When giving a talk about advanced biofuels, more often than not, my first question to a group, "What do you think of when you hear the word, biofuel?" elicits "corn ethanol" as the primary response (if they have any idea about biofuels at all).

That answer certainly makes sense where the predominant biofuel *is* corn ethanol; however, it also illustrates the general public's limited understanding of key elements of a successful transition toward a renewable energy future, the multiplicity of feedstocks to make a diverse array of fuels using a variety of conversion technologies.

Thus, my excitement and eager anticipation have led to a text that covers a wide range of renewable energy topics including renewable transportation fuels; a text that mixes the science and technology, the research and development, with practical aspects of deployment, and, a text about biofuels that recognizes the importance of exploring all three legs of the sustainability stool: environmental, social, and economic.

Too often we forget that if everyone along the value chain cannot make a living, environmentally beneficial practices are doomed to fail. The world we live in today, our global society, relies on business interactions, even if they are financed by governments. We cannot rely on altruism, good works or even on enlightened public policies to effect a successful transition from fossil fuels to renewables. Economic incentives with entrenched policy preferences, in coal, gas, and oil underpin the world's energy systems. As so many involved with renewable energy have found out the hard way, competition in price and performance is key to a true transition to a more lasting sustainable energy future. Perhaps enlightened policy will follow.

Contributors to this text include some of the best-known names in the field. They span the topic from agricultural and forestry considerations related to production of feedstock through the science and engineering of conversion technologies to economic and policy issues which are crucial to assuring environmentally conscientious implementation choices.

The more you learn about energy alternatives, the more you know what you don't know. This text will provide valuable, practical information that can be applied to implementing renewable energy projects.

But it is not just for those doing hands-on implementation or for students in an academic setting. This text will help civic leaders, economic development professionals, farmers, growers, investors, fleet managers, reporters, and the general public engage in informed discussions. It will enhance understanding of the complexities related to choosing among competing feedstocks, technologies, and products. And I hope that it will provide a foundation to guide knowledgeable policy development.

I have been watching the advanced biofuels industry evolve since 2007. I know of no other text that focuses on bio-based energy as a transition to a truly sustainable energy future or that combines scientific detail with practical information for implementation along with attention to the big picture issues and benefits.

The world is changing fast. The questions answered in this text were not even asked a half decade ago: What is the difference between biodiesel and renewable diesel? What sugars derive from cellulose and hemicellulose? How should we use comparative life cycle analyses?

The information in this text was not a part of anyone's basic education. It fills a need not only for current students, but should also serve as a basic resource for anyone interested in an organized introduction to the language, feedstocks, technologies, and products in the bio-based renewable energy world, including transportation fuels.

We need a text that discusses the challenges of transitioning away from fossil fuel, not as barriers to change, but as issues worthy of serious study, research, and reflection. We need a text that provides a framework for elucidating the practical challenges, the benefits, opportunities, and potential pitfalls of moving to a renewable energy future. This is that text.

We know so much more about bio-based energy than we did 10 years ago. We owe it to ourselves to base our choices for tomorrow on that hard-won knowledge, not on uninformed fears and glib proclamations. This text is one tool to successfully engage in that effort.

Joanne Ivancic
Executive Director
Advanced Biofuels USA

Preface

Bioenergy: Biomass to Biofuels, a title like this comprising of all the possible related topics (biodiesel, biogas, gasification, ethanol, waste oil, wood, grass, oil-seed fuels, algal fuels, fuel sustainability, energy return, economics, etc.) could possibly be put together only with contributions from topic-specific highly accomplished experts engaged in education and practice. I tried to accomplish this colossal task by bringing together the leading bioenergy/biofuel experts nationwide from universities, the department of energy, nonprofit organizations, and businesses, who graciously contributed the respective chapters per their expertise. Also complementing the chapters, presented are the first-hand service learning experiences of students in the form of case studies and comments from participating in a well-established university level program. These participants actually partnered with bio-energy-related for-profit and nonprofit businesses, farms, and other community partners to take the learning experience to the next level by complementing the in-class learning with hands-on experience.

In a way the debut of this book was formalized back in 2010, although at that time I was not sure about the direction and shape that this book would eventually take. Joanne Ivancic (Executive Director, *Advanced Biofuel USA*) has been observing the development of advanced biofuels research and financing for more than 10 years. She noticed my Biomass to Biofuels program at the University of Vermont in Biofuels Digest[1], planned for that summer, and when the course was completed and relisted for the next offering, she asked me about the broad outreach of the course and future directions. I shared with her about anticipating to write a book to help develop and run programs like mine elsewhere to include materials, lesson plans, procedures, and tips for setting up the bioenergy-related community relationships (more in this interview[2] reproduced below).

This book is designed to be used by a reader of any background exploring bioenergy, biomass, or biofuels from any angle (definitions of these terms are provided in the Part I introduction). As described in *How to use this book (and specifically in the introductions to respective parts)*, it is organized into seven parts: *Part I: Bioenergy: Biomass to Biofuels: an overview; Part II: Wood and Grass Biomass as Biofuels; Part III: Biomass to Liquid Biofuels; Part IV: Gaseous fuels and Bioelectricity; Part V: Conversion Pathways for Cost-effective Biofuel Production; Part VI: Biofuels Economics, Sustainability, Environmental and Policy; and Part VII: Quizzes*. Each part starts with a part introduction and presents topic-specific chapters—each a self-contained information package making it possible for the reader to pick up any topic without going through a particular order in this book. The service learning project case studies under Parts II to V complement the chapters as they are built on hands-on experience working with the bioenergy farms, businesses who actually encounter real-world problems and shared the experience with student interns. This service learning component can be incorporated in any bioenergy-related teaching course designed anywhere by engaging the local bioenergy community partners.

Advancement in the bioenergy area has received quite a bit of criticism with the foremost argument being it is unsustainable option—due to the "food versus fuel" debate, land use, and carbon issues. Evidently, new cutting edge technologies have already paved new ways towards the sustainability

---

[1]http://www.biofuelsdigest.com/.
[2]http://advancedbiofuelsusa.info/university-of-vermonts-biomass-to-biofuels-course-brings-real-world-to-academia-and-visa-vers.

of renewable fuel options described in several chapters in this book. The above-stated criticism is also addressed by two bioenergy economists in this book: a well-referenced report from the Farm Foundation found that while some of the same factors that drove commodity and food price increases in 2008 are at work today, new and very different factors have also emerged[3]. One of the authors of this report, Prof. Wallace Tyner at the Department of Agricultural Economics at Purdue University, has authored the chapter *Biofuel Economics and Policy: The Renewable Fuel Standard the Blend Wall, and Future Uncertainties* in this book that provides an indication of what might happen under a wide range of possible changes to the current renewable fuel standards (RFS), which is important to the biofuels sector; however, because of the blend wall and perhaps other issues, the RFS has now come under increased attack. The extension Agriculture economist, Prof. Bob Parsons, Professor at the Department of Community Development and Applied Economics, University of Vermont, in his chapter asserts, "There are economic associations with the quantities and prices of various inputs, substitutes, complements, production scale, and consumption of direct and byproducts that place different but important perspectives on the use of biofuels."

A healthy dose of skepticism helps move an area forward in the right direction. Some of the bioenergy critics arguments revolve around the issues related with feedstock markets causing biofuel production to be uneconomical, and some of these groups have been making demonstrations against *all* bioenergy options with the primary intent towards marginalizing the first generation biofuels (food vs fuel debate and land use). There is another angle to this debate. As Ron Kotrba, editor of *Biodiesel* magazine[4] (May/June 2014 issue), quoted in his conversation with the National Biodiesel Board's chair, "there's a lot of talk about foregoing first-generation biofuels for next generation biofuels, and the most certain way to kill next-generation biofuels is to take a pause in current generation biofuels." In the same issue, Ron provides a convincing view (especially to the ethanol critics) in *Time has come Today* through the example of, "the co-location of biodiesel and ethanol production at ethanol plants benefiting from the sharing of the existing infrastructure to share process essentials, an in-house feedstock in distillers corn oil and use of ethanol rather than methanol for biodiesel reactions", and Ron further makes the point in *The myths & reality of biodiesel feedstock availability*, "conversely the most certain way to advance next generation biofuels and the feedstocks from which they will be manufactured is to continue to grow existing-generation biofuels, and on any given day, feedstocks could be tight, but over the course of a year markets basically adjust… markets have not indicated to us that we are anywhere close to those limits (of biofuel feedstock)."

Bioenergy: Biomass to Biofuels is a journey towards a new era of "growing fuel"—on the lines of perfecting agriculture. It is about time to channelize our positive efforts based on cutting edge technologies with all the information at our fingertips unlike the days of Rudolf Diesel who demonstrated his engine at the Paris World Exposition in the year 1900 by using "peanut oil"— way before the use of fossil fuels replaced the fuel choice of its inventor. We are over a century behind now and have a lot of catching up to do!

– **Anju Dahiya**

---

[3]http://www.farmfoundation.org/news/articlefiles/1742-FoodPrices_web.pdf
[4]http://www.biodieselmagazine.com/.

Advanced Biofuels USA Interview[5] reproduced here:

UNIVERSITY OF VERMONT'S BIOMASS TO BIOFUELS COURSE BRINGS REAL WORLD TO ACADEMIA AND VISA VERSA
SUBMITTED BY JOANNE ON JUNE 20, 2011—8:51 AM

by Joanne Ivancic (Advanced Biofuels USA) Whoever thought of inviting "degree and non-degree seeking students, farmers, budding entrepreneurs, teachers (interested in developing curriculum, or projects at school or college levels) and others" to a university course on biomass to biofuels?

Anju L. Dahiya, Ph.D., did in 2009 as research scientist at the Gund Institute for Ecological Economics; Rubenstein School of Environment and Natural Sciences; and president of General Systems Research LLC. And she got the students she hoped for.

A surprisingly diverse student population is not Dahiya's only innovation. The second year the course was offered with Continuing Education, as a biofuels instructor in the Plant and Soil Science Department, she incorporated a service learning component to bring in even more real-world experience.

THE BEGINNINGS

Prior to responding to a request for proposals from the Vermont Sustainable Jobs Fund, Dahiya had been thinking about how academic research, particularly her on-going algae-to-biodiesel research, fit into the biodiesel industry's needs; or into the broader developing biofuels industry.

Seizing the early 2009 $20,000-grant opportunity, she integrated these thoughts into action. She proposed a course that would provide a wide range of biofuels-related science and technology topics, with environmental, economic, social, and other issues related to biofuels. She planned to build into the lesson plans hands-on work in the field and tours to bioenergy businesses. To enhance the real-world value of the course, she enlisted the Continuing Education Department in encouraging students not seeking degrees to participate.

Her objective for students was that once they completed the course, they would know not just information about biofuels and the developing bioenergy industry, but they would understand real-life problems and situations encountered in the bioenergy industry; and they would also learn what they need to know to develop a biofuels or bioenergy business.

IMPLEMENTATION

As part of her work in the Plant and Soil Science Department at the University of Vermont in Burlington, and as president of General Systems Research, a local company conducting algae research, she had met many Vermonters involved in biofuels and bioenergy businesses and organizations. She tapped into those resources for experts to complement the university faculty. As it turned out, she instigated another innovation as she developed this course. The group of lecturers morphed into a consortium of experts once the grant was awarded and work intensified on the specifics.

---

[5]http://advancedbiofuelsusa.info/university-of-vermonts-biomass-to-biofuels-course-brings-real-world-to-academia-and-visa-versa.

They divided the subject matter into general areas of liquid biofuels (seed-based biodiesel, bioethanol, and algal biofuels), solid biofuels (from woody biomass and grasses), and biogas (particularly from manure and other farm waste).

The teaching team now includes ten university faculties along with four instructors from other organizations including three farms. A number of guest lecturers from community organizations and businesses will also speak.

During the second year, in addition to the classroom lectures, field trips, and an imaginary exercise in identifying and addressing a biofuels business challenge, students were required to participate in a service learning component. In collaboration with the Service Learning Office at the University of Vermont which contributed $500 for expenses, students volunteer with a "community partner" or other entity who has agreed to participate in this element of the biomass to biofuels course.

Instead of just imagining a business situation, students work with their community partners to identify a business project that they can work on over a 6-week period. They begin with a set of questions to get them to learn details about the business they chose. Their experience volunteering with the business or organization then informs their perspective and participation in the class lectures and field trips. In addition, they are required to write a report describing the project, the goals, work they did, what they learned and the results of their work. This is a major part of their evaluation with students taking the course for credit graded on somewhat stricter criteria than nondegree students.

In the past years, students have worked on farm-based biogas projects, with a wood chip gasifier, with oil-based biodiesel, with a hydrological compost heating system, etc. This past spring, two teachers took the course: one to bring the information and the learning/teaching experience back to his classroom; the other, a retiree who wants to set up a biofuels business. Farmers planning to develop biofuels businesses at their farms have also participated, as has someone interested in creating a wood pellet mill who used this course to understand and sort out problems while creating a business plan. Some of the class-related service learning has resulted in formal internships for students. Some are working on grant proposals of their own based on their work for this course.

An online discussion forum was also created to enable discussions in specific topic areas related to the coursework and service learning experiences. In addition, faculty from other parts of the university who are not formally involved in the course are available when questions arise either from class and online discussions or from the service learning experiences.

SHARING AND OUTREACH

Now that the beta testing is completed, Dr Dahiya is ready to increase her outreach for the course encouraged by student evaluations of the course including phrases like: "Course was helpful for me I learned a ton of new information—also gave me contacts for later in life." "I feel reinforced with knowledge." "This course has inspired me to protect biomass to make sure it is used conscientiously as biofuels." "Broadened my perspective of biofuels." "Will carry this knowledge with me in the future." "Everything I needed at this time."

When she started preparing the grant, Dahiya felt she was exploring new territory. As with so many other biofuels projects, this had never been done before, as far as she could tell. She wanted to bring together all the possible biofuels-related know-how she could to help students find direction in this area of interest. She now feels that her goals have been well met.

Lately, Dahiya has heard from many teachers in Vermont who would like to develop something like this for the secondary school level; another teacher who contacted her is putting together a similar college-level course.

Dahiya anticipates writing a book about the course including materials, lesson plans, procedures, and tips for setting up the community relationships. In the meantime, she will be happy to consult with others who want to use this as a model course to expand the understanding and development of biofuels and bioenergy among traditional and nontraditional university students. She can be reached through the University of Vermont Web site for the course. READ MORE[6] (Revised 6/27/11).

---

[6]http://advancedbiofuelsusa.info/university-of-vermonts-biomass-to-biofuels-course-brings-real-world-to-academia-and-visa-versa.

Acknowledgments

With a span of over two years of work to shape this work, it has been a tremendous learning experience and is my hope that this book deserves merit due to contributions from extremely talented and highly accomplished bioenergy experts and dedicated bioenergy students nationwide who kindly accepted my request to contribute and made this book happen—my special thanks and deep appreciation goes out to each one of the contributors for their dedication, hard work in writing, and revising the chapter texts.

My extra special thanks to Ellen Kahler and Netaka White (Vermont Sustainable Jobs Fund) for supporting the early stages of Biomass to Biofuels program at the University of Vermont that guided this book immensely. I am most grateful to my biomass to biofuels students for their help with the book design selection from the two available samples, and voting for the best design: Nicholas Allgaier, Caleb Atwood, Ethan Bellavance, Samuel Grubinger, Elias Lichtenstein, James MacLeish, Adam Riggen, Gregory Schab, Heather Snow, Grant Troester, and Frank Geier.

I want to thank Sharon Lezberg, (Environmental Resources Center University of Wisconsin, Madison) for introducing the team leaders to bioenergy program. Robert Brown (Iowa State University) for directing me to a topic-specific expert. Rebecca Theller (Purdue University) for providing the originals; Ray Albrecht (Northeast US Region, National Biodiesel Board); Trish Cozart and Kristi Theis, (National Renewable Energy Laboratory) for helping with the factsheets.

I wish to acknowledge numerous reviewers who kindly evaluated chapters and offered constructive criticism, and volunteers who commented, gave feedback and provided their support in a timely manner that helped me pull together this work: Kristin Lewis (John A. Volpe National Transportation Systems Center); Semih Eser (Pennsylvania State University); Rich Altman, David Ripplinger (North Dakota State University); Zhiguang Zhu (Virginia Tech); Ron Wilson and Gene Simpson (Auburn University), Bin Yang (Washington State University); Kelly Donovan (University of Pittsburgh); Tobias Kind (UC Davis Genome Center, metabolomics group); Melanie Thom (Baere Aerospace Consulting, Inc.); Netaka White and Sarah Galbraith (Vermont Sustainable Jobs Fund); Guy Roberts (Avatar Energy); Matt Sayre, Elizabeth Hayward, and Lucy Singer (University of Vermont (for sorting some of the technical issues)). Current and former affiliates at University of Vermont: Michelle Smith, Tom Joslin, Ariadne Brancato, Deandra Perrucio, Rich Smith, Gregory Schab, Chris Smith, and Jason McCune-Sanders. And many others not listed here.

My acknowledgements would be incomplete without mentioning Prof. John Todd (one of the pioneers in the emerging field of ecological design and engineering) for his continuous support and inspiration over many years for the biomass to biofuels course, my other projects including this book. I would also like to thank Prof. Deborah Neher, Chair, Department of Plant and Soil Science at University of Vermont (whom I fondly refer as a million-dollar-smile professor), who kindly supported a proposal for a larger bioenergy project at a time when my vision about this book was shaping up. Dr Scott Costa (may he rest in peace – would be thrilled to see this book and would have definitely asked me if I included a chapter on biodiesel from dung beetles – Sorry Scott, you would have been that contributor). I also want to thank Prof. Ann Hazelrigg, Prof. Ernesto Mandez, Dr Joel Tilley, Colleen Armstrong, and Prof. Alan McIntosh, for cheering me up for this book project and their good wishes that helped materialize this book.

The preliminary design of these book cover was made by Benjamin Kaufman and with my help further redesigned by the publisher's design team. I appreciate permission to include pictures and acknowledge support from Bioenergy Technologies Office staff, Earthrise staff, Gulf publication staff, Myriant Corporation staff, and Robert Henrikson and David Schwartz, Algae Industry Magazine.

I want to express my gratitude to the Elsevier team—Kattie Washington, Tiffany Gasbarrini, Raquel Zanol, Poulouse Joseph, and the design and production team for their complete support, patience, and understanding.

My thanks to so many other professors and intellectuals I got a chance to work with over the last decades. Also my gratitude to H.S. Mann and B.S. Gahlot, for all their support.

I'm grateful to my friends and family, my dear mom, my dear brother, my dear parents in-laws, for all their support, and extremely grateful to my dear husband and my dear son for their unwavering support and encouragement and above all patience that helped me bring together this work.

– **Anju Dahiya**

How to Use this Book: Helpful Suggestions

The leading bioenergy, biomass to biofuels experts from universities (including extensions), Department of Energy National laboratories, National Biodiesel Board, and other nonprofits, representing all the four regions in the Unites States (Midwest, Northeast, West, and South) have literally poured in their years of education and research experience into the chapters presented in this book. Each chapter is written in best possible comprehensive way to help a reader of any background to grasp the concepts, learn about the state-of-the-art technology and be able to take the learning to the next level. I have carefully presented these contributed book chapters and organized the topics in accordance to the design I had earlier implemented for a university level program[1] on "bioenergy: biomass to biofuels" (developed with support from Department of Energy and Vermont Sustainable Jobs Fund) that received overwhelmingly encouraging response from diverse sectors in the state (academic, non-profit, and government). It is hoped that this book will be used in facilitating similar programs and help the participants and individuals exploring bioenergy nationwide and elsewhere.

Participants in this program have included degree and nondegree seeking students at undergraduate to doctoral levels (from several different backgrounds—agriculture, economics, engineering, mathematics, environmental, policy, energy, etc.), researchers, biofuel farmers, budding entrepreneurs, teachers (for curriculum/projects at school/college levels), engineers, retired personnel exploring biofuel businesses, and others.

Helpful suggestions are presented here for students, researchers, instructors/experts (teachers, professors, guest speakers), and individuals exploring bioenergy, biofuels, planning to use this book as a teaching or learning tool. A brief overview of the seven parts in this book is presented in the end. First, an overview of the book presented below followed by the respective suggestions.

OVERVIEW OF THE BOOK

- **Part introductions:** this book is organized into seven parts (each indicated with a roman numeral) that cover all the possible bioenergy topics (wood, grass, oil seeds, algal fuels, waste oil, biodiesel, biogas, bioheat, gasification, ethanol, and related sustainability, economics, environmental, policy etc.).

  Part I Bioenergy—Biomass to Biofuels: an Overview

  Part II Wood and Grass Biomass as Biofuels

  Part III Biomass to Liquid Biofuels

  Part IV Gaseous Fuels and Bioelectricity

  Part V Conversion Pathways for Cost-effective Biofuel Production

  Part VI Biofuels Economics, Sustainability, Environmental and Policy

  Part VII Quizzes

  As reflected in the table of contents, at the beginning of each part, a 'part-introduction' is provided to give an overview of each of the chapters included in that part along with the suggestions to most efficiently pair a chapter/topic of interest with other related chapters included in other parts of this book for a comprehensive understanding of a particular bioenergy option.

[1]http://go.uvm.edu/7y1rr.

- **The structure of a chapter**: each chapter is a unique and self-contained information package on the respective topic; first, because it is contributed by a leading expert in the area, and second, the contributing expert has organized the chapter to first give a good introduction of the topic followed by the specifics, along with easy-to-follow examples that describe the concepts further, and additional up-to-date resources and references to explore the topic in greater depth. This pattern makes it possible for the reader to pick up any topic of interest without going through a particular order in this book.
- **Service learning projects and case studies** contributed by the participants of the bioenergy—biomass to biofuels program present their real-world experience from partnering/interning with bioenergy businesses are included in parts II to V. These studies complement the main chapters and also provide additional backgrounds for particular topics. By including the first-hand project descriptions, I have tried to provide a model of participatory learning built on hands-on experience working with the bioenergy businesses who actually encounter the real-world problems and can share the experience with student interns. This component can be incorporated in any bioenergy related teaching course designed anywhere by engaging the local bioenergy community partners—the effort would greatly enhance student learning, jump start their career, and help in the community partners providing an internship experience.
- **Supplementary materials** are provided at the book Web site at: http://booksite.elsevier.com/9780124079090. Some of the materials were originally intended to be included in this book, but due to page limits and with good intentions of including all bioenergy related possible topics, we decided to post it online.

Useful suggestions provided below are intended to help students effectively use this book as reading material for their courses, and also for instructors planning on designing such courses or for enhancing their ongoing courses.

HELPFUL SUGGESTIONS FOR BOTH THE STUDENTS AND THE INSTRUCTORS ENGAGED IN BIOENERGY COURSES

- **Scan the "table of contents"** and identify the topic of your interest (fuel type: solid, liquid, gaseous; or economics, policy, etc.) and then go through the respective part introduction to look for the related supplementary chapters in other parts. For instance, for the biodiesel topic see part II, for the feedstock option, go through the part introduction (not chapter) and pair it with the respective fuel conversion pathway in Part V. It can be further paired with the related economics and policy related information presented in Part VI to understand the "cradle to grave" picture if needed or if you want to explore a topic in greater depth.
- **Use case studies as additional topics** as some of the bioenergy topics are also covered through the case studies that provide additional topics with excellent background descriptions. For example, "compost heat" under Part III gives a good background on the Jean Pain system.
- **Refer to the glossary** provided in Part I for the terms unclear in the text. If you do not see the term you are looking for then please check the book Web site for supplementary material. If you still do not see it, do a quick search on the Internet and continue reading your chapter or vice versa. (*We would highly appreciate your dropping a message through the book message board at the book Web site or via email address provided there.* Efforts will be made to add the information as part of the supplementary material in a reasonable period of time).

ADDITIONAL SUGGESTIONS FOR AN EXPERT (TEACHER, PROFESSOR, GUEST SPEAKER) PLANNING TO USE MATERIAL PROVIDED IN THIS BOOK AS A TEACHING TOOL OR IN PRESENTATIONS OF ANY KIND

- **Chapters as readings and discussion material to supplement a lecture or talk**: the topic-specific chapters could be suggested as readings for student participants. The students can be directed to either read the specified topic(s) to come prepared for a lecture or to read those chapters after a class to strengthen their knowledge of a particular topic and to actively participate in the class or forum discussions, and for completing the class assignments or the project reports.
- **Readings for field trips**: if your teaching module consists of field trips, the students could be directed to read the topic-specific chapters before heading out to a related field trip.
- **Student Projects**: depending on the time commitment and allowed credits, student projects may range from self-studies, hands-on under supervision, or as broad as partnering with a local bioenergy business or an organization engaged in a particular bioenergy related topic to do service learning projects. This book contains examples of topic specific projects done by students in a College level program that students can be directed to use for developing their own similar hands on projects and even form a partnership (or internship) with local community partners engaged in bioenergy areas.
- **Quiz questions** on the topics included in this book are presented along with the respective answers at the end of the book. These questions could be given as tests after a topic-specific lecture or field trip. Additional suggested quizzes are available online as supplementary material.

BRIEF OVERVIEW OF SEVEN PARTS

See the respective part introductions for details as presented at the beginning of a part. A chapter in itself is a comprehensive package on the respective topic. For further understanding, the related additional details of the biomass conversion pathways common for producing different solid, liquid, or gaseous fuels (e.g., *chemical/thermal/enzymatic treatments; hydrolysis, fermentation; pyrolysis, hydroprocessing* etc.) are presented in Part V, and the related economics and policy or environmental issues in Part VI.

PART I BIOENERGY—BIOMASS TO BIOFUELS: AN OVERVIEW

In this part, the Chapter *Introduction to Bioenergy* provides a general overview of bioenergy. Additionally the biodiesel introduction contributed by National Biodiesel Board and a glossary of bioenergy-related terms with brief definitions are provided for referring while reading the chapters.

PART II WOOD AND GRASS BIOMASS AS BIOFUELS

The respective three chapters cover all the possible aspects of wood and grass energy, and the two service learning case studies describe the use of wood and grass for the production of pellets through the formation of a co-op, and present an assessment of a 130-acre property for on-site related biomass resources.

PART III BIOMASS TO LIQUID BIOFUELS

Nine chapters in this part present a variety of liquid-biofuel topics ranging from feedstock options, types of fuels, oil seed-based biodiesel, life cycle assessment with energy return on assessment calculations, straight vegetable oil, cellulosic ethanol, bioheat as heating oil, and algae biofuel. Related service learning projects and case studies include the cost analysis of oil seed production for biodiesel, the thermal switch from wood to oil heat, and ethanol distillery. Related biomass conversion pathways are presented in Part V.

PART IV GASEOUS FUELS AND BIOELECTRICITY

Four chapters cover the topics biogas (anaerobic digestion) and gasification in great detail. The related conversion pathways in Part V are available for additional information. Related service learning projects present the effect of daily variation in food waste on biogas production during anaerobic digestion, the potential for anaerobic digestion in meeting statewide energy needs, and biomass gasification as a strategy for rural electrification lessons from the field.

PART V CONVERSION PATHWAYS FOR COST-EFFECTIVE BIOFUEL PRODUCTION

Ten chapters on conversion pathways topics include transesterfication at different scales with different types of feedstocks (e.g., oil-seeds, waste vegetable oil, algae), ethanol production, pyrolysis, mechanical, chemical, and biological conversions (e.g., pretreatment, hydrolysis, enzymatic, saccharification, fermentation), and technologies to integrate with biorefineries (e.g., indirect liquefaction, hydroprocessing). The case study examples include the processes used in biodiesel production, wastewater treatment for biogas, and the fungal breakdown of lignocellulosic biomass.

PART VI BIOFUELS ECONOMICS, SUSTAINABILITY, ENVIRONMENTAL AND POLICY

Nine chapters in this part cover a wide range of related topics including bioenergy, biomass to biofuels related economics, sustainability, environmental policy, and some other issues. Some of those are contributed by the National Biodiesel Board. Entrepreneurial opportunities are also discussed here.

PART VII QUIZZES

Lastly, self-test quizzes are provided as multiple choice questions along with respective answers.

It is hoped that the broad array of topic-specific chapters in this book contributed by the respective leading experts in the areas will provide the reader with insightful information and guide in such a way that the reader is able to grasp the concepts and able to find the additional sources needed to take the learning to the next level.

For additional information on the topics covered in the book, visit the companion site: http://booksite.elsevier.com/9780124079090

BIOENERGY— BIOMASS TO BIOFUELS: AN OVERVIEW

ANJU DAHIYA[1,2]

[1] *University of Vermont, USA,* [2] *GSR Solutions, USA*

The terms *bioenergy*, *biofuels*, and *biodiesel* are sometimes used interchangeably. The reader may find a variety of definitions of these and other related terms elsewhere in the literature which may be applicable in different contexts and viewpoints as described below. Depending on a region or country of interest, a reader is encouraged to explore the corresponding definitions and fuel specifications.

Part 1 of this book presents introductions to "bioenergy," biodiesel, fuel specifications, and a rich glossary of terms based on the editor's experience with a University level bioenergy program.

Chapter 1 (Introduction to Bioenergy), as the title suggests, provides a general overview of "bioenergy." It discusses the energy products and feedstock (wood, agricultural sources of bioenergy, waste—manure; liquid biofuels such as biodiesel and advanced fuels—algae, etc.) currently considered feasible, or plausibly so in near future resulting from the current momentum in technology, policy, and economic change. It is hoped that the reader will

FIGURE 1

The "Bioenergy—Biomass to Biofuels" program creation.[1]

Picture courtesy: A. Dahiya.

gain an appreciation of the scope of bioenergy, biomass supply, impacts, and related issues from this chapter.

Chapter 2 (Introduction to Biodiesel and Glossary of Terms) prepared by the National Biodiesel Board provides a state-of-the-art introduction to biodiesel, the biodiesel standards, and the BQ-9000 fuel quality program.

Chapter 3 presents bioenergy - biomass to biofuels terms and terms and definitions related with bioenergy, and a quick reference list of conversion factors used for bioenergy feedstock.

TERMS, DEFINITIONS AND FUEL SPECIFICATIONS IN THIS BOOK

The definition of a term may vary as per different sources. The context of its use is important. Following are some of the key terms as defined by different organizations and experts.

Bioenergy: As defined in Chapter 1 (Introduction to Bioenergy) of this book, "*bioenergy is the renewable energy derived from recently living biological material called biomass.*" The U.S. Department of Energy's Oak Ridge National Laboratory's (ORNL) Bioenergy Feedstock Network defines bioenergy as, "*useful, renewable energy produced from organic matter—the conversion of the complex carbohydrates in organic matter to energy.*" Food and Agriculture Organization of the United Nations (FAO)[2,3] defines bioenergy as, "*energy from biofuels.*"

Biofuels: The U.S. Office of Energy Efficiency and Renewable Energy (EERE)[4] defines the term biofuels as, "*Biomass converted to liquid or gaseous fuels such as ethanol, methanol, methane, and hydrogen.*" ORNL defines biofuels as, "*Fuels made from biomass resources, or their processing and conversion derivatives,*" whereas FAO defines biofuels as, "*Fuel(s) produced directly or indirectly from biomass.*" The first chapter of this book states that "*Biofuels refers to solid, liquid, and gas fuels,*" accordingly Chapter 26 states the

---

[1]http://go.uvm.edu/7y1rr
[2]http://www.fao.org/energy/befs/definitions/en/.
[3]ftp://ftp.fao.org/docrep/fao/007/j4504e/j4504e00.pdf.
[4]http://www.energy.gov/eere/bioenergy/full-text-glossary.

term, "*biofuels* are widely used to address different types of fuels derived from living matter, e.g., ethanol, biodiesel, biogas."

Biodiesel: ORNL defines biodiesel as, "*Fuel derived from vegetable oils or animal fats.*" The U.S. National Biodiesel Board[5] (Chapter 2) defines biodiesel as, "*a drop-in diesel alternative, made from domestic, renewable resources such as plant oils, animal fats, used cooking oil, and even new sources such as algae.*" EERE's definition is "*A biodegradable transportation fuel for use in diesel engines that is produced through the transesterification of organically derived oils or fats.*" "Fatty acid esters" are termed as biodiesel in the chapter titled Biodiesel Production. Chapter 27 of this book states that "biodiesel" is generally used for mono-alkyl esters of long-chain fatty acids as an alternative to petroleum diesel that can be used as blends in diesel engines.

Biomass: Note that many definitions of terms depend on how the term "biomass" is defined. EERE defines biomass as, "*an energy resource derived from organic matter.*" FAO defines biomass as, "*a material of biological origin excluding material embedded in geological formations and transformed to fossil.*" The first chapter of this book defines biomass as "*recently living biological material and animal wastes.*"

Fuel Specifications: Depending on the regulations in a region or a country, fuel specification standards may vary. For instance, as described in Chapter 28, the U.S. specification for biodiesel is ASTM D6751, and the European specification is EN 14214. The key difference between the two is that the former applies to the fatty acids extraction via any type of alcohol (e.g., methanol, ethanol), whereas the latter applies to the extraction of fatty acids only via methanol.

The modified definitions of these terms are often used in deriving specific terms, for example, the chapter 4 on wood energy defines, wood bioenergy as "*energy created from the direct or indirect conversion of biomass from trees and woody shrubs.*" Similarly, definitions of bioenergy options from first, second, and third-generation of biofuels can be derived.

FIRST, SECOND, AND THIRD GENERATIONS OF BIOFUELS

Biofuels are categorized in the first chapter as follows:

First-generation biofuels—produced from oils, sugars, and starches originating in food crops.

Second-generation biofuels—produced from nonfood crops such as perennial grasses and woody materials, and nonfood portions of food crops.

Third-generation biofuels—produced from algae that can produce multifold times higher yields.

According to the Biomass Research and Development Board, Office of Energy Efficiency & Renewable Energy, Department of Energy 1 described the three generations of feedstocks as follows:

• First generation feedstocks include corn for ethanol and soybeans for biodiesel. These feedstocks are currently in use and their yields have been increasing.

---

[5]http://www.biodiesel.org/.

- Second generation feedstocks consist of the residues or "left-overs" from crop and forest harvests. They show much promise for near-term adoption with the development of cellulosic conversion technologies.
- Third generation feedstocks are crops which require further R&D to commercialize, such as perennial grasses, fast growing trees, and algae. They are designed exclusively for fuels production and are commonly referred to as "energy crops". They represent a key long-term component to a sustainable biofuels industry.

INTRODUCTION TO BIOENERGY

Carol L. Williams[1], Anju Dahiya[2,3], Pam Porter[4]

[1] *Department of Agronomy, University of Wisconsin, Madison, WI, USA;* [2] *University of Vermont, USA; GSR Solutions, USA;* [3] *GSR Solutions, Burlington, VT, USA;* [4] *Environmental Resource Center, University of Wisconsin–Madison, Madison, WI, USA*

PURPOSE

This chapter provides a comprehensive overview of bioenergy and its related issues. The geographic focus is North America (the United States, in particular), although much of the information presented is applicable to other regions and countries, with the exception of policy. The discussion is limited to energy products and feedstocks (e.g., wood, agricultural sources, waste, manure, advanced fuels, algae) that are currently considered feasible or plausibly will be in the near future because of the current momentum in technology, policy, and economic change. From this chapter, the reader will gain an appreciation of the scope of bioenergy, biomass supply, impacts, and related issues.

INTRODUCTION

Interest in renewable energy is increasing globally in response to the growing demand for energy and environmental protection. Energy providers are thus seeking new technologies and new resources to fulfill these needs. *Biomass*—recently living biological material and animal wastes—has been used since early history to cook and heat spaces where humans live and labor. Today, biomass has an expanded role in global energy consumption. *Bioenergy*—energy produced from biomass—is a promising solution to environmental challenges and a driver of economic development from local to global levels (Coleman and Stanturf, 2006; Kleinschmidt, 2007).

To meet bioenergy demand, energy providers must continuously secure a sufficient and reliable supply of biomass at prices allowing them to operate profitably. As attention focuses on the sustainability of resource use, biomass producers (e.g., farmers) must balance market pressures for ever-increasing supply at low prices with demands for nonmarket values, such as soil conservation, water quality protection, and biodiversity enhancement. Hence, solutions are necessary that will reduce the potential trade-offs between economic development and resource conservation, as well as the competition for land resources.

To understand the full scope of bioenergy, it is necessary to know and understand the various forms of bioenergy, the types of biomass materials and their sources, and the types of co-products and byproducts resulting from conversion of biomass materials. At the conclusion of this chapter, readers

Bioenergy. http://dx.doi.org/10.1016/B978-0-12-407909-0.00001-8

should be able to answer the following questions: What are the major forms of bioenergy and what are they used for? What are the major drivers of bioenergy development? What is the difference between "biomass" and "feedstock"? What are the key differences between woody and nonwoody biomass? What are the primary sources for biomass? What are the emerging feedstock sources for the future? What are advanced biomass sources? What are the advantages and disadvantages associated with different types of biomass? What are the social, sociological, and economic impacts of bioenergy?

BIOENERGY DEFINED

Bioenergy is renewable energy derived from recently living biological material, or biomass. Fossil carbon sources of energy, such as coal and petroleum, are not sources of bioenergy because these materials are the result of geological processes that transformed plants living many thousands of years ago. Bioenergy is a form of renewable energy because the energy contained in biomass is energy from the sun captured through natural processes of photosynthesis. As long as the quantity of biomass used is equal to or less than the amount that can be regrown, it is potentially renewable indefinitely. Forms of bioenergy include power, heat, and solid, liquid, and gas fuels. Uses of these various forms of bioenergy include industrial, residential, and commercial applications (Figure 1.1).

Biopower is electricity generated from combustion of biomass, either alone or in combination with coal, natural gas, or other fuel (termed *co-firing*). Most biopower plants are direct-fired systems where biomass feedstocks are burned in a boiler to produce high-pressure steam that runs turbines connected to electric generators. The electricity produced can be distributed for industrial, residential, or commercial use (Figure 1.1). The steam generated from combustion of biomass can also be used to directly power mechanical processes in industrial settings (Figure 1.1). Technical challenges in biopower generation involve feedstock quality, boiler chemistry, ash deposition and ash disposal. However, these challenges are being resolved as technology advances. Boiler efficiency in co-firing may be reduced slightly compared to the efficiency of firing 100% fossil fuel feedstocks due to the moisture content of

FIGURE 1.1

Energy products and their end uses.

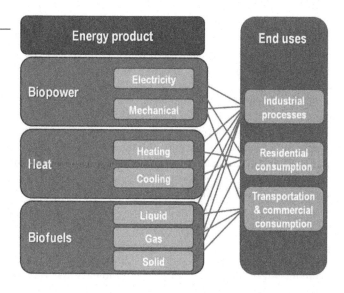

biomass (Prochow et al., 2009). Biomass co-fired in a coal power plant can assist the utility in meeting renewable energy standards and help cut pollution from coal burning.

Biofuel refers to solid, liquid, and gas fuels. Solid fuels are typically used for space heating via combustion. Liquid and gas fuels are used for transportation and industrial processes (Figure 1.1). Liquid and gas biofuels are produced through fermentation, gasification, pyrolysis, and torrefaction, described below. Bioethanol and biodiesel are major forms of biofuel. Biofuels produced from oils, sugars, and starches originating in food crops are known as *first-generation biofuels*. First-generation biofuels are produced through relatively simple and established technologies. Conversion technologies still under development allow the creation of *second-generation biofuels*, also known as *advanced biofuels*, which are produced from nonfood crops such as perennial grasses and woody materials, and the nonfood portions of food crops. *Third-generation biofuels* are produced from algae (Goh and Lee, 2010; Lee and Lavoie, 2013). Feedstock materials and conversion technologies are further discussed below.

Heating and cooling (i.e., thermal energy) can be derived from steam and waste heat generated in biopower generation and biofuel conversion processes. Combined heat and power, also known as *co-generation*, is the simultaneous production of electricity and heat from a single fuel source, including biomass.

Bioenergy production and use occurs along a continuum of scale. At one end of the spectrum is household use of biomass for heat and cooking. In the middle section of the spectrum are numerous production and distribution arrangements and organization types, including mixed-scale operations, cooperatives, and community-based or small-scale distributed energy projects. At the other end of the bioenergy production/use spectrum are large energy companies who own the means of biomass production and conversion, as well as the means of distribution of bioenergy products. At this end of the spectrum utility, companies take advantage of economies of scale to acquire vast quantities of biomass either from their own lands or lands they lease, or from large-scale suppliers, and then convert the biomass into bioenergy to make it available for thousands of consumers. There are advantages and disadvantages associated with different scales of bioenergy production and use. Decision-makers should regard the tradeoffs among these benefits and disadvantages, particularly in relation to public costs and benefits (Elghali et al., 2007).

BIOENERGY DEVELOPMENT AND DRIVERS

Much of current interest in bioenergy revolves around issues of *energy security, energy independence*, and perceived opportunities for economic growth and development. Energy security is reliable access to affordable energy. Energy independence is state/national self-reliance for energy production/supply. Below, then, is a discussion of policy and other drivers of bioenergy development in response to these issues.

POLICY

Interest in ethanol as a liquid transportation fuel, although used in the United States since at least 1908 with the Ford Model T, grew in the late twentieth century as a result of oil supply disruptions in the Middle East and environmental concerns over the use of lead as a gasoline octane booster. Ethanol production in the United States soon grew with support from federal and state ethanol tax subsidies and mandated use of high-oxygen gasoline. Additional incentives in the 1980 and 1990s, and passage of the Clean Air Act Amendments of 1990 (Waxman, 1991), further incentivized expanded U.S. ethanol production. Political instability in major oil producing nations has led to increased interest in energy independence as a means of reducing widespread economic risks associated with price shocks due to supply disruptions.

Today, nearly all ethanol production in the United States uses corn grain in fermentation processes creating first-generation biofuel (Baker and Zahniser, 2006). However, bioenergy development is focused primarily on advanced biofuels and biopower projects, as well as next-generation biomass crops. Policy at national and state levels provides major incentives for these development agendas.

In his 2006 State of the Union Address, President George W. Bush rolled out the Advanced Energy Initiative, which included increased research funding for cutting-edge biofuel production processes. In early 2007, he announced the "Twenty-in-Ten" initiative, a plan to reduce gasoline consumption by 20% in 10 years (Bush, 2007). Congress responded in December 2007 by passing a Renewable Fuel Standard (RFS) as part of the Energy Independence and Security Act of 2007 (EISA, 2007). The RFS requires annual production of 36 billion gallons of biofuels by 2022 and includes specific provisions for advanced biofuels, paving the way for advanced technologies. Many state governments have adopted similar policy initiatives and programs.

In 2007, the Bush Administration proposed a farm bill that included funds for new renewable energy and energy efficiency-related spending at the U.S. Department of Agriculture (USDA), including support for cellulosic ethanol projects. In May 2008, Congress passed the 2008 Farm Bill, titled the Food, Conservation and Energy Act of 2008 (FCEA, 2008) with mandatory funding for bioenergy activities. These and similar provisions will be reconsidered in subsequent farm bills.

GOVERNMENT RESEARCH PROGRAMS

Additional major influences on bioenergy development are new or recently expanded federally funded or sponsored research initiatives, programs, and offices. Mostly notable are those within, administered by, or in partnership with the U.S. Department of Energy (DOE) (2011) and the USDA. Chief among the research centers are the National Renewable Energy Lab (NREL), Idaho National Laboratory, Sandia National Labortories, and the DOE's suite of three bioenergy research centers: Oak Ridge National Laboratory and collaborators, the Great Lakes Bioenergy Research Center and collaborators, and the Joint BioEnergy Institute led by the Lawrence Berkeley National Laboratory.

The DOE's Office of Energy Efficiency and Renewable Energy sponsors energy initiatives and leads 10 programs, including a biomass research, development, and demonstration program. The USDA leads in bioenergy research and development through its Economic Research Service and Agricultural Research Service. Through its Farm Service Agency, the USDA administers important programs incentivizing bioenergy crop production, such as the Biomass Crop Assistance Program.

MARKET INERTIA

Although U.S. agricultural and energy policies have aimed to encourage the development of cellulosic bioenergy from sources such as perennial grasses (EISA, 2007), its development has been sluggish. As a result, potential multiple benefits have been foregone. Cellulosic bioenergy is faced with a "chicken and egg" dilemma: investors in biomass conversion technologies and infrastructure are reluctant to engage until there is sufficient biomass supply, and biomass producers are unwilling to invest in new crops and production systems until there is sufficient demand. Overcoming this market inertia may require novel intervention (i.e., nongovernmental) for reducing risk and uncertainty in enterprise development and biomass supply (McCormick and Kaberger, 2007; Taylor et al., 2013).

FEEDSTOCKS

Bioenergy *feedstocks* are biomass-derived materials that are converted to energy through the application of microbial activity, heat, chemicals, or a combination of these processes. Biomass materials usually must be at least minimally processed to be ready for conversion into bioenergy. That is, biomass does not usually exist in a form that can be converted directly into energy without some alteration. Combustion of fuelwood for household use is an exception. Once processed, biomass is considered feedstock.

The *bulk density* of biomass is relatively low (McKendry, 2002). Bulk density is the weight of biomass per volume of biomass. Low bulk density of biomass means it takes up space in transport vehicles that are otherwise equipped to handle heavier loads, which means more hauling trips are necessary compared to materials with greater bulk density. Therefore, it is frequently necessary to process biomass into forms that are economical to transport. A first step is *aggregation*—the process of gathering up harvested biomass into easily handled units such as bales (Figure 1.2). Low bulk density also translates into low energy density, particularly compared to other sources of energy such as coal.

FIGURE 1.2

Baling of biomass is a form of aggregation that makes handling more efficient.

(Source: Photo: Williams, 2012.)

FIGURE 1.3

Pelletized biomass. Cattail (*Typha* spp.) biomass has been milled and pressed into a relatively high-density solid fuel.

(Source: Photo: Williams, 2013.)

Therefore, biomass may also be densified. *Densification* is the application of pressure and other processes to create solid, condensed feedstock (Tumuluru et al., 2010). Increased energy density improves conversion efficiency and therefore reduces costs associated with conversion (Stephen et al., 2010). *Pelletization*—densification into pellets—is a common method of increasing bulk density of biomass (Figure 1.3).

As biomass moves from where it was grown to where it is converted, it passes through various stages of processing and handling. Each stage adds value to what started as a relatively low-value material. The sequence of processes is known as a supply chain, sometimes also called a value chain. Because of low bulk density and subsequent low energy density of biomass, optimal conversion facility size frequently depends on biomass haulage costs, and feedstock supply/value chains trend toward smaller, more distributed, and more localized facilities (Jack, 2009; Searcy et al., 2007).

BIOMASS MATERIALS AND SOURCES

There are three main types of biomass materials from which bioenergy feedstocks are derived: lipids, sugars/starches, and cellulose/lignocellulose (Figure 1.4). Lipids are energy-rich, water-insoluble molecules such as fats, oils, and waxes. Lipids are a feedstock source derived from nonwoody plants and algae. Soyabean (*Glycine max*), oil palm (*Elaeis guineensis* and *Elaeis oleifera*), and various seed crops such as sunflower (*Helianthus annus*) are common agricultural sources of oils for biodiesel. Sugars and starches are the carbohydrates typically found in the edible portions of food crops, such as corn (*Zea mays*) grain, which are sources for first-generation biofuels. Cellulosic/lignocellulosic

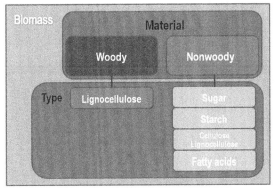

FIGURE 1.4

Types of biomass derived from woody and nonwoody plants.

biomass is composed of complex carbohydrates and noncarbohydrate molecules typically found in the leaves and stems of plants. Cellulose/lignocellulose is of little or no food value to humans. Advanced biofuels therefore offer an opportunity to take these relatively low-value materials and use them in the production of high-value energy products (Clark et al., 2006).

There are two broad categories of plants from which cellulosic/lignocellulosic feedstocks are derived: woody and nonwoody (Figure 1.4). In addition to cellulose, many plants also contain hemicellulose and lignin. Hemicellulose is a large, complex carbohydrate molecule that helps cross-link cellulose fibers in plant cell walls. Lignin is a noncarbohydrate polymer that fills the spaces between cellulose and hemicellulose. When cellulose, hemicellulose, and lignin are present together, they are referred to as *lignocellulose*. Hemicellulose can be broken down into fermentable sugar and then converted into ethanol and other fuels. Lignin is difficult to convert into other usable forms and is therefore considered a byproduct. As technologies for transforming lignin improve, new markets for its use may emerge.

The majority of biomass for bioenergy feedstocks comes from three sources: forests, agriculture, and waste (Figure 1.5). Algae is an important emerging source of bioenergy feedstocks. Agroforestry and nonforest conservation lands, such as grasslands and savannahs, are also potential sources of bioenergy feedstocks. Forests provide woody materials (Figure 1.5); agriculture and waste sources can provide both woody and nonwoody biomass for bioenergy production (Figure 1.5). Each of these sources has limitations of biomass availability and quality, as well as issues of accessibility. Additionally, these sources typically have competing uses for their biomass, which may affect price and availability (Suntana et al., 2009).

FOREST-BASED FEEDSTOCKS

Woody biomass from forests is the original source of bioenergy (Demirbas, 2004). It remains the most important source of fuel for cooking and space heating throughout the world, particularly among subsistence cultures (Cooke et al., 2008). Few extended-rotation forests (i.e., growth–harvest cycles of decades), whether public or private, are or will likely be managed specifically to provide biomass for bioenergy (Hedenus and Azar, 2009). Instead, biomass for bioenergy is typically a co-product of forest management activities (e.g., fuel hazard removal) or commercial activities emphasizing higher value materials such as merchantable wood.

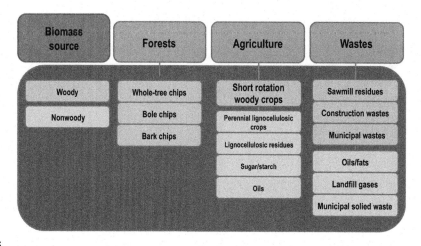

FIGURE 1.5

Sources of biomass.

In general, only wood that is not merchantable as lumber or pulp is used in bioenergy production. There are two main ways low-grade wood is removed from forests for bioenergy use: as bark and as wood chips. Bark is typically burned to firewood kilns at mills, or it is sold in higher value markets such as landscaping materials. Although bark has a high energy density (more than wood chips), it has high silica and potassium content, which affects its quality as a feedstock (Lehtikangas, 2001). Wood chips, however, can be used directly as a solid fuel (for combustion) or they can be refined and densified into pellets.

There are three main types of wood chips: mill chips, whole-tree chips, and bole chips (Figure 1.5). Mill chips are produced from waste wood (off-cuts and slabs from sawing logs into lumber). Because logs are debarked before sawing, mill chips are usually very clean. Whole-tree chips originate from managed forests with little commercial value for lumber and where removal of trees could improve future commercial timber value. Whole-tree chips are produced by either chipping the entire low-grade tree or from only the tops and limbs severed from logs. Although a majority of whole-tree chips are generated from forest management activity, they are also produced from land clearing and land-use conversion projects that make way for roads, parking lots, buildings, and open spaces, for example. The felled trees are typically chipped onsite. Bole chips are produced from low-grade or pulp logs, usually from managed forests. The difference between whole-tree chips and bole chips is that bole chips do not include branches or foliage.

AGRICULTURE-BASED FEEDSTOCKS

Agriculture is a source of sugars, starches, lipids, nonwoody cellulosic materials, and woody materials (i.e., lignocellulosic biomass; Figure 1.5). Agriculture-based biomass comes from crops grown specifically for bioenergy production (i.e., dedicated bioenergy crops), as well as agricultural residues. Agricultural residues are nonedible, cellulosic materials that remain after harvest of edible portions of crops. Dedicated bioenergy crops include annual crops grown for their sugars, starches, or oils, and perennial herbaceous nonfood plants grown for their cellulose. Agricultural residues include plant leaves and stems. Some annual crops, such as corn, can be dedicated bioenergy crops for both their grain and their cellulosic residues.

Most of the world's first-generation bioethanol is made from feedstocks derived from annual food crops. Annual row crops are grown and harvested in a single year and must be planted every year. Sugarcane (*Saccharum officinarum*) and corn are the primary feedstock sources for first-generation bioethanol. However, bioethanol is also produced from cereal crops, sugar beets (*Beta vulgaris*), potatoes (*Solanum tuberosum*), sorghum (*Sorghum bicolor*), and cassava (*Manihot esculenta*) as well. Sugarcane is the primary feedstock in Brazil, and corn grain is the primary feedstock in the United States. These two feedstock sources are converted into approximately 62% of the world's bioethanol (Kim and Dale, 2004).

The primary agricultural sources of lipids for first-generation biodiesel are annual row crops soyabean, palm, and oilseed rape (or rapeseed, *Brassica napus*). Soyabean is the primary feedstock source for biodiesel produced in the United States, Europe, Brazil, and Argentina, which are world leaders in biodiesel production (Bergmann et al., 2013). Palm, a tropical plant, is the primary feedstock source in Southeast Asia (e.g., Malaysia and Indonesia), while oilseed rape is grown in Europe, Canada, the United States, Australia, China, and India (Rosillo-Calle et al., 2009). Inedible oil crops are being examined for commercial potential in second-generation biodiesel production, including castor (*Rincinus communis*) and Camelina (*Camelina sativa*; Atabani et al., 2012). For more information on biodiesel feedstocks and production technologies, see Salvi and Panwar (2012).

Perennial crops are the primary sources of lignocellulosic biomass for second-generation biofuels. They have received considerable attention because they are not food crops, and they provide both long-term yield potential and environmental benefits not usually achieved in annual row crop agriculture (Sanderson and Adler, 2008, and references therein). These potential environmental benefits include wildlife habitat, soil erosion prevention, and water quality improvement (Glover et al., 2010). Perennial crops live for more than one growing season and do not have to be planted every year. Perennial crops include herbaceous plants (i.e., plants lacking permanent woody stems) and woody plants. Perennial grasses in particular are of considerable value in advanced biofuels, as are fast-growing trees such as hybrid poplars (*Populus* spp.) and willows (*Salix* spp.). Whether they are herbaceous or woody, perennial dedicated bioenergy crops are typically grown with some amount of agronomic intensity (e.g., inputs of fertilizer and pesticides), which is why they are considered as crops.

Governments and academic researchers have embarked on rigorous evaluations of a variety of perennial herbaceous plants as candidates for advanced biofuels (Lewandowski et al., 2003). Switchgrass has been extensively studied for second-generation biofuels, particularly in North America (Lewandowski et al., 2003; McLaughlin and Kszos, 2005; Wright and Turhollow, 2010), and miscanthus has been widely evaluated in Europe (e.g., Christian et al., 2008). Hence, these two bioenergy crops are further detailed here.

Switchgrass is a C4 grass that has evolved as a component of diverse tall grass prairie ecosystems in the eastern two-thirds of the United States (Parrish and Fike, 2005). It has been used since arrival of Europeans to graze ruminant livestock, and over time has been intentionally managed and improved for forage. Recently, switchgrass has been scrutinized for bioenergy purposes (Wright and Turhollow, 2010). Switchgrass as managed forage and in bioenergy research is typically grown as a pure-grass sward (i.e., *monoculture*), although interest is high in its use in *polycultures*, which are diverse plant mixtures that may include different plant functional groups (e.g., grasses, forbs, and legumes, Tilman et al., 2006). Polycultures are receiving research and development attention for their biomass yields as well as wildlife habitat and environmental benefits compared to monocultures (Adler and Sanderson, 2008; Tilman et al., 2006, 2009). There are numerous switchgrass varieties and cultivars, each having different responses to characteristics of location (e.g., soil, day length) and mineral fertilization (Casler et al., 2007; Fike et al., 2006; Virgilio et al., 2007). Choice of variety or cultivar will depend on characteristics of the location where it will be grown (e.g., growing season) and the management that will be used. Management of switchgrass in monocultures can be quite different than polycultures in which it is a component. Whether in monoculture or polyculture, switchgrass is grown from seed. It is harvested once or twice annually, depending on location. Reported yields for switchgrass vary according to cultivar, location, fertilizer use, and other factors, generally ranging from 5.3 Mg DM/ha/year to 21.3 Mg DM/ha/year (Fike et al., 2006; Lemus et al., 2002; Lewandowski et al., 2003).

Perennial C4 grasses of the genus *Miscanthus* originate in the tropics and subtropics of East Asia. Due to their high yields and wide climatic adaptability, they have received much attention as potential bioenergy crops (Lewandowski et al., 2000). *Miscanthus x giganteus* (hereafter, miscanthus) is a hybrid of *Miscanthus sinensis* and *Miscanthus sacchariflorus*; it is the frontrunner in bioenergy crop research, development, and production in Europe (Lewandowski et al., 2003) and has recently begun receiving attention in the United States (e.g., Heaton et al., 2004, 2008). Miscanthus is a sterile hybrid that does not produce seeds and instead reproduces vegetatively from rhizomes. It is grown in monocultures that are established through manual or mechanical planting of rhizomes or rhizome pieces (Lewandowski et al., 2000). Miscanthus plantations, therefore, are monocultures of clones (i.e., genetically identical plants). The rhizomes are grown in nursery fields, where they are mechanically collected and divided just before planting in fields for biomass production (Lewandowski et al., 2000). Winter kill can be a problem in miscanthus cultivation (Heaton et al., 2010; Lewandowski et al., 2000). Miscanthus is harvested only once per year, usually after senescence (Lewandowski et al., 2000). Biomass yields of miscanthus vary widely depending on location, use of irrigation, and harvest timing. Yield reports range from 7 to 40 Mg DM/ha/year (Lewandowski et al., 2000, 2003; Price et al., 2004). For comprehensive overviews of miscanthus improvements, agronomy, and biomass characteristics, see Lewandowski et al. (2000) and Jones and Walsh (2001).

Crop residues also are an important source of cellulosic feedstocks (Figure 1.4). Crop residues include materials left in fields after crops have been harvested. For example, the stems, leaves, and stalks of corn leftover after harvest of grain, referred to as stover, can be used in the production of cellulosic ethanol. Their use potentially limits the impacts of biofuels on food security (Kim and Dale, 2004). Global residue biomass is estimated as 3.8 billion Mg/year (Lal, 2005); however, the availability of residues differs by location (region, country, and within countries) according to climate and soil variations, affecting the growth suitability of particular crops. For example, rice straw is readily available in Asia, whereas stover (corn residue) is available in the United States, Mexico, and Europe (Kim and Dale, 2004). The amount of residue that is potentially available differs widely among crops (Lal, 2005; see further discussion below). The use of agricultural residues must be carefully planned and managed due to their important role in soil erosion control and maintenance of soil quality, and their use as forage, fodder, and bedding for livestock (Lal, 2005).

WASTE-BASED FEEDSTOCKS

Waste-based biomass includes organic materials leftover from industrial processes, agricultural liquid and solid wastes (e.g., manure), municipal solid wastes, and construction wastes (Figure 1.5). Many industrial processes and manufacturing operations produce residues, waste, or co-products that can be potentially used for bioenergy. The major sources of nonwoody wastes include waste paper, liquid leftover from paper production (i.e., black liquor), and textile manufacturing. Major sources of woody waste materials include used pallets, sawmill byproducts such as sawdust and shavings, cutoffs from furniture manufacturing, and composite wood products containing nonwood resins, adhesives and/or fillers. Conversion technologies for these wastes are potentially the same as for virgin wood (Antizar-Ladislao and Turrion-Gomez, 2008).

Agricultural wastes include byproducts of agro-industrial processes and manure from livestock. Agro-industrial processes such as animal processing, grain milling, starch production, and sugar production result in byproducts that may be used as bioenergy feedstocks. Bagasse, the fibrous material leftover from sugarcane and sorghum crushing in sugar production, for example, is sometimes used as a fuel source for heat in sugar mills but it can also be converted to bioethanol (Botha and Blottnitz, 2006). Animal processing generates large quantities of feathers, bones, and other materials. These animal byproducts are a potential source of diseases that have public and/or animal health risks (e.g., bovine spongiform encephalopathy), so rigorous protocols must be followed to eliminate the possibility of spread of disease. Accordingly, animal byproducts are used as feedstocks in anaerobic digesters that kill potential pathogens and produce biogas (i.e., methane). Biogas is a substitute for propane, kerosene, and firewood to produce heat and electric power. It is also compressed and liquefied for use as a transportation fuel.

Manure can be used as a fertilizer on agricultural fields, but manure application is a highly regulated agricultural activity and its disposal can present challenges to farm profitability. In some circumstances, manure cannot be applied directly to fields because the ground is frozen, or the amount of manure available exceeds the amount that can be put onto fields without endangering nearby water resources with contamination. Use of manure as a bioenergy feedstock is therefore an opportunity for turning a potentially expensive liability into a benefit. Livestock manure is converted into biogas via anaerobic digestion.

Municipal solid waste is a major source of biomass. Also called trash and urban solid waste, municipal solid waste is predominantly household or domestic waste. Municipal solid waste includes biodegradable waste, such as kitchen food waste and food packaging; clothing and toys; recyclable materials such as paper, plastics, and metals; appliances and furniture; and debris. Most municipal solid waste is diverted to landfills, but in some locations it is incinerated to make electricity. Portions that are not incinerated can be converted to syngas through gasification. Syngas can be co-fired in boilers with coal, for example, to produce electricity.

Construction waste consists of wood, plastic, and metal debris. Although plastic and metal may be used in combustion for production of power, for example, only woody construction wastes are feedstocks for bioenergy. Construction waste varies greatly in composition and by location. Currently, the primary conversion technology for construction waste is combustion for heat, steam, and biopower, although as lignocellulosic material it can potentially be used in biological and other conversion technologies for biofuels (Antizar-Ladislao and Turrion-Gomez, 2008).

AGROFORESTRY FEEDSTOCKS

Agroforestry is the intentional integration of perennial nonfood crops with food crops on the farm. This may occur as *alley cropping*—the planting of trees or shrubs in rows of wide spacing that allow for the planting of crops in between rows of woody crops (Garrett et al., 2009; Headlee et al., 2013). Alternatively, fast-growing, intensively managed woody crops may be grown in monoculture as part of a diverse farm enterprise (Dickmann, 2006). Regardless of the production system, agroforestry is an emerging source of lignocellulosic feedstocks for second-generation biofuels. Short-rotation woody crops (SRWC) typically grow to harvestable size in less than 15 years—depending on species and management, possibly in as little as 3 years (Volk et al., 2004). Globally, *Eucalyptus* species are the most extensively planted species, whereas other hardwoods predominate in temperate regions (Rockwood et al., 2008). In temperate regions, SRWCs include hybrid poplars (*Populus* spp.), willows (*Salix* spp.), and maples (*Acer* spp.). Most SRWCs are shade intolerant, which makes them suitable to the openness of farm fields. Many temperate SRWCs have the ability to coppice (sprout new growth from stumps) when harvested. Hence, coppicing will produce harvestable biomass on shorter rotations.

A potentially important source of whole-plant biomass for biorefineries is Jatropha (*Jatropha curcas*), an oil-bearing tree that can be grown in agroforestry systems (Achten et al., 2007). Jatropha is native to Mexico, Central America, and parts of South America. It is drought resistant, easily propagated, and performs well in a wide variety of soils, including degraded lands. Jatropha contains inedible oil and is toxic to humans and animals. The biodiesel production with Jatropha results in valuable byproducts, such as seedcake and husks. These characteristics make it an attractive biorefinery feedstock candidate (Manurung et al., 2009). A biorefinery typically integrates conversion processes so that whole-plant feedstocks may be used to produce energy, chemicals, and other value-added products.

BIOMASS FROM CONSERVATION LANDS

To avoid potential competition with production of food and forage on prime agricultural lands, government authorities and researchers are considering the potential benefits and risks associated with periodic harvest of biomass from conservation lands, such as those set aside in agriculturally dominated landscapes for purposes of soil conservation, water quality improvement, wildlife habitat, hunting access, or other nonagricultural purpose (Adler et al., 2009; Fargione et al., 2009; Rosch et al., 2009). Conservation lands, whether privately or publically owned, typically require management for maintaining cover types and various conservation goals. Harvest may be a viable habitat management tool (Figure 1.6). Biomass resulting from habitat management actions has the potential to be used in a variety of biomass conversion technologies. Land managers, therefore, may be able to offset management costs with sale of biomass resulting from periodic management actions. However, the long-term impacts of removal of biomass from conservations lands, such as nutrient loss and soil compaction, are currently unknown. Greater scrutiny is therefore necessary to understand impacts of biomass harvest on lands set aside for wildlife and other resource management goals.

ADVANCED FUELS FROM ALGAE

Challenges in meeting the demands for bioenergy include competition for land, water, and other resources needed to produce plant-based feedstocks (Dale et al., 2011). As a biofuel feedstock, algae

FIGURE 1.6

Harvest of grassland biomass for habitat management on public conservation land in Wisconsin.

(Source: Photo: Williams, 2012.)

provides a very attractive alternative option (DoE, 2010). Algae does not compete with food, land, and water resources. The algae fuel yield is also estimated to be 100 times more than other biofuel sources. The low-temperature fuel properties and energy density of algae fuel make it suitable as jet fuel, home heating oil, and general transportation fuel for colder regions. In addition, it ensures a continuous supply, can capture waste CO_2 for generating biomass, can manage farm nutrients runoff, and can treat wastewater.

NREL's Aquatic Species Program (active from 1978 to 1996) demonstrated biofuel production from algae (Sheehan et al., 1998), but a cost-effective technology for larger-scale production of biofuel has yet to be found. For comprehensive review of microalgae cultivation for biofuels, see the U.S. Department of Energy Biomass publication, "National Algal Biofuels Technology Roadmap" (DoE, 2010), which laid the groundwork for identifying challenges that will likely need to be surmounted for algae to be used in the production of economically viable, environmentally sound biofuels. This document resulted from inputs of over 200 scientists, engineers, industry representatives, research managers, and other stakeholders.

Critical hurdles in terms of algae biofuel production and big economic barriers in the production of algal-based drop-in biofuel include the cost-efficiency involved in the development and production of algae biomass for biofuel (Dahiya et al., 2012), which is currently limited due to the availability of *robust oleagenous algae strains* that can grow in nonsterile environments, such as dairy farm manure wastewater (Dahiya, 2012) and other lignocellulosic materials (e.g., residues: corn stover, straw, manure, rural food industry wastewater, molasses, bagasse, wood chips/ bark, grasses, etc.) to efficiently use low-cost nutrient sources of mainly carbon, nitrogen, and phosphorus. Most of the sterile photo-bioreactors have produced significantly more volumetric algal biomass than the open raceway ponds (Figures 1.7). However, leading studies (Huntley and Redalje, 2007; Chisti, 2007; Schenk et al., 2008; DoE, 2010) have pointed out that large-scale sterile reactors

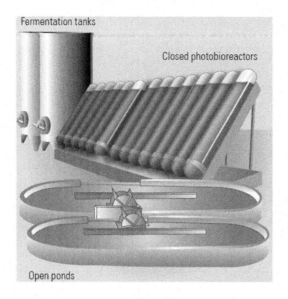

FIGURE 1.7

Algae cultivation systems.

(Credit: Bioenergy Technologies Office, DoE, 2010)

are not cost-effective due to infrastructure development and that economic solutions will have to include a nonsterile open ponds option, which requires consideration of the contamination issues.

Micro- and macroalgae are therefore being explored as commercially viable feedstocks for third-generation biofuels. Low-cost throughput feedstocks for growing algae (e.g., CO_2, wastewater containing nutrients for algae, low-cost carbon sources from lingo-cellulosic wastes including corn stover and bagasse, grasses) and the cost-effectiveness of commercial-scale systems would be the determining factors in developing advanced biofuels.

BIOMASS SUPPLY AND AVAILABILITY

In 2005, the USDA and DOE released a joint report on the feasibility of biomass feedstock supply to fulfill the renewable fuel goals of the Biomass Research and Development Technical Advisory Committee, a panel established by the U.S. Congress to guide future direction of federally funded biomass research and development. The report, "Biomass as Feedstock for a Bioenergy and Bioproducts Industry: The Technical Feasibility of a Billion-Ton Annual Supply" (Perlack et al., 2005), found that there were sufficient land resources in the United States for fulfilling the goal of replacing 30% of current U.S. petroleum consumption with biofuels by 2030. Moreover, the Billion-Ton Study concluded that the 1 billion dry tons of biomass feedstock needed annually could be sustainably produced with "modest changes" in land use and management practices within U.S. agricultural lands and forestlands. In 2011, the Billion Ton Study was updated to account for changes in underlying assumptions (including economic restrictions) and changes in analytical methods. The U.S. Billion Ton Update generally supports the findings of the 2005 report but modifies the magnitudes of specific resources (Downing et al., 2011). Forest and crop residues were determined to be less than the 2005 estimates, but energy crop potential was determined to be significantly greater than the 2005 estimate. The Billion Ton Update concluded that biomass resources could be sustainably increased from the then-current 473 billion dry tons to nearly 1.1 billion dry tons by 2030—enough to replace approximately 30% of current petroleum consumption. Individual states have conducted or are conducting similar studies to determine the potential inventory of their resources (e.g., Willyard and Tilkalsky, 2006).

In 2008, Sandia National Laboratory and General Motors' Research and Development Center conducted a joint biofuel systems analysis to assess the feasibility, impacts, limitations, and enabling factors of large-scale production of biofuels in the United States. The findings of the analysis are reported in the "90-Billion Gallon Biofuel Deployment Study" (Sandia National Labs, 2009). According to the report, 90 billion gal/year of biomass-derived ethanol can be produced and distributed, with 15 billion gal/year from corn grain ethanol and the balance from cellulosic ethanol. The production of 45 billion gal/year of cellulosic ethanol requires 480 million tons of biomass, of which 215 million tons was projected to come from dedicated perennial energy crops, requiring 48 million acres of planted cropland from what is now idle, pasture, or nongrazed forest.

The greatest supply of U.S. forest residues is the northwest, the extreme northeast, and across the southern tier of states east of the Mississippi River (Figure 1.8). Crop residues are primarily available within the Corn Belt, and among counties along the Mississippi River in the central

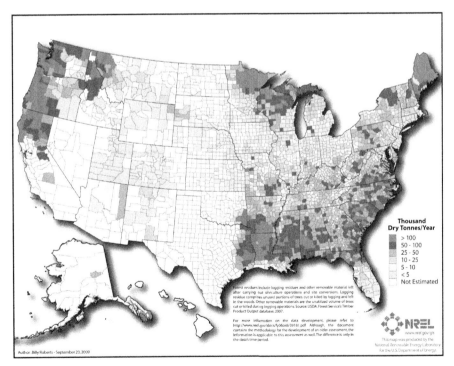

FIGURE 1.8

Distribution of forest residues for potential use in bioenergy production.

(Source: National Renewable Energy Laboratory.)

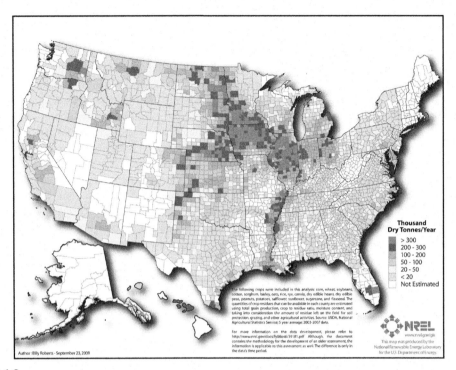

FIGURE 1.9

Distribution of crop residues for potential use in bioenergy production.

(Source: National Renewable Energy Laboratory.)

Mississippi Valley (Figure 1.9). Dedicated perennial feedstocks will likely be distributed in regions across the country based on climate, soils, and agronomic factors (Figure 1.10). According to the Oak Ridge National Laboratory, hybrid poplars are suitable throughout the United States, except in southeastern states where other hardwood species are suitable for growth. Areas most suitable for the growth of switchgrass and reed canary grass are the Plains states and Texas, Corn Belt states, and the Upper Midwest. The most suitable area for growth of miscanthus and other tropical grasses is in the southeastern states. According to analysis by NREL, counties with the greatest potential biomass supply are capable of producing greater than 500,000 tons annually. These counties have high amounts of forest residues, wood and paper industry wastes, and urban/construction wastes. Counties with high amounts of agricultural residues are also among those with highest biomass supply, potentially producing 250,000–500,000 tons of biomass annually.

OVERVIEW OF CONVERSION TECHNOLOGIES

To make use of the energy available in biomass, it is necessary to use technology to either release the energy directly, as in direct combustion for heat, or to transform it into other forms such as solid,

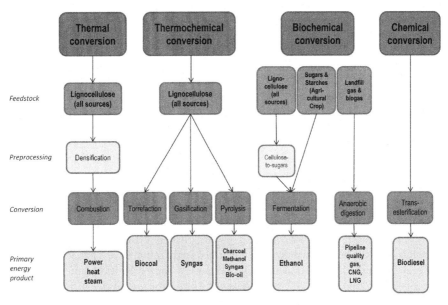

FIGURE 1.10

Conversion pathways.

liquid, or gaseous fuel. There are three main types of conversion technologies currently available: thermal, chemical, and biochemical (Figure 1.10). These technologies may also be used in combination. As implied by its name, thermal conversion processes use predominantly heat to convert biomass into other forms. Thermal conversion processes include combustion, pyrolysis, torrefaction, and gasification. Pyrolysis is the decomposition of biomass at high temperatures in the absence of oxygen. Torrefaction is pyrolysis at low temperature. Gasification is the conversion of solid biomass into various gases using heat and varying amounts of oxygen. Chemical conversion involves the use of chemical agents to convert biomass into liquid fuels. Biochemical conversion involves the use of enzymes of bacteria or other microorganisms to break down biomass through the processes of anaerobic digestion, fermentation, or composting. Although relevant technologies exist (and continue to be developed), some are not yet cost effective, particularly for the large-scale conversion of cellulosic biomass (Committee on America's Energy Future, 2009).

In the biological conversion of lignocellulosic feedstocks, pretreatment is required. Pre-treatments break down cellulose and hemicellulose into sugars and separate lignin and other plant constituents from fermentable materials. Pretreatment technologies are physical, biological, and combinatorial; the form of pretreatment used will depend on the nature of the feedstock. Physical pretreatment includes gamma-ray exposure; chemical pretreatment methods include the use of acids, alkali, and ionic liquids; and biological methods include use of microorganisms to degrade lignin and hemicellulose (Zheng et al., 2009). For more information on pretreatment technologies and related developments in agronomic qualities of bioenergy crops, see Coulman et al. (2013) and Sticklen (2006).

CO-PRODUCTS AND BYPRODUCTS

The production of bioenergy typically results in the production of nonenergy co-products (i.e., products for which there is an economic use) and byproducts (i.e., products for which there does not yet exist an economic use). All of the conversion pathways can result in co-product creation, but primary co-products are discussed here, including distillers grains, bagasse, glycerol, and biochar.

The primary co-products of corn grain ethanol are cereal byproducts of the distillation process, notably dried distillers grains (DDGs) and wet distillers grains (WDGs). Both DDGs and WDGs can be used for animal feed, mostly cattle, but transportation costs are less for DDGs than distillers grains that have not been dried. Most DDGs and WDGs are used in the dairy and feedlot industries. For more information on use of DDGs and WDGs to feed cattle, see Schingoeth et al. (2009).

Bagasse is the fibrous material left after sugarcane or energy cane has been crushed and the juice extracted. It can be used directly as a feedstock (i.e., burned directly in biopower combustion), but it can also be used in cellulosic biofuel production. Bagasse is sometimes used as a primary fuel for heat and electricity use in sugar mills. Glycerol, also known as glycerin, is the primary co-product of biodiesel production. It is a liquid co-product of the transesterification of oils and fats. Glycerol is used in a range of products including pharmaceuticals, cosmetics, and in the food industry as a sweetener, emulsifier, and solvent. Approximately 950,000 tons of glycerol are produced in the United States annually. However, because of the growth of biodiesel production, there is currently more glycerol produced than market demand. As a result, a substantial amount of glycerol is considered surplus (Pinki et al., 2010).

A primary co-product of gasification and pyrolysis is biochar, or black carbon. Biochar is a form of charcoal that is most commonly used as a fertilizer or soil amendment. The amount of biochar produced varies according to temperature of thermal conversion process, with cooler temperatures resulting in more biochar. There is currently a great deal of scientific interest in the potential use of biochar to sequester carbon in soils (e.g., Laird, 2008).

There are three primary wastes resulting from bioenergy production: fly ash (combustion), wastewater (from anaerobic digestion and fermentation), and gaseous effluent (from combustion, gasification, and pyrolysis). Fly ash and waste water are discussed here.

Fly ash is fine particulate matter resulting from combustion of solid fuel, which is either carried into the air or may fall to the bottom of certain furnaces/boilers. It is composed of substantial amounts of silica. It is an air pollutant and known health hazard regulated by the U.S. Environmental Protection Agency, which issues specifications for its collection and removal. Less than half of the fly ash produced in the United States is recycled, most often as an amendment to Portland cement (Berry and Malhotra, 1980).

Waste water is a major challenge in the production of first-generation bioethanol and biodiesel. Grains are crushed and then mixed with water for fermentation to produce bioethanol. After fermentation, water is removed from the alcohol through distillation. The leftover water contains proteins, residual sugars, enzymes, and dead yeast cells. Water is used in the biodiesel process to remove methanol and glycerol contaminants. Bioethanol and biodiesel effluents must be treated before being reused or being discharged to nearby surface waters (which is an activity subject to regulation). The primary treatment option for biofuel effluents is biodigestion (i.e., anaerobic digestion).

SOCIAL, ECONOMIC, AND ENVIRONMENTAL IMPACTS

There are many potential benefits of bioenergy, as well as many potential negative effects. Planners, policymakers, and other decision-makers are frequently interested in impacts to communities, economies, and the environment. Hence, these issues are discussed below.

COMMUNITY IMPACTS OF BIOENERGY

In rural areas across the country, community leaders are reconsidering traditional drivers of economic activity in search of sustainable, diversified, and environmentally friendly options. Bioenergy and renewable energy may be viable economic development options for communities that can grow dedicated energy crops and develop energy industries to process those crops into power or fuel. The development of a bioenergy industry may be particularly well suited for local economies (given adequate investors) in that the costs of transporting bioenergy crops makes local processing necessary. Thus, economic activity and economic benefit may stay local, although local net benefits are not always guaranteed when balanced against negative impacts to community life and well-being, such as increased truck and/or train traffic (i.e., noise, air quality, traffic safety), and odors and noise from the biomass conversion facility (Bain, 2011; Selfa et al., 2011). Economic benefits must also be weighed against impacts to water supply and other resources. Each community and situation is different, and local decisions around the choice of energy crops, processing systems, and markets will define the economic benefits, while state and federal policy can provide incentives and influence outcomes.

FOOD SECURITY

Chief among the concerns over impacts of bioenergy is food security, or lack thereof. These concerns stem largely from a mid-2008 crisis when an unexpected rise in grain prices created supply shortages in some countries (Nonhebel, 2012). Critics were quick to label biofuels as the leading cause of the crisis, but years later the cause is still being debated. Many analysts have concluded that tight interlinkages of global commodity supply and trade, price speculation, and other factors were as much if not more to blame than biomass production (Godfray et al., 2010; Mittal, 2009; Mueller et al., 2011). Nonetheless, in an increasingly globalized world economy, bioenergy is inextricably connected to complex webs of price/supply fluctuations of agricultural commodity crops. Sustainability analyses of bioenergy are thus converging on a multitactic approach for resolving the multiple challenges of providing food, energy, and environmental protection for the world population (Groom et al., 2008; Reijnders, 2006; Tilman et al., 2009). For more consideration of food-insecurity bioenergy challenges, see Bryngelsson and Lindgren (2013), Foley et al. (2011) and Tilman et al. (2009).

ECOLOGICAL AND ENVIRONMENTAL IMPACTS OF BIOENERGY

Bioenergy can have positive and negative effects on ecosystems and species within them, and geophysical systems such as water and climate. The overall net impact can also be positive or negative depending on the particular system or project under consideration. Specific impacts and net impacts depend on the feedstock type, biomass production system, conversion technology, transportation/distribution system, and use or disposal of co-products and byproducts, respectively. Following is a brief overview of major ecological and environmental impacts of bioenergy, including land use and land use change, greenhouse gas emissions and climate change, wildlife and biodiversity, invasive and transgenic plants, marginal lands, and water quantity and quality.

LAND USE AND LAND USE CHANGE

Land use decisions are affected by many factors, including public policy and the prices of agricultural and forest commodities and petroleum (Figure 1.11). Profitability of specific land uses, noneconomic benefits of alternative land use, and opportunity costs associated with competing land uses also influence decision-making. Decision-makers who choose to produce bioenergy feedstocks must consider land use and land use change. Land use is the management of land resources for economic benefit and includes tillage, maintenance, and harvest activities as well as conservation practices. Land use change (LUC) includes conversion of native ecosystems into agriculture use (i.e., land cover change), as well as switching from one crop type to another (or switching from one forest type to another; Walsh et al., 2003). Land use and LUC associated with bioenergy feedstock production can increase or decrease the direct, indirect, and nonfinancial benefits of native and managed ecosystems. Native ecosystems and managed ecosystems provide many benefits that indirectly affect human well-being and livelihoods. Water and nutrient cycling are two examples of the benefits ecosystems provide but for which there is presently no market value.

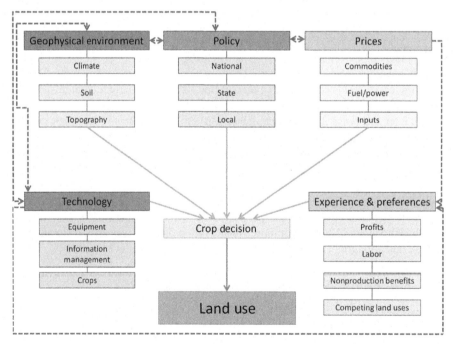

FIGURE 1.11

Factors and their interactions influencing land use decisions.

(Source: Williams, 2011.)

GREENHOUSE GASES AND CLIMATE CHANGE

Greenhouse gases (GHGs) are gases in the atmosphere that absorb and emit thermal radiation in a process known as the greenhouse effect—the mechanism by which solar radiation is captured and earth is warmed to an extent necessary for supporting life. The primary GHGs are water vapor, carbon dioxide, methane, nitrous oxide, and ozone. These gases differ in their abundances in the atmosphere as well as their warming power. The science of global climate change indicates an overall warming trend for the earth as a whole in association with rising levels of GHGs.

According to climate change researchers, the power plant, industrial processes, and transportation sectors are the biggest contributors to annual GHG emissions. These contributions are a result of combustion of fossil fuels, most notably coal, natural gas, and petroleum, which releases the carbon that was once stored in these nonrenewable energy sources into the atmosphere as carbon dioxide and thereby enriches the greenhouse effect. Bioenergy has the potential to be carbon neutral by balancing the amount of carbon released in use of bioenergy products with an equivalent amount put into and stored in soils, plant, and animal tissues, or other material such as the ocean floor. That is, bioenergy is part of a global carbon cycle where plants take up atmospheric carbon dioxide and convert it into plant tissue (i.e., they sequester carbon). Carbon is released back to the atmosphere when the plant biomass is burned directly or after it has been converted into a fuel and used (i.e., carbon emissions), and from there it is available again for plant uptake (i.e., sequestration). However, when taking a life-cycle assessment approach, bioenergy may or may not prove to be carbon neutral. The neutrality of bioenergy with regard to other GHGs such as methane and nitrous oxide is also questionable. Of importance in evaluating the carbon-neutrality of a bioenergy crop are land use change and feedstock production systems.

WILDLIFE AND BIODIVERSITY

Bioenergy and biomass crops are often promoted by environmentalists and government leaders as having the potential to provide wildlife habitat and to support biodiversity. Several important studies provide evidence of the negative impacts of first-generation biofuel on wildlife and biodiversity (e.g., Brooke et al., 2009; Meehan et al., 2010) while others provide evidence that biopower and second-generation biofuels have positive effects on wildlife and biodiversity (e.g., Robertson et al., 2008). However, policies to outline environmental standards for bioenergy production are lacking, and financial programs for compensating land owners and farmers for habitat- and biodiversity-protecting land practices are also lacking. Most significantly, land conversion could decrease native habitats, reduce biodiversity, and decrease ecosystem services (Fargione et al., 2010). Key issues are habitat loss and fragmentation with expanding corn and soybean acres; loss of Conservation Reserve Program land; persistence of pesticides in the environment associated with conventionally managed row crops; timing of the harvest of perennial crops and forests; and impacts to water quality associated with agricultural run-off (Figure 1.12).

INVASIVE AND TRANSGENIC PLANTS

There is growing concern about the invasiveness of plants used for biomass production as the traits of ideal biomass crops are also commonly found among invasive plant species (Raghu et al., 2006). Invasive species are of concern due to adverse environmental and ecological impacts, the economic costs associated with lost productivity of natural ecosystems and the services they provide, and the costs

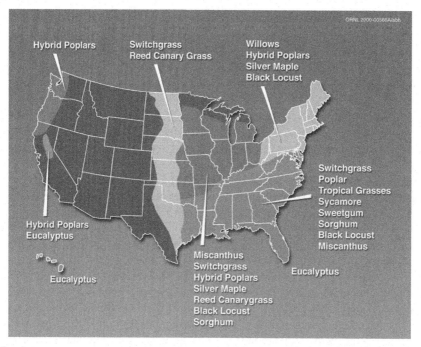

FIGURE 1.12

Distribution of perennial dedicated crops for potential use in bioenergy production.

(Source: Oak Ridge National Laboratory.)

associated with invasive species control (Pimentel et al., 2005). The invasiveness issue is particularly acute for wildlife and biodiversity managers in public agencies and nongovernment organizations (Smith et al., 2013). Warm-season grass monocultures, for example, are seen as providing very little value as wildlife habitat (Fargione et al., 2009; Hartman et al., 2011). Genetic contamination is also a concern. Although it is native to North America, many switchgrass ecotypes and improved germplasms are being introduced to new locations; subsequent out-cross with local ecotypes could erode native biodiversity at local and regional levels (Kwit and Stewart, 2012). The use of transgenic plants, particularly SRWC in agroforestry, is also of concern to managers of ecological systems and wildlife species (Hinchee et al., 2009). There are no easy solutions for these challenges, and many decision-makers must seek to balance the benefits of biofuels and biomass production with known and potential risks. For further overviews of invasive plants and biofuels, see Gordon et al. (2011) and Smith et al. (2013).

SUSTAINABILITY CHALLENGES

It is beyond the scope of this chapter to discuss the complexity of specific challenges to sustainability of bioenergy. Hence, this section offers a broad, integrated definition of sustainability and a statement about its importance. A brief survey of major sustainability challenges is then given, including yield gaps, farmer risk and knowledge, marginal lands, water quality and quantity, and rural development and social justice.

SUSTAINABILITY DEFINED

There are many definitions of sustainability, each supporting various principles and concepts. Essentially, however, sustainability can be described as both a set of goals and practices/behavior that support such goals. As a set of goals, sustainability describes desired conditions of the environment and human well-being as a result of interaction with the environment, now and in the future. As practices and behaviors, sustainability describes human actions that support and enhance the environment and human benefits. Sustainability is important because the choices and actions of today affect everything in the future. Sound decisions at present may prevent undesirable outcomes in the future.

Bioenergy is frequently evoked as an important tool in improving environmental conditions, as well as human lives and livelihoods (Domac et al., 2005; Faaij and Domac, 2006; Tilman et al., 2009). However, much remains to be understood about the impacts of bioenergy on the environment and human society. Ultimately, however, the sustainability of bioenergy will depend on the goals defined; when, where, and by whom those goals are defined; what actions and behaviors people are willing and able to adopt to support those goals; and the ability of science to assist human knowledge of connections between the many aspects of bioenergy and sustainability goals. In the meantime, governments, international agencies, and nongovernment organizations at different levels have produced white papers and various guidelines in an effort to encourage sustainable practices in biomass production (Hull et al., 2011; UNEP, 2009; RSB, 2011).

YIELD GAPS

To meet competing land uses for food, feed, fiber, and fuel, many experts assert that crop yields will need to increase. Although it is thought there is sufficient supply of biomass from forest, agricultural, and waste resources for meeting current and future demands for bioenergy, there is no scientific consensus regarding the simultaneous fulfillment of food and feed needs. The yield potential of biomass sources is constrained by plant genetics and the environments in which plants grow (specifically solar radiation, temperature, and water supply).

The climatic and genetic upper yield boundary of crops (yield potential) far surpasses the average yield (that which farmers actually produce, as an average). Yield gaps—the difference between average and potential yields—result from environmental stresses (e.g., insufficient water supply) and effects of management (planting dates, tillage methods, pest management, etc.). Efforts to decrease yield gaps have focused on the decreasing yield gap between average yield and attainable yield through management and technology improvements (e.g., addressing factors that limit yield, such as water and nutrient availability); and overcoming the limits of plant traits (using plant breeding or biotechnology) to improve tolerance to environmental stress. Biotechnology is thought to have a significant role in yield improvements of bioenergy crops, including annual and perennial crops and nonforest biomass sources of woody materials, such as poplar (*Populus* spp.). For more on yield potentials and yield gaps, see Foley et al. (2011) and Lobell et al. (2009).

FARMER KNOWLEDGE AND RISK

Bioenergy feedstock production involves risk. The agricultural production of biomass perhaps involves more risk than forest-based production because of seasonal weather uncertainties (e.g., flood, drought, hail), fluctuating yields (i.e., interannual variation in yield), and rapidly fluctuating prices.

Farm profitability is directly related to how well these risks can be minimized or absorbed. The USDA's Risk Management Agency has identified five primary sources of risks in agriculture: production, marketing, financial, legal, and human resources (ERS, 2013). Agricultural production of biomass crops presents acute challenges in each of these categories. Traditional strategies for managing risks include crop insurance, revenue insurance, production contracting, and investments in technology. These strategies are not yet applicable in dedicated cellulosic bioenergy crop production.

For biomass producers, knowledge is perhaps the most viable tool currently available for managing risk associated with perennial biomass crop production. However, information about crop varieties and cultivars, crop suitability, agronomic practices and associated technologies, and opportunities for value-added preprocessing is either unknown or not yet widely distributed. That is, farmer knowledge regarding agronomic and financial management of second-generation bioenergy cropping is limited.

MARGINAL LANDS

Definitions of marginality with regard to productivity of arable lands varies greatly, but in general marginal lands are those that have one or more characteristics not conducive to annual crop production. Characteristics such as steep slopes, shallow soils, excessive wetness, or drought proneness generally have negative effects on the profitability of agricultural use. Therefore, marginal lands usually are of fairly low value (i.e., comparatively low price per acre for rent or taxation purposes). Row-crop production on marginal lands is associated with land degradation and decreased productivity over time as a result (Pimentel, 1991). Hence, the production of perennial crops is seen as a potential source of resource protection and income for farmers.

It has been suggested that marginal lands be targeted for production of biomass for bioenergy, not only for meeting renewable energy goals, but also as a potential means for avoiding land use conflicts contributing to food insecurity (Achten et al., 2012; Campbell et al., 2008). The 2007 U.S. Census of Agriculture, for example, identifies approximately 12 million hectares of idle lands, land in cover crops for soil improvement, and fallow rotations as potentially available for biomass production (COA, 2009). Conversion of steep or wet land currently in food crop production (i.e., row crop) to less intensive bioenergy crops such as high diversity, low-input perennial mixes is thought to have the potential to generate more ecosystem services (Tilman et al., 2009). However, some researchers caution that conversion of marginal lands, particularly that in set-aside programs (i.e., currently idle or planted in perennial cover), to more intensively managed bioenergy cropping systems could lead to permanent land degradation and net increases in greenhouse gases and food insecurity (Bryngelsson and Lindgren, 2013; Zenone et al., 2013). Government, academic, and private sector research is needed to assess whether and to what degree marginal lands can be relied upon for meeting future bioenergy demand, while policy makers and other decision-makers address questions of whether and to what degree marginal lands should be relied upon for bioenergy needs.

WATER QUANTITY AND QUALITY

Bioenergy production affects water availability and quality through water use in biomass production and water use during feedstock conversion. Emission of air pollutants from biopower combustion and burning of biofuels also potentially impacts water quality, mostly via precipitation. Many bioenergy feedstocks have relatively high water requirements when grown at commercially viable levels. In areas where there is sufficient precipitation, yields may be close to yield potential (in the case of crops). However, in areas lacking sufficient precipitation, irrigation is required for achieving commercially viable yields (Service, 2009).

The use of irrigation water has the potential to depress groundwater supplies, divert water used to grow food crops to the growth of energy crops, and contribute to soil salinization. In both irrigated and rainfed systems, crops that remain longer on the landscape (i.e., perennial) and crops that provide good ground cover increase structural filtering of precipitation, thus promoting more rainfall interception and retention and better control of sediment run-off. Grasslands, once established, may provide better water services than traditional annual row crops because of greater soil coverage and root density. In addition to sediment run-off, pesticides and fertilizers associated with conventional row crop production threaten water resources. One of the top pollutants to aquatic ecosystems, as recognized by the U.S. EPA, is nitrogen. Nitrogen in excess can be harmful to both ecosystem and human health. Nitrogen and other agronomic pollutants, such as phosphorous, can enter water resources through runoff and leaching. Nutrient pollution is a leading cause of water quality impairment in lakes and the Gulf of Mexico (Costello et al., 2009). The major concern is the promotion of algal growth and subsequent dissolved oxygen depletion, fish mortality, clogged pipelines, and reduced recreational values.

Biopower generation requires use of water to produce high-pressure steam to drive turbines and low-pressure steam to deliver heat/cooling in centralized heating/cooling districts. Some biopower plants also use water for cooling, particularly in the co-firing of coal or natural gas with biomass. Water is also required during fermentation of ethanol and in posttransesterification of biodiesel. Note, however, that water use for ethanol production is lower than that of many other industrial processes. Effluents from these industrial processes must be treated before being discharged to nearby surface waters. Water use for fermentation may deplete surrounding surface or groundwater supplies and limit their availability for other uses including drinking water, wildlife habitat, and recreation.

RURAL DEVELOPMENT AND SOCIAL JUSTICE

In rural areas, community leaders are reconsidering traditional drivers of economic activity in search of sustainable, diversified, and environmentally friendly options. Bioenergy may be a viable economic development option for communities that can grow dedicated energy crops and develop energy industries to process those crops into power or fuel. The development of a bioenergy industry may be particularly well suited for local economies (given adequate investors) in that the costs of transporting bioenergy crops makes local processing necessary. Thus, economic activity and economic benefit may stay local, although local net benefits are not always guaranteed when balanced against negative impacts to community life and well-being, such as food insecurity, increased truck and/or train traffic (i.e., noise, air quality, traffic safety), and odors and noise from the biomass conversion facility (Selfa et al., 2011). Economic benefits must also be weighed against impacts to water supply. Each community and situation is different, and local decisions around the choice of energy crops, processing systems, and markets will define the economic benefits, while state and federal policy can provide incentives and influence outcomes. An additional issue, however, is rural self-determination and empowerment. Government policies tend to overlook social considerations in biofuel development strategies (Mol, 2007; Rossi and Hinrichs, 2011), leading to macrolevel goals that adversely affect local-level realities. For comprehensive reviews of rural development and social justice issues of biofuels, see Dale et al. (2013), and Van der Horst and Vermeylen (2011).

CONCLUSION

Bioenergy is renewable energy derived from recently living biological material, or biomass. Forms of bioenergy include power, heat, and solid, liquid, and gas biofuels. Interest in bioenergy is increasing in response to concerns about energy security, energy independence, and environmental and climate impacts associated with the use of nonrenewable energy resources. Policy, government programs, and sponsored research are major drivers of bioenergy development. Biomass materials, after pre-processing into suitable forms for various conversion technologies, provide feedstock for a variety of bioenergy products and end uses. The majority of biomass for bioenergy comes from forests, agriculture, and waste. Algae and nonforest conservation lands are important emerging sources of bioenergy feedstocks. Major feedstock types include sugars and starches from agricultural crops, and cellulosic/lignocellulosic materials from forests, agriculture, agroforestry, and industrial and construction wastes.

Pressure is increasing for bioenergy to meet energy demands, reduce greenhouse gas emissions, improve soil and water quality, and provide economic development and other socioeconomic benefits. Whether and to what degree bioenergy delivers on the simultaneous demands depends on science, policy, and socioeconomic dynamics from local to global levels. Expansion of biomass production, processing into feedstocks, handling, transportation, and storage, if done sustainably, may provide supply/value chains that support renewable energy goals while enhancing rural livelihoods. Biomass supply is potentially available to meet renewable energy goals. However, land availability, competing land uses, yield potential, yield gaps, producer profitability, and other important constraints influence the potential supply of biomass. Ultimately, decision-makers at all levels must balance the costs, benefits, advantages, and disadvantages among specific biomass types and production systems to make informed decisions with regard to the desired goals for the present and the future.

ACKNOWLEDGMENTS

This material is based upon work supported by the National Institute of Food and Agriculture, U.S. Department of Agriculture, under Agreement No. 2007-51130-03909 to Carol L. Williams and Pam Porter. Any opinions, findings, conclusions, or recommendations expressed in this publication are those of the author(s) and do not necessarily reflect the view of the U.S. Department of Agriculture.

Anju Dahiya gratefully acknowledges support from the Department of Energy, Vermont Sustainable Jobs Fund, National Science Foundation, and Vermont EPSCoR.

REFERENCES

Achten, W.M.J., Mathijs, E., Verchot, L., Singh, V.P., Aerts, R., Muys, B., 2007. *Jatropha* biodiesel fueling sustainability? Biofuels, Bioproducts and Biorefining 1, 283–291.
Achten, W.M.J., Trabucco, A., Maes, W.H., Verchot, L.V., Aerts, R., Mathijs, E., Vantaomme, P., Singh, V.P., Muys, B., 2012. Global greenhouse gas implications of land conversion to biofuel crop cultivation in arid and semi-arid lands – lessons learned from *Jatropha*. Journal of Arid Environments, 1–11.
Adler, P.R., Sanderson, M.A., Weimer, P.J., Vogel, K.P., 2009. Plant species composition and biofuel yields of conservation grasslands. Ecological Applications 19, 2202–2209.

Antizar-Ladislao, B., Turrion-Gomez, J.L., 2008. Second-generation biofuels and local bioenergy systems. Biofuels, Bioproducts and Biorefining 2, 455–469.

Atabani, A.E., Silitonga, A.S., Badruddin, I.A., Mahlia, T.M.I., Masjuki, H.H., Mekhilef, S., 2012. A comprehensive review on biodiesel as an alternative energy resource and its characteristics. Renewable and Sustainable Energy Review 16, 2070–2093.

Bain, C., 2011. Local ownership of ethanol plants: what are the effects on communities? Biomass and Bioenergy 35, 1400–1407.

Baker, A., Zahniser, S., 2006. Ethanol reshapes the corn market. Amber Waves 4, 30–35. U.S. Department of Agriculture, Economic Research Service. Available at: http://www.agclassroom.org/teen/ars_pdf/social/amber/ethanol.pdf.

Bergmann, J.C., Tupinamba, D.D., Costa, O.Y.A., Aleida, J.R.M., Barreto, C.C., Quirino, B.F., 2013. Biodiesel production in Brazil and alternative biomass feedstocks. Renewable and Sustainable Energy Reviews 21, 411–420.

Berry, E.E., Malhotra, V.M., 1980. Fly ash for use in concrete – a critical review. American Concrete Institute Journal Proceedings 77, 59–73.

Botha, T., Blottnitz, H.V., 2006. A comparison of the environmental benefits of bagasse-derived electricity and fuel ethanol on life-cycle basis. Energy Policy 34, 2654–2661.

Brooke, R., Fogel, G., Glaser, A., Griffin, E., Johnson, K., 2009. Corn Ethanol and Wildlife: How Increases in Corn Plantings Are Affecting Habitat and Wildlife in the Prairie Pothole Region. National Wildlife Federation, Washington, DC. Available at: http://hdl.handle.et/2027.42/62096.

Bryngelsson, D.K., Lindgren, K., 2013. Why large-scale bioenergy production on marginal is unfeasible: a conceptual partial equilibrium analysis. Energy Policy 55, 454–466.

Bush, G.W., 2007. Twenty in Ten: Strengthening America's Energy Future. State of the Union Address. Available at: http://georgewbush-whitehouse.archives.gov/stateoftheunion/2007/initiatives/energy.html.

Campbell, J.E., Lobell, D.B., Genova, R.C., Field, C.B., 2008. The global potential of bioenergy on abandoned agricultural lands. Environmental Science and Technology 42, 5791–5794.

Casler, M.D., Vogel, K.P., Taliaferro, C.M., Ehlke, N.J., Berdahl, J.D., Brummer, E.C., Kallenbach, R.L., West, C.P., Mitchell, R.B., 2007. Latitudinal and longitudinal adaptation of switchgrass populations. Crop Science 47, 2249–2260.

Christian, D.G., Riche, A.B., Yates, N.E., 2008. Growth, yield and mineral content of *Miscanthus x giganteus* grown as a biofuel for 14 successive harvests. Industrial Crops and Products 28, 320–327.

Chisti, Y., 2007. Biodiesel from microalgae. Biotechnology Advances 25, 294–306.

Clark, J.H., Budarin, V., Deswaarte, F.E.I., Hardy, J.J.E., Kerton, F.M., Hunt, A.J., Luque, R., Macquarrie, D.J., Milkowski, K., Rodriguez, A., Samuel, O., Tavener, S.J., White, R.J., Wilson, A.J., 2006. Green chemistry and the biorefinery: a partnership for a sustainable future. Green Chemistry 8, 853–860.

COA, 2009. 2007 Census of Agriculture. AC-07-A-51. U.S. Department of Agriculture, Washington, DC. Available at: http://www.agcensus.usda.gov/Publications/2007/Full_Report/usv1.pdf (accessed 08.09.13.).

Coleman, M.D., Stanturf, J.A., 2006. Biomass feedstock production systems: economic and environmental benefits. Biomass and Bioenergy 30, 693–695.

Committee on America's Energy Future, 2009. America's Energy Future: Technology and Transformation. National Academy of Science, National Academy of Engineering, National Research Council, Washington, DC. Available at: http://www.nap.edu/catologue/12710.html.

Cooke, P., Kohlin, G., Hyde, W.F., 2008. Fuelwood, forests and community management – evidence from household studies. Environment and Development Economics 13, 103–135.

Costello, C., Griffin, W.M., Landis, A.E., Matthews, H.S., 2009. Impact of biofuel crops production on the formation of hypoxia in the Gulf of Mexico. Environmental Science and Technology 43, 7985–7991.

Coulman, B., Dalai, A., Heaton, E., Lee, C.P., Lefsrud, M., Levin, D., Lemaux, P.G., Neale, D., Shoemaker, S.P., Singh, J., Smith, D.L., Whalen, J.K., 2013. Developments in crops and management systems to improve lignocellulosic feedstock production. Biofuels, Bioproducts, and Biorefining 7, 582–601.

Dahiya, A., 2012. Integrated approach to algae production for biofuel utilizing robust algae species. In: Gordon, R., Seckbach, J. (Eds.), The Science of Algal Fuels: Cellular Origin, Life in Extreme Habitats and Astrobiology, vol. 25. Springer, Dordrecht, pp. 83–100.

Dahiya, A., Todd, J., McInnis, A., 2012. Wastewater treatment integrated with algae production for biofuel. In: Gordon, R., Seckbach, J. (Eds.), The Science of Algal Fuels: Cellular Origin, Life in Extreme Habitats and Astrobiology, vol. 25. Springer, Dordrecht, pp. 447–466.

Dale, V.H., Kline, K.L., Wright, L.L., Perlack, R.D., Downing, M., Graham, R.L., 2011. Interactions among bioenergy feedstock choice, landscape dynamics, and land use. Ecological Applications 21, 1039–1054.

Dale, V.H., Efroymson, R.A., Kline, K.L., Langholtz, M.H., Leiby, P.N., Oladosu, G.A., David, M.R., Downing, M.E., Hilliard, M.R., 2013. Indicators for assessing socioeconomic sustainability of bioenergy systems: A short list of practical measures. Ecol Indicators 26, 87–102.

Dickmann, D.I., 2006. Silviculture and biology of short-rotation woody crops in temperate regions: then and now. Biomass and Bioenergy 30, 696–705.

Demirbas, A., 2004. Combustion characteristics of different biomass fuels. Progress in Combustion and Energy Science 30, 219–230.

DoE: Algae Roadmap Publication, 2010. In: Algal Biofuels Technology Roadmap Workshop. US Department of Energy, Office of Energy Efficiency & Renewable Energy (EERE), Office of the Biomass Program. December 9–10, 2008. UMD. http://www1.eere.energy.gov/bioenergy/pdfs/algal_biofuels_roadmap.pdf.

Domac, J., Richards, K., Risovic, S., 2005. Socio-economic drivers in implementing bioenergy projects. Biomass and Bioenergy 28, 97–106.

Downing, M., Eaton, L.M., Graham, R.L., Langholtz, M.H., Perlack, R.D., Turhollow Jr., A.F., Stokes, B., Brandt, C.C., 2011. U.S. Billion-Ton Update: Biomass Supply for a Bioenergy and Bioproducts Industry. U.S. Department of Energy.

EISA, 2007. Energy Independence and Security Act. One hundred tenth Congress of the United States of America. Government Printing Office, Washington, DC. Available at: http://www.gpo.gov/fdsys/pkg/BILLS-110hr6enr/pdf/BILLS-110hr6enr.pdf.

Elghali, L., Clift, R., Sinclair, P., Panoutsou, C., Bauen, A., 2007. Developing a sustainability framework for the assessment of bioenergy systems. Energy Policy 35, 6075–6083.

ERS, 2013. Risk in Agriculture. Economic Research Agency, US. Department of Agriculture. Available at: http://www.ers.usda.gov/topics/farm-practices-management/risk-management/risk-in-agriculture.aspx.

Faaij, A.P.C., Domac, J., 2006. Emerging international bio-energy markets and opportunities for socio-economic development. Energy for Sustainable Development X, 7–19.

Fargione, J.E., Cooper, T.R., Flashpohler, D.J., Hill, J., Lehman, C., McCoy, T., McLeod, S., Nelson, E.J., Oberhauser, K.S., Tilman, D., 2009. Bioenergy and wildlife: threats and opportunities for grassland conservation. BioScience 59, 767777.

Fargione, J.E., Plevin, R.J., Hill, J.D., 2010. The ecological impact of biofuels. Annual Review of Ecology, Evolution, and Systematics 41, 351–377.

Fike, J.H., Parrish, D.J., Wolf, D.D., Blasko, J.A., Green Jr., J.T., Rasnake, M., Reynolds, J.H., 2006. Switchgrass production for the upper southeastern USA: influence of cultivar and cutting frequency on biomass yields. Biomass and Bioenergy 30, 207–213.

Foley, J.A., Ramankutty, N., Brauman, K.A., Cassidy, E.S., Gerber, J.S., Johnston, M., Mueller, N.D., O'Connell, C., Ray, D.K., West, P.C., Balzer, C., Bennett, E.M., Carpenter, S.R., Hill, J., Monfreda, C., Polasky, S., Rockstrom, J., Sheehan, J., Siebert, S., Tilman, D., Zaks, D.P.M., 2011. Solutions for a cultivated planet. Nature 478, 337–342.

FCEA, 2008. Food, Conservation, and Energy Act. One hundred and tenth Congress of the United States. Government Printing Office, Washington, DC. Available at: http://www.gpo.gov/fdsys/pkg/PLAW-110publ234/html/PLAW-110publ234.htm.

Garrett, H.E., McGraw, R.L., Walter, W.D., 2009. Alley cropping practices. In: Garrett, H.E. (Ed.), American Agroforestry: An Integrated Science and Practice, second ed. American Society of Agronomy, Madison, Wisconsin, USA, pp. 133–162.

Godfray, H.C.J., Beddington, J.R., Crute, I.R., Haddad, L., Lawrence, D., Muir, J.F., Pretty, J., Robison, S., Thomas, S.M., Toulmin, C., 2010. Food insecurity: the challenge of feeding 9 billion people. Science 327, 812–818.

Goh, C.S., Lee, K.T., 2010. A visionary and conceptual macroalgae-based third-generation bioethanol (TGB) biorefinery in Sahab, Malaysia as an underlay for renewable and sustainable development. Renewable and Sustainable Energy Reviews 14, 842–848.

Glover, J.D., Culman, S.W., DuPont, S.T., Broussard, W., Young, L., Mangan, M.E., Mai, J.G., Crews, E., DeHaan, L.R., Buckley, D.H., Ferris, H., Turner, R.E., Reynolds, H.L., Wyse, D.L., 2010. Harvested perennial grasslands provide ecological benchmarks for agricultural sustainability. Agriculture, Ecosystems and Environment 137, 3–12.

Gordon, D.R., Tancig, K.J., Onderdonk, D.A., Gantz, C.A., 2011. Assessing the invasive potential of biofuel species proposed for Florida and the United States using the Australian Weed Risk Assessment. Biomass and Bioenergy 35, 74–79.

Groom, M.J., Gray, E.M., Townsend, P.A., 2008. Biofuels and biodiversity: principles for creating better policies for biofuel production. Conservation Biology 22, 602–609.

Hartman, J.C., Nippert, J.B., Orozco, R.A., Springer, C.J., 2011. Potential ecological impacts of switchgrass (*Panicum virgatum* L.) biofuel cultivation in the Central Great Plains, USA. Biomass and Bioenergy 35, 3415–3421.

Headlee, W.L., Hall, R.B., Zalensy Jr., R.S., 2013. Establishment of alleycropped hybrid aspen "Crandon" in Central Iowa, USA: effects of topographic position and fertilizer rate on aboveground biomass production and allocation. Sustainability 5, 2874–2886.

Heaton, E., Voigt, T., Long, S.P., 2004. A quantitative review comparing the yields of two candidate C4 perennial biomass crops in relation to nitrogen, temperature and water. Biomass and Bioenergy 27, 21–30.

Heaton, E.A., Dohleman, F.G., Long, S.P., 2008. Meeting US biofuel goals with less land: the potential of *Miscanthus*. Global Change Biology 14, 2000–2014.

Heaton, E.A., Dohleman, F.G., Miguez, A.F., Juvik, J.A., Lozovaya, V., Widholm, J., Zabotina, O.A., McIssac, G.F., David, M.B., Voigt, T.B., Boersma, N.N., Long, S.P., 2010. Miscanthus: a promising biomass crop. Univ. of Illinois at Urbana-Champaign. Advances in Botanical Research 56, 76–137.

Hedenus, F., Azar, C., 2009. Bioenergy plantations or long-term carbon sinks? – A model based analysis. Biomass and Bioenergy 33, 1693–1702.

Hinchee, M., Rottmann, W., Mullinax, L., Zhang, C., Chang, S., Cunningham, M., Pearson, L., Nehra, N., 2009. Short-rotation woody crops for bioenergy and biofuels applications. In Vitro Cellular Development Biology – Plant 45, 619–629.

Hull, S., Arntzen, J., Bleser, C., Crossley, A., Jackson, R., Lobner, E., Paine, L., Radloff, G., Sample, D., Vandenbrook, J., Ventura, S., Walling, S., Widholm, J., Williams, C., 2011. Wisconsin Sustainable Planting and Harvest Guidelines for Nonforest Biomass. Wisconsin Department of Agriculture, Trade, and Consumer Protection. Available at: http://datcp.wi.gov/uploads/About/pdf/WI-NFBGuidelinesFinalOct2011.pdf.

Huntley, M.E., Redalje, D.G., May 2007. CO_2 mitigation & renewable oil from photosynthetic microbes: a new appraisal. Mitigation and Adaption Strategies for Global Change 12 (4), 573–608 (36).

Jack, M.W., 2009. Scaling laws and technology development strategies for biorefineries and bioenergy plants. Bioresource Technology 100, 6324–6330.

Jones, M.B., Walsh, M. (Eds.), 2001. Miscanthus for Energy and Fibre. James 7 Jams Ltd, London.

Kim, S., Dale, B.E., 2004. Global potential bioethanol production from wasted crops and crop residues. Biomass and Bioenergy 26, 361–375.

Kleinschmidt, J., 2007. Biofueling Rural Development: Making the Case for Linking Biofuel Production to Rural Revitalization. Policy Brief no. 5. Carsey Institute, University of New Hampshire, Durham.

Kwit, C., Stewart, C.N., 2012. Gene flow matters in switchgrass (*Panicum virgatum* L.), a potential widespread biofuel feedstock. Ecological Applications 22, 3–7.

Laird, D.A., 2008. The charcoal vision: a win-win-win scenario for simultaneously producing bioenergy, permanently sequestering carbon, while improving soil and water quality. Agronomy Journal 100, 178–181.

Lal, R., 2005. World crop residues production and implications for its use as a biofuel. Environment International 31, 575–584.

Lee, R.A., Lavoie, J.-M., 2013. From first- to third-generation biofuels: challenges of producing a commodity from a biomass of increasing complexity. Animal Frontiers 3, 6–11.

Lehtikangas, P., 2001. Quality properties of pelletised sawdust, logging residues and bark. Biomass and Bioenergy 20, 351–360.

Lemus, R., Brummer, E.C., Moore, K.J., Molstad, N.E., Burras, C.L., Barker, M.F., 2002. Biomass yield and quality of 20 switchgrass populations in southern Iowa, USA. Biomass and Bioenergy 23, 433–442.

Lewandowski, I., Clifton-Brown, J.C., Scurlock, J.M.O., Huisman, W., 2000. Miscanthus: European experience with a novel energy crop. Biomass and Bioenergy 19, 209–227.

Lewandowski, I., Scurlock, J.M.O., Lindvall, E., Christou, M., 2003. The development and current status of perennial rhizomatous grasses as energy crops in the US and Europe. Biomass and Bioenergy 25, 335–361.

Lobell, D.B., Cassman, K.G., Field, C.B., 2009. Crop yield gaps: their importance, magnitudes, and causes. Annual Review of Environment and Resources 34, 179–204.

Manurung, R., Wever, D.A.Z., Wildschut, J., Venderbosch, R.H., Hidayat, H., van Dam, J.E.G., Liejenhorst, E.J., Broekhuis, A.A., Heeres, H.J., 2009. Valorization of *Jatropha curcas* L. plant parts: nut shell conversion to fast pyrolysis oil. Food and Bioproducts Processing 87, 187–196.

McCormick, K., Kaberger, T., 2007. Key barrier for bioenergy in Europe: economic conditions, know-how, and institutional capacity, ad supply chain co-ordination. Biomass and Bioenergy 31, 443–452.

McKendry, P., 2002. Energy production from biomass (part 1): overview of biomass. Bioresource Technology 83, 37–46.

McLaughlin, S.B., Kszos, L.A., 2005. Development of switchgrass (*Panicum virgatum*) as a bioenergy feedstock in the United States. Biomass and Bioenergy 28, 515–535.

Meehan, T.D., Hurlbert, A.H., Gratton, C., 2010. Bird communities in future bioenergy landscapes of the upper midwest. Proceedings of the National Academy of Sciences 107, 18533–18538.

Meinhausen, M., Meinhausen, N., Hare, W., Raper, S.C.B., Frieler, K., Knutti, R., Frame, D., Allen, M.R., 2009. Greenhouse-gas emission targets for limiting global warming to 2 °C. Nature 458, 1158–1163.

Mittal, A., 2009. The 2008 Food Price Crisis: Rethinking Food Security Policies. G-24 Discussion Paper Series, Research papers for the Intergovernmental Group of Twenty-Four on International Monetary Affairs and Development. United Nations, New York and Geneva. Available at: http://www.g24.org/Publications/Dpseries/56.pdf (accessed 08.09.13.).

Mol, A.P.J., 2007. Boundless biofuels? Between environmental sustainability and vulnerability. Sociologia Ruralis 47, 297–315.

Mueller, S.A., Anderson, J.E., Wallington, T.J., 2011. Impact of biofuel production and other supply and demand factors on food price increases in 2008. Biomass and Bioenergy 35, 1623–1632.

Nonhebel, S., 2012. Global food supply and the impacts of increased use of biofuels. Energy 37, 115–121.

Parrish, D.J., Fike, J.H., 2005. The biology and agronomy of switchgrass for biofuels. Critical Reviews in Plant Sciences 24, 423–459.

Perlack, et al., 2005. Biomass as Feedstock for a Bioenergy and Bioproducts Industry: The Technical Feasibility of a Billion-ton Annual Supply. DOE/GO-10200502135. US DOE, Oak Ridge, TN. Available at:http://www1.eere.energy.gov/biomass/pdfs/final_billionton_vision_report2.pdf.

Pimentel, D., 1991. Ethanol fuels: energy security, economics, and the environment. Agricultural and Environmental Ethics 4, 1–13.

Pimentel, D., Zuniga, R., Morrison, D., 2005. Update on the environmental and economic costs associated with alien-invasive species in the United States. Ecological Economics 52, 273–288.

Pinki, A., Saxena, R.K., Sweta, Y., Firdas, J., 2010. A greener solution for a darker side of biodiesel: utilization of crude glycerol in 1,3-propanediol production. Journal of Biofuels 1, 83–91.

Price, L., Bullard, M., Lyons, H., Anthony, S., Nixon, P., 2004. Identifying the yield potential of *Miscanthus x giganteus*: an assessment of the spatial and temporal variability of *M x giganteus* biomass productivity across England and Whales. Biomass and Bioenergy 26, 3–13.

Prochow, A., Heiermann, M., Plochl, M., Amon, T., Hobbs, P.J., 2009. Bioenergy from permanent grassland – a review: 2. Combustion. Bioresource Technology 100, 4945–4954.

Raghu, S., Anderson, R.C., Daehler, C.C., Davis, A.S., Wiedenmann, R.N., Sumberloff, D., Mack, R.N., 2006. Adding biofuels to the invasive species fire? Science 313, 1742.

Reijnders, L., 2006. Conditions for the sustainability of biomass based fuel use. Energy Policy 34, 863–876.

Robertson, G.P., Dale, V.H., Doering, O.C., Hamburg, S.P., Melillo, J.M., Wander, M.M., Parton, W.J., Adler, P.R., Barney, J.N., Cruse, R.M., Duke, C.S., Fearnside, P.M., Follett, R.F., Gibbs, H.K., Goldemberg, J., Mladenoff, D.J., Ojima, D., Palmer, M.W., Sharpley, A., Wallace, L., Weathers, K.C., Weins, J.A., Wilhelm, W.W., 2008. Sustainable biofuels redux. Science 322, 49–50.

Rockwood, D.L., Rudie, A.W., Ralph, S.A., Zhu, J.Y., Winandy, J.E., 2008. Energy product options for *Eucalyptus* species grown as short rotation woody crops. International Journal of Molecular Science 9, 1361–1378.

Rosch, C., Skarka, J., Raab, K., Stelzer, V., 2009. Energy production from grassland – assessing the sustainability of different process chains under German conditions. Biomass and Bioenergy 33, 689–700.

Rosillo-Calle, F., Pelkmans, L., Walter, A., 2009. A Global Overview of Vegetable Oils, with Reference to Biodiesel. Report for the IEA Bioenergy Task Force 40. International Energy Agency, Paris, France.

Rossi, A.M., Hinrichs, C.C., 2011. Hope and skepticism: farmer and local community views on the socio-economic benefits of agricultural bioenergy. Biomass and Bioenergy 35, 1418–1428.

RSB, 2011. Consolidated RSB EU RED Principles & Criteria for Sustainable Biofuel Production. RSB-STD-11-001-01-001 (Ver 2.0). The Round Table on Sustainable Biofuels, Lausanne, Switzerland. Available at: http://rsb.org/pdfs/standards/RSB-EU-RED-Standards/11-05-10-RSB-STD-11-001-01-001-vers-2-0-Consolidated-RSB-EU-RED-PCs.pdf (accessed 08.09.13.).

Salvi, B.L., Panwar, N.L., 2012. Biodiesel resources and production technologies – a review. Renewable and Sustainable Energy Reviews 16, 3680–3689.

Sanderson, M.A., Adler, P.R., 2008. Perennial forages as second generation bioenergy crops. International Journal of Molecular Sciences 9, 768–788.

Sandia National Labs, 2009. 90-Billion Gallon Biofuel Deployment Study. U.S. Department of Energy Publications 84. U.S. Department of Energy, Washington, DC. Available at: http://digitalcommons.unl.edu/cgi/viewcontent.cgi?article=1083&context=usdoepub.

Searcy, E., Flynn, P., Ghafoori, E., Kumar, A., 2007. The relative cots of biomass energy transport. Applied Biochemistry and Biotechnology 136–140, 639–652.

Schenk, P.M., Thomas-Hall, S.R., Stephens, E., Marx, U.C., Mussgnug, J.H., Kruse, O., Hankame, B., 2008. 2nd generation biofuels: high-efficiency microalgae for biodiesel production. BioEnergy Research vol. 1, 20–43.

Schingoeth, D.J., Kalscheur, K.F., Hippen, A.R., Garcia, A.D., 2009. The use of distillers products in dairy cattle diets. Journal of Dairy Science 92, 5802–5813.

Selfa, T., Kulcsar, L., Bain, C., Goe, R., Middendorf, G., 2011. Biofuels Bonanza? Exploring community perception of the promises and perils of biofuels production. Biomass and Bioenergy 35, 1379–1389.

Service, R., 2009. Another biofuels drawback: the demand for irrigation. Science 326, 516–517.

Sheehan, J., Dunahay, T., Benemann, J., Roessler, P., 1998. A Look Back at the U.S. Department of Energy's Aquatic Species Program-Biodiesel from Algae. National Renewable Energy Program.

Smith, A.L., Klenk, N., Wood, S., Hewitt, N., Henriques, I., Yan, N., Bazley, D.R., 2013. Second generation biofuels and bioinvasions: an evaluation of invasive risks and policy responses in the United States and Canada. Renewable and Sustainable Energy Reviews 27, 30–42.

Stephen, J.D., Mabee, W.E., Saddler, J.N., 2010. Biomass logistics as a determinant of second-generation biofuel facility scale, location and technology selection. Biofuels, Bioproducts and Biorefining 4, 503–518.

Sticklen, M., 2006. Plant genetic engineering to improve biomass characteristics for biofuels. Current Opinion in Biotechnology 17, 315–319.

Suntana, A.S., Vogt, K.A., Turnblom, E.C., Upadhye, R., 2009. Bio-methanol potential in Indonesia: forest biomass as a source of bio-energy that reduces carbon emissions. Applied Energy 86, 5215–5221.

Taylor, C.M., Pollard, S.J.T., Angus, A.J., Rocks, S.A., 2013. Better by design: rethinking interventions for better environmental regulation. Science of the Total Environment 447, 488–499.

Tilman, D., Hill, J., Lehman, C., 2006. Carbon-negative biofuels from low-input high-diversity grassland biomass. Science 314, 1598–1600.

Tilman, D., Socolow, R., Foley, J.A., Hill, J., Larson, E., Lynd, L., Pacala, S., Reilly, J., Searchinger, T., Somerville, C., Williams, R., 2009. Beneficial biofuels – the food, energy and environment trilemma. Science 325, 270–271.

Tumuluru, J.S., Wright, C.T., Kenny, K.L., Hess, J.R., 2010. A Technical Review on Biomass Processing: Densification, Preprocessing, Modeling and Optimization. ASABE Paper NO. 1009401. American Society of Agricultural and Biological Engineers, St. Joseph, Michigan.

UNEP, 2009. Towards Sustainable Production and Use of Resources: Assessing Biofuels. United Nations Environment Programme. Available at: http://www.unep.org/pdf/biofuels/Assessing_Biofuels_Full_Report.pdf (accessed 08.09.13.).

U.S. Department of Energy, 2011. U.S. Billion-Ton Update: Biomass Supply for a Bioenergy and Bioproducts Industry. RD Perlack and BJ Stokes (leads). ORNL/TM-2011/224. Oak Ridge National Laboratory, Oak Ridge, TN.

Van der Horst, D., Vermeylen, S., 2011. Spatial scale and social impacts of biofuel production. Biomass and Bioenergy 35, 2435–2443.

Virgilio, N.D., Monti, A., Venturi, G., 2007. Spatial variability of switchgrass (Panicum virgatum L.) yield as related to soil parameters in a small field. Field Crops Research 101, 232–239.

Volk, T.A., Verwijst, T., Tharaken, P.J., Abrahamson, L.P., White, E.H., 2004. Growing fuel: a sustainability assessment of willow biomass crops. Frontiers in Ecology and the Environment 2, 411–418.

Walsh, M.E., De La Torre Ugarte, D.G., Shapouri, H., Slinksky, S.P., 2003. Bioenergy crop production in the United State. Environmental and Resource Economics 24, 313–333.

Waxman, H.A., 1991. An overview of the Clean Air Act amendments of 1990. Environmental Law 21, 1721–1766.

Williams, C., Porter, P., 2011. Introduction to Bioenergy. Module 1. In: Lezberg, S., Mullins, J. (Eds.), Bio- energy and Sustainability Course. On-line curriculum. Bioenergy Training Center. http://fyi.uwex.edu/biotrainingcenter.

Willyard, C., Tilkalsky, S., 2006. Bioenergy in Wisconsin: The Potential Supply of Forest Biomass and its Relationship to Biodiversity. State of Wisconsin, Department of Administration, 48 pp.

Wright, L., Turhollow, A., 2010. Switchgrass selection as a "model" bioenergy crop: a history of the process. Biomass and Bioenergy 34, 851–868.

Zheng, Y., Pan, Z., Zhang, R., 2009. Overview of biomass pretreatment for cellulosic ethanol production. International Journal of Agricultural and Biological Engineering 2, 51–68.

Zenone, T., Gelfand, I., Chen, J., Hamilton, S.K., Robertson, G.P., 2013. From set-aside grassland to annual and perennial cellulosic biofuel crops: effects of and use change on carbon balance. Agricultural and Forest Meteorology 182–183, 1–12.

INTRODUCTION TO BIODIESEL AND GLOSSARY OF TERMS

National Biodiesel Board, USA

- **What is biodiesel?**—Biodiesel is a drop-in diesel alternative, made from domestic, renewable resources such as plant oils, animal fats, used cooking oil, and even new sources such as algae. Biodiesel contains no petroleum, but can be blended with petroleum diesel. Biodiesel blends can be used in compression-ignition (diesel) engines with little or no modifications. Biodiesel is cleaner burning, simple to use, biodegradable, nontoxic, and essentially free of sulfur and aromatics. Biodiesel is America's advanced biofuel.

- **What biodiesel is not**—Biodiesel is not raw vegetable oil. Fuel-grade biodiesel must be produced to strict industry specifications (ASTM D6751) in order to ensure proper performance. Only biodiesel that meets the specification and is registered with the Environmental Protection Agency is a legal motor fuel.

  - **Biodiesel is not the same as ethanol**. Biodiesel is made from a variety of materials for use in diesel engines, with different properties and benefits than ethanol.

- **Biodiesel, n.**—A fuel comprised of mono-alkyl esters of long chain fatty acids derived from vegetable oils or animal fats, meeting ASTM D 6751, designated B100. (i.e., the pure fuel).

- **Biodiesel blend, n.**—A blend of biodiesel fuel with petroleum-based diesel fuel designated BXX, where XX is the volume percent of biodiesel (i.e., B5 = 5 percent biodiesel blended with 95 percent petroleum diesel; B20 = 20 percent biodiesel blended with 80 percent petroleum diesel, etc.).

- **Feedstocks**—The raw materials used to make biodiesel fuel. Common biodiesel feedstocks in the U.S. include:
  - **Vegetable oils** from soybeans, canola, camelina, sunflower, cottonseed
  - Dry distillers grain (**DDG**) **corn oil** left over from ethanol production process
  - **Used cooking oil**/yellow grease
  - **Animal fats** including beef tallow, pork lard, poultry fat
  - **Future feedstocks** including algae, pennycress, jatropha, brown grease, halophytes, low ricin castor oil, and others

- **Transesterification**—Biodiesel is made through a chemical process called transesterification whereby the glycerin is separated from the fat or vegetable oil. The process leaves behind two products—methyl esters (the chemical name for biodiesel) and glycerin (a valuable byproduct usually sold to be used in soaps and other products).

Bioenergy. http://dx.doi.org/10.1016/B978-0-12-407909-0.00002-X

- **Advanced Biofuel (EPA Definition 40 CFR 80.1401)**—**Advanced Biofuel** means renewable fuel, other than ethanol-derived from cornstarch, that has **lifecycle greenhouse gas emissions** that are at least **50 percent less** than baseline lifecycle greenhouse gas emissions (i.e., diesel fuel).
  - Biodiesel is America's first domestically produced, commercially available advanced Biofuel and meets EPA requirements for inclusion and use under the new Renewable Fuel Standard (RFS-2).
  - US biodiesel reduces lifecycle carbon emissions by 57 to 86 percent compared to petroleum diesel.
- **Energy balance**—The arithmetic balancing of energy inputs versus outputs for an object, reactor, or other processing system; it is positive if energy is released, and negative if it is absorbed.
  - **Biodiesel has the highest positive energy balance (5.54:1) of any commercially available fuel**, returning 5.54 units of renewable energy for every one unit of fossil energy needed to produce it. Since petroleum diesel has a negative energy balance of 0.88, every gallon of biodiesel used has the potential to extend our petroleum reserves by over four gallons.
- **Cetane number**—A measurement of the combustion quality of diesel fuel during compression ignition. It is a significant expression of the quality of a diesel fuel. Biodiesel generally has a higher cetane number (average over 50) compared to diesel fuel (average 42–44), making biodiesel a cleaner burning fuel.
- **Viscosity**—Viscosity is a measurement of how resistant a fluid is to attempts to move through it. A fluid with a low viscosity is said to be "thin," while a high-viscosity fluid is said to be "thick." Properly processed biodiesel has a viscosity that is in the same range as conventional diesel; however, the viscosity of the raw, unprocessed vegetable oil is much thicker than diesel.
- **Cold flow properties**—Three important cold weather parameters that define operability for diesel fuels and biodiesel:
  - **Cloud point**—Temperature where crystals first appear
  - **Cold filter plugging point**—The lowest operating temperature in which a vehicle will operate
  - **Pour point**—Lowest temperature where fuel is observed to flow
  - Users of a B20 blend with #2 diesel will usually experience an increase of these properties beginning at approximately 2–10 °F.
  - Similar precautions employed for petroleum diesel are needed for fueling with B20 blends during cold weather.
- **Lubricity**—**Lubricity** is the measure of the reduction in friction of a lubricant. In a modern diesel engine, the fuel is part of the engine lubrication process. Diesel fuel naturally contains sulfur compounds that provide good lubricity, but because of regulations in the United States, sulfur must be removed.
  - **Biodiesel has excellent lubricity properties, and even in low blends of 1–2% it can completely replace the lubricity that is lost in today's Ultra Low Sulfur Diesel (ULSD).** Without biodiesel added, ULSD has a lower lubricity and requires lubricity-improving additives to prevent excessive engine wear.
- **ASTM**—ASTM International, formerly known as the American Society for Testing and Materials (ASTM), is a globally recognized leader in the development and delivery of international voluntary consensus standards. Working in an open and transparent process and using ASTM's advanced electronic infrastructure, ASTM members deliver the test methods, specifications, guides, and practices that support industries and governments worldwide.

The ASTM Standards for Biodiesel Are As Follows:

- **ASTM D6751** is the approved standard for B100 for blending up to B20, in effect since 2001.
  - Performance-based standard: Feedstock and process neutral.
- **ASTM D975**—Covers petrodiesel and blends up to five percent biodiesel maximum for on/off road engines; B5 is now fungible with diesel fuel.
- **ASTM D396**—Covers heating oil and blends up to five percent biodiesel; B5 is now fungible with petro-based heating oil.
- **ASTM D7467**—Covers blends containing six to 20 percent biodiesel for on/off road engines.
  - Designed so that if B100 meets D6751 and petro diesel meets D975, then B6 to B20 blends will meet their specifications; Important quality control is at B100 level.
- **BQ-9000**®—BQ-9000 is a cooperative and voluntary fuel-quality management program for the accreditation of laboratories, producers, and marketers of biodiesel fuel. The program is a unique combination of the ASTM standard for biodiesel, ASTM D6751, and a quality-systems program that includes storage, sampling, testing, blending, shipping, distribution, and fuel-management practices. **More than 81 percent of the biodiesel in the U.S. is now supplied by BQ-9000 certified suppliers**. For more information, visit www.bq-9000.org.
- **RFS-2—Renewable Fuel Standard (RFS)**. The original RFS program was created under the Energy Policy Act (EPACT) of 2005, and established the first renewable fuel volume mandate in the United States.

Under the Energy Independence and Security Act (EISA) of 2007, the RFS-2 program was expanded in several key ways:

- EISA expanded the RFS program to include **diesel and biodiesel**, in addition to gasoline.
- EISA increased the volume of renewable fuel required to be blended into transportation fuel **from 9 billion gallons in 2008 to 36 billion gallons by 2022**.
- EISA established new categories of renewable fuel, and set separate volume requirements for each one (Biodiesel qualifies for the RFS-2 under the categories of **Biomass-Based Diesel and Non-Cellulosic Advanced Biofuel**).
- EISA required EPA to apply **lifecycle greenhouse gas performance** threshold standards to ensure that each category of renewable fuel emits fewer greenhouse gases than the petroleum fuel it replaces.
- **EPACT Credits**—EPACT stands for the **Energy Policy Act** of 1992, later amended by EPACT of 2005. Under Title III of EPACT 1992, 75 percent of a federal fleet's covered light-duty vehicle acquisitions in US metropolitan areas must be alternative-fueled vehicles.
- Agencies also receive credits for each light-, medium-, and heavy-duty AFV they acquire each year and for the use of B20 biodiesel blends. Simply switching over a diesel vehicle fleet to **use B20 biodiesel represents one of the most economical options for EPACT compliance for fleets** because no vehicle modifications or special equipment are needed in order to use biodiesel. Visit http://www.fleet.wv.gov for more information.
- **CAFE** (corporate average fuel economy) **Credits**—The U.S. EPA and U.S. Department of Transportation finalized new fuel-efficiency standards on August 28, 2012. The new corporate average fuel economy (CAFE) standards will increase fuel economy to the equivalent of 54.5 mpg for cars and light-duty trucks by model year (MY) 2025.

- Currently the EPA provides greenhouse gas (GHG) emissions incentives in the form of CAFE credits to automakers that produce vehicles capable of operating on biofuels, specifically B20 biodiesel blends. However, those incentives under the light-duty vehicle GHG program are set to expire after MY2015.

BIODIESEL STANDARDS

ASTM D6751 is the approved standard for B100 for blending up to B20, in effect since 2001
- Performance-based standard: feedstock and process neutral.

D975—Covers petrodiesel and blends up to five percent biodiesel maximum for on/off road engines; B5 is now fungible with diesel fuel.

D396—Covers heating oil and blends up to five percent biodiesel; B5 is now fungible with petro-based heating oil.

D7467—Covers blends containing six to 20 percent biodiesel (B6–B20) for on/off road engines.
- Designed so that if B100 meets D6751 and petro diesel meets D975, then B6 to B20 blends will meet their specifications.

- Important quality control is at B100 level.

BQ-9000 FUEL QUALITY PROGRAM

The biodiesel industry has an excellent fuel quality management program called BQ-9000 (www.bq-9000.org)

- The BQ-9000 program helps biodiesel producers, marketers, and laboratories put quality management systems in place to ensure that only the highest quality biodiesel meeting ASTM specifications gets put into your customers' fuel tanks.

- This quality-control system covers everything from biodiesel manufacturing, sampling, testing, blending, storage, shipping, and distribution.

- The process yields an ASTM Grade Fuel, produced and supplied by BQ-9000 certified companies.

- Many OEMs are now either requiring or strongly encouraging their customers to source their fuel from BQ-9000 certified suppliers.

ACKNOWLEDGMENTS

This factsheet is published with permission from the National Biodiesel Board http://www.biodiesel.org/. The editor is grateful to Ray Albrecht, P.E. (Technical Representative, Northeast US Region) and Jessica Robinson, Director of Communications for their time and effort in making it available for this book.

BIOENERGY: BIOMASS TO BIOFUELS GLOSSARY OF TERMS AND CONVERTION FACTORS

US Department of Energy[1,2]

[1] *Oak Ridge National Laboratory, USA;* [2] *Office of Energy Efficiency & Renewable Energy, USA*

GLOSSARY OF TERMS

Acid hydrolysis: The treatment of cellulosic, starch, or hemicellulosic materials using acid solutions (usually mineral acids) to break down the polysaccharides to simple sugars.

Aerobic fermentation: Fermentation processes that require the presence of oxygen.

Agricultural residue: Agricultural crop residues are the plant parts, primarily stalks and leaves, not removed from the fields with the primary food or fiber product. Examples include corn stover (stalks, leaves, husks, and cobs), wheat straw, and rice straw.

Alcohol: An organic compound with a carbon bound to a hydroxyl (hydrogen and oxygen or –OH) group. Examples are methanol, CH_3OH, and ethanol, CH_3CH_2OH.

Aldehyde: Any of a class of highly reactive organic chemical compounds characterized by the common group CHO and used in the manufacture of resins, dyes, and organic acids.

Algae: Simple photosynthetic plants containing chlorophyll, often fast growing and able to live in freshwater, seawater, or damp oils. May be unicellular and microscopic or very large, as in the giant kelps.

Alkali: Soluble mineral salts with characteristically "basic" properties, a defining characteristic of alkali metals.

Anaerobic Digestion: Decomposition of biological wastes by micro-organisms, usually under wet conditions, in the absence of air (oxygen), to produce a gas comprising mostly methane and carbon dioxide.

Archaea (formerly Archaebacteria): A group of single-celled microorganisms. A single individual or species from this domain is called an archaeon (sometimes spelled "archeon"). They have no cell nucleus or any other organelles within their cells.

Aromatic: A chemical that has a benzene ring in its molecular structure (benzene, toluene, xylene). Aromatic compounds have strong, characteristic odors.

Ash content: Residue remaining after ignition of a sample determined by a definite prescribed procedure.

B20: A mixture of 20% biodiesel and 80% petroleum diesel based on volume.

Barrel of Oil Equivalent: (BOE) The amount of energy contained in a barrel of crude oil, i.e., approximately 6.1 GJ (5.8 million Btu), equivalent to 1700 kWh. A "petroleum barrel" is a liquid measure equal to 42 US gallons (35 Imperial gallons or 159 L); about 7.2 barrels are equivalent to one tonne of oil (metric).

Batch process: Unit operation where one cycle of feedstock preparation, cooking, fermentation, and distillation is completed before the next cycle is started.

Bioenergy: Useful, renewable energy produced from organic matter. The conversion of the complex carbohydrates in organic matter to energy. Organic matter may either be used directly as a fuel or processed into liquids and gases.

Bioproduct: Materials that are derived from renewable feedstocks. Examples include paper, ethanol, and palm oil.

Biorefinery: A facility that processes and converts biomass into value-added products. These products can range from biomaterials to fuels such as ethanol or important feedstocks for the production of chemicals and other materials. Biorefineries can be based on a number of processing platforms using mechanical, thermal, chemical, and biochemical processes.

British thermal unit (Btu): The amount of heat required to raise the temperature of one pound of water 1 °F under 1 atm of pressure and a temperature of 60–61 °F.

Capital cost: The total investment needed to complete a project and bring it to a commercially operable status. The cost of construction of a new plant. The expenditures for the purchase or acquisition of existing facilities.

Catalyst: A substance that increases the rate of a chemical reaction without being consumed or produced by the reaction. Enzymes are catalysts for many biochemical reactions.

Cetane (also called Hexadecane): An alkane hydrocarbon with the chemical formula C16H34. Consists of a chain of 16 carbon atoms, with three hydrogen atoms bonded to the two end carbon atoms, and two hydrogens bonded to each of the 14 other carbon atoms. Cetane is often used as a shorthand for cetane number, a measure of the detonation of diesel fuel. Cetane ignites very easily under compression; for this reason, it is assigned a cetane number of 100, and serves as a reference for other fuel mixtures.

Cetane number: A measurement of the combustion quality of diesel fuel during compression ignition. It serves as an expression of diesel fuel quality among a number of other measurements that determine overall diesel fuel quality (often abbreviated as CN).

Char: The remains of solid biomass that has been incompletely combusted (e.g., charcoal, if wood is incompletely burned).

Cord: A stack of wood consisting of 128 cubic feet (3.62 cubic meters). A cord has standard dimensions of 4 × 4 × 8 feet, including air space and bark. One cord contains about 1.2 U.S. tons (oven-dry) (i.e., 2400 pounds or 1089 kg).

Corn stover: The refuse of a corn crop after the grain is harvested.

Cracking: A reduction of molecular weight by breaking bonds, which may be done by thermal, catalytic, or hydrocracking. Heavy hydrocarbons, such as fuel oils, are broken up into lighter hydrocarbons such as gasoline.

Cropland: Total cropland includes five components; cropland harvested, crop failure, cultivated summer fallow, cropland used only for pasture, and idle cropland.

Dehydration: The removal of the water from any substance.

Dehydrogenation: The removal of hydrogen from a chemical compound.

Denaturant: A substance that makes ethanol unfit for consumption.

Dewatering: The separation of free water from the solids portion of spent mash, sludge, or whole stillage by screening, centrifuging, filter pressing, or other means.

Digester: A biochemical reactor in which anaerobic bacteria are used to decompose biomass or organic wastes into methane and carbon dioxide.

Disaccharides: The class of compound sugars that yields two monosaccharide units upon hydrolysis; examples are sucrose, maltose, and lactose.

Discount rate: A rate used to convert future costs or benefits to their present value.

Distillate: The portion of a liquid that is removed as vapor and condensed during a distillation process.

Distillation: The process by which the components of a liquid mixture are separated by boiling and recondensing the resultant vapors. The main components in the case of alcohol production are water and ethanol.

Distillers dried grains (DDG): The dried grain byproduct of the grain fermentation process, which may be used as a high-protein animal feed.

Downdraft gasifier: A gasifier in which the product gases pass through a combustion zone at the bottom of the gasifier.

Drop-in fuel: A substitute for conventional fuel that is completely interchangeable and compatible with conventional fuel. A drop-in fuel does not require adaptation of the engine, fuel system, or the fuel distribution network and can be used "as is" in currently available engines in pure form and/or blended in any amount with other fuels.

Drying: Moisture removal from biomass to improve serviceability and utility.

Dry ton: 2000 pounds of biomass on a moisture-free basis.

E-10: A mixture of 10% ethanol and 90% gasoline based on volume. In the United States, it is the most commonly found mixture of ethanol and gasoline.

E-85: A mixture of 85% ethanol and 15% gasoline based on volume.

Effluent: The liquid or gas discharged after processing activities, usually containing residues from such use. Also discharge from a chemical reactor.

Emissions: Waste substances released into the air or water.

Energy crop: A commodity (crop) grown specifically for its fuel value. These include food crops such as corn and sugarcane and nonfood crops such as poplar trees and switchgrass.

Enzymatic hydrolysis: Use of an enzyme to promote the conversion, by reaction with water, of a complex substance into two or more smaller molecules.

Enzyme: A protein or protein-based molecule that speeds up chemical reactions occurring in living things. Enzymes act as catalysts for a single reaction, converting a specific set of reactants into specific products.

Ester: A compound formed from the reaction between an acid and an alcohol. In esters of carboxylic acids, the $-COOH$ group of the acid and the $-OH$ group of the alcohol lose a water molecule and become a $-COO$ linkage.

Ethanol (CH_3CH_2OH): A colorless, flammable liquid produced by fermentation of sugars. Ethanol is used as a fuel oxygenate; the alcohol found in alcoholic beverages.

Externality: A cost or benefit not accounted for in the price of goods or services. Often refers to the cost of pollution and other environmental impacts.

Fatty acid: A fatty acid is a carboxylic acid (an acid with a $-COOH$ group) with long hydrocarbon side chains.

Feedstock: Any material used directly as a fuel, or converted to another form of fuel or energy product. Bioenergy feedstocks are the original sources of biomass. Examples of bioenergy feedstocks include corn, crop residue, and woody plants.

Fermentation: A biochemical reaction that breaks down complex organic molecules (such as carbohydrates) into simpler materials (such as ethanol, carbon dioxide, and water). Bacteria and yeasts can ferment sugars to ethanol.

Fixed carbon: The carbon that remains after heating in a prescribed manner to decompose thermally unstable components and distill volatiles. Part of the proximate analysis group.

Flash point: The temperature at which a combustible liquid will ignite when a flame is held over the liquid; anhydrous ethanol will flash at 51 °F.

Fluidized bed: A gasifier or combustor design in which feedstock particles are kept in suspension by a bed of solids kept in motion by a rising column of gas. The fluidized bed produces approximately isothermal conditions with high heat transfer between the particles and gases.

Fly ash: Small ash particles carried in suspension in combustion products.

Forest land: Land at least 10% stocked by forest trees of any size, including land that formerly had such tree cover and that will be naturally or artificially regenerated. Forest land includes transition zones, such as areas between heavily forested and non-forested lands that are at least 10% stocked with forest trees and forest areas adjacent to urban and built-up lands.

Forestry residues: Includes tops, limbs, and other woody material not removed in forest harvesting operations in commercial hardwood and softwood stands, as well as woody material resulting from forest management operations such as pre-commercial thinnings and removal of dead and dying trees.

Fossil fuel: A carbon or hydrocarbon fuel formed in the ground from the remains of dead plants and animals. It takes millions of years to form fossil fuels. Oil, natural gas, and coal are fossil fuels.

Fungi: Plant-like organisms with cells with distinct nuclei surrounded by nuclear membranes, incapable of photosynthesis. Fungi are decomposers of waste organisms and exist as yeast, mold, or mildew.

Fuel cycle: The series of steps required to produce electricity. The fuel cycle includes mining or otherwise acquiring the raw fuel source, processing and cleaning the fuel, transport, electricity generation, waste management, and plant decommissioning.

Fuel handling system: A system for unloading wood fuel from vans or trucks, transporting the fuel to a storage pile or bin, and conveying the fuel from storage to the boiler or other energy conversion equipment.

Fuel treatment evaluator (FTE): A strategic assessment tool capable of aiding the identification, evaluation, and prioritization of fuel treatment opportunities.

Furnace: An enclosed chamber or container used to burn biomass in a controlled manner to produce heat for space or process heating.

Galactose: A six-carbon sugar with the formula $C_6H_{12}O_6$. A product of hydrolysis of galactan found in the hemicellulose fraction of biomass. Galactan is polymer of galactose with a repeating unit of $C_6H_{10}O_5$.

Gasification: Any chemical or heat process used to convert a feedstock to a gaseous fuel.

Gasifier: A device that converts solid fuel to gas. Generally refers to thermochemical processes.

Gigawatt (GW): A measure of electrical power equal to one billion watts (1,000,000 kW). A large coal or nuclear power station typically has a capacity of about 1 GW.

Glucose ($C_6H_{12}O_6$): A six-carbon fermentable sugar.

Glycerin ($C_3H_8O_3$): A liquid by-product of biodiesel production. Used in the manufacture of dynamite, cosmetics, liquid soaps, inks, and lubricants.

Green diesel: A diesel fuel substitute made from renewable feedstocks by using traditional distillation methods. It is also known as renewable diesel.

Green gasoline: A liquid identical to petroleum-based gasoline, but is produced from biomass such as switchgrass and poplar trees. In the United States, it is still in the development stages. It is also known as renewable gasoline.

Grid: An electric utility's system for distributing power.

Hardwood: One of the botanical groups of dicotyledonous trees that have broad leaves in contrast to the conifers or softwoods. The botanical name is angiosperms; hardwood has no reference to the actual hardness of the wood. Short-rotation, fast growing hardwood trees are being developed as future energy crops.

Hectare: Common metric unit of area, equal to 2.47 acres. 100 hectares = 1 square kilometer.

Hemicellulose: Consists of short, highly branched chains of sugars. In contrast to cellulose, which is a polymer of only glucose, a hemicellulose is a polymer of five different sugars. It contains five-carbon sugars (usually D-xylose and L-arabinose), six-carbon sugars (D-galactose, D-glucose, and D-mannose), and uronic acid. The sugars are highly substituted with acetic acid. The branched nature of hemicellulose renders it amorphous and relatively easy to hydrolyze to its constituent sugars compared to cellulose. When hydrolyzed, the hemicellulose from hardwoods releases products high in xylose (a five-carbon sugar). The hemicellulose contained in softwoods, by contrast, yields more six-carbon sugars.

Horsepower (electrical horsepower; hp): A unit for measuring the rate of mechanical energy output, usually used to describe the maximum output of engines or electric motors. 1 hp = 550 foot-pounds per second = 2545 Btu per hour = 745.7 W = 0.746 kW.

Hydrocarbon (HC): An organic compound molecule that contains only hydrogen and carbon.

Hydrocarbon emissions: In vehicle emissions, these are usually vapors of hydrogen-carbon compounds created from incomplete combustion or from vaporization of liquid gasoline. Emissions of hydrocarbons contribute to ground-level ozone.

Hydrocracking: A process in which hydrogen is added to organic molecules at high pressures and moderate temperatures; usually used as an adjunct to catalytic cracking.

Hydrogenation: Treatment of substances with hydrogen and suitable catalysts at high temperature and pressure to saturate double bonds.

Hydrolysis: A chemical reaction that releases sugars that are normally linked together in complex chains. In ethanol production, hydrolysis reactions are used to break down the cellulose and hemicellulose in the biomass.

Indirect liquefaction: Conversion of biomass to a liquid fuel through a synthesis gas intermediate step.

Industrial wood: All commercial round wood products, except fuel wood.

Inoculum: Microorganisms produced from a pure culture; used to start a new culture in a larger vessel than that in which they were grown.

Joule: Metric unit of energy, equivalent to the work done by a force of 1 N applied over a distance of 1 m (= $1 \text{ kg m}^2/\text{s}^2$). One joule (J) = 0.239 calories (1 calorie = 4.187 J).

JP-8 (or JP8 for "Jet Propellant 8"): A kerosene-based jet fuel, specified in 1990 by the U.S. government, as a replacement for the JP-4 fuel; the U.S. Air Force replaced JP-4 with JP-8 completely by the fall of 1996 to use a less flammable, less hazardous fuel for better safety and combat survivability. The U.S. Navy uses a similar formula, JP-5.

Kilowatt (kW): A measure of electrical power equal to 1000 W 1 kW = 3412 Btu/ h = 1.341 horsepower.

Kilowatt hour (kWh): A measure of energy equivalent to the expenditure of 1 kW for 1 h. For example, one kWh will light a 100-W light bulb for 10 h 1 kWh = 3412 Btu.

Landfill gas: Biogas produced from the natural degradation of organic material in landfills.

Life-cycle assessment (LCA): The investigation and evaluation of the environmental impacts of a given product or service caused or necessitated by its existence. Also known as life-cycle analysis, ecobalance, and cradle-to-grave analysis.

Lignocellulose: Refers to plant materials made up primarily of **lignin, cellulose,** and **hemicellulose.**

Megawatt (MW): A measure of electrical power equal to one million watts (1000 kW). See also watt.

Methane (CH_4): The major component of natural gas. It can be formed by anaerobic digestion of biomass or gasification of coal or biomass.

Methanol (wood alcohol) (CH_3OH): An alcohol formed by catalytically combining carbon monoxide with hydrogen in a 1:2 ratio under high temperature and pressure.

Microorganism: Any microscopic organism such as yeast, bacteria, fungi, etc.

Mill/kWh: A common method of pricing electricity in the United States. Tenths of a U.S. cent per kilowatt hour.

Moisture-free basis: Biomass composition and chemical analysis data are typically reported on a moisture free or dry weight basis. Moisture (and some volatile matter) is removed prior to analytical testing by heating the sample at 105 °C to constant weight. By definition, samples dried in this manner are considered moisture free.

Monoculture: The cultivation of a single species crop.

Monosaccharide: A simple sugar such as a five-carbon sugar (xylose, arabinose) or six-carbon sugar (glucose, fructose). Sucrose, on the other hand, is a disaccharide, composed of a combination of two simple sugar units, glucose and fructose.

Municipal solid waste (MSW): Any organic matter, including sewage, industrial, and commercial wastes, from municipal waste collection systems. Municipal waste does not include agricultural and wood wastes or residues.

Nitrogen oxides (NO_x): A product of photochemical reactions of nitric oxide in ambient air; the major component of photochemical smog.

Nonrenewable resource: A resource that cannot be replaced after use. Although fossil fuels like coal and oil are, in fact, fossilized biomass resources, they form at such a slow rate, that in practice, they are nonrenewable.

Organic compound: Compound containing carbon chemically bound to hydrogen. Often contains other elements (particularly O, N, halogens, or S).

Oxygenate: A compound which contains oxygen in its molecular structure. Ethanol and biodiesel act as oxygenates when they are blended with conventional fuels. Oxygenated fuel improves combustion efficiency and reduces tailpipe emissions of CO.

Particulates: A fine liquid or solid particle such as dust, smoke, mist, fumes, or smog, found in air or emissions.

Petroleum: Substance comprising a complex blend of hydrocarbons derived from crude oil through the process of separation, conversion, upgrading, and finishing, including motor fuel, jet oil, lubricants, petroleum solvents, and used oil.

Photosynthesis: A complex process used by many plants and bacteria to build carbohydrates from carbon dioxide and water using energy derived from light. Photosynthesis is the key initial step in the growth of biomass and is depicted by the equation:

$$CO_2 + H_2O + light + chlorophyll = (CH_2O) + O_2$$

Pilot scale: The size of a system between the small laboratory model size (bench scale) and a full-size system.

Polymer: A large molecule made by linking smaller molecules ("monomers") together.

Process development unit: An experimental facility that establishes proof of concept, preliminary process economics, and engineering feasibility for a pilot or demonstration plant.

Process heat: Energy, usually in the form of hot air or steam, needed in the manufacturing operations of an industrial plant.

Producer gas: Fuel gas high in carbon monoxide (CO) and hydrogen (H_2), produced by burning a solid fuel with insufficient air or by passing a mixture of air and steam through a burning bed of solid fuel.

Proof: The ethanol content of a liquid at 60 °F, stated as twice the percent by volume of the ethyl alcohol.

Proximate analysis: The determination, by prescribed methods, of moisture, volatile matter, fixed carbon (by difference), and ash. Does not include determinations of chemical elements or determinations other than those named. The group of analyses is defined in ASTM D 3172.

Pyrolysis: The breaking apart of complex molecules by heating in the absence of oxygen, producing solid, liquid, and gaseous fuels.

Quad: One quadrillion Btu (10^{15} Btu) = 1.055 EJ (EJ), or approximately 172 million barrels of oil equivalent.

Refuse-derived fuel (RDF): Fuel prepared from municipal solid waste. Noncombustible materials such as rocks, glass, and metals are removed, and the remaining combustible portion of the solid waste is chopped or shredded. RDF facilities process typically between 100 and 3000 tons of MSW per day.

Residues, biomass: By-products from processing all forms of biomass that have significant energy potential. For example, making solid wood products and pulp from logs produces bark, shavings and sawdust, and spent pulping liquors. Because these residues are already collected at the point of processing, they can be convenient and relatively inexpensive sources of biomass for energy.

Return on Investment (ROI): The interest rate at which the net present value of a project is zero. Multiple values are possible.

Rotation: Period of years between the establishment of a stand of timber and the time when it is considered ready for final harvest and regeneration.

Saccharide: A simple sugar or a more complex compound that can be hydrolyzed to simple sugar units.

Saccharification: A conversion process using acids, bases, or enzymes in which long-chain carbohydrates are broken down into their component fermentable sugars.

Softwood: Generally, one of the botanical groups of trees that in most cases have needle-like or scale-like leaves; the conifers; also the wood produced by such trees. The term has no reference to the actual hardness of the wood. The botanical name for softwoods is gymnosperms.

Sustainable: An ecosystem condition in which biodiversity, renewability, and resource productivity are maintained over time.

Tar: A liquid product of thermal processing of carbonaceous materials.

Therm: A unit of energy equal to 100,000 Btu (= 105.5 MJ); used primarily for natural gas.

Thermochemical conversion: The use of heat to chemically change substances to produce energy products.

Timberland: Forest land that is producing, or is capable of producing, crops of industrial wood and that is not withdrawn from timber utilization by statute or administrative regulation. Areas qualifying as timberland are capable of producing more than 20 cubic feet per acre per year of industrial wood in natural stands. Currently, inaccessible and inoperable areas are included.

Ton (tonne): One US ton (short ton) = 2000 pounds. One Imperial ton (long ton or shipping ton) = 2240 pounds. One metric tonne (tonne) = 1000 kg (2205 pounds). One oven-dry ton or tonne (ODT, sometimes termed bone-dry ton/tonne) is the amount of wood that weighs one ton/tonne at 0% moisture content. One green ton/tonne refers to the weight of undried (fresh) biomass material–moisture content must be specified if green weight is used as a fuel measure.

Total solids: The amount of solids remaining after all volatile matter has been removed from a biomass sample by heating at 105 °C to constant weight. (Source: Ehrman, T. Standard Method for Determination of Total Solids in Biomass. NREL-LAP-001. Golden, CO: National Renewable Energy Laboratory, October 28, 1994.)

Transesterification: A process that includes chemical reactions of alcohols and triglycerides contained in vegetable oils and animal fats to produce biodiesel and glycerin.

Triglyceride: A combination of glycerol and three fatty acids. Most animal fats are comprised primarily of triglycerides.

Volatile: A solid or liquid material that easily vaporizes.

Waste streams: Unused solid or liquid by-products of a process.

Watt: The common base unit of power in the metric system. One watt equals 1 J per second, or the power developed in a circuit by a current of 1 A flowing through a potential difference of 1 V. One Watt = 3.412 Btu/hr.

Wood: A solid lignocellulosic material naturally produced in trees and some shrubs, made of up to 40–50% cellulose, 20–30% hemicellulose, and 20–30% lignin.

Credit: The glossary of terms is based on the compilation by Oak Ridge National Laboratory and the Office of Energy Efficiency & Renewable Energy.

REFERENCES

Oakridge National Laboratory, Bioenergy Feedstock Information Network. https://bioenergy.ornl.gov/main.aspx (accessed 15.06.2014.).

Office of Energy Efficiency & Renewable Energy. http://www.energy.gov/eere/bioenergy/full-text-glossary#B (accessed 15.06.14.).

QUICK REFERENCE LIST OF CONVERSION FACTORS USED FOR BIOENERGY FEEDSTOCK

Bioenergy Feedstock Information Network

Oak Ridge National Laboratory

This is a quick-reference list of conversion factors used by the Bioenergy Feedstock Development Programs at Oak Ridge National Laboratory. It was compiled from a wide range of sources, and it is designed to be concise and convenient rather than all inclusive. Most conversion factors and data are given to only three significant figures. Users are encouraged to consult other original sources for independent verification of these numbers. The following are links to Web sites we have found useful (many universities worldwide maintain good guides and conversion calculator pages):

- U.S. National Institute of Standards and Technology (NIST)
- Centre for Innovation in Mathematics Teaching, University of Exeter, U.K.
- Department of Geological Sciences, University of Michigan
- Convertit.com Measurement Converter

Energy contents are expressed here as Lower Heating Value (LHV) unless otherwise stated (this is closest to the actual energy yield in most cases). Higher Heating Value (HHV, including condensation of combustion products) is greater by between 5% (in the case of coal) and 10% (for natural gas), depending mainly on the hydrogen content of the fuel. For most biomass feedstocks, this difference appears to be 6–7%. The appropriateness of using LHV or HHV when comparing fuels, calculating thermal efficiencies, etc. really depends upon the application. For stationary combustion where exhaust gases are cooled before discharging (e.g., power stations), HHV is more appropriate. Where no attempt is made to extract useful work from hot exhaust gases (e.g., motor vehicles), the LHV is more suitable. In practice, many European publications report LHV, whereas North American publications use HHV.

ENERGY UNITS
QUANTITIES

- 1.0 joule (J) = 1.0 Newton (N) applied over a distance of 1.0 meter (= 1.0 kg m^2/s^2)
- 1.0 J = 0.239 calories (cal)
- 1.0 cal = 4.187 J
- 1.0 gigajoule (GJ) = 10^9 J = 0.948 million Btu = 239 million cal = 278 kilowatt-hours (kWh)
- 1.0 British thermal unit (Btu) = 1055 J (1.055 kJ)

- 1.0 Quad = one quadrillion Btu (10^{15} Btu) = 1.055 exajoules (EJ), or approximately 172 million barrels of oil equivalent (boe)
- 1000 Btu/lb = 2.33 GJ per tonne (GJ/t)
- 1000 Btu/US gallon = 0.279 megajoules per liter (MJ/l)

POWER

- 1.0 watt (W) = 1.0 J/s = 3.413 Btu/h
- 1.0 kilowatt (kW) = 3413 Btu/h = 1.341 horsepower (hp)
- 1.0 kWh = 3.6 MJ = 3413 Btu
- 1.0 hp = 550 foot-pounds per second = 2545 Btu/h = 745.7 W = 0.746 kW

ENERGY COSTS

- $1.00 per million Btu = $0.948/GJ
- $1.00/GJ = $1.055 per million Btu

SOME COMMON UNITS OF MEASURE

- 1.0 U.S. ton (short ton) = 2000 pounds (lb)
- 1.0 imperial ton (long ton or shipping ton) = 2240 lb
- 1.0 metric tonne (tonne) = 1000 kg = 2205 lb
- 1.0 U.S. gallon = 3.79 liter (L) = 0.833 imperial gallon
- 1.0 imperial gallon = 4.55 L = 1.20 U.S. gallon
- 1.0 L = 0.264 U.S. gallon = 0.220 imperial gallon
- 1.0 U.S. bushel = 0.0352 m^3 = 0.97 UK bushel = 56 lb, 25 kg (corn or sorghum) = 60 lb, 27 kg (wheat or soybeans) = 40 lb, 18 kg (barley)

AREAS AND CROP YIELDS

- 1.0 hectare (ha) = 10,000 m^2 (an area 100 × 100 m, or 328 × 328 ft) = 2.47 acres
- 1.0 km^2 = 100 ha = 247 acres
- 1.0 acre = 0.405 ha
- 1.0 U.S. ton/acre = 2.24 t/ha
- 1.0 metric tonne/hectare = 0.446 ton/acre
- 100 g/m^2 = 1.0 tonne/ha = 892 lb/acre
  - For example, a "target" bioenergy crop yield might be 5.0 U.S. tons/acre (10,000 lb/acre) = 11.2 tonnes/hectare (1120 g/m^2)

BIOMASS ENERGY

- **Cord:** a stack of wood comprising 128 cubic feet (3.62 cubic meters); standard dimensions are 4 × 4 × 8 feet, including air space and bark. One cord contains approximately 1.2 U.S. tons (oven-dry) = 2400 lb = 1089 kg

- 1.0 metric tonne **wood** = 1.4 cubic meters (solid wood, not stacked)
  - Energy content of **wood fuel** (HHV, bone dry) = 18–22 GJ/t (7600–9600 Btu/lb)
  - Energy content of **wood fuel** (air dry, 20% moisture) = about 15 GJ/t (6400 Btu/lb)
- Energy content of **agricultural residues** (range due to moisture content) = 10–17 GJ/t (4300–7300 Btu/lb)
- Metric tonne **charcoal** = 30 GJ = 12,800 Btu/lb (but usually derived from 6 to 12 t air-dry wood, i.e., 90–180 GJ original energy content)
- Metric tonne **ethanol** = 7.94 petroleum barrels = 1262 L
  - Ethanol energy content (LHV) = 11,500 Btu/lb = 75,700 Btu/gallon = 26.7 GJ/t = 21.1 MJ/L. HHV for ethanol = 84,000 Btu/gallon = 89 MJ/gallon = 23.4 MJ/L
  - Ethanol density (average) = 0.79 g/mL (=metric tonnes/m^3)
- Metric tonne **biodiesel** = 37.8 GJ (33.3–35.7 MJ/L)
  - Biodiesel density (average) = 0.88 g/ml (=metric tonnes/m^3)

FOSSIL FUELS

- **Barrel of oil** equivalent (boe) = approximately 6.1 GJ (5.8 million Btu), equivalent to 1700 kWh. "Petroleum barrel" is a liquid measure equal to 42 U.S. gallons (35 imperial gallons or 159 L); about 7.2 barrels oil are equivalent to one tonne of oil (metric) = 42–45 GJ.
- **Gasoline**: U.S. gallon = 115,000 Btu = 121 MJ = 32 MJ/L (LHV). HHV = 125,000 Btu/gallon = 132 MJ/gallon = 35 MJ/L
  - Metric tonne gasoline = 8.53 barrels = 1356 L = 43.5 GJ/t (LHV); 47.3 GJ/t (HHV)
  - Gasoline density (average) = 0.73 g/ml (=metric tonnes/m^3)
- **Petro-diesel** = 130,500 Btu/gallon (36.4 MJ/L or 42.8 GJ/t)
  - Petro-diesel density (average) = 0.84 g/ml (=metric tonnes/m^3)
- Note that the energy content (heating value) of petroleum products per unit mass is fairly constant, but their density differs significantly—hence the energy content of a liter, gallon, etc. varies between gasoline, diesel, and kerosene.
- Metric tonne **coal** = 27–30 GJ (bituminous/anthracite); 15–19 GJ (lignite/subbituminous) (the above ranges are equivalent to 11,500–13,000 Btu/lb and 6500–8200 Btu/lb)
  - Note that the energy content (heating value) per unit mass varies greatly between different "ranks" of coal. "Typical" coal (rank not specified) usually means bituminous coal, the most common fuel for power plants (27 GJ/t).
- **Natural gas**: HHV = 1027 Btu/ft^3 = 38.3 MJ/m^3; LHV = 930 Btu/ft^3 = 34.6 MJ/m^3
  - Therm (used for natural gas, methane) = 100,000 Btu (=105.5 MJ)

CARBON CONTENT OF FOSSIL FUELS AND BIOENERGY FEEDSTOCKS

- **Coal** (average) = 25.4 metric tonnes carbon per terajoule (TJ)
  - 1.0 metric tonne **coal** = 746 kg carbon
- **Oil** (average) = 19.9 metric tonnes carbon/TJ
- 1.0 U.S. gallon **gasoline** (0.833 imperial gallon, 3.79 L) = 2.42 kg carbon
- 1.0 U.S. gallon **diesel/fuel oil** (0.833 imperial gallon, 3.79 L) = 2.77 kg carbon

- **Natural gas (methane)** = 14.4 metric tonnes carbon/TJ
- 1.0 cubic meter **natural gas (methane)** = 0.49 kg carbon
- Carbon content of **bioenergy feedstocks**: approximately 50% for woody crops or wood waste; approximately 45% for graminaceous (grass) crops or agricultural residues

Source Credit: Bioenergy Feedstock Information Network, Oak Ridge National Laboratory.

WOOD AND GRASS BIOMASS AS BIOFUELS

2

ANJU DAHIYA[1,2]

[1] *University of Vermont, USA,* [2] *GSR Solutions, USA*

The term wood refers to a *solid lignocellulosic material naturally produced in trees and some shrubs, made of up to 40–50% cellulose, 20–30% hemicellulose, and 20–30% lignin.* As defined by the U.S. Office of Energy Efficiency and Renewable Energy[1]. FAO[2] defines wood energy as the "*energy derived from wood fuels corresponding to the net calorific value of the fuel*" and energy grass as, "*herbaceous energy crop.*" These two solid biomass energy options: wood and grass, and their use as biofuels are addressed in greater details in Part II. The related biomass conversion processes (gasification, pyrolysis, chemical/biological pretreatment, hydrolysis, fermentation, etc.) are described in Part V.

Chapter 4 (Wood Bioenergy) defines wood energy as, "energy created from the direct or indirect conversion of biomass from trees and woody shrubs." It describes wood energy sources followed by wood bioenergy uses in the United States and worldwide, and the values and benefits of wood bioenergy. The chapter also describes managing wood biomass

---

[1]http://www.energy.gov/eere/bioenergy/full-text-glossary.
[2]ftp://ftp.fao.org/docrep/fao/007/j4504e/j4504e00.pdf.

FIGURE 1

Woody and grass biomass for fuel.

Picture Courtesy: A. Dahiya.

for bioenergy, harvesting, transporting, storing, and the utilization of woody biomass. The wood bioenergy processes are presented including "direct combustion" (for cooking, heat, electricity using chords or pellets), "liquid and gaseous biofuel conversion," "biochemical technological processes" (including the anaerobic or aerobic digestion and fermentation of lignocellulosic feedstock), and "thermochemical processes" (pyrolysis, gasification, liquefaction, hydrothermal upgrading process). The chapter ends with the description on economics and sustainability of woody bioenergy.

Chapter 5 (Perennial Grass Biomass Production and Utilization) first introduces the grass energy as, "grasses used to produce biomass are a type of *herbaceous biomass* which includes feedstock that come from non-woody plants grown directly for energy or from plant parts or residues." Then it describes the four major elements of a biomass system (feedstock supply, biomass conversion, distribution, and end use or utilization) for developing and evaluating a perennial grass biomass system. Different energy grasses described include switchgrass, giant miscanthus, reed canarygrass used for biomass conversion to solid fuels in the form of pellets, briquettes, and cubes, and liquid fuels such as ethanol, methanol, alkanes, and other advanced biofuels. The description of feedstock quality characteristics for grass biomass is followed by the environmental impact of grass biomass, and the chapter ends with economic considerations for grass biomass and future directions.

Chapter 6 (Wood- and Grass energy-related Service Learning Projects & Case Studies) presents an overview of wood and grass energy related projects undertaken by students as part of the University level Bioenergy—Biomass to Biofuels program followed by the description of detailed projects as three case studies examples that describe the use of wood and grass for production of pellets, and provide an assessment of 130-acre property for on-site biomass resources, and the exploration of compost heat. Details are provided at the beginning in the chapter description.

WOOD BIOENERGY

4

William G. Hubbard

Southern Regional Extension Forestry, The University of Georgia, Athens, GA, USA

INTRODUCTION TO WOOD BIOENERGY

Energy created from the direct or indirect conversion of biomass from trees and woody shrubs is known as wood bioenergy. This energy can be derived by directly burning the wood or burning the wood in combination with coal or other fuel sources. It can also be derived indirectly through a number of thermal and/or chemical conversion processes. Wood bioenergy is one of many energy sources that is considered to be renewable or "nondepletive" in nature. Woody biomass that is consumed produces carbon dioxide and other emissions that can be used when young trees are planted and managed. This "carbon neutral" cycle is one prominent reason why wood bioenergy is considered to be a viable alternative to traditional depletive sources within energy discussions in the United States and around the world.

Wood bioenergy resources come from wood biomass that is created in trees and woody shrubs due to growing processes. In its simplest form, wood biomass is a combination of water, cellulose, hemicellulose, and lignin. Wood biomass is produced on all trees and performs several critical functions, including mechanical support and the means by which a trees' xylem and phloem cells carry water and nutrients to various parts of the plant. It is formed through the photosynthetic process whereby sunlight, water, and carbon dioxide combine to make energy for the plant. Wood biomass consists of the accumulated organic material in the trees trunk, bark, branches, leaves, and roots.

WOOD ENERGY SOURCES: A VAST AND RENEWABLE RESOURCE

Much of the feedstock for a wood energy economy comes directly or indirectly from trees on forestland. Forests are defined as lands that contain at least 10% tree cover (Smith et al., 2009). Over a third of all land in the United States is covered in forests. This accounts for approximately 750 million acres, or 3 million square kilometers. From a global perspective, these figures vary by country, with Russia, Brazil, and Canada holding close to 4 billion acres of forest (14 million square kilometers) and some countries containing virtually no forest cover (Figures 4.1) (Food and Agriculture Organization of the United Nations, 2012). These forests are extremely diverse, representing a wide range of species and ecosystems that are dependent upon various climatic, physiographic, and local environmental conditions.

Ownership of these forests varies drastically by country and even regionally within a country. In the United States, for example, a vast majority of the forestland in the east is privately owned, whereas

Bioenergy. http://dx.doi.org/10.1016/B978-0-12-407909-0.00004-3

Region/subregion	Forest area	
	1 000 ha	% of total forest area
Eastern and Southern Africa	267 517	7
Northern Africa	78 814	2
Western and Central Africa	328 088	8
Total Africa	**674 419**	**17**
East Asia	254 626	6
South and Southeast Asia	294 373	7
Western and Central Asia	43 513	1
Total Asia	**592 512**	**15**
Russian Federation	809 090	20
Europe excl. Russian Federation	195 911	5
Total Europe	**1 005 001**	**25**
Caribbean	6 933	0
Central America	19 499	0
North America	678 961	17
Total North and Central America	**705 393**	**17**
Total Oceania	**191 384**	**5**
Total South America	**864 351**	**21**
World	**4 033 060**	**100**

FIGURE 4.1

Credit—Food and Agriculture Organization of the United Nations.

most of the forestland in the west is publically owned by agencies such as the U.S. Forest Service, U.S. Bureau of Land Management, and the U.S. National Park Service (Figure 4.2) (Nelson et al., 2010). Private ownership includes individuals, corporations, investment firms, and nongovernment organizations. These lands vary in size and their management varies by ownership objective. Depending on markets and landowner interests, forest owners may wish to invoke a form of forest management known as plantation management, or they may leave it to nature to control. In the southeastern United

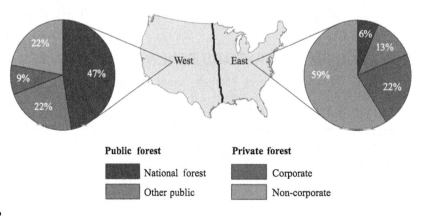

FIGURE 4.2

Credit—U.S. Forest Service, Northern Research Station.

FIGURE 4.3

Credit—U.S. Natural Resource Conservation Service.

States, for example, there are more than 45 million acres of planted pines that have been stocked for the timber market (Figure 4.3) (Smith et al., 2009). These forest owners plant trees with the expectations that they will be used for products such as pulp and paper, lumber, utility poles, chips, or wood biomass. Many landowners wait years for their trees to reach merchantable size and even longer if markets are not desirable. Unlike agricultural crops, however, forest owners can keep their trees "on the stump" until markets improve. This often leads to the trees growing larger and into more valuable product classes, such as high quality lumber, veneer, or utility poles.

In the United States and many other countries, professional forest management and policies are in place to ensure sustainable management of public and private forests. Most forestland in the United States is either artificially regenerated or regenerates naturally following harvest. Voluntary "best management practices" are followed to ensure water quality is not degraded by forest management. In addition, various forestry techniques are used to ensure forests can withstand insect and disease outbreaks as well as other naturally or human-caused environmental stressors.

Many forests in the United States and around the world, however, are considered to be "overstocked", or unhealthy—that is, growing too densely for optimal healthy conditions. Overstocked stands tend to be wildfire hazards, stagnant growers, and breeding grounds for undesirable insects, diseases, and nonnative plants (Figure 4.4). This overstocked and undermanaged state of the forest has led to much devastation. The Yellowstone fire in the western United States in 1988 burned close to 800,000 acres (3200 square kilometers). This fire, on public lands, may have been prevented or reduced drastically if active management such as controlled burning, thinning trees, or even limited commercial harvesting had been allowed.

Another forest health issue in many parts of the world is one known as "hi-grading." This is the practice of taking only the biggest and/or best trees from the forest and leaving the sick, small, and unhealthy trees to remain to grow and regenerate the forest. This has a negative effect on forest and ecosystem composition, structure, and function. It also limits the owner's ability to achieve long-term economic sustainability. Higraded forests, especially in hardwood regions of the world, are often best

FIGURE 4.4

Credit—http://www.bugwood.org.

managed by clearing the sick and small trees and beginning a regime of active reforestation and management. Wood biomass markets in these areas provide an economic incentive for landowners to improve their forests.

WOOD BIOENERGY USES IN THE UNITED STATES AND WORLDWIDE

Wood has been used as an energy source for eons. The utilization of wood for heating and cooking not only dates to precivilized conditions, but it also continues to be used for these purposes to this day in many developed and developing nations. In fact, there has been resurgence in the use of wood for heating in many European countries due to clean air regulations. This has primarily been met through the production and shipment of wood pellets to Europe from the United States and Canada. Many forest companies in the United States use the wood chips, slabs, shavings, and sawdust from their mill operations in combined heating and power (CHP) systems. This reduces or eliminates their reliance on external energy supplies and in some cases offers them the opportunity to sell energy to others. In the United States, wood-based bioenergy accounts for close to 23% of the total of renewable energy sources (Figure 4.5) (U. S. Energy Information Administration, 2012).

VALUES AND BENEFITS OF WOOD BIOENERGY

There are many values and benefits associated with using a natural resource such as wood to create energy. As mentioned earlier, an environmental advantage associated with wood energy when it displaces traditional coal or oil energy is the fact that it is not removing carbon from long-term storage sites within the earth. A growing body of science supports the fact that society is impacting global temperatures, partly due to the burning of fossil fuels (Stocker et al., 2014). Wood bioenergy offers a cleaner alternative when managed appropriately. Actively managed forests and woodlands can use carbon emitted from burning processes. Wood bioenergy can ultimately offset excess carbon and other particulates that are contributing to global climate change.

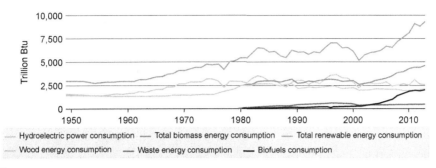

FIGURE 4.5

Credit—U.S. Energy Information Administration.

In addition to the environmental benefits of using wood bioenergy, there are a number of social, political, and economic benefits. Due to recent global conflicts, energy security and self-sufficiency looms large on many country's political agendas. In countries such as the United States and Canada, where forest resource supplies are abundant and well-managed, the opportunity to supply a portion of total energy needs through wood bioenergy reduces the dependency on foreign supplies. Rural communities, landowners, and allied professionals with access to forest resources also stand to benefit from the increased use of wood bioenergy. Biomass operations in the forest add jobs and support local economies. Traditional forest operations supply pulpwood and sawtimber to mills from forests. Prices and desired quantities for these products fluctuate based on the demand in the open market. Wood biomass operations offer landowners, timber harvesters, and others an added product that can benefit them, the land and forest, and the communities and businesses that produce and use the energy, while enhancing the forest in some cases and using a traditionally wasted supply in others.

WOOD ENERGY SOURCES

Wood energy can come from a variety of biomass sources that originate on the land, the mill, or the landfill. Much of the current woody biomass used comes from what is known as logging residues and thinnings. This includes several tons of whole trees, limbs, leaves, branches, and tops per acre (Figure 4.6). These resources are often left to rot on the harvest site but can be used for a variety of wood bioenergy products, such as pellets or liquid fuels. Although scientific studies are underway to determine the short-term and long-term environmental impacts of removing these materials from the site, the current consensus within the forestry profession is that adverse effects can be minimized as long as acceptable guidelines and recommendations are followed.

Other sources of wood bioenergy come from thinnings associated with reducing wildfire risks and harvesting debris from natural disasters. These efforts typically involve the removal of live, damaged, or dead trees that would tend to lead to further catastrophes. Although not obvious to the untrained eye, situations like the Yellowstone catastrophe (Figure 4.7) could have been minimized if those forests had been thinned appropriately. Disasters such as hurricanes in the south and ice storms in the northeast often lead to wildfire, insect, and disease hazards due to the extreme amount of dead, dying, and damaged trees.

FIGURE 4.6

Credit—Jeremy Stovall, Stephen F. Austin
University.

FIGURE 4.7

Credit—Billy Humphries.

Additional sources of wood energy can come from wood manufacturing waste or urban/
landscape waste typically sent to landfills. Wood manufacturing waste includes sawdust, chips,
slabs, and other material not used in the production of paper, lumber, or furniture. Many forestry
companies are able to convert much of this waste into energy that powers their mills and production
plants. Some are entirely self-sufficient, and some companies have excess electrical power that they
sell to others. Woody biomass can also be used from construction debris and landscaping materials.
Additional treatment is necessary in these cases to remove nonwoody materials such as concrete,
metal, plastic, etc.

A final source of wood bioenergy comes directly from forests managed specifically for energy
wood. These forests are actually plantations called short rotation woody crop systems (SRWCs).
SRWCs often consist of fast-growing trees with desirable energy traits, such as hybrid poplar, eucalyptus, and sweetgum. Specialized harvesting equipment is designed to cut and chip the trees for easier
transportation and storage (Figure 4.8).

FIGURE 4.8

Credit—U.S. Forest Service.

MANAGING WOOD BIOMASS FOR BIOENERGY

Farmers, forest owners, loggers, and others who are interested in actively pursuing wood biomass production for bioenergy systems will appreciate the fact that many current forest management production systems need only be modified slightly, if at all. Silviculture, or forest management, involves the purposeful manipulation of a specified landscape to reach one or more objectives. Many times, these objectives involve the production of one or more forest product types, such as pulpwood or sawtimber. Landowner objectives might also include improving habitat for wildlife, recreation activities, aesthetics/beauty purposes, or improving the land for future generations. Whatever the case, landowners will work with the landscape that they have within their control to progress from a current situation to a future desired situation. Foresters and other professionals advise the landowners and are responsible for various activities that help the landowners achieve their objectives.

Silvicultural operations that involve landowner objectives range in intensity levels. Some systems, such as the highly managed pine plantation systems in the southern United States, involve an agricultural approach where the most productive seedlings are planted on the most productive sites. These seedlings are planted in rows and number in the hundreds of trees to the acre. The site is prepared much like an agricultural field and fertilizer is applied where needed. The forests are monitored and entered as frequently as necessary to ensure maximum production of wood for various products. Forest treatments include thinning the stand to remove trees for marketable products and to ensure trees in the remaining stands grow into higher valued product classes, such as those used for sawtimber for lumber and furniture. These plantations are mostly pine throughout the United States; however, some hardwood plantations are predominant in various regions, which is at least in part due to the increase in woody biomass for energy demands. These stands are then harvested when they reach a marketable age. This age varies based on growing conditions, geography, and markets. The forests can be completely harvested as in the situation of a clearcut, or they can be partially harvested through one or more different systems known as seed tree, shelter wood, gap cuts, or group cuts. Each of these systems has advantages and drawbacks. Each of these systems can produce some level of woody biomass for energy use.

In addition to the silvicultural system, landowners are presented with options based on the local and regional physiology of their property. Forests are highly unique systems made up of one or more tree species and include a diversity of biota and abiota. Forests are described by their dominant species or species mix. Forest trees are dichotomized as either hardwood or softwood species. Generally speaking, hardwood trees are broad-leaved trees and softwood trees have needles. Examples of hardwood tree species groups include oaks, hickories, pecans, and poplar trees. Examples of softwood tree species groups include pines, firs, and cedars. Hardwood trees generally grow slower than softwood trees, but advances are being made in the areas of SRWCs with several hardwood species. Biomass and associated bioenergy properties vary between species types, with softwoods generally having more desirable bioenergy traits and characteristics.

HARVESTING, TRANSPORTING, AND STORING WOODY BIOMASS FOR BIOENERGY

Harvesting, transportation, and storage systems for woody biomass consists of equipment and activities that are designed with landowner objectives, resource efficiency, and environmental impact in mind. Variations range from conventional and large-scale systems to newer, small-scale systems. Forestry operations in many parts of the North America are large-scale and mechanized to take advantage of economies of scale. Minimum profitable acreage size is dependent on a variety of factors, including the amount and quality of expected wood product, access to roads and markets, and terrain and landscape features. To be profitable, harvesting professionals and loggers must invest heavily in mechanized equipment. It is expensive to operate and it is often desirable to woody biomass in "one-pass" systems. In this system, machines such as feller/bunchers, skidders, harvesters, and chippers are strategically operated on a tract of land so as to maximize product selection (Figure 4.9). This may include separating small-, medium-, and large-sized trees as well as collecting tops, limbs, and branches for biomass purposes. The goal is to collect and separate at one time so as to minimize

FIGURE 4.9

Credit—U.S. Forest Service.

machine movement, product transportation, and environmental impact. Two-pass methods involve recovery of the primary product first, then collection of the forest residues on the second pass. This has not proven to be economically feasible as of yet.

Small-scale systems including modified tractors, small tracked skidder steers, and even animal power such as horses and mules remain an option due to lower upfront costs and the ability to retrieve smaller diameter or irregularly shaped trees where access may be a problem. There is also the potential for fewer negative impacts on the harvest site, as well as more manageable workforce and training costs for operators. Often, however, small-scale operations such as these lack experienced operators and their safety records are questionable due to the relative difficulty for oversight from the Occupational Safety and Health Administration. Small-scale woody biomass harvesting systems remain a viable alternative and continued research in this area is revealing promising results.

Transporting and storing woody biomass for bioenergy also presents challenges due to the condition, size, and weight of the woody biomass. Biomass must first be preprocessed to ready it for transportation. This is due to the size of the biomass material, its moisture and grit content, and quality aspects. Equipment such as in-woods chippers, shredders, and grinders serve to break down or comminute the material. Size of the comminuted particle is important for a variety of reasons. It influences drying and loading activities as well as the utility in its current state to be directly loaded into boilers or other bioenergy processing machinery. Particles should not be too wet, large, or small as they will not burn or flow efficiently in energy conversion machinery. Woody biomass can also be bundled in a noncomminuted form in the forest and transferred to a bioenergy processing plant in what are known as bundler and composite residue logs for further treatment. Finally, woody biomass can be transported in its raw form as limbs, branches, stems, and leaves. Neither of these last two options are economically popular operations in the United States or Canada.

Woody biomass is then transported via trucks or trailers to intermediate or final bioenergy processing points. Biomass that is transported in its raw, uncomminuted form is going to have a low bulk density problem due to the amount of air space mixed in with tops, branches, and stems. Biomass transported in comminuted forms, while having lower bulk density, has durability and longevity issues during storage. Bundled materials and whole logs are typically transported via tractor-trailers or log trailers while chips, ground material, and sawdust is transported via chip vans or open-top bulk vans. Material must be transported in a manner that meets state and federal regulations to reduce or eliminate flying debris while in transit.

A final concept of importance here relates to the storage of woody biomass. Supply and demand typically dictate the amount of woody biomass that is necessary for conversion to energy products at any given time and place. In many cases, the harvesting, processing, transportation, and use of the woody biomass occur relatively quickly with no need to store for any lengthy period of time. In some cases, there will be the need to store materials in either comminuted or unconsolidated forms. When this is the case, there are advantages to keeping the biomass in an unconsolidated form. In these circumstances, leaves and needles are allowed to dry and fall off, thereby reducing ash content, which can be an issue in some cases. The drying that occurs in unconsolidated forms allows for a drop in moisture content and an increase in heating capacity. Storing chips or comminuted materials for long periods of time has its drawbacks. High temperatures result in dry matter loss from fungi and mold, as well as the potential for self-ignition. Care should be taken to keep storage piles and temperatures from getting too high.

USING WOODY BIOMASS

Woody biomass is an extremely common yet complex source of renewable energy. The components of cellulose, hemicellulose, lignin, and minerals vary by species type and can be burned directly for heat and power generation, or converted to energy indirectly via a variety of mechanical, thermal, chemical, and biological processes. The cellulose and hemicellulose in wood are made up of more complex bonded sugar molecules than many perennial crops, which contain simple starches such as corn and sorghum. Therefore, the processes used to convert them to energy sources are more difficult and costly under current conditions.

In many situations, forest and wood processing residues are the primary feedstock for wood energy today. In other situations, it may be whole trees from current forestry operations or even biomass from SRWC energy plantations. Forest processing residues are those typically left on the forested tract following a harvest for pulpwood harvest. Wood or mill processing residues are those residues resulting from conversion to lumber, paper, or other wood product and most likely physically situated at the point of conversion. Three major types of fuel can be used from woody biomass: solid, liquid, and gaseous. Solid woody biofuels include wood, charcoal, biochar, torrified biomass, and biocoke (in combination with coal). Liquid fuels from woody biomass examples include lignocellulosic ethanol, butanol, methanol, biodiesel, and pyrolysis oils as well as nonfuel chemicals such as acetic acid, ethyl acetate, and ethylene. Gaseous fuels include biogas, syngas, producer gas, and substitute natural gas (Basu, 2013). The following sections review the many processes that are available to convert woody biomass to wood energy.

DIRECT COMBUSTION OPTIONS
WOOD FOR HEAT, COOKING, AND ELECTRICITY

Wood has been used for thousands of years to heat homes and cook meals. Wood can be converted into charcoal by heating the wood in the absence of oxygen—a process known as pyrolysis. Charcoal is still used for cooking in many countries around the world, and many forms of pyrolysis are in practice today in the wood bioenergy industry.

Wood can be co-fired or burned directly in a variety of forms in homes, at traditional or retrofitted power generation facilities, or in boiler systems of varying sizes at institutions such as schools or hospitals. In order for wood to be used directly for heat or electricity, its form is often mechanically altered and dried appropriately. Wood can be burned as cordwood pieces (large pieces that fit into fireplaces and furnaces) or can be mechanically reduced to sawdust, slabs, chips, or chunks. Often, the wood is a residue of other industrial processes, such as slabs from lumber, so it comes in this form already. As noted earlier, much of the forest industry in the southern United States is self-sufficient with regards to their heat and power needs because of the ability to use excess wood from their operations. Wood can also be mechanically reduced and reformed for ease of transportation, storage, and co-firing. This is the case for the increasing interest in wood pellets.

WOOD PELLETS

A surging demand in Europe for wood pellets has highlighted the continued interest and utility of wood in its basic form for energy. Wood pellets are made from shavings, sawdust, chips, slabs, or other wood matter. The wood is ground, pulverized, or broken down with other mechanical means to small, grain-sized particles. The process also includes drying mechanisms to reduce excess moisture content from 50% or more in the raw material to around 10%. The material is then heated, pressed, and molded using the natural based lignin found in the wood cells as a binding agent. The final wood pellets are the approximate diameter of a pencil and less than an inch in length (Figure 4.10). These pellets can then be burned in residential or commercial institutions using a variety of technologies, such as home heating stoves or institutional boiler systems. They can also be co-fired with coal at large generation plants to meet environmental regulations. Another potential for these pellets is in combined heat and power systems. Pellets can also be converted into gases, liquids, or oils based on one or more processes outlined in future sections of this chapter. A torrefied version of the wood pellet has also been developed, which burns more effectively alongside crushed coal; however, the technology has not proven to be commercially viable yet.

Wood pellet production continues to increase dramatically due to worldwide interest in the product. Much of this interest is derived from environmental regulations in Europe based on the desire to reduce dependence on fossil fuels. Pellets are also exported from North America, namely from the United States and Canada. Policy and public opinion will dictate the future growth of this industry, but industry experts expect these numbers to continue to grow as other countries around the globe look for environmentally friendly fossil fuel substitutes.

Wood for heating and electricity through the generation of steam in boilers is being used more within the CHP industry. Commercial-scale CHP systems function well where there are plentiful forests with timber supplies that are not currently being used for traditional markets. They also find successful niches in areas where the population centers are located near the wood supply. Interest in this form of utilizing wood energy will likely continue to vary with the cost, availability, and high load heating demand for fossil fuels in a particular region (Jacobson, 2013). The Fuels for Schools programs, which started in Vermont and now has expanded to many nonschool sites across the United States, are excellent examples of using wood to heat and power educational institutions in light of pricey traditional fuels.

FIGURE 4.10

Credit—U.S. Forest Service.

LIQUID AND GASEOUS BIOFUEL CONVERSION OPTIONS

The exploration to complement or replace the current fossil fuel and first-generation biofuels in the United States and elsewhere includes the creation of second-generation or advanced fuels from lignocellulosic sources, such as those found in wood. There are two major routes to convert biomass to useable fuels: biochemical and thermochemical. Biochemical conversion processes center around digestion (anaerobic and aerobic) and fermentation. Thermochemical routes focus on using heat, pressure, and varying degrees of oxygen and inclusion of chemical reactants.

Advanced biofuel production from wood commonly relies on hydrolysis of the cellulose and hemicellulose in the wood. Unlike hydrolysis of the starch in corn, however, the cellulosic bonds in wood are much stronger. This requires various chemical (e.g., acids) or enzymatic pretreatments to ready the feedstock for conversion to sugars and ultimately alcohols and other usable biofuels.

BIOCHEMICAL TECHNOLOGICAL PROCESSES

Anaerobic digestion takes place in the absence of oxygen and uses bacteria to decompose the woody biomass. The resulting bioenergy products include methane, or biogas, which can be collected and utilized. The type and quality of the product is highly dependent on a number of factors, including temperature retention time, chemical composition of the influent, presence of toxicants, and collection systems. Aerobic digestion, on the other hand, uses ambient oxygen and decomposes woody biomass, releasing heat, carbon dioxide, and noncompostable solids. Digestion systems that capitalize on this decomposition process either with or without the presence of oxygen have not been used on any commercial scale to date within the woody bioenergy realm.

Fermentation of lignocellulosic feedstocks first involves the separation of the wood into cellulose, hemicellulose, and lignin, followed by enzymatic reactions mixed in water to convert the sugars, which can be distilled and used as an energy product. These enzymatic reactions can be yeast, fungus, or bacteria-based. As these microorganisms digest the sugar, they produce ethanol, carbon dioxide, hydrogen, and other products. Byproducts of these processes, including the residual lignin, can be used as boiler fuel for electricity and steam production. Carbon dioxide, another byproduct of the process, can be sold to the beverage industry.

THERMOCHEMICAL PROCESSES
PYROLYSIS

Wood that is subject to rapid thermal decomposition in the absence of oxygen is known as pyrolysis. This process produces bio-oil liquids, gases, and char. The wood is first comminuted and then delivered to a high-heat reactor with a temperature of approximately 932°F (500 °C). At this temperature, the wood turns into a vapor. Following this transformation, it is cooled, condensed, and recollected in the liquid bio-oil, gases, and char forms. The term *fast pyrolysis* comes from the fact that the time between heating and cooling is decreased substantially, resulting in a higher amount of the desired bio-oil liquid product (Jacobson, 2013). An example of a pyrolysis process is the production of charcoal for cooking.

GASIFICATION

Gasification occurs when wood is subject to high temperatures in a furnace. These temperatures range between 1112 and 1832°F (600–1000 °C); this is a special combustion process due to the inclusion of a limited amount of oxygen and/or steam (Basu, 2013). At these high temperatures, the wood is converted directly into biogas. The gas includes hydrogen, methane, carbon monoxide, and carbon dioxide. This gas can be compressed for storage and transportation or, as often is the case, integrated with other conversion processes to create various biobased syngases. There are several types of gasification processes, including the fluidized bed reactor, entrained flow gasifier, moving gasifiers, and hybrid or novel gasifiers (Phillips, 2012). Byproducts of this process include ash, tars, char, and other hydrocarbons (Hubbard et al., 2007).

LIQUEFACTION

This process is similar to the naturally occurring developing of crude oil, except for the fact that it occurs in a matter of a few minutes rather than millions of years. After subjecting the biomass to intense heat and pressure, the wood turns into hydrocarbon oils and byproducts. Direct liquefaction is also called thermal depolymerization; it has been shown to be successful in the production of liquid oils. A newer, indirect liquefaction procedure has been successful in producing syngas, ethanol, and methanol (Hubbard et al., 2007).

FISCHER–TROPSCH PROCESS

The Fischer–Tropsch process was first demonstrated in Germany in the 1920s. It converts carbon monoxide and hydrogen into oils or fuels that can be substituted for petroleum products. The reaction uses a catalyst based on iron or cobalt and is fueled by the partial oxidation of coal or wood-based fuels such as ethanol, methanol, or syngas, typically coming from an adjacent gasifier. This process can produce "green diesel" or syngas, depending on the temperature and level of oxygen involved in the process (Hubbard et al., 2007).

HYDROTHERMAL UPGRADING PROCESS

The hydrothermal upgrading process (HTU) process converts a large variety of biomass feedstocks into a liquid fuel that can be upgraded to a high-quality diesel fuel. The HTU process heats the feedstock in water to 572–662° F at 100–180 bars of pressure for 5–20 min, which facilitates the removal of oxygen. A majority of the oxygen is removed in equal parts of carbon dioxide and water (Hubbard et al., 2007).

ECONOMICS OF WOODY BIOENERGY

There are several factors that influence the economic viability of a woody biomass to bioenergy operation, including the costs associated with growing, transporting, and storing the feedstock and costs associated with the conversion, transportation, and storage of the converted product. Specific costs vary based on the type of woody bioenergy product desired, the specific feedstock being used, the conversion process, and the scale of operation planned and implemented. Much of the economic success of any fuel alternative relates to the economies of scale and the ability to produce and transport fuels at marketable prices.

Wood delivered in its raw form to a cogeneration facility, or scraps and sawdust used in-house at a sawmill or pulpmill for energy purposes, will have an entirely different set of supply, demand, and price structures than wood converted to liquid or gas fuels via one of several thermal, chemical, or biological processes. Two of the woody bioenergy sectors receiving the greatest attention these days are wood for electricity and heat production and wood for liquid fuels. Some general economic concerns for each of these are discussed below.

ECONOMICS OF WOOD FOR ELECTRICITY AND HEAT PRODUCTION

With the potential for an increase in renewable energy portfolios and renewable fuel standards across the country, it behooves policymakers and energy companies to investigate the economics of generating heat and electricity from renewable energy sources, such as wood. As has already been mentioned, wood pellet consumption in Europe has increased dramatically since 2010. Given the current policy and economic conditions in many countries in Europe at the moment, it is economical for energy companies to purchase pellets from the United States and Canada. This involves a variety of agreements, including those with forest landowners, forest industry, pellet manufacturing companies, and trucking and shipping interests. There are several long-term supply contracts in existence at this time, which ensures this form of woody bioenergy will be around for the foreseeable future. The economics of mass production of pellets for heating and electricity in North America has not been seen yet due to the lack of policies and regulations. However, several cogeneration facilities in North America are using a mix of wood and coal feedstocks.

ECONOMICS OF WOODY BIOENERGY PRODUCTION

Several economic factors influence the ultimate success of any woody bioenergy production system, including those related to the feedstock production or procurement, as well as the harvesting, transportation, conversion, and deployment of the woody biomass and woody energy product. The factors will vary by the source of the woody biomass (urban wood waste, SRWCs, conventional timber harvesting materials, etc.) and the desired end energy product under consideration (e.g., solid, liquid, or gas fuels).

The economics of wood procured from forest residues varies based on ownership characteristics and objectives. Many private landowners do not expect to receive much for tops, branches, and other residues following a conventional harvest for pulpwood or sawtimber. In some instances, landowners may actually pay to have these residues removed. In the case of forest thinnings, local markets and the size and type of timber removed will dictate prices paid or received for removal. Precommercial thinnings in plantations, for example, are designed to encourage growth in the remaining stand. There is usually little to no current market value for the trees removed. In the case of natural stands, timber removed at various times during the life of the stand may or may not have commercial value. Many forested stands are overly stocked and unhealthy, as well as fire-, insect-, and disease-prone. Thinnings of various natures can improve the stand and offer local biomass resources for energy production systems.

Urban wood waste is also a resource that can be procured for little to no cost depending on the quality, location, and other factors. Many large landscaping and land clearing firms are often happy to provide these resources as energy wood. In these situations, hauling and landfill costs are avoided. Energy plantation wood and trees diverted from conventional pulpwood and sawtimber harvesting systems would logically be more expensive to acquire and would require higher energy prices or policy requirements in the marketplace that are currently not in place.

In addition to the economics related to feedstock production or procurement, there are several other variables to take into account. Harvesting costs, for example, refer to equipment, personnel, and business expenses associated with removing and readying the feedstock for transportation to the energy production plant. Some urban wood waste may need processing to remove incompatible materials, such construction debris. Finally, special equipment is needed for harvesting woody energy crops such as willow or poplar.

Following harvesting or procurement, the feedstock needs to be prepared for transportation. Depending on the source, the feedstock may be chipped onsite, bundled, dried, or further refined. Several studies have examined the cost effectiveness of chipping and drying onsite versus hauling to woody bioenergy plants for conversion. Results are varied and depend on several factors, including the feedstock source, the haul distance, and the woody bioenergy product under consideration.

Trucking and transportation costs need to be accounted for and may include transportation to a centralized feedstock collection and conversion site or transportation to the final destination. These costs vary based on hauling distance and time, fuel costs, tolls, taxes, labor, and a number of other factors. They also vary based on the feedstock hauled and the form that the feedstock is in (e.g., comminuted vs. whole tree).

Once the material reaches the woody bioenergy plant, costs include storage, further refinement, pretreatment, and other preprocessing requirements. Cost curves for various woody bioenergy products have been developed for different parts of the country. Converting conventional coal plants to accept woody biomass in the furnace, for example, is a costly venture that requires heavy investments and/or low interest loans, grants, or subsidies. The construction of cellulosic ethanol facilities requires an economies-of-scale approach that can take millions of dollars and several years to build. Profitable conversion of woody biomass to liquid fuels such as ethanol under current market, policy, and social situations is proving to be a challenging undertaking.

Woody biomass has rural economic development impacts as well. The growing, managing, harvesting, transporting, processing, and marketing of woody biomass to an energy resource can have significant positive effects on job growth and retention, personal income, tax incomes, rural spending, and a host of other direct and indirect features of the rural economy. Woody biomass can be grown in many parts of the United States and abroad. Unlike coal and fossil fuels, woody biomass is tended and used locally in many circumstances. These lead to more financial resources being kept and spent locally.

SUSTAINABILITY OF WOODY BIOENERGY

In addition to financial and economic sustainability studies relating to the viable production of energy from woody biomass sources, there has been considerable research and discussion with regard to the environmental, social, and economic sustainability of woody biomass to bioenergy systems. Sustainability discussions are complex and multifaceted. Research on the impact of managing and harvesting woody feedstocks on forest ecosystems, wildlife, water, biodiversity, air, and other natural resource amenities is being undertaken by both public and private entities.

Impacts are a function of system intensity, geography, scale, and other factors. Intensive short-rotation woody plantations such as eucalyptus, for example, may affect local water supplies if not managed properly. Wildlife and biodiversity can be positively or negatively affected as well. In cases where thinning overstocked forests provides feedstocks, the sunlight created by the forest openings provides for plant growth that can benefit deer and other browsing animals. Certain animals, however, require mature forests with little to no disturbance. Their habitat can be negatively affected by collecting excessive woody biomass from old-growth forests.

Social and economic development concerns are also being addressed through research and analysis. Quality of life, energy independence, jobs, wages, and payroll are all affected when a woody biomass to bioenergy production system is initiated. Economic impact tools, such as input–output (I-O) models and community sustainability indices, have been used to estimate tradeoffs and benefits under various scenarios. Community impacts are based on the scale of the plant built and the type of fuel products produced.

SUMMARY

In conclusion, the future potential and sustainability of woody biomass systems for alternative energy will rely on the continuation of thoughtful intercourse, research, study, policy implementation, and public acceptance. Woody biomass for energy is already a socially acceptable and feasibly economic practice in many parts of the world today. Solid wood consumption in the energy sector is on the rise in Europe due to social and political mandates requiring a reduction in pollutants that affect air and water. Inexpensive sources of nonrenewable energy, such as oil from fracking ventures and natural gas, will become more expensive as the readily available supplies become more scare and environmentally monitored. Woody bioenergy captures energy from the sun in a manner that, once stored in the tree, can be released years or decades later with minimal negative impacts on the environment.

REFERENCES

Basu, P., 2013. Biomass Gasification, Pyrolysis and Torrefaction Practical Design and Theory (pp. 1 Online Resource). Retrieved from. http://alias.libraries.psu.edu/eresources/proxy/login?url=http://www.sciencedirect.com/science/book/9780123964885.

Food and Agriculture Organization of the United Nations, 2012. State of the World's Forests. Food and Agriculture Organization of the United Nations, Rome (pp. v.).

Hubbard, W., Biles, L., Mayfield, C., Ashton, S. (Eds.), 2007. Sustainable Forestry for Bioenergy and Bio-based Products: Trainers Curriculum Notebook. Southern Forest Research Partnership, Inc, Athens, GA.

Jacobson, M., 2013. Wood-Based Energy in the Northern Forests. Springer, New York.

Nelson, M.D., Liknes, G.C., Butler, B.J., 2010. Map of Forest Ownership of the Conterminous United States [Scale 1:7,500,000] (Res. Map NRS-2). USDA Forest Service, Newton Square, PA. Northern Research Station Retrieved from. http://www.nrs.fs.fed.us/pubs/rmap/rmap_nrs2.pdf.

Phillips, J., 2012. Different Types of GasIfiers and Their Integration with Gas Turbines. Department of Energy. Retrieved from. http://www.netl.doe.gov/File Library/research/coal/energy systems/gasification/gasifipedia/1-2-1.pdf.

Smith, W.B., Miles, P.D., Perry, C.H., Pugh, S.A., 2009. Forest Resources of the United States, 2007. Forest Service, U.S. Dept. of Agriculture, Washington, D.C.

Stocker, T.F., Qin, D., Plattner, G.-K., Tignor, M., Allen, S.K., Boschung, J., Nauels, A., Xia, Y., Bex, V., Midgley, P.M. (Eds.), 2014. Climate Change 2013-The Physical Science Basis: Working Group I Contribution to the Fifth Assessment Report of the Intergovernmental Panel on Climate Change. Cambridge University Press, Cambridge, UK.

U.S. Energy Information Administration, 2012. Annual Energy Review. Retrieved January 17, 2014, 2014, from. http://www.eia.gov/beta/MER/index.cfm?tbl=T10.01-/?f=A&start=1949&end=2013&charted=6-14-15-11-12-13.

PERENNIAL GRASS BIOMASS PRODUCTION AND UTILIZATION

5

Sidney C. Bosworth

Plant and Soil Science Department, University of Vermont, Burlington, VT, USA

INTRODUCTION

Grasses used to produce biomass are a type of *herbaceous biomass,* which includes feedstock that come from nonwoody plants grown directly for energy or from plant parts or residues. Herbaceous biomass crops, including both annuals and perennials, are usually harvested on an annual basis, and can be broken down into three types depending on their potential utilization: sugar/starch crops, oil-seed crops, and fiber/cellulosic crops. Sugar and starch biomass crops are most often converted into ethanol and other alcohol products, and are the same crops generally grown for grains for feed and food or for sugars for sweeteners. The most common annual crop used for biomass in the U.S. is corn (*Zea mays*), although other crops, such as sugar beets (*Beta vulgaris*) or sweet sorghum (*Sorghum bicolor*), have much potential. Sugarcane (*Saccharum* spp.), is a perennial crop that is widely used for ethanol production in Brazil. Oil crop seed are used to make biodiesel and include such crops as canola, sunflowers, and soybeans. Fiber/cellulosic crops include many grasses and other herbaceous plants. At a developed stage of maturity, the quantity of cellulose, hemicellulose, and lignin in grasses can make up 65 to 70% of the dry matter. Usually the whole, aboveground portions of the plants are harvested and used for direct combustion for heat and/or power, cellulosic conversion to ethanol, thermo-chemical processes for fuel supplements, or anaerobic digestion for methane.

It was estimated in a 2005 U.S. Department of Energy (US Dept of Energy, 2005) report that 0.91 billion metric tons (1 billion U.S. tons) of biomass feedstock would have to be produced annually to meet a Federal goal of replacing 30% of current U.S. petroleum consumption with biofuels by 2030. This would include wood and nonwoody biomass sources. Based on their analysis, the United States has the potential to produce that much biomass annually and still continue to meet food, feed, and export demands. The sources of this biomass would include crop residues, grains, animal manures, processed residues, and perennial crops (primarily perennial grasses and woody crops such as willows). Approximately 38% would be derived from perennial crops requiring 22.2 million hectares (55 million acres) of cropland, idle cropland, and/or cropland pasture. Much of these perennial crops will be the perennial grasses, and this will be the focus of this chapter.

DEVELOPING AND EVALUATING A PERENNIAL GRASS BIOMASS SYSTEM

Regardless of biomass type, the four major elements of a biomass system include feedstock supply, biomass conversion, distribution, and end use or utilization (Figure 5.1). For any biomass system to be

FIGURE 5.1

The four elements of a biomass system.

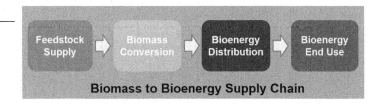

economically viable and ecologically sustainable, these four major elements of the system must be carefully evaluated and efficiently managed. Depending on the particular system, each element may vary in their relative influence on the success or failure of the system. The success of a grass biomass system is highly dependent on all four elements.

FEEDSTOCK SUPPLY

Feedstock supply includes the production, harvest, transportation, and storage of the biomass prior to the conversion process. Biomass feedstock availability must be reliable and meet the minimum quality requirements for the particular energy conversion product (which will be discussed later in this chapter). Total annual biomass requirements will largely be determined by land availability and biomass yields, and a critical amount of land suited for biomass production must be located within an economical distance for transportation to the conversion site, which is usually considered less than 80 km.

Biomass yield can vary greatly across locations and is highly influenced by site characteristics, latitude, temperature, crop species, soil fertility levels, and harvest management. Yield can also vary from year to year depending on growing conditions. Therefore, a robust and resilient biomass system should probably be designed to account for 20% more land than actually needed to offset poor growing years. This would require alternative markets when biomass supplies are in excess beyond the normal storage capacity. The biomass producer would need to receive a decent price for their product that would offset their production, harvest, and transportation costs.

Plant species used for perennial herbaceous biomass

Almost any plant could be used for biomass; however, there are specific criteria that have narrowed the list to a few species that appear to be most suitable for economical and sustainable biomass production. According to Heaton et al. (2004), the ideal crop utilized for biomass fuel should possess a sustained capacity to convert solar energy into harvestable feedstock with maximal efficiency and minimal inputs, while having a minimal impact on the environment. The authors outline several factors to consider in selecting a species and cultivar that would be most suited as an ideal biomass crop:

- A highly favorable energy balance, with a system requiring low energy inputs with relatively high outputs
- Maximum utilization of light, resulting in high yields and high efficient use of inputs
- A harvested biomass that is relatively dry; therefore, minimizing the requirement of additional energy to artificially dry the feedstock
- A high water-use efficiency (unit of yield per unit of water to growth the crop)
- A high efficiency of nitrogen and other nutrients uptake and utilization

- A plant with C_4 photosynthesis, since this is the most efficient type in converting CO_2 into harvestable biomass
- A minimal use of cultivation and need for pesticides
- A noninvasive species or one with minimal risk to spread into environmentally sensitive areas
- Other environmental benefits, such as wildlife habitat, soil and water conservation, nutrient runoff mitigation, and carbon sequestration
- A feedstock with the quality characteristics suited for its end use, whether that be a solid fuel (for combustion or gasification) or a liquid fuel.

Most suitable species are perennial grasses adapted to marginal soils and often having stands that last at least 10 years. Long-term stands help spread the costs of establishment over a longer period. With a long-term perennial sod crop, there is less soil disturbance, which generally results in a more environmentally friendly ecosystem. These grasses also need to be resilient to variable weather conditions and relatively tolerant to diseases and insect pests.

High-yield potential is one of the most important criteria for crop selection. Yield is the most important variable in affecting the cost per unit of biomass; therefore, high yields are important to assure reasonable returns to biomass producers. For most biomass needs, these grasses are often harvested only once or twice per year. Most of the highest yielding grasses have tropical or sub-tropical origins referred to as "warm season" grasses and possessing the C_4 pathway for photo-synthesis (Table 5.1). This is in contrast to "cool season" grasses that have the C_3 photosynthetic pathway. C_4 grasses generally have a higher growth rate during the summer months and are more efficient in utilizing carbon during photosynthesis. C_4 grasses are also more efficient in utilizing soil nitrogen and other nutrients compared to C_3 grasses. However, in higher latitude regions of North America or Europe that have a short growing season and frigid winters, only limited C_4 grasses are adapted and some C_3 grasses may be better suited for biomass production.

The following are a brief description of three grass crops that appear to have the greatest potential for biomass production in the United States.

Switchgrass (*Panicum virgatum*) is a native warm season grass (C_4) that is often associated with the tall grass prairie regions of the United States (Figure 5.2). It is one of the leading perennial biomass crops in the United States, due to its high-yield potential and adaptability to marginal soils (Mitchell et al., 2013). It is also grown for pasture and hay for livestock, and is an excellent crop for soil and water conservation and wildlife habitat. There are over 70 years of research and plant breeding with

Table 5.1 Major Differences in C_3 and C_4 Grasses

Cool Season (C_3)	Warm Season (C_4)
• CO_2 converted to a 3 carbon acid (3 phosphoglyceric acid)	• Co_2 converted to a 4 carbon acid (oxaloacetate)
• Has photorespiration	• No photorespiration
• Opt. temp. 18−24 °C, 72 °F	• Opt. temp. 32−35 °C, 90 °F
• Matures in June/July	• Matures in July/August/September
• More mesophyll cells	• More vascular bundles
• Relatively lower fiber	• Relatively higher fiber
• Responds to high N inputs	• More efficient in N utilization
• Moderate to high N content	• Low to moderate N content

FIGURE 5.2

Switchgrass seedhead (a) and stands (b) of two cultivars, 'Cave-N-Rock' on the left and 'Kanlow' on the right (University of Vermont).

this species, primarily for its use as a pasture and hay crop, but more recently for its biomass potential (Mitchell et al., 2008).

Switchgrass is most adapted to the Central and Eastern portions of the U.S. between 70° W and 103° W (Figure 5.3). There are two basic ecotype groups of switchgrass, *upland* and *lowland*. As the name implies, upland ecotypes are usually found in upland sites not usually prone to flooding, whereas, lowland ecotypes are adapted to floodplains. In regions below 38.2° N, such as the lower Midwest or Southern U.S., lowland cultivars generally yield better and mature later in the season compared to

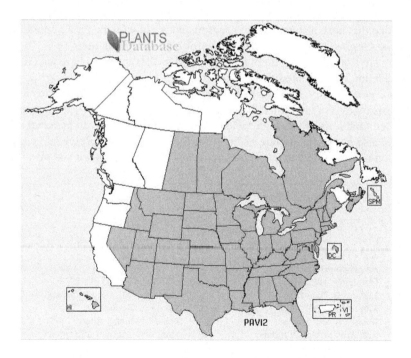

FIGURE 5.3

Switchgrass has a wide range of adaptation (U.S. Department of Agriculture).

upland ecotypes. They are taller with thicker stems (Figure 5.2). Upland cultivars are most adapted to Northern regions of the U.S., such as the Northern Great Plains, the upper Midwest, the Northeast, as well as the Canadian providences of Ontario and southern Quebec (Parrish and Fick, 2005).

Yields vary across the U.S. depending on site location, temperature, precipitation, length of growing season, soil fertility, cultivar, and age of stand. Yields have been reported as high as 39 MT/ha (10.4 U.S. tons/acre). Modeling studies using data from numerous trials across the U.S. (Wullschleger et al. 2010) have shown that the greatest yield potential of switchgrass (as high as 23 Mt/ha) is in the middle latitudes (35° N to 40° N) of the Eastern U.S. (75° W to 98° W). Beyond this region, yield potential declines. In Vermont, a region representing the outer limits for switchgrass adaptability, yields of the best adapted cultivars were reported to be approximately 11 Mt/ha (5 tons/acre) of dry biomass for stands 3 years old or older (Bosworth and Kelly, 2013).

Switchgrass can also be grown in mixtures with other native warm season grasses, such as big bluestem (*Andropogon gerardii*) and Indiangrass (*Sorghastrum nutans*) (Figure 5.4). These three species have grown together in the North American tall grass prairie for thousands of years. As a mixture, the stand can be more resilient to changes in soil conditions, weather, disease, or insect pests compared

FIGURE 5.4

Big bluestem (a), Indiangrass (b), and a polyculture mix of switchgrass and big bluestem (c) (University of Vermont).

to a monocrop. Due to seed quality challenges, a slow establishment rate, and variable yields, it is less likely that either big bluestem or Indiangrass would be grown as a biomass monoculture, but they do complement a mixture with switchgrass (Bosworth, 2013). Yields vary by cultivar and location.

Giant miscanthus (*Miscanthus x giganteus*) is another grass species that has gained much interest in recent years. Originally from Asia, giant miscanthus is a C_4 warm season grass that was introduced to the U.S. and Europe in the 1930s as an ornamental plant due to its straight, tall stems and silver seed head. After the 1970s oil crisis, Europeans began to evaluate giant miscanthus for its biomass potential. It is now commercially used in Europe for bedding, heat, and electricity. It was not until 2001 that research began in the U.S. at the University of Illinois to evaluate this grass for biomass potential in the Midwest. It was soon discovered that it could yield two times more biomass than traditional switchgrass cultivars with very few inputs. The only other energy crops that have comparable yields are sugarcane and energy cane (*Saccharum* spp.), but giant miscanthus has a much broader growing range. It is estimated that with these high-yield potentials, giant miscanthus could meet the 30% U.S. biofuel goals by 2030 without bringing new land into production or displacing the food supply. Most of the focus with this grass in the United States has been as a source of cellulosic ethanol. However in Europe, researchers in Germany reported that optimally harvested miscanthus had very good combustion quality characteristics if used as a solid fuel, reporting low moisture, ash, and mineral content (Heaton, 2010).

Giant miscanthus is a cross of *Miscanthus sacchariflorus*, a tetraploid species with *Miscanthus sinensis*, a diploid, resulting in a sterile triploid. Its only means of reproduction is by rhizome cuttings, which makes establishment costs about three times higher than those grasses that propagate by seed. However, with proper management, stands should last up to 20 years; therefore, the higher cost of establishment can be offset by a long stand life of high yields (Heaton et al., 2004).

Adaptation and site selection are extremely important (Figure 5.5). Giant miscanthus has a higher demand for water compared to switchgrass or the other tall prairie grasses and can show poor growth on droughty soils. It requires at least 32 inch of water during the growing season. Giant miscanthus is also not as tolerant to northern climates as the tall prairie grasses, although it has been successfully grown for at least five years in small research plots in the Champlain Valley of Vermont. However, in the higher elevations of Vermont, it has shown severe winter injury and poor yields (Figure 5.5). In Michigan and Wisconsin, it was reported that overwinter survival was poor if the stand was cut in the first year (Heaton et al., 2014). In Vermont, two-year-old stands that were cut in late October either did not survive or were severely injured compared to stands cut one month later (Bosworth et al., 2013).

Reed Canarygrass (*Phalaris arundinacea*) is a cool season, C_3, perennial grass that is commonly grown as a hay or pasture grass in the Northeast and upper Midwest. It is most adapted to wet soils, but can be fairly drought tolerant on upland soils. It tolerates a wide range of soil pH from 4.9 to 8.2. It is a sod-forming grass spreading by rhizomes. It has high biomass yield potential when compared to other cool season grasses, and has been evaluated in Europe for many years.

As a cool season grass, reed canarygrass begins growth in the spring, heads out much earlier than switchgrass or miscanthus (Figure 5.6), and is at full maturity by early August. For biomass yield and feedstock quality, it is important that reed canarygrass be harvested in August when the stand is most mature. If the harvest is delayed until mid- to late fall, the reproductive stems will tend to lodge and new growth comes up around the fallen material. The new growth is primarily leaf material, resulting in much higher ash, mineral, and nitrogen content compared to the mature, reproductive tillers.

Reed canarygrass must be amended with N fertilizer to assure optimum yields. In Vermont, it was found that for optimum biomass yield, reed canarygrass needed about 84 kg/ha (75 lb/a) applied when

FIGURE 5.5

All three images are of two year old stands of Miscanthus in Vermont with (a) growing in a moist soil producing optimum growth, (b) growing on a well drained, dry site, and (c) growing in a higher elevation subject to winter injury and loss (University of Vermont).

FIGURE 5.6

Reed canarygrass in head (a). The stands of grass (b) show the difference in phenology of the cool season reed canarygrass on the left with that of the warm season switchgrass on the right.

the grass was just initiating growth in April/May. Generally, reed canarygrass has a higher ash, mineral, and nitrogen content than the warm season grasses. However, site selection, careful fertility management, and a timely harvest can minimize these characteristics.

BIOMASS CONVERSION

There are several potential avenues for biomass conversion from grass feedstock (US Dept. of Energy 2010). Grasses can be used as a solid fuel, burned directly either alone or cofired with another fuel source. Grasses can also be used to produce synthetic fuels, either by thermochemical or biochemical methods. The best use may depend on the area of the country where grass biomass is grown and the costs of other competing fuels. Sections of the Midwest or the Southeast that have large areas of noncrop land suitable for grass biomass production could be suitable for a liquid fuel processing plant, since there is enough land within a reasonable distance to support a steady supply of feedstock. In the Northeast where there can be significant distances between agricultural tracts of land, smaller plants or mobile processors may be better suited.

Solid Fuels—Direct combustion is the most energy efficient utilization of grass feedstock. Normally, grass is harvested in a dry form (10–12% moisture content) either as hay or chopped directly in the field. It can be burned as whole bales, chopped and burned, or powdered and burned without any additional processing or densification. Whole bale burners can be found in Europe and are often associated with large, industrial-size boilers. There are less processing costs associated with these systems; however, the storage facility to hold this bulky material will be greater.

Densification of grass feedstock involves grinding and pressing the grass into a more dense form. Densification provides a more uniformly dense feedstock of a set physical size. This improves handling, storage, and transportation, as well as a better control over the combustion process. The most common forms are pellets, briquettes, and cubes (Figure 5.7). Pellets are the smallest and densest forms, but cost more to process. Briquettes and cubes are less dense than pellets, and are best suited for large biomass combustion systems. They are more economical to make since the material does not need to be ground as finely as pellets. Often, briquettes are split into 30 cm wide pieces referred to as "pucks," making them easier for storage and auguring (Cherney et al., 2014).

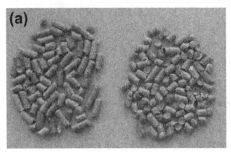

FIGURE 5.7

Grass pellets (a) of switchgrass, left, and reed canarygrass, right, with a diameter of 7 mm and a density of 610 kb/m³. Grass briquettes (b) of switchgrass with a diameter of 56 mm and a density of 560 kb/m³ (University of Vermont).

The process of densification involves several steps. In most cases, the form of the harvested feedstock is hay, which can be harvested and stored as a large, round, or rectangular package, or the conventional small square bale. Hay is often fed into a tub grinder, which breaks the particles into small enough pieces to easily feed into a grinder or hammer mill. For wood, there is an additional drying process during this step; however, grass hay is usually dry enough without this process. The particle size from a tub grinder is usually adequate for making briquettes, which is one reason the cost is less. For pellets, the chopped material requires additional grinding with a hammer mill, and the material must be small enough to pass through a 3.5 mm or smaller mesh screen. It is also important to separate air from the particles, which is usually accomplished by passing the chopped or ground material through a cyclone that is mounted above the hammer mill and pellet mill (Cherney et al., 2014).

Ground material is then pressed through a die creating pressure and frictional heat. The compaction process physically crushes and eliminates any pore space and a combination of heat, pressure, and moisture conditions will cause some of the components of the feedstock to soften. This allows the material to "flow," increasing physical contact and strengthening the connections between particles. Once the pellet is formed, it is usually run through a conveyor to screen for fines and allow it to cool quickly to near-ambient temperature. If cooled too slowly, it can absorb moisture, which reduces its strength and stability.

The size of the briquette or pellet depends on the perforation diameter and thickness of the die. Typically, pellets are approximately 20 mm in length and 6–8 mm in diameter.

Liquid Fuels—Grass biomass has the potential to be converted to liquid fuels, such as ethanol, methanol, alkanes, and other advanced biofuels. A large portion of mature grass is made up of cell wall material, mostly consisting of the complex carbohydrates cellulose and hemicellulose, as well as lignin. Biochemical conversion is a process that uses enzymes plus heat and other chemicals to breakdown the cellulose and hemicellulose into sugars and other intermediary chemicals. The overall process is broken down into several steps (DOE, 2010):

1. *Feedstock*—A cost-effective feedstock supply of specific quality and composition is the first requirement. The form may vary according to the needs of the processing plant, and the production, harvesting, handling, and transportation systems must be cost-effective to keep input prices competitive.
2. *Pretreatment*—The feedstock is usually heated and treated with either an acid or base to break down the fibrous cell walls, making the cellulose easier to hydrolyze.
3. *Hydrolysis*—The material is then treated with either an enzyme or other catalyst to break down cellulose and hemicellulose into simple sugars, such as glucose and xylose.
4. *Fermentation*—Microbes are added in anaerobic conditions to convert the sugars into alcohol or other fuels.
5. *Chemical Catalysis*—As an alternative process to microbial fermentation, the sugars can also be converted to other fuels utilizing chemical catalysis.
6. *Product Recovery*—The fuel products are separated from any water, solvents, and residual solids in the mix
7. *Product Distribution*—There are various options for the final liquid fuel products. Some may be sent to blending facilities, while other products may be sent to traditional refineries.
8. *Heat and Power*—The remaining solids from the process are made up primarily of lignin, which can then be burned for generating heat and power for the fermentation process.

A biomass conversion plant can vary in size from a permanent site, such as an industrial size cellulosic ethanol plant, to a mobile unit that can travel from farm to farm, such as a briquette machine pulled by a trailer. The final costs of the conversion product must be competitive with other comparable fuel sources, and are most affected by initial and maintenance costs of the equipment and labor. Efficiency of processing is critical and is often measured by output per unit of time. Variation in feedstock characteristics can have a large influence on processing efficiency. For example, the ability to pelletize grasses efficiently depends largely on grass species and time of harvest. An effective conversion processing plant must be able to adjust to the highly variable nature of perennial grass feedstock.

Feedstock quality characteristics for grass biomass

Feedstock quality is very important and can influence energy conversion efficiency and cost, wear and corrosion of equipment, and the potential for undesirable emissions and particulate matter which can reduce air quality. The quality characteristics needed for a grass biomass feedstock will depend somewhat on its end use, particularly if a solid fuel for combustion or a liquid fuel such as ethanol.

Moisture content is very important, particularly for combustion utilization (Ciolkosz, 2010). A feedstock that is too moist (greater than 12% moisture content) will not burn as efficiently, reducing the total energy content of the biomass product. Grasses harvested as hay will usually have a moisture content of about 10–12% in storage. Compared to wet-wood biomass, such as cordwood or woodchip, grasses would have an advantage in energy efficiency. However, compared to wood pellets that are typically 6–8% moisture content, grasses would need to be dried some before they are densified into pellets if their standards are going to be consistent with wood (Ciolkosz, 2010).

Ash content and certain minerals of the ash can have a significant effect on the fuel properties of grasses used for combustion. Ash is the mineral residue left over after complete combustion. Ash content can vary widely in grasses, ranging from less than 1.0% to over 10% on a dry matter basis. Soil contamination can increase ash content even more so with some reports up to 20%. It is not only the quantity of ash that can influence the combustion performance of a feedstock, but rather the composition of the ash (specific minerals) that has the greatest influence. Within the ash, it is the alkali metals potassium (K) and sodium (Na) that combine with silica (Si) and sulfur (S) during combustion that causes melting or sintering at low-burn temperatures to form clinkers, an agglomeration of metals (Figure 5.8). In addition, chlorine (Cl) acts as a catalyst for this process (Cherney and Verma, 2013).

FIGURE 5.8

"Clinkers" caused by the melting of ash add "from grass biomass." (University of Vermont).

The alkali metals influence the softening and melting temperatures of ash, increasing corrosion as well as slagging and fouling of furnaces (Cherney and Verma, 2013). Of the alkali metals, K is the most important because of its relatively large content found in plants. Na is usually found in a relatively small amount. Grasses can range from 0.2% to over 3.0% K on a dry matter basis but typically for grasses harvested as biomass, it will be below 1.0%. Plants will take up K at levels beyond what is required for normal plant metabolism and growth; therefore, soils high in K can often result in plants that are higher in K content. However, other factors also influence K levels, particularly stage of maturity at harvest, species, and harvest management. Generally, the older the plant at harvest, the more diluted the mineral content. This is why it is recommended to delay the biomass harvest of perennial grasses until late fall or even the next spring. An optimum or acceptable range of K content for a biomass feedstock depends on the type of boiler or furnace used for combustion.

Silica combines with alkali metals to form alkali silicates that melt at lower temperatures. Perennial grasses tend to have a high concentration of Si; however, Si content is influenced by soil type, water uptake, and plant species. Silica moves as a soluble nutrient in water; therefore, plants growing in moist soils will generally have a higher Si content. Also, some grass species are accumulators of Si. Warm season grasses, such as switchgrass, have a lower intake of water and, therefore, tend to have a lower Si content. Soil contamination of harvested biomass feedstock will also increase the Si levels.

Sulfur reacts with the alkali metals to form alkali sulfates that stick on grates and other heat transfer surfaces. The upper limit for biomass feedstock is usually about 0.2% of dry matter. Another concern with sulfur is the potential for air pollution. High S can result in sulfur oxide (SO_x) emissions, which are attributed to acid rain.

Chlorine facilitates the chemical reactions of alkali metals with Si or S, and when combusted, Cl produces gases that corrode boiler heat exchangers and exhaust flues. The standards for Cl in biomass feedstock for combustion is usually less than 300 ppm. Grasses will vary in Cl content depending on soil levels, as well as fertilizer or manure applications. Chlorine will also leach from plant residue; therefore, harvest management can greatly vary the levels found in the final feedstock.

High nitrogen content can increase the production of nitrogen oxide (NO_x) emissions when the feedstock is combusted at very high temperatures. This is a broad category that includes nitric oxide (NO) and nitrogen dioxide (NO_2), both of which contribute to particulate matter, smog, and acid rain. Grass species, nitrogen fertilization, stage of maturity at harvest, and in-field harvest management will all influence the final feedstock content of N. Generally, it would be desirable to have a grass feedstock less than 1.5% N (Cherney et al., 2014).

For grass pellets, the amount of fines or dust is critical. Fines are very small particles that either occur during the grinding and densification process or result from the breakdown of the pellet during transportation and/or storage. Too many fines reduces efficiency of combustion (BERC, 2011).

One of the challenges with grasses utilized for biomass as a solid fuel is its wide degree of variation in quality. Grass species, soil and growing conditions, and harvest management can all have a huge impact on ash, mineral, and nitrogen content (Bosworth et al., 2013). The data presented in Table 5.2 shows that species and the type of soil both had an influence on ash and nitrogen content. All of these were harvested within a day of each other in early October. Delaying the harvest until later in the fall or even waiting until the following spring can usually result in lower ash and mineral content (Figure 5.9). Utilizing grass biomass for residential uses, such as pellets for stoves, will require high standards involving consistent and careful production practices using dedicated grass species known to accumulate lower levels of minerals. It is likely this will be too expensive at today's prices

Table 5.2 Ash and Nitrogen Content of Warm Season Grasses Harvested in October From Two Sites in Vermont. The dry site was a well-drained, coarse-textured soil and the wet site was a somewhat poorly drained, fine-textured soil.

	Ash Content		Nitrogen Content	
	Dry Soil	Wet Soil	Dry Soil	Wet Soil
	% of dry matter		% of dry matter	
Switchgrass[a]	3.6	4.8	0.69	0.61
Big bluestem[b]	3.0	4.4	0.50	0.47
Indiangrass[c]	3.9	6.3	0.58	0.69
Giant miscanthus[c]	3.9	4.1	1.22	1.03
Reed canarygrass	3.9	6.5	1.23	1.15

[a]Average of four cultivars.
[b]Average of two cultivars.
[c]Single cultivar.
Source: Adapted from Bosworth and Kelly (2013)

compared to other forms of energy. On the other hand, mid- to large industrial size boiler systems have been developed that can handle a wide range in ash and mineral content.

For the biochemical conversion of grass feedstock into liquid fuels, with the end use being a fermentable product such as ethanol, complex carbohydrates in the cell wall (cellulose, hemicellulose) must be broken down into simple sugars. This is achieved utilizing enzymes (biocatalysts) in addition to heat and other chemicals, such as lime, acids, or ammonia. Two components that reduce the efficiency of the digestion are lignin and ash content. Both components are indigestible, and lignin can also limit the digestion of cellulose and hemicellulose. New technology may help solve the lignin issues. First, research to identify a new generation of enzymes and enzyme production technology that more effectively hydrolyzes the structural carbohydrates looks promising. In addition, the lignin left over after conversion can be utilized as a combustion fuel for providing heat and power for the fermentation process.

FIGURE 5.9

An adapted variety of switchgrass grown on a well drained soil (a), and a windrow of switchgrass (b) cut in late fall but not harvested until the next spring allowing for leaching of minerals (University of Vermont).

ENVIRONMENTAL IMPACT OF GRASS BIOMASS

There are many environmental implications for using perennial grasses for biomass production, both positive and potentially negative. The following are a summary of the major issues (Cherney, 2014):

Greenhouse Gas Emissions—In an ideal situation, the atmospheric CO_2 that is fixed during the growth of the grass for biomass feedstock balances the amount of CO_2 that is released back into the atmosphere when it is burned, so that the net gain is at least neutral. For grasses such as switchgrass and giant miscanthus that require very few inputs, this can be achieved when the energy conversion is for direct combustion. In fact, there is some evidence that during the early years of growth as the root systems develop, there is actually a net positive balance (more CO_2 is sequestered than lost) with these grasses. However, if the system is mismanaged, such as overapplication of nitrogen fertilizer or use of an inefficient conversion technology, than many of the gains in CO_2 sequestration can be lost.

Energy Conversion Efficiency—Grass biomass that is used for direct combustion is one of the most efficient renewable energy systems. The energy-out: energy–in ratio (EOIR) is an important criteria to evaluate any renewable energy system. Pelleted grasses used for combustion have a ratio of 14:1 compared to biodiesel of 3:1 or corn ethanol at 1.2:1. This means that it takes far less energy to produce a unit of energy from the grass system. Most of the energy input for perennial grasses is in the establishment phase (which can be spread out over the life of the stand), during harvest and transportation, and in the energy conversion phase. If grasses are converted to cellulosic ethanol, the EOIR is far less than direct combustion.

Protection of Soil and Water Resources—Perennial sod grasses provide excellent cover and soil holding capabilities which prevent soil erosion, improve water infiltration, and reduce nutrient runoff and leaching. In addition, these crops require very few pesticides. Initial establishment may be enhanced with herbicide weed control but once the stand is established, there is very little need for further products.

Wildlife Nesting—The once per year, delayed harvest for grass biomass is very favorable for grassland birds, allowing them to complete their nesting.

Suitable for Noncrop Land—The grass species of interest will grow on marginal soils not suited for crop production; therefore, they should not interfere with our existing food or feed production.

Air Emissions—There is a higher risk of increased NO_x and SO_2 levels with grasses compared to wood; therefore, crop practices that reduce nitrogen and sulfur uptake by the grass plants, as well as properly designed furnaces that utilize scrubbers, will be important to minimize these risks.

Invasive Plants—When evaluating potential species for biomass production, it is always critical to assess their ability to escape and spread across a wide range of environments.

ECONOMIC CONSIDERATIONS FOR GRASS BIOMASS

Grass biomass is not a high-return crop. Keeping input costs to a minimum but also assuring optimum yields will be a key to a viable production system. If fertilizer inputs are kept relatively low, then the major costs for biomass production include the cost of establishment and annual harvest costs. Establishment costs include land preparation, seed costs, lime and fertilizer, and herbicides Bosworth, 2013. For biomass crops started from seed, the cost can range from $400 to $500 per acre. For giant miscanthus, which has to be seeded vegetatively, the costs can be $1500–$2000 per acre.

Table 5.3 Cost of Production for Switchgrass Based on Five Different Yield Scenarios

Grass Biomass Cost Scenarios Over a 10 Year Period					
Establishment Cost ($/a):				$350 – $400	
Annual Maintenance and Harvest Cost ($/a):				$170 – $270	
Scenario:	1	2	3	4	5
Biomass Yield (tons/acre)					
Establishment year	0	0	1.5	1.5	1.5
Year 2	2.0	2.0	2.5	2.5	2.5
Years 3–10	4.0	5.0	4.0	5.0	6.0
Total yield (10 years)	34	42	36	44	52
Average yield/year	3.4	4.2	3.6	4.4	5.2
Annual Average Cost					
Cost, $/acre	261	261	261	261	286
Cost, $/ton	77	62	73	59	55

Yield is the most sensitive variable influencing economic returns. Table 5.3 illustrates how different yield scenarios can influence the costs of production. In scenarios 1 and 2, there was no harvest in the seeding year. This is not unusual for switchgrass, which can be very slow to establish. A site with high yield potential, such as in scenario two, can compensate over time for the lack of a harvest the first year. Scenario 4 and 5 have the highest total yields and, therefore, the lowest costs per unit. The only difference is the extra cost in scenario 5 for adding 50 lbs. of N fertilizer. This clearly had a cost benefit since this scenario had the overall lowest cost per unit.

THE FUTURE

As a renewable energy resource, there is much potential for the use of perennial grasses for biomass. On the production side, there is much research in the genetic improvement of biomass yields and quality traits. Though still in the development stage, newly developed switchgrass hybrids have shown significantly higher yields compared to conventional ecotypes. Improvements in nutrient- and water-use efficiency will also improve the EOIR for these crops.

Much of the effort in liquid fuels research is in the development of efficient enzymes and enzyme production systems for the hydrolysis process. The U.S. Department of Energy has focused much toward identifying highly productive, naturally occurring enzymes and utilizing molecular biology techniques to increase their efficiency. There is also much research on the development of more efficient fermentation microorganisms.

Newer combustion technologies, especially out of Europe, are improving the abilities of these boiler systems to handle grass feedstock with a wide range of ash and mineral composition.

REFERENCES

BERC, 2011. Technical Assessment of Grass Pellets as Boiler Fuel in Vermont. Biomass Energy Resource Center, Montpelier, VT. URL: http://www.vsjf.org/assets/files/RFPs/VT%20Grass%20Pellet%20Feasibility%20Study %202010.pdf.

Bosworth, S., 2013. Establishing warm season grasses for biomass production. In: The Vermont Crops and Soils Home Page, Plant and Soil Science Dept., University of Vermont. http://pss.uvm.edu/vtcrops/articles/ EnergyCrops/Establishing_Warm_Season_Grasses_Biomass_Production_UVMEXT.pdf.

Bosworth, S., Kelly, T., 2013. Evaluation of warm season grass species and cultivars for biomass potential in Vermont 2009–2012. In: The Vermont Crops and Soils Home Page, Plant and Soil Science Dept., University of Vermont. http://pss.uvm.edu/vtcrops/articles/EnergyCrops/Vermont_WSG_Biomass_Report4.2013revised.pdf.

Cherney, J.H., 2014. Benefits of grass biomass. Bioenergy Information Sheet #2, Cornell University Cooperative Extension, Cornell University, Ithaca, NY.

Cherney, J.H., Verma, V.K., 2013. Grass pellet Quality Index: a tool to evaluate suitability of grass pellets for small scale combustion systems. Appl. Energy 103, 679–684. http://dx.doi.org/10.1016/j.apenergy.2012.10.050.

Cherney, J.H., Paddock, K.M., Kiraly, M., Ruestow, G., 2014. Grass Pellet Combustion – Summary of NYS Studies. Bioenergy Information Sheet #30. Cornell University. http://forages.org/files/bioenergy/Bioenergy_ Info_Sheet_30.pdf.

Ciolkosz, D., 2010. Characteristics of Biomass as a Heating Fuel. Renewable and Alternative Energy Factsheet. Penn State Biomass Energy Center, The Pennsylvania State University.

Heaton, Emily, 2010. Giant miscanthus for biomass production. AG201 Factsheet, Iowa State University Extension, Ames, Iowa.

Heaton, E.A., Clifton-Brown, J., Voigt, T.B., Jones, M.B., Long, S.P., 2004. *Miscanthus* for renewable energy generation: European union experience and projections for Illinois. In: Mitigation and Adaptation Strategies for Global Change, vol. 9. Kluwer Academic Publishers, The Netherlands, pp. 433–451.

Heaton, E.A., Boersma, N., Caveny, J.D., Voigt, T.B., Dohleman, F.G., 2014. Miscanthus (*Miscanthus X Giganteus*) for Biofuel Production. In eXtension website: http://www.extension.org/pages/26625/miscanthus-miscanthus-x-giganteus-for-biofuel-production#.U3dQvSigf8e.

Mitchell, R.B., Vogel, K.P., Sarath, G., 2008. Managing and enhancing switchgrass as a bioenergy feedstock. Biofuels, Bioproducts, and Biorefining 2, 530–539.

Mitchell, R., Vogel, K., Schmer, M., 2013. Switchgrass (*Panicum irgatum*) for Biofuel Production. In eXtension website: http://www.extension.org/pages/26635/switchgrass-panicum-virgatum-for-biofuel-production#.U3dMiCigf8d.

Parrish, D.J., Fike, J.H., 2005. The biology and agronomy of switchgrass for biofuels. Critical Reviews in Plant Sciences 24, 423–459.

US Department of Energy, 2005. Biomass Program Factsheet, EE 0830. Energy Efficiency and Renewable Energy Information Center, US Department of Energy.

US Department of Energy, 2010. Biomass as Feedstocks for a Bioenergy and Bioproducts Industry: The Technical Feasibility of a Billion Ton Annual Supply. US Department of Energy.

Wullschleger, S.D., Davis, E.B., Borsuk, M.E., Gunderson, C.A., Lynd, L.R., 2010. Biomass production in switchgrass across the United States: database description and determinants of yield. Agronomy Journal 102, 1158–1168.

WOOD AND GRASS ENERGY SERVICE LEARNING PROJECTS AND CASE STUDIES

Anju Dahiya[1,2]

[1] *University of Vermont, USA;* [2] *GSR Solutions, USA*

The pencil sized pellets produced from wood and grass feedstocks are now a mature technologically feasible heating source. As per the World Bioenergy Association's global bioenergy statistics report (WBA, 2014), around 18.5 million tons of wood pellets were produced globally in 2012 and of that volume approximately 5.91 million metric tons were produced in the Americas. According to a wood energy expert, a Southern Regional Extension Forester, Dr. William Hubbard[1] (see Chapter 4 on wood bioenergy), "there has been resurgence in the use of wood for heating in many European countries due to clean air regulations. This has primarily been met through the production and shipment of wood "pellets" to Europe from the US and Canada." In terms of energy return on investment, a grass energy expert and professor from the Northeast, Dr. Sid Bosworth[2] (see Chapter 5 on grass energy) asserts, "pelleted grasses used for combustion have a ratio of 14:1 compared to biodiesel of 3:1 or corn ethanol at 1.2:1, and this means that it takes far less energy to produce a unit of energy from the grass system."

Students in the established Biomass to Biofuels program[3] have found the wood and grass energy topics intriguing. This chapter presents the solid biofuel-related service learning projects undertaken by students over many years. Similar projects can be undertaken in any location in partnership with the local solid biofuel-based businesses.

In the 2013 spring semester, in response to a lecture on woody biomass, a student commented in the class forum, "*using wood as a biofuel seems to most people to be a very 'stone age' idea, there is a huge technology gap between what it used to be, even 60 years ago, to what it is today. This technology gap seems to be rather large though, the amount of money and distribution of knowledge would be necessary for this idea to catch on seems too high. For this reason, I do not know if using wood as a biomass will ever be the solution to our problem of finding a suitable alternative energy source.*" A year ago in the 2012 semester, apparently this thought was put into practice by a couple of students who implemented their technology skills and partnered as interns with the Biomass Energy Resource Center (BERC)[4]—a nonprofit working toward advancing the use of community-scale biomass energy throughout North America and beyond. BERC maintains a database of existing biomass heating projects and the boilers in United States and Canada. These students undertook the projects, "*Biomass Energy Resource*

---

[1]http://www.extension.org/pages/68832/bill-hubbard#.U5xc5fkVJp8
[2]http://www.uvm.edu/~pss/?Page=faculty/facultybosworth.php
[3]http://go.uvm.edu/7y1rr
[4]http://www.biomasscenter.org/

Bioenergy. http://dx.doi.org/10.1016/B978-0-12-407909-0.00006-7

Center–Wood Energy", and *"Biomass Energy Resource Center—Heating,"* and contributed in updating the database with the current manufacturers and distributers, and completing the new information about community scale projects. Afterward both students continued working as BERC interns.

During the 2011 spring semester, another Biomass to Biofuels student, Tom Tailer, the Executive Director of the Vermont Sustainable Heating Initiative (VSHI), undertook an ambitious project as part of his service learning requirements. He compared two existing pellet manufacturing companies to establish a base for establishing a pellet facility at a county level and found that the co-op model made the most sense for a county as part of his service learning project. As a result, Tom developed a business plan to raise $14,250,000 for the development of a wood pellet manufacturing company—a nonprofit to serve as a base for the research, education, and production of biomass derived fuels. This case study is presented below. The following year a 2012 batch student partnered with VSHI to conduct surveys and help move this work forward.

In the 2011 version of the course, another innovative service learning project on solid biomass use was undertaken by a couple of first year undergraduate students (Erick Crockenberg and Tad Cooke), *"Feedstock Development and Analysis for On-Farm Compost-Based Heating and Methane Capture Systems"* in partnership with a nonprofit CompostPower and the University of Vermont's Center for Sustainable Agriculture to analyze the potential thermal output of several composted feedstocks typically available to Vermont farmers. They wanted to take the "compost heat recovery from woody biomass" concept developed by a French farmer Jean Pain in the 1970s to a newer level for which they explored three separate mixtures of feedstocks containing varying proportions of woody feedstock (woodchips) and agricultural feedstock (cow manure, hay, wood shaving bedding). Eric and Tad used the course project opportunity to build a framework for their grant application to study compost as an energy source for diversified farms. They received the award ($76,000+ via the Clean Energy Fund) and eventually the duo installed a heat capturing system designed to improve the potential of hoop houses for Vermont farms[5] (Figure 6.1).

A similar approach was taken in 2012 by two students whose project *"Cedar Creek Farm: Creating Heat cost-efficiently for running greenhouses—a great benefit to the farmer"* helped a local

FIGURE 6.1

Erick Crockenberg (right) and Tad Cooke prepare two growing beds in the greenhouse where warm air from a compost pile will be delivered to grow plants year-round.

Source: Photo: Sally McCay.

[5]http://prezi.com/jh0sij4gz1e2/copy-of-clean-greenhouse-energy/.

farmer build a cost-effective greenhouse from the alder trees invading in his farm's backyard, and designed a heat capture system from compost. When Ted and Erick presented their work in the 2013 batch of Biomass to Biofuels class, their story inspired a student, Heather Snow (founder of Addie's Acres), to undertake a similar service learning study but with a twist—in a real-farm setting hosting an abundance of woody material. Heather focused on the feasibility of solid biomass-based compost heating systems for greenhouse-based food production at her farm. Heather's project titled, "*Exploration of the Jean Pain Composting Method for Greenhouse Heating*" is presented in this chapter. Heather has provided a comprehensive description of Jean Pain's system as background, its application on her farm, and has also included a success story from a similar project as a case study that also became her inspiration. In 2013 besides Heather, another student (a former Vermont State Legislator on Education Committee, and Peace Corps Volunteer, taught family planning and composting) also focused on *smart greenhouses*, while designing closed-loop mobile greenhouses powered by bioenergy. In 2014 he connected an educational component to his project for designing a mobile classroom for elementary schools and farmers markets teaching about composting and saving energy.

During the 2011 version of the course, while Eric and Ted were busy developing their compost heating concepts, and Tom was building a pellet co-op, other students had undertaken innovative projects related with other aspects of wood and grass energy. Four students (Lance Nichols, Alex Kozlowski, Dan Treadway, and Sam Quinn) grouped together to work on "*Retrofitting Trinity Campus for Woody Biomass.*" This group studied the Trinity campus energy situation with the existing two 6.3 MMBTU natural gas boilers that provide heat and hot water to a fraction of the campus, and based on heating requirements, a cost analysis, fuel sources, and other factors determined that during the winter months, which require the most heat input, only 20–40% of the capacity of a single boiler was utilized. They figured that if the system in place were expanded to supply all other buildings on Trinity campus, the heat input requirements would be roughly 3–4 MMBTU. A single boiler then would still be sufficient enough to provide heat and hot water and would only be utilizing 50–70% of the maximum capacity. The boilers are only utilized to provide heat and hot water, not electricity production; therefore it is important to consider this when choosing a system because burning biomass for electricity production is not as efficient (15–25% efficient) as using biomass for heating (65–85% efficient). Accordingly, the student's team recommended Trinity campus not to install a biomass boiler/gasification system, as the return on investment was far too great for this to be economically feasible, besides there were also many other problems associated with installing a biomass system in an urban area.

Earlier in 2012, a biomass to biofuels student pursued "*Role of biomass in Vermont's Energy Future: A position proposal for The Vermont Sierra Club*" to formulate a position in regard to biomass use in Vermont and for the National Sierra Club to consider revisiting its position on the matter. He presented this proposal to the local chapter of the club. In the same spring semester, another biomass to biofuels student, a Former Director of Residential Energy Services, Ron McGarvey, had undertaken a service learning project, "*Assessment of 130 acre property for on-site biomass resources to meet the energy needs of the community,*" on a property consisting of forested, grassy, and wetland areas. He concluded that among various resources, grass could be harvested and pelletized on-site and burned in modified space heating equipment, and this bioenergy resource was judged to be able to make the greatest contribution to current energy use as described in the case study below.

Tom and Ron have been presenting their projects in the follow-up versions of the Biomass to Biofuels course. Compared to the idea of the Stone Age usage of wood, the concept of pellets was well received in the class and the technologically advanced palletization process of biomass such as wood and grass was appreciated. A 2013 biomass to biofuels student, an Engineering Energy Consultant, Ethan Bellavance, undertook a service learning project, and partnered with his own medium-sized commercial dairy farm to quantify the economics of reducing the farm's thermal heating costs by using harvested biomass from the farm to replace heating oil. His study is presented in Chapter 15. According to Ethan (as posted in the class forum), *"Grass energy truly has to continue to go through an economic vetting process before it hits the main stream like wood pellets have. I see this long term as being a more environmentally/land sustainable option than the use of wood pellets but there are issues that currently need to be addressed: (1) establishment of a Market for this product; (2) clear economic picture for growers/processors; (3) Standardization and collaboration within the industry to have universal ash contents and burning characteristics. Without these questions answered I believe that grass energy will continue to be an economically viable option for a niche market that has the land and wherewithal to make this come to pass, but it will not be able to break out of that niche into the broader market until we answer some of these tough questions."*

Note: it was not possible to include multiple projects undertaken by students over the years; however, effort has been made to present some to demonstrate how the students started the studies and what their conclusions were. These case studies are presented to inspire the readers and give ideas how similar studies can be incorporated as service learning projects in their areas.

REFERENCE

WBA GBS report, 2014. World Bioenergy Association (WBA) Global Bioenergy Statistics. www.worldbio energy.org.

THE VERMONT BIOMASS ENERGY CO-OP

6A

Tom Tailer

2011 Biomass to Biofuels Course Student, University of Vermont, Burlington, VT, USA

INTRODUCTION

As part of this course (Biomass to Biofuels course), I developed a business plan to produce local wood pellets in Vermont. At the same time the Vermont Sustainable Heating Initiative (VSHI)[1] was working on a feasibility study of pelletizing in Chittenden County, Vermont, with the Chittenden County Regional Planning Commission and the Biomass Energy Resource Center (BERC). Part of my work for this course was a survey of what Vermonters want in a biomass utility. Most want a locally owned, locally run biomass utility that would promote environmental sustainability and social justice issues. Biomass-derived energy has the potential for both great economic gain and environmental benefit to Vermont. It also has the potential to do great harm if not developed thoughtfully. Can we affordably heat our homes and businesses with solid biofuels? The answer is Yes! VSHI has placed 22 pellet stoves into low income homes, both to save these homeowners money and to save the Vermont Fuel Assistance Program money. VSHI has also developed a comprehensive biomass energy plan for Vermont. Vermont must use its biomass resources wisely both to provide for its citizens today and to protect the interest of those who will live here after them. If done well Vermonters will save $100,000,000 a year while keeping $100,000,000 in the local economy, and it will be a sustainable business by providing both steady jobs and long-term stewardship of the environment all across Vermont.

COMMUNITY PARTNER

The Vermont Sustainable Heating Initiative (VSHI).

EXECUTIVE SUMMARY OF THE BUSINESS PLAN FOR THE VERMONT BIOMASS ENERGY CO-OP

The purpose of this business plan is to raise $12,000,000 for the development of a cooperative wood pellet manufacturing company. This co-op will serve as a base for research, education, and production of biomass-derived fuels. The co-op will sell wood pellets to its members, provide pellets to the Vermont Fuel Assistance Program (LIHEAP), and serve other state needs in regard to research and

---

[1]Tom Tailer is a founding member of VSHI. As of May 2014, VSHI is continuing to pursue the development of a wood pellet co-op in Vermont. If you are interested in current VSHI activities, please visit our website, www.sustainableheatingvt.org.

development of the biofuels industry. This effort will be part of a larger effort to develop many regional pellet co-ops in Vermont, to serve all Vermonters.

BIOMASS FEEDSTOCK FOR PELLET PRODUCTION

There is enough feedstock to produce over 75,000 tons of high-quality pellets in the area surrounding Chittenden County, Vermont (see BERC 2011 study[2]). The feedstock will be primarily green soft-wood, because it is an underutilized resource. Forest products have a range of economic values. Veneer lumber and saw logs have the highest value. Hardwood for cordwood sales generates more income for harvesters than selling hardwood as feedstock. This leaves softwood, an underutilized resource, because it currently is of low-economic value to harvesters.

Chittenden County is currently developed into urban, suburban, and rural areas. There may be significant possible biomass feedstock for pellet production from developed areas. This would come from primarily suburban areas of the county based on ease of road access, significant biomass that needs to be removed, and the inability of home owners to dispose of unwanted biomass. Currently home owners must pay to dispose of unwanted biomass. There are free drop off places but for large logs they often must pay to have them removed. Chittenden County has 539 square miles of land area or 334,960 acres of land. Roughly 150,000 acres are suburban in nature. This should generate between 20,000 and 60,000 tons of low-grade feedstock a year. These calculations are being left out of the BERC study on purpose. There is no comparable study that has done this type of analysis. By creating a biofuel co-op it is hoped that the community will be more interested in turning their low-quality biomass into fuel. The calculations by BERC are based only on conventional feedstock that is available from land parcels 40 acres or more in size and are within a 50 mile radius of the county.

In phase two and three of the development of a biomass energy co-op, feedstock from nonwoody sources will be introduced. Once local economic viability is developed for traditional pellet energy systems, new technology will be explored.

TWO CASE STUDIES THAT INFORMED THE BUSINESS PLAN FOR THE VERMONT BIOMASS ENERGY CO-OP

In this case study, I will compare two existing pellet manufacturing companies to establish a base for a pellet facility in Chittenden County, Vermont. The two companies are the Vermont Wood Pellet Company in Vermont and the Show Me Energy Co-op in Missouri. I selected the Vermont Wood Pellet Company because it is the only currently operating pellet company in Vermont. It is also similar in scale to VSHI's proposed facility in Chittenden County, and produces pellets with a product model that is similar to what will work in Chittenden County. Show Me Energy Co-op has been chosen because it is a co-op model corporation that is successfully making pellets. Also, it is one of the only facilities that is specializing in grass pellets. Grass energy is a developing energy market with significant system advantages.

The Vermont Wood Pellet Company is located near Rutland, Vermont. They started with a very small operation producing 10,000 tons of pellets a year. Since starting operations they have scaled

[2]http://www.ccrpcvt.org/library/energy/VSHI_PelletManufacturing_report_201108.pdf.

production up to over 20,000 tons/year. Show Me Energy is a co-op that is related to a for-profit business that sells energy crop seeds. Their plant can produce ~100,000 tons/year. They built 3/4 of the facility and then started a fund drive to finish the facility as a producer co-op. "Memberships are $2500 each with a minimum purchase of two memberships. Each membership requires the delivery of 500–1000 pounds of biomass to the processing plant in Centerview." Thus by becoming a co-op they raised the capital to finish the plant and locked in feedstock suppliers. The $5000 buy-in limited the co-op members to serious investors. Local landowners and farmers were their target members.[3]

The Vermont Wood Pellet Company, under the direction of Chris Brooks, has put together an operation using mostly used equipment that is relatively easy to repair. They located their facility where wood pallets were once manufactured and had the cost advantages of using a "brown field" site in an already industrial zoned location.

FINANCING

The Vermont Wood Pellet Company's financing came from a local partner who had significant capital to invest. Show Me Energy Co-op's financing was done by spending, "six and a half million in the first phase and $12 million in the second phase.[3]" This comes to $120/ton capacity capital investment in the first phase.

FEEDSTOCK ANALYSIS

The Vermont Wood Pellet Company harvests within a 15 mile radius of Rutland, Vermont. Thus the community they serve also provides the feedstock. Their product is a softwood pellet that has a very high energy value per pound. This is potentially a problem for people who do not realize their pellet stove can overheat, if using their product on the stove's highest setting (communications with Andy Boutin of Pellergy, April 2011). One reason to make a softwood pellet though is because there is already a market for low-value hardwood as firewood. Currently there is almost no market in Vermont for low-value softwood. McNeil Generating Station in Burlington, Vermont, is the largest consumer of low-quality softwood in Chittenden County, but is limited by the number of trucks that can travel into Burlington to deliver fuel. In a timber harvesting operation, where hardwood is being harvested, softwood not suitable for lumber can be cut and diverted to a pellet mill. One drawback of softwood pelletization is the amount of green wood needed. To produce one ton of hardwood pellets takes about 1.2–1.5 tons of green wood. To produce one ton of softwood pellets requires about 2.4 tons of green wood (Communications with Andy Boutin of Pellergy, April 2011). This means that there is more energy invested in trucking a softwood feedstock to the mill; thus placement of the pellet mill in an area with significant sources of softwood feedstock is important (Communications with Chris Brooks August 2010).

Show Me Energy Co-op uses feedstock from all sources within a 100 mile radius. One of their major providers is the United States Mint where they dispose of ground up paper money. They also use corn stover and other "grasses," including "switchgrass, seed waste from processing plants, even coffee and tea grounds." Their feedstock is sorted at their yard by energy density, moisture content, and ash content. The co-op members are then paid based on what they have delivered. They also must

---

[3]http://www.hpj.com/archives/2007/jan07/jan29/ShowMeEnergyCooperativebegi.cfm.

deliver the feedstock on a fixed date schedule. If they miss the date or just drop it off they receive less money. This is set up to keep the plant running at capacity.

ENERGY DENSITY OF PELLETS

The energy density of softwood pellets is higher than that of hardwood pellets because of the saps and oils in the woody part of the trees. They can range from 8200 to 9000 BTU/lb (Communications with Andy Boutin and Bob Garrit, April 2011). The high energy density is significant if the pellets need to be trucked to where they are consumed. The Vermont Fuel Dealers Association has expressed concern that the energy density of pellets is significantly lower than that of fuel oil. Fuel dealers must make more trips to deliver a million BTU of pellets than fuel oil. Thus the highest possible energy density of the pellets increases the EROEI or Energy Return on Energy Investment of the entire system when energy allocation includes transportation of the fuel. A 10 ton truckload of fuel oil has 2.3 times the energy of a 10 ton truckload of pellets (Presentation given by the VT Fuel dealers Association, Spring 2011). One option is to cut the pellet feedstock with lower energy density materials to decrease the energy per pound slightly so as not to damage the burn units. Show Me Energy Co-op produces a lower energy density pellet with a very high ash content. They claim less than 8% ash; however, testing done by my group showed ash content up to 12%. I am guessing that the energy density is less than 8000 BTU/lb.

PELLET DATA

The analysis of the Vermont Wood Company pellets posted online is 8600 to 9000 BTU/lb with an ash content of 0.23% and a moisture content of 4.74%. This is a very low ash content and a standard moisture content in the pellet industry. Show Me's pellet data are not available, but they are intentionally producing a high ash content pellet. This utility pellet will not work well in most available pellet systems as the ash will fuse with the burn pot.

SALES

The Vermont Wood Pellet Company sells primarily in a 35 mile radius. However, there are some pellet consumers who like to buy "designer pellets" and will pay to have them trucked to Boston. They sell pellets in the standard 40 lb bags in one ton lots, as well as half ton bags and one ton bags. Purchasing pellets in larger bags can diminish packaging waste, because these bags are reusable. The larger bags only cost ~ $12 each. Some of VT Wood Pellet's sales are now by bulk truck with air delivery systems. Show Me Energy Co-op sells to local consumers who have systems that can burn the high ash content pellets. They also sell to a local coal burning power plant that use the pellets to lower the plant's sulfur emissions (5% pellets and 95% coal is the mix[4]).

CONCLUSIONS

The type of pellet that is made by the Vermont Wood Pellet Company is suitable for the Vermont market and feedstock supply. The pellets made by Show Me Energy Co-op are substandard in quality.

---

[4]http://www.hpj.com/archives/2007/jan07/jan29/ShowMeEnergyCooperativebegi.cfm.

The co-op nature of Show Me works well, in that they have a stable feedstock supply. However a co-op of consumers, as opposed to producers, will allow for capital to be raised and the local sustainability of the "harvest" to be ensured. It is from researching these two companies that I have come to the conclusion that a consumer co-op that produces high-quality pellets is the best business model.

A SURVEY OF WHAT VERMONTERS WANTED AS A BIOMASS UTILITY

One of the deeper understandings I have learned in this course is that the concept of sustainability is based on the ability to evolve and adapt. Within the scope of solid biofuels this will require that the community is continually involved in a discussion of what sustainability means. Thus the development of a woody biofuels processing facility will start with a discussion in the community of what sustainability means in the contexts of woody biomass. The factors of feedstock supply, processing, fuel quality, transportation distance of the feedstock, and the pellet product are all addressed in the business plan. The reason to establish a pellet mill in Chittenden County, Vermont, is that this is where the highest density of potential consumers live. The BERC study shows that there is a sufficient feedstock supply to produce 28,000 tons of pellets a year. The harder factors of feedstock supply are related to the community's understanding of the biomass energy, its limits, its potential, and how best to use it. These factors will need to be evaluated to bring a community of possible investors in the co-op to consensus. To start this process I created a Survey Monkey with some of these questions. As of May 5, 2011, 74 people completed the survey. The following are the learning points from this survey:

- There is significant local understanding of the issues related to biofuels.
- There is interest in producing high-quality pellets at a reasonable price.
- Pellet production should be managed for higher levels of environmental protection than required by the state of Vermont.
- The greatest interest is in a co-op.
- Of the respondents 80% were supportive of either 2% or 5% of pellet production supporting LIHEAP. This is a strong support for a social mission. (Almost 60% supported a 5% support to LIHEAP, and between 67% and 84% supported some form of educational relationship with local schools and universities.)
- Of the respondents 46% were interested in investing in a co-op.
- Of the respondents 43% were interested in investing at $200 per share to minimize the amount of money borrowed.

ASSESSMENT OF 130 ACRE PROPERTY FOR ON-SITE BIOMASS RESOURCES TO MEET THE ENERGY NEEDS OF THE COMMUNITY

6B

Ron McGarvey

2012 Biomass to Biofuels Course Student, University of Vermont, Burlington, VT, USA

INTRODUCTION

An assessment was made of the potential for on-site biomass resources to meet the energy needs of Rock Point, Burlington, VT, a 130 acre property consisting of forested, grassy and wetland areas. The assessment identified wood, grass and food waste as potential on-site biomass energy resources. Among these resources, grass, which could be harvested and pelletized on-site and burned in modified space heating equipment, was judged to be able to make the greatest contribution to current energy use at Rock Point.

Rock Point, Burlington, Vermont, is a rocky peninsula into Lake Champlain that consists of 130 acres of generally undeveloped forest, meadows, and wetlands. The property is owned by the Episcopal Diocese of Vermont and is the location for the Diocese administrative headquarters; the Rock Point School, a residential high school; the Bishop Booth Conference Center; and four residential structures.

The Diocese has chosen to maintain the majority of Rock Point in an undeveloped and natural state and there are no specific plans to actively manage the natural environments. The Diocese has also made a commitment to make Rock Point an example of energy conservation and efficiency; Bishop Thomas Ely, "…to make Rock Point, by the year 2015, a model of energy conservation and efficiency in Vermont and beyond."

To support the diocese goals for energy conservation and efficiency this project has been undertaken to evaluate the feasibility of using biomass resources at Rock Point to supply some portion of the nonelectrical energy needs of the buildings and facilities. The major elements of this evaluation are:

- Identification of current energy uses, by energy type and by building
- Estimation of the biomass resources currently available at Rock Point
- Assessment of the feasibility of these on-site biomass resources to meet some portion of the energy needs at Rock Point.

ENERGY USE AT ROCK POINT

Energy is used at Rock Point for space heating, electrical applications, e.g., lighting, drive motors, pumps, and refrigeration equipment, and some limited transportation (Table 6B.1). Transportation

Table 6B.1 The Energy Use, by Type and by Building, Identified by Building Energy Audits

Building	Square Footage	Heating Fuel	Annual Energy Use by Building Beating Energy Use	Heating in MMBTU	Electrical Use
Rock Point School	27,000	Natural gas	21,152 CCF	2115 MMBTU	135,777 KWH
Bishop Booth CC					62300 KWH
Butterfield	6390	Natural gas	4812 CCF	481 MMBTU	
Van Dyck	6000	Natural gas	1230 CCF	123 MMBTU	
Kerr	2500	Natural gas	1806 CCF	181 MMBTU	
Diocese office	4115 2057 heated	Propane	1217 gallons	112 MMBTU	17,439 KWH
Dean of student house	1754	Natural gas	1252 CCF	125 MMBTU	10,000 KWH
Headmaster's house	2900	Propane Wood	1056 gallons 2 cords	97 MMBTU 40 MMBTU	6716 KWH
Bishop's house	5050	Fuel oil	2780 gallons	375 MMBTU	8921 KWH
Property manager's house	1968	Fuel oil	589 gallons	80 MMBTU	−13,184 KWH
		Wood	3 cords	60 MMBTU	
			Total	3789 MMBTU	254,337 KWH

MMBTU, million BTUs; KWH, kilowatt hours. CCF, 100 cubic feet (C for Centum)

energy use is limited to the vehicles that residents and employees use to come and go. There are no through roadways or outside traffic; therefore, there is limited transportation energy use associated with any of the activities at Rock Point.

BIOMASS RESOURCES AT ROCK POINT

The biomass resources potentially available at Rock Point consist of wood from the forested areas; grasses from the meadows or "old fields"; and food waste from the commercial kitchens of the Rock Point School and the Bishop Booth Conference Center. There is an additional biomass resource that, although technically not on Rock Point, may be available, i.e., wood chip heated hot water from the wood chip fired boilers at Burlington High School, which is located adjacent to the Rock Point property.

1. Wood: As shown on the table of current energy uses, wood is being used as a supplemental heating fuel for two of the residences as cut/split logs, and given that the majority of the property is forested, wood could be a major source of biomass for use in heating the buildings. However, cut/split logs for heating the other buildings, i.e., the school, conference center, and diocese office would not be practical due to the labor required to maintain a consistent heating system. Wood applications for these commercial-scale buildings could be realized by using wood chip systems, but this would require an aggressive forest management program to produce enough wood chips, plus burner or boiler conversion equipment to accommodate wood chips, and wood chip storage capacity at each building. Wood pellets are also not attractive due to the additional cost for pellet-making equipment; the increased labor involved in pellet making (on-site); or the increased transportation of wood biomass to and from a pellet-making facility.

Although wood is an available biomass resource, the major constraint that limits the ability to use a relatively large quantity of on-site wood is the designation of the forested areas at Rock Point as a "sustainable forest community" by the City of Burlington, and the Diocese's desire to maintain the property in its natural state.

Given these constraints it is not recommended that on-site wood biomass be considered as a primary energy source for space heating. However, to the extent that deadfalls and diseased or damaged trees must be removed, and that there are residents or students who are willing to undertake the maintenance and material handling involved, cut/split logs could be a source of supplemental heating for the residences.

2. Grass: There are approximately 15 acres of "old fields" or meadows at Rock Point that are currently populated by various species of grass. About three to five of these acres have been used for the "solar orchard" leaving 10 acres available for grasses. Currently these meadows are brush-hogged every year to limit forest encroachment, and the cut grass is left in the fields. Several samples of the current grasses were examined by Prof. Sid Bosworth and he identified the primary species as reed canary and broom grass, with some additional invasive species.

To determine the potential of this grass as a fuel source, Adam Dantzscher, Renewable Energy Resources, Inc., Bennington, estimated that if the grass is harvested at full stand, the remaining 10 acres of available meadow could be harvested and pelletized to produce 2 tons of grass pellets per acre, for a total of 20 tons. Prof. Bosworth independently confirmed that this level of biofuel production was feasible given these types of grass and the marginal, nonmanaged site.

Based on a heating value of 7900 BTU/lb, for grass at 12% moisture, Prof. Bosworth provided a calculation that burning the 20 tons of pellets would generate the heat equivalent to 2000 gallons of fuel oil or the equivalent of 270 MMBTU. This would be enough to displace all of the fossil fuel currently being used to heat three of the four residences or some combination of residences and the diocese office.

In addition to projecting the quantity of potential grass biofuel, Adam Dantzscher also estimated the costs to harvest the grass and densify it using his mobile palletizing equipment:

Cost to harvest 10 acres	$800
Cost the pelletize cut grass	$2800

The other costs associated with the use of grass pellets for heating would be the conversion of existing space heating equipment, boilers or furnaces, to being able to burn pellets and the addition of storage capacity at the locations selected for using pellets. Pellergy in Barre, Vermont, identified that the cost of a replacement boiler or furnace, designed to burn pellets, would be in the order of $9000–$11,000, installed. Therefore, for an investment of about $14,000 ($4000 to harvest and make grass pellets and $10,000 to convert existing equipment to burn pellets), it would be possible to displace the equivalent of 2000 gallons of fuel oil or 270 MMBTU (million BTU).

3. Food/organic wastes: Both the Rock Point School and the Bishop Booth Conference center have commercial kitchens for the preparation of meals for students and conference center attendees. It is estimated that during the school year the school generates approximately 100 gallons of organic waste per week and that operation of the conference center kitchen during the summer months produces a comparable amount.

The most feasible option for using these organic wastes as an energy source would be conversion through a biodigester to produce methane, which could be used to displace some current natural gas uses such as cooking, water heating, and space heating. The quantity of methane produced by this amount of food waste was estimated using an on-line biodigester (www.electrigaz.com). The quantity of methane estimated to be produced would not justify the cost of electrical generation equipment, but the methane could supplement or displace some current low volume uses of natural gas, e.g., chemistry lab.

4. Burlington High School—Wood Chips: Another possible biomass resource that may be available is wood chip-heated hot water from the Burlington High School. The high school has a wood chip-fired boiler plant located at the rear of the high school, which is a distance of approximately 100 yards from the Rock Point School. The high school boiler plant has a capacity of 10 million BTU per hour, which is more than the high school currently utilizes. The wood chips used at the high school are sustainably harvested, delivered approximately twice a week during the heating season, and the high school has on-site storage for approximately 90 tons of chips.

The best option for utilizing this existing wood chip facility would require the construction of insulated hot water or steam supply and return piping from the high school to the Rock Point School, a distance of approximately 100 yards. The cost to construct this type of insulated piping is not known, but given the current low cost of natural gas, which the Rock Point School currently uses, it seems that this biofuel option would not be cost-effective at this time.

RECOMMENDATIONS

There are several sources of biomass available on-site at Rock Point and several options for their utilization to reduce the use of fossil fuels. The option that has the best potential to meet the largest portion of heating energy needs is the harvesting of grass from the current meadow areas, densifying the cut grass into pellets, and burning the grass pellets in the existing heating equipment that has been modified to burn grass pellets. The estimated 20 tons of grass pellets available from the 10 acres of meadow represent the heating potential of providing 270 MMBTU or 7% of the total heating energy use at Rock Point.

The food wastes from the school's and conference center's commercial kitchen represent another potential biofuel in the form of methane from a biodigester. If students at Rock Point School were interested, this could be a very interesting science project to pursue on a small scale.

Lastly, the use of wood chip heated hot water from the Burlington High School power plant, although not cost effective at this time, may have potential for future consideration if natural gas prices increase.

EXPLORATION OF THE JEAN PAIN COMPOSTING METHOD FOR GREENHOUSE HEATING

6C

Heather M. Snow

2013 Biomass to Biofuels Course Student, University of Vermont, Burlington, VT, USA

INTRODUCTION

Jean Pain (c.1920–1980) was a French farmer, forester, researcher, and inventor who developed techniques for composting forest residues while also capturing and using the heat produced from the biological activity within the pile. With an experimental pile of finely shredded brushwood compost that was 100 yd^3 in size, he is reported to have yielded an average of 1 gallon/min of 60 °C water for 18 months. In terms of energy, this is the equivalent of 620 MBTU of captured heat (Gorton). However, replication of Pain's methods has not been found in the scientific literature. This is likely because Pain was meticulous in his methods. With his homemade chipper, he was able to easily run his materials through it twice and he chose relatively small brush and branches to chip as well. He also soaked the material in water for a significant period of time. With such meticulous methods, it would be a daunting task for a person to replicate and produce 100 yd^3 to test. Pain's energy was clearly boundless.

This exploration of Jean Pain's method will focus specifically on the use of compost to heat greenhouses. In the cold Northeast, four-season food production is only possible with the aid of greenhouses, and the cost of heating those structures can be significant because of the need to allow the penetration of light means that they are poorly insulated. It is desirable that we find affordable, sustainable, and local resources that can displace our reliance on fossil fuels for food production.

COMMUNITY PARTNERS

The community partners were (1) Addie's Acres farm; (2) University of Vermont (UVM) Horticulture Farm (Project Leaders Colleen Armstrong, Greenhouse Manager and Project Advisor; Former Biomass to Biofuels course students, Tad Cooke and Erick Crockenberg); and (3) Compost Power Network (Gorton and McCune-Sanders, 2011)

OBJECTIVE

The objective was to complete a feasibility study considering the economics, energy, and logistics of implementing a John Pain system on my farm based on the available resources found there. An exploration of good greenhouse design that is somewhat capable of self-regulating and maintaining its temperature should be done in combination with an exploration of the Jean Pain method for best results. I will use the system already implemented at the Slade greenhouse at the UVM (see the link to this study in the references) as a case study in addition to drawing on knowledge that is available in the literature, corresponding with those implementing another system at the Common Ground Horticulture Farm, and through the experiences of my community partners. I will provide assistance in implementing the system at the Horticulture Farm as time allows.

SHORT HISTORY OF ADDIE'S ACRES

Addie's Acres farm sets on the remaining 21.8 acres of a 125 acre parcel that my grandparents purchased in the town of Sunderland, Vermont, in the late 1940s. The property is evenly split between pasture and woodland, and spruces that were once established as part of a Christmas tree planting program now tower over the house and garage. Though the majority of the farm infrastructure built during my grandparents' time is in a state of disrepair, there are endless possibilities for the standing lumber that remains. Since I purchased the property in 2012, I have found myself tasked with reimagining the potential of the land and would like to use the wisdom of permaculture design to achieve my goal of creating a highly diversified yet integrated system. The definition of permaculture according to David Holmgren (2002), the co-originator of the concept, is "consciously designed landscapes which mimic the patterns and relationships found in nature, while yielding an abundance of food, fiber, and energy for provision of local needs." It takes into account the relationship of different elements to others in the system. It recognizes that every element should have at least three functions, and every essential function should be met by multiple elements. That is how smart design and resilience manifests itself in a system. By taking locational and functional relationships between elements into account, energy can be saved and harnessed in a more efficient manner. As Permaculture instructor Keith Morris emphasizes, it is important to have a well thought out, long-term plan for a site such that each element is placed in its ideal location the first time instead of placing elements blindly and later having to rearrange or redesign them as a result of poor foresight. I hope to use this project as an opportunity to explore the optimal design characteristics for a greenhouse that could be sited on my farm in the future capable of winter production. Ideally it would have an efficient and sustainable heating system that would utilize local resources. With the use of my brother's wood chipper and the abundance of woody material on the property, I believe this could be achieved with the help of a few outside resources.

BACKGROUND

Some of the earliest work and best known example of heating greenhouses with compost-generated heat was initiated by the New Alchemy Institute (NAI) in Massachusetts in 1983 with a 700 ft^2 prototype. The NAI experimented and published widely on appropriate technology, ecology, solar energy, bioshelters, solar greenhouses, integrated pest management in greenhouses, organic farming, and sustainable agriculture. UVM greenhouse manager Colleen Armstrong had ties with the NAI compost heat project in the 1980s and was uniquely suited to help with the advising process at both the Slade project and with Tad and Erick's Horticulture Farm project. At the NAI, compost was actually being cured within the greenhouse instead of being placed just outside the structure, so some of their findings were unique to that variation of the Jean Pain method. The NAI did identify several problems with the concept by 1987. They recognized that it was an experimental technology that was risky and suggested that it only be considered in situations where each operation, both greenhouse and composting, made sense in their own right. Labor requirements are high for a small operation if specialized composting equipment and machinery are not readily available or affordable. They also determined that the composting component needs to be based on the CO_2 production instead of on the amount needed to heat the greenhouse if the mound is to be placed within the structure. For the climate of southern New England, half a cubic yard of compost is needed to heat each square foot of greenhouse space. However, this would provide six times the amount of CO_2 needed for the optimal atmosphere enrichment of plants. Additionally, the amount of ammonia released would be 50 times that needed for optimal plant growth. If the composting component is instead sized to meet CO_2 needs instead, only 15% of the greenhouse's energy demands would be met (Diver, 2001).

Good greenhouse design is key to managing heat needs. Greenhouses should always be oriented toward the sun, so facing south for those of us here in the Northern Hemisphere. The attachment of a greenhouse to a home can work to couple the heating and cooling needs of each. A greenhouse should have an awning to block the direct summer sun from above yet allow the low winter sun to penetrate the area. Otherwise, excessive measures will have to be taken in the summer to avoid "cooking" the plants. The use of bricks, rocks, or some form of thermal mass such as black water barrels on the far wall or in the floor can store the sun's heat and radiate it back to the greenhouse during the night. Vents can also be installed to allow heat to circulate into the house. In the summer, enabling cool air to be vented from the basement can help to cool the greenhouse when it becomes too hot. Gray water from showers and cooking can also be channeled into a storage area in the greenhouse where remaining heat can be captured instead of going straight down the drain after use. After all, a significant amount of energy goes into the heating of water. That wastewater can later be used to water the plants in the greenhouse. Of course, care should be taken as to what chemicals are being used in the water and whether the plants being grown are meant for human consumption.

A greenhouse needs to be heated at the rate at which it is losing heat in order to maintain some determined temperature above freezing. Heat is lost as a result of conduction, infiltration, and radiation. The different materials used in the construction of a greenhouse conduct heat to the colder outside at different rates with metals such as aluminum conducting heat at a relatively fast rate. Wood is less conductive but more prone to rot, making it generally unsuitable for greenhouse construction. A greenhouse covered in a single layer of polyethylene loses heat at a rate of 1.1 $Btu/ft^2/h$ when the outside temperature is 1 °F colder than inside, but only 0.7 $Btu/ft^2/h$ when a second layer is added (Nelson, 2012). By creating this dead air space insulation, a 40% reduction in heat loss is achieved. In order to maximize this capacity, the spacing between the two layers should not exceed 18″ because air currents could be established in such a space. Additionally, it is important to never let the two layers touch as all insulation value will be lost. Infiltration is characterized by the physical exchange of air between the greenhouse and the outside. It is assumed that the volume of air inside a double-layer film plastic greenhouse can be lost as often as once every 60 min while an old, poorly maintained glass greenhouse might experience the same exchange as often as every 15 min. In a structurally tight glass greenhouse that exchanges its air once every 30 min, about 10% of its total heat loss is the result of infiltration. Radiation is the passage of heat from warmer objects to colder objects without warming the air significantly. Polyethylene greenhouses can lose significant amounts of energy as a result of radiation unless a film of moisture forms on the plastic to form a barrier.

PLAN OF WORK

Week of: 2/24: Background research; compile information and lots of reading!; 3/3: Continue research; analyze logistics of implementation at my farm; document with pictures; start paper; 3/10: Continue paper; visit Slade's Jean Pain system and write case study findings; 3/17: Meet with UVM Greenhouse Manager Colleen Armstrong to discuss greenhouse heating systems and Tad and Erick's project; 3/24: Continue paper, edit; contact partners with any nagging questions; 3/31: Do more research on greenhouse efficiency; 4/7: Make PowerPoint; practice presentation; anticipate questions; 4/14: Finalize paper and presentation; final editing; 4/21: Present findings.

RESULTS
ECONOMIC ANALYSIS

According to a 2011 analysis done by Environmental Engineer Jason McCune-Saunders:

Greenhouse heating with Pain's system compares favorably to propane for greenhouse sizes of 100 m^2 and larger, particularly when the compost product is utilized. At 100 m^2, the system yields an ROI of 36% with mechanical salvage and compost value; this increases to 59% for a 300 m^2 greenhouse... Depending on efficiency of use, compost powered process heat becomes advantageous when competing energy costs rise above $0.07–$0.12 per kWh, and may generate significant savings under the right conditions.

Although the assumptions of this model including labor rates, fuel costs, cost of feedstock and mechanical components, value of finished compost, heat demand and storage capacity, longevity of heat production, and highly simplified thermal characteristics are subject to change and are potentially flawed, they do warrant further exploration of this technology. For this reason, McCune-Saunders and the Compost Power Network have partnered with the Slade greenhouse at UVM for further analysis of the system's potential.

FUTURE DIRECTIONS

When considering the heating efficiency of composting wood chips, energy output should be compared between the number of Btu that would have otherwise been captured from burning the wood. If heat is needed only at select times when temperatures dip below freezing, it might make more sense to simply burn the wood instead of composting it during that time. However, when that is done, high-quality compost is not also yielded. The compost system seems to make a lot of sense for Vermont's extended cold period. It is also convenient in the fact that it would not require daily inputs and maintenance as a wood-burning stove would. It is a one-time investment of energy in the fall to construct the system followed by a deconstruction phase in the summer when the compost is harvested and the heat is no longer necessary. Determining the size of the pile needed to adequately provide heat for a given area over a given period of time as well as the ideal compost recipe will be areas for further exploration for the Compost Power Network in the future. Being able to closely monitor temperature and flow of the system implemented at the Slade greenhouse over the next two years will help them to gain further insight. The system that Tad Cooke and Erick Crockenberg are implementing at the Common Ground Horticulture Farm will be experimenting with drawing both heat and CO_2 from the compost into the growing beds with the hope of improving crop productivity. According to the duo, between 10 and 50 g of CO_2 can be obtained from each kilogram of material being composted. They had hoped to implement their project in mid-March, but delays have pushed construction to the end of April. They now hope to have the system up and running by mid May. The stock for their compost will be dairy manure and bedding which will be stored in a covered bunker on the north wall of the greenhouse. It will be interesting to see how much crop growth will be affected by the CO_2 additions. It is truly unfortunate that their project has been delayed so much since the heating of the greenhouse will not be as crucial at this point with the coming summer temperatures. I guess that is just the reality of dealing with administrative and academic constraints.

BENEFITS TO COMMUNITY PARTNERS

Addie's Acres farm benefited significantly from this project in not only exploring different means of sustainably heating a greenhouse with compost, but also in designing a greenhouse in such a way as to minimize its actual heating needs. This document will prove to be a useful source of reference within the next 10 years as the farm develops and could be useful to others who are also considering building a greenhouse in the cold northeast where heating in the winter can be a significant cost. The case study for the Slade greenhouse provides a concise and complete document that could be distributed to others who are interested in learning about and applying the Jean Pain method. (Unfortunately, I was not as much use to Tad and Erick's project at the Horticulture Farm as I would have liked to have been as a result of administrative delays in approval that pushed the construction of their compost heat system to the end of April instead of the middle of March as initially hoped.) Had I been able to participate in its construction, I would have a better idea of the amount of labor, materials, hardware, and tools necessary for such a project while also providing service.

REFERENCES

"Compost Power: Using Compost Power to Heat a Greenhouse." UVM Office of Sustainability. Google: http://www.uvm.edu/sustain/clean-energy-fund/cef-projects/compost-power-using-compost-power-to-heat-a-greenhouse.

Diver, S., January 2001. "Compost Heated Greenhouses." Appropriate Technology Transfer for Rural Areas (ATTRA). Google. http://www.clemson.edu/sustainableag/CT137_compost_heated_greenhouse.pdf.

Gorton, S., McCune-Saunders, J., March 2011. Woody Biomass Energy Research Symposium for the Northern Forest: Design and Feasibility Analysis of Biothermal Energy and Compost Generation from Forest and Agricultural Feedstocks. University of Vermont. http://www.uvm.edu/~cfcm/symposium/?Page=Gorton.html.

Holmgren, D., 2002. Permaculture: Principles & Pathways beyond Sustainability. Holmgren Design Services, Australia.

McCune-Saunders, W.J., Rizzo, D.M., 2011. "Examining the Potential of Heat Extraction from Wood-based, Static Compost piles." Department of Civil & Environmental Engineering, University of Vermont, Burlington, VT.

Nelson, P., 2012. Greenhouse Operation and Management, seventh ed. Prentice Hall, Boston.

Pain, I., Pain, J., 1980. The Methods of Jean Pain or "Another Kind of Garden", seventh ed. translated from French.

BIOMASS TO LIQUID BIOFUELS

3

ANJU DAHIYA[1,2]

[1] *University of Vermont, USA,* [2] *GSR Solutions, USA*

As quoted in the preface, **the most certain way to advance next generation biofuels and the feedstock from which they will be manufactured is to continue to grow existing-generation biofuels**[1]. This part (Part III) covers liquid biofuels of both existing and next generations.

Part III contains nine chapters on biofuel feedstock and fuel options including bioenergy crops, on-farm oilseed-based biodiesel production, straight vegetable oil (SVO), cellulosic ethanol, bioheat, and algae. For additional insight, the biofuel conversion processes and renewable hydrocarbon liquid fuel diesel are covered in parts IV, V, and VI.

Chapter 7 (Bioenergy Crops) presents the rationale of planting biofuel crops (sugar and starch crops, cellulosic crops, woody crops, oilseed crops, crop residue and organic waste) from economics and most importantly, a farmer's perspective. It discusses the detailed crop-by-crop analysis, the crop handling and logistics, and the sustainable management for biomass production.

Chapter 8 (On-Farm Oil-Based Biodiesel Production) addresses oil-seed crop species and the variety selection, related planting considerations, pest management, harvesting, cleaning, drying and storage of seed, oil extraction and by-products, challenges and opportunities

---

[1]Kotrba, in Biodiesel magazine May/June issue, 2014. http://www.biodieselmagazine.com/.

FIGURE 1

Biomass to liquid biofuels.

Picture courtesy: A. Dahiya.

with farm-based fuel production. Additionally, two case studies are presented, which highlight a 13,000 gallon per year and a 100,000 gallon per year processing capacity. Examples are provided, which have been shown to result in fuel production costs of \$0.60–2.52 per gallon at regional level, and energy return on an investment of 2.6–5.9:1 and a potential net carbon emission reduction of up to 1420 lbs per acre of production.

Chapter 9 (Life Cycle Assessment: The Energy Return on Invested of Biodiesel) discusses Energy Return on Invested (EROI) approaches and controversies and provides the EROI calculation-related data and methods that readers can apply in their EROI-related studies, and reports on a life cycle assessment on biodiesel produced from oilseed grown on five farms and from reclaimed vegetable oil. The demonstrated EROI estimates from a state are compared to those reported by the United States Department of Agriculture.

Chapter 10 (Energy Management during Field Production Practices) explores bioenergy crops related field operations including tractor, tractor ballasting/slip/tire inflation, tractor transmission, no-till seeding, fertilizer, and other cultural and technological issues affecting energy use. It ends with an excellent list of resources a reader may explore further for in-depth knowledge.

Chapter 11 (Straight Vegetable Oil as a Diesel Fuel?) discusses the viability of fueling vehicles with SVO, or waste oils from cooking and other processes without intermediate processing, and how SVO and waste oils differ from biodiesel (and conventional diesel) in some important ways and are generally not considered acceptable vehicle fuels for large-scale or long-term use.

Chapter 12 (Cellulosic Ethanol—Biofuel beyond Corn) introduces cellulose as ethanol feedstock, and describes the challenges in the production of cellulosic ethanol, related plant biotechnology, and processes involved (pretreatment, hydrolysis, fermentation).

Chapter 13 (Bioheat) first distinguishes the difference between bioheat (also known as renewable heating oil), biodiesel, and petroleum. Properties of heating oil and bioheat are examined next. Three steps to successful bioheat management are described along with the oil filtration, the fuel additive treatment, and preventative maintenance.

Chapter 14 (Advanced Biofuel Production from Algae Biomass Cultivation) describes the use of algae biomass as a sustainable feedstock for biofuel. The brief historical perspective follows different biofuel options from algae. The start-to-finish process of liquid biofuel from algae involved is presented: (1) algae strain selection (different algae type); (2) algae cultivation (algae growth systems—photobioreactors, open raceways, and fermenters); (3) biomass harvesting; (4) algae oil extraction (mechanical, chemical—transesterification, enzymatic, supercritical fluid, etc.). The challenges in upscaling of algal biofuel operations from bench to commercial scales are described in the end followed by the life cycle analysis, economic and environmental effects.

Chapter 15 (Liquid Biofuels-Related Service Learning Projects and Case Studies) Related service learning projects and case studies presented include the cost analysis of oil seed production for biodiesel, the thermal switch from wood to oil heat, and a mobile ethanol distillery as examples.

BIOENERGY CROPS

Dennis Pennington
Michigan State University Extension

INTRODUCTION

The Energy Independence and Security Act (EISA) of 2007 set mandates for the amount of renewable fuels to be incorporated into the transportation fuel supply stream over the next 15 years. Oil companies will be required to blend 36 billion gallons by 2022. Starch-based products like corn grain will account for 15 billion gallons, while the remaining 21 billion gallons will come from biomass-derived fuels.

This policy has created a flurry of activity and investment in research of biomass crops. It will take a concerted effort to produce, collect, and deliver the quantity of biomass needed to achieve these goals. Scientists across the country are evaluating various crops to determine what might be the best fit for their soils, climates, and production systems. Sustainable production practices are being evaluated and best management practices written in order to mitigate any environmental risks. This chapter covers crop-by-crop production information and discusses some of the handling and logistical issues that need to be addressed (Figure 7.1).

Readers should finish this chapter with a basic understanding of current and future opportunities for bioenergy crop production and some of the economic factors driving it.

ECONOMICS

There are many factors for a farmer to consider before dedicating land to bioenergy crop production. Questions worth asking before planting a biofuel crop include:

1. Do I have access to a reliable market?
2. Do I need to invest in additional equipment or labor to plant, harvest or handle biomass?
3. What is my cost of production?
4. Which energy crop species fit my situation?
5. What are the potential yield and price I need for a biofuel crop to be at least as profitable as my current crop?
6. Do I need to develop an exit strategy or timeline in consideration of the above listed items?

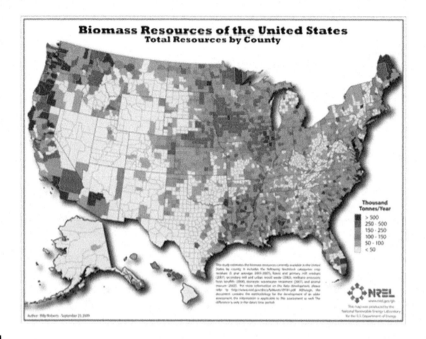

FIGURE 7.1

Biomass resources of the United States, national renewable energy laboratory, Billy Roberts, September 23, 2009.

The type of crop grown will affect the yearly cost of production and rate of return. Comparing annual with perennial crops can be tricky. Estimating the cost of production for annual crops like corn is relatively easy. It is more difficult for perennial crops, because you have to amortize the establishment costs over the life of the crop. Most of the costs for perennial crops are incurred up-front and are not recouped until later in the life of the crop. Additional consideration should be given to the timing of cash flows. For example, woody biomass, including poplar and willow trees, is harvested every 3–7 years. That means paying establishment and maintenance costs for a crop you will not see any income from for at least 3 years.

COMPARATIVE PROFITABILITY

Comparative crop profitability is based on partial enterprise budgets of annualized expenses. This provides a performance baseline for cash flow and potential net revenue. When conducting this kind of analysis we have to make some assumptions. It isn't realistic to know the exact yields, costs, and management of every farm. Once we have the models built, assumptions can be challenged and changed to fit each individual farm. The model provides a place to start in making comparisons. Table 7.1 summarizes the estimated enterprise production costs for continuous corn and

Appendix 1

Continuous corn plus stover

Annual - high input

	Quantity	unit	Price /unit		Yearly Total per acre	
Revenue sources						
Grain yield	135	bu	$	4.50	$	607.50
Stover (38%)	1.4	ton	$	60.00	$	84.00
Total revenue					$	691.50
Cash expenses						
Planting material						
Seed	0.4	80K kernel	$	184.00	$	69.00
Fertilizer[1]						
Nitrogen	173	lbs	$	0.46	$	79.51
P_2O_5	53	lbs	$	0.62	$	33.15
K_2O	118	lbs	$	0.63	$	74.40
Lime	0.3	ton	$	25.60	$	6.40
Pest control						
Lexar	0.8	gal	$	42.07	$	31.55
Machine costs						
Chisel plow	1	acre	$	14.17	$	14.17
Soil finish	1	acre	$	11.10	$	11.10
Planter-no till (corn)	1	acre	$	16.00	$	16.00
Bulk spreader	1	acre	$	3.75	$	3.75
Sprayer	1	acre	$	5.60	$	5.60
Post-harvest						
Combine, 6 row head	1	acre	$	29.11	$	29.11
Baling - lg round	2	bale	$	8.35	$	15.59
Bale-to-storage	2	bale	$	3.10	$	5.79
Drying[2]	135	bu	$	0.04	$	5.40
Marketing	135	bu		5%	$	30.38
Trucking (corn, 20mi)	135	bu	$	0.20	$	27.00
Trucking (stover, 20mi)	1	ton	$	3.00	$	4.20
Total cash expenses					$	462.09
Rev above expenses					$	229.41

	$ 30.00	60	90
	$30/ton	$60/ton	$90/ton
	187.41	229.41	271.41

1. Assumes no N contribution from previous crop and soil test indicating 50 lbs/acre available phosphorus and 150 lbs/acre available potassium. Refer to Michigan State University Extension bulletin E-2567, 2000, "Tri-State Fertilizer Recommendations for Corn, Soybeans, Wheat & Alfalfa" for more detail.

2. Assumes drying to 13% moisture from 23% pts

SOURCE:

MICHIGAN STATE
U N I V E R S I T Y
EXTENSION

Extension Bulletin E-3084, Profitability of Converting to Biofuel Crops
Web: http://bioenergy.msu.edu/economics/index.shtml

Table 7.1 Estimated Enterprise Production Costs for Continuous Corn and Five Perennial Biofuel Crops

Crop	Fertilizer Applied?	Avg. Annual Yield over 10 Years	Cost per acre (annualized)
Corn	Yes	135 bu Grain 1.4 ton stover	$462
Switchgrass	Yes	4 ton	$167
Grass mix	No	3.5 tons	$123
Native prairie	No	2.1 ton	$131
Miscanthus	Yes	10 tons	$412
Poplar	No	5 tons	$267

five alternative biofuel crops. A copy of these costs can be found at http://bioenergy.msu.edu/bioenergy/economics.

Miscanthus spp. production expenses were estimated using a custom planting rate of $900 per acre to purchase and plant rhizomes. This would require large nurseries, as was the case in several of the miscanthus BCAP projects. In spring, 2010 rhizomes in the United States cost $2.25 each. In contrast, in Europe at the same time, rhizomes cost $0.05 each. It is expected that the price of rhizomes will decrease significantly as more growers begin to propagate the crop themselves. In spring, 2013, rhizomes cost $0.27 each in the United States.

These perennial biofuel crops are evaluated on a 10-year replanting cycle. The grass crops take 2–3 years to produce mature biomass yields, and poplar grows for the full 10 years with only one harvest. During this establishment period, input costs (especially for weed control) are relatively high. We convert all costs over a 10-year period to an annual amortized basis so that profitability can be compared with that of continuous corn. Table 7.2 presents the per-acre annualized net return to land, labor and management, assuming biomass prices at $30/ton, $60/ton and $90/ton.

At $60 per ton for biomass, corn grain plus stover harvest has the highest return to land, labor, and management. At $90 per ton all crop systems are profitable.

Table 7.2 Annualized Net Return to Land, Labor and Management[a] ($/acre)

Crop System	$30/ton	$60/ton	$90/ton
Corn + stover	187.41	229.41	271.41
Switchgrass	−63.00	46.00	155.00
Grass mix	−82.00	−14.00	55.00
Native prairie	−82.00	−14.00	55.00
Misc	−184.00	30.00	243.00
Poplar	−146.00	−35.00	76.00

[a]*Land and management costs are assumed equal across all crops and are not included in the analysis.*

A CROP-BY-CROP ANALYSIS
SUGAR AND STARCH CROPS

Many of the sugar and starch crops that are candidates to produce biofuels are those currently being grown for sweeteners or for feed and food. Corn grain is currently the primary feedstock being used for ethanol production in the United States, due to the ease of converting corn to ethanol, the abundance of corn already in production and the existing infrastructure to produce, transport, and store corn.

Corn is converted to ethanol through the biochemical process of fermentation, using microorganisms that convert the sugars into alcohol. Most corn grain is converted to ethanol using a dry mill process that produces animal feed—and sometimes CO_2—in addition to ethanol. For non-technical summaries of the process of making ethanol, and issues and opportunities associated with ethanol production, see the National Sustainable Agriculture Information Service (ATTRA) publication "Ethanol Opportunities and Questions" (Morris and Hill, 2006) and the Extension publication "Corn for Biofuel Production" (Hay, 2010).

Corn grain derived ethanol has created new wealth and opportunity in the rural Midwest. However, environmental, economic, and social sustainability concerns have also been raised. There is debate over the impact corn ethanol has had on net energy balance, greenhouse gas emissions, water quality, market volatility, risk for farmers and competition for use of feedstock, as well as on local communities.

Ethanol can also be made from grain sorghum, wheat, sugarcane, and sugar beets. In Brazil, sugarcane is used extensively as a bioenergy feedstock. Sugar beets and sweet sorghum can be grown in temperate regions as far north as Canada.

CELLULOSIC CROPS

Plants grown for cellulose include herbaceous and woody perennials. Leaves, stems, and stalks—the cellulosic parts of plants, can be burned directly to generate electricity, or can be converted to liquid fuels, energy gases, and chemicals through a variety of conversion technologies.

Perennial crops grown for bioenergy production offer environmental benefits to soil, water and wildlife resources. One of the main advantages of these crops is that they re-grow from the root system every year, and thus do not need to be replanted annually. Perennial crops maintain cover on the land, provide habitat, and usually require fewer fertilizer and water inputs. These crops can be planted on less productive or marginal agricultural landscapes. Perennial crops under consideration as bioenergy feedstock include switchgrass and other native prairie grasses, *Miscanthus* spp., reed canary grass and tropical maize.

At present, the major concern about perennial bioenergy crops is economic; farmers cannot anticipate market demand or estimate the time for a return on their investment, because, as yet, there are few established market for these crops. Additionally, environmental concerns about these crops have also been raised because of their limited production history in the United States. There is more to learn about their input requirements, potential pests, and susceptibility to disease. Some of these species, such as Miscanthus, may cause controversy because they are non-native.

WOODY CROPS

Perennial non-forest woody crops, also called fast growing, short-rotation crops (SRC), provide multiple products and benefits. For example, SRCs such as hybrid poplar and willow yield pulpwood products as well as energy. These crops are typically harvested then chipped on-site. The chips are often combined with some other feedstock during densification (e.g., pelletizing).

Perennial woody crops are usually established as a monoculture, but can be grown in mixed plantings as well. These short-rotation crops provide long-term yield potential and environmental benefits like wildlife habitat, soil erosion prevention, and water-quality improvement.

Producing woody biomass crops causes some social and economic concerns, such as competition with traditional forest product industries for land and wood products, along with questions about whether the financial returns will be worth the investment.

OILSEED CROPS

Oilseed crops are grown for their seed, which is harvested and crushed to extract oil. This oil can then be converted into biodiesel through a process called transesterification. There are a number of oilseed crops being evaluated for bioenergy potential. Many of these crops have special adaptations that make them regionally viable based on climatic and soil conditions. Some characteristics include salt tolerance, cold hardiness, drought resistance, and low fertility requirements.

Because a number of oilseed crops, including soybean, canola, and sunflower, are grown and used for food grade vegetable oil, there are concerns that they may be diverted to fuel use at the expense of food production. Indeed, several oilseed importing European countries have determined that oilseed production for export often diverts land from food production. Sustainability standards are under development to remedy this situation. Land use change and competition with food crops may be a factor in determining what crops will be eligible for tax credits and incentives for biofuel production in the United States.

CROP RESIDUE AND ORGANIC WASTE

In annual cropping systems, aboveground, non-grain portions of plants—including stalks, leaves, chaff, and husks—are left in the field after harvest. These residue materials are composed of ligno-cellulose and may be used as a bioenergy feedstock, burned directly for biopower, or converted to a liquid transportation fuel or gas.

There are economic concerns about the use of crop residue as a bioenergy feedstock. In the Midwest, corn stover (stalks and leaves) is the leading candidate for production of cellulosic ethanol because it is abundant and relatively inexpensive. However, the farmer incurs an additional cost in terms of equipment, time, and fuel used to harvest stover. The increased use of some crop residues for biofuels could reduce their availability for more traditional uses. For example, straw from small-grain crops is typically used as animal bedding, either on-farm or sold as a cash crop. Diverting this form of crop residue to the biofuel market could decrease the supply of animal bedding forcing livestock farmers to find alternative sources for bedding.

Removal of crop residues for biofuel production can have environmental consequences. When left in the field, these residues reduce erosion and increase soil organic matter (SOM) the soil. On the other hand, excessive residue cover may interfere with spring planting and hinder soil warming and drying. It will be important to remove residue at sustainable rates so that it can simultaneously provide an extra income for farmers, improve spring planting, and still prevent erosion and maintain SOM.

The following crop-by-crop analysis gives a brief synopsis of yield potential, production and management practices, and estimated energy potential. For a quick "at a glance" view, please see the crop comparison matrix in Appendix 2.

Appendix 2

"At a Glance" Cellulosic Energy Crop Comparison Matrix

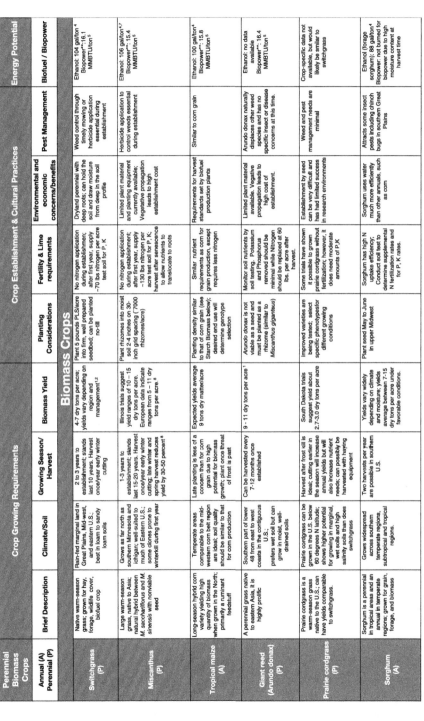

Perennial Biomass Crops / Annual (A) Perennial (P)	Crop Growing Requirements				Crop Establishment & Cultural Practices				Energy Potential
	Brief Description	Climate/Soil	Growing Season/ Harvest	Biomass Yield	Planting Considerations	Fertility & Lime requirements	Environmental and economic concerns/benefits	Pest Management	Biofuel / Biopower
Biomass Crops									
Switchgrass (P)	Native warm-season grass; grown for hay, forage, wildlife cover, biofuel crop	Rain-fed marginal land in Great Plains, Mid-west, and eastern U.S., best in loam to sandy loam soils	2 to 3 years to establishment; stands last 10 years. Harvest once/year early winter cutting	4-7 dry tons per acre; yields vary depending on region and management[1,2]	Plant 5 pounds PLS/acre into a firm, well prepared seedbed; can be planted no-till	No nitrogen application during establishment; after first year, supply ~70 lbs nitrogen per acre test soil for P, K	Dryland perennial with deep roots; can hold the soil and draw moisture from deep in the soil profile	Weed control through timely mowing and herbicide application essential during establishment	Ethanol: 104 gal/ton.[4] Biopower**:16.1 MMBTU/ton[5]
Miscanthus (P)	Large warm-season grass, native to Asia; natural hybrid between M. saccchariflorus and M. sinensis with nonviable seed	Grows as far north as southern Minnesota and Michigan; well-suited to much of Eastern U.S.; some clones prone to winterkill during first year	1-3 years to establishment; stands last 15-20 years. Harvest once/year early winter cutting; late winter and spring harvest reduces yield by 30-50 percent[3]	Illinois trials suggest yield ranges of 10-15 dry tons per acre. European data indicate ranges from 5-11 dry tons per acre.[3]	Plant rhizomes into moist soil 2-4 inches on 30-inch grid spacing (~7000 rhizomes/acre)	No nitrogen application during establishment; after first year, supply ~130 lbs nitrogen per acre test soil for P, K; harvest after senescence to allow nutrients to translocate to roots	Limited plant material and planting equipment currently available; Vegetative propagation leads to high establishment cost	Herbicide application to control weeds essential during establishment	Ethanol: 106 gal/ton.[4,7] Biopower**: 15.4 MMBTU/ton[5]
Tropical maize (A)	Long-season hybrid corn variety yielding high quantity of biomass when grown in the North; primarily a ruminant feedstuff	Temperate areas comparable to the mid-western corn belt region are ideal; soil quality should be similar to that for corn production	Late planting is less of a concern than for corn grain due to high potential for biomass growth; plant once frost is past	Expected yields average 9 tons dry matter/acre	Planting density similar to that of corn grain (see Starch Biomass below); desired and use will determine genotype selection	Similar nutrient requirements as corn for grain production, except requires less nitrogen	Requirements for harvest standards set by biofuel production plants	Similar to corn grain	Ethanol: 100 gal/ton[4] Biopower**: 15.8 MMBTU/ton[5]
Giant reed (Arundo donax) (P)	A perennial grass native to eastern Asia, it is highly prolific	Southern part of lower 48 from east to west coasts in the contiguous U.S.; prefers wet soil but can grow in moist, well-drained soils	Can be harvested every 7-12 months once established	9 - 11 dry tons per acre[?]	Arundo donax is not viable as a seed and must be planted as a rhizome (similar to Miscanthus giganteus)	Monitor soil nutrients by soil testing. Potassium and Phosphorus removed should be minimal while Nitrogen should be replaced at 60 lbs. per acre after harvest.	Limited plant material available. Vegetative propagation leads to high cost of establishment.	Arundo donax naturally displaces other weed species and has no specific insect or disease concerns at this time.	Ethanol: no data available Biopower**: 16.4 MMBTU/ton[5]
Prairie cordgrass (P)	Prairie cordgrass is a warm-season grass native to the U.S.; can have yields comparable to switchgrass.	Prairie cordgrass can be grown in the U.S. below 60 degrees N latitude; shows higher potential for growing in marginal, wet soils and high salinity soils than does switchgrass	Harvest after frost kill is ideal; cutting earlier in the season will increase annual yields but will also increase nutrient needs; can possibly be harvested with haying equipment	South Dakota trials suggest yield about 2.7-3.0 dry tons per acre	Improved varieties are being tested; select specific phenotype for different growing conditions	Some trials have shown it possible to grown prairie cordgrass without fertilization; however, it does need moderate amounts of P,K	Establishment by seed can be very difficult and has had limited success in research environments	Weed and pest management needs are minimal	Crop-specific data not available, but would likely be similar to switchgrass
Sorghum (A)	Sorghum is a perennial in tropical areas and an annual in temperate regions; grown for grain, forage, and biomass	Grows widespread across southern temperate regions, subtropical and tropical regions.	Two harvests per year are possible in southern U.S.	Yields vary widely depending on climate and moisture; yields average between 7-15 dry tons per acre under favorable conditions.	Plant seed May to June in upper Midwest	Sorghum has high N uptake efficiency; Conduct soil test to determine supplemental N fertilization rates and for P, K rates.	Sorghum uses water much more efficiently than other annuals, such as corn	Attracts some insect pests including chinch bugs in southern Great Plains	Ethanol (forage sorghum): 88 gal/ton[4] Biopower: not burned for biopower due to high moisture content at harvest time

(Source: Pennington, D., C. Gustafson, J. Hay, R. Perrin, G. Wyatt, and Zamora, D. 2011. Bioenergy Crop Production and Harvesting: Module 2 in S. Lezberg and J. Mullins (eds.). Bioenergy and Sustainability Course. On-line curriculum. Bioenergy Training Center)

Annual (A) Perennial (P)	Brief Description	Climate/Soil	Growing Season/ Harvest	Biomass Yield	Planting Considerations	Fertility & Lime requirements	Environmental and economic concerns/benefits	Pest Management	Biofuel / Biopower
Woody Biomass Crops									
Willow (P)	Willow is a fast-growing shrub	Temperate areas, midwestern and northern states and Canada are good growing regions	Harvest every 3 to 4 years; can be harvested with modified forage harvesting equipment	Yields of 4 to 6 dry tons per acre are realistic in the Midwest	Plant cuttings or rooted saplings in twin rows about 2.5 ft apart; 5 ft between twin rows.	Apply 100 lbs of nitrogen after establishment year and again after each harvest on 3-4 year cycles.	Trees require long-term rotations. Yields are currently similar and sometimes lower than other perennial biomass crops	May have minor insect problems from Japanese and willow leaf beetles.	Ethanol: 105 gal/ton[4] Biopower**: 16.9 MMBTU/ton[5]
Poplar (P)	Fast growing tree grown for pulpwood and other uses	Widely grown; range from southern states to much of Canada; Western poplar species grown in western states.	Trees grow 3-4 years before first cutting and 3 years between subsequent cuttings.	Yields of 3 to 6 dry tons per acre per year are realistic for the Midwest	Planted with cuttings or bareroot saplings. 12 feet apart with 8 to 12 ft rows. Plant in May to June in upper Midwest	Low fertilizer requirement. In the range of 50 lb/ac per year	Trees require long term rotations. Yields are currently similar and sometimes lower than other perennial biomass crops	Some insect pressure. Deer browsing when trees are small can slow growth.	Similar to Poplar.
Starch Crops									
Corn grain (A)	The grain portion of the corn crop. It is about half the above-ground biomass	Widely adapted throughout the U.S. and Mexico as far north as southern North Dakota, high yields in well-drained prairie soils	Annual, planted in the spring and harvested in late summer or fall	Grain yield range of 50 to 250 bushels per acre or 1.4 to 7 tons silage per acre; national average 152 bu/acre in 2010; current conversion to ethanol ~2.8 gal per bu	Planting is in 20 to 40 inch rows with 30 most common; 20,000 to 35,000 plants per acre at 2-4 inches deep	High N requirement. Sidedress nitrate test should be done to determine N credit. Soil test for P, K application.	High N rates have lead to leaching and runoff of N and P, causing ground and surface water contamination	Attracts many insect and disease pests; control with GMO seed, pesticides and herbicides	Ethanol: 124 gal/ton[4] Biopower**: 14 MMBTU/ton
Crop Residues									
Corn stover (A)	Above-ground biomass left after corn grain harvest, including husks, leaves, and cobs	Widely adapted as far north as southern North Dakota and South Dakota, Mexico; prefers well-drained highly fertile soils	Days to maturity varies from 90 to 120 days; an annual planted and harvested in mid- to late fall	Above-ground biomass yield average 4.2 tons per acre; harvest efficiency 37-50 percent[6]	Plant March to May in the upper Midwest; can be planted no-till	High N requirement. Sidedress nitrate test should be done to determine N credit. Soil test for P, K application.	Soil erosion and organic matter concerns should determine the amount of stover left on surface after harvest; harvested amount needs to be limited on most soils	Attracts many insect and disease pests; control with GMO seed, pesticides and herbicides	Ethanol: 113 gal/ton[4] Biopower**: 15.7 MMBTU/ton[5]
Wheat straw (A)	Above-ground biomass left after wheat grain harvest, including stems, leaves	Widely adapted across the U.S.; major growing regions include, North and South Dakota, Kansas, Oklahoma, Montana, Minnesota, Washington	Spring and winter varieties available; spring wheat is planted in early spring and harvested in summer, winter wheat is planted in the fall and harvested the following summer	Wheat straw yields vary with climate and variety selection; yields generally range from 0.75 to 1.5 dry tons/acre	Winter wheat should be planted after Hessian Fly-free date; to prevent disease and winter kill, limit growth prior to first fall frost	Wheat has moderate N requirements; test soil for K	Harvest of wheat straw impacts soil and water and results in nutrient removal, increased erosion, soil organic matter losses, etc.	Insect and disease issues are common, some may be controlled with pesticides	Ethanol: 96 gal/ton[4] Biopower**: 14.9 MMBTU/ton[5]
Sugar Crops									
Sugarcane Bagasse	The fiber left after sucrose is squeezed from the stems; burned to produce heat and power for ethanol production	Tropcial and subtropical areas	See sugarcane, below	8 to 10 tons of bagasse per acre	Plant stem cuttings	Needs large amounts of nutrients to produce high yields	Bagasse burning dramatically increases the net energy balance		Ethanol: 111 gal/ton[4] Biopower**: 16.1 MMBTU/ton[5]
Sugarcane (P)	A tall, perennial grass native to Asia. sucrose is squeezed and directly fermented into ethanol	Grown in subtropical and tropical areas. Currently Brazil is the world's largest producer of sugarcane.	Harvest 12 to 24 months after planting; can be harvested 2-10 times between replantings; harvested by hand, mechinization becoming more common	Yields vary by planting date and location; 12 to 15 dry tons/acre is characteristic, which corresponds to 2-6 tons per acre of sugar	Planting is mainly done using stem cuttings, at 6000 to 8000 cuttings per acre	With such high yields sugar cane also has high nutrient requirements	This is a heavily intensive crop with high fertilizer and pesticide requirements; Bagasse can be burned in CHP process, dramatically reducing the energy input for processing	Insects and diseases can be controlled with rotations and other cultural practices and pesticide applications; weed control prior to canopy closure is important	Ethanol: 16.77 gal/ton[1] Biopower ** (total plant): 30.7 MMBTU/ton; Biopower: (Bagasse only): 16.5 MMBTU/ton

Annual (A) Perennial (P)	Brief Description	Climate/Soil	Growing Season/ Harvest	Biomass Yield	Planting Considerations	Fertility & Lime requirements	Environmental and economic concerns/benefits	Pest Management	Biofuel / Biopower
Sugar beets (A)	Grown for the sucrose stored in the large tap root	Grown in temperate regions of the northern and and southern hemispheres	Sugar beet is a biennial, yet the root is harvested in its first year of growth; beet tops and pulp used as cattle feed	Sugar beet generally has a higher sugar content than sugarcane but lower tons per acre yield.	Plant seeds in spring; seeds are very small, requiring only about 1 lb per acre	Nitrogen fertilizer is typically needed; soil test for other nutrients	Currently sugar beets are used for edible sugars, yet ethanol production plants are in planning stages in the U.S.	GMO sugar beets are on the market; these reduce weed control costs	Ethanol: 22 gal/ton[1]
Sweet sorghum (A)	Sweet sorghum is a variety of sorghum that has a high concentration of soluble sugars in the juice	Unlike sugarcane, sweet sorghum can be grown in temperate regions	Annual in temperate regions	Yields similar to sugarcane, with similar to slightly higher sugar content; 5–18 dry tons/acre[1] with brix sugar content of 14-20 percent	Plant seed with similar spacing, depth and population density as corn	Nutrient studies have shown mixed results in the Midwest with moderate N applications	Benefit of high sugar content in stalk; to maximize ethanol yield, juice extraction must be at least 50% of juice from stalk.		Ethanol (from juice): 24-32 gal/ton[1] Ethanol (total plant): 130 gal/ton
Oil Seed Crops									
Canola/rapeseed (A)	Canola is a Brassica species developed for its high quality oil	Can be grown from Oklahoma to North Dakota, in the Southern U.S. and Pacific Northwest; prefers well drained soils	Spring and winter varieties; harvested either after windrowing or direct harvest; small seeds have special harvest considerations	Yield of 40 bu/acre are common; oil content is ~40 percent; biodiesel yield is 1 gal biodiesel per gal of vegetable oil	Plant winter varieties in fall and spring varieties in spring; avoid soil crusting	Similar to other small grains, with special attention to sulfur requirements	Canola will produce a high quality oil which has edible and industrial uses	Birds can become problem at harvest time	Biodiesel: 272 gal/ton Biopower**: 32.5 MMBTU/ton
Mustard (A)	Mustards are Brassica species similar to canola.	Can be grown in wide area; tends to be more drought tolerant than canola	Spring and winter varieties; harvested either after windrowing or direct harvest; small seeds have special harvest considerations	Yields approximately 400 – 920 lbs. per acre in Montana[4] for ~ 61 gal of oil per acre. Seeds have an oil concentration slightly lower than its canola relatives. Biodiesel yield is 1 gal biodiesel per gal of vegetable oil.	Planting in fall for winter varieties and in spring for spring varieties. Avoid soil crusting.	Similar to small grains	Mustards produce high quality oil with low cloud point, desirable for biodiesel production; mustard meal is not as desirable as other oil seed meal		Biodiesel: 272 gal/ton Biopower**: 32.5 MMBTU/ton
Soybeans (A)	Soybeans are native to China and grown for their protein (meal) and oil	Soybeans are grown from Texas to Canada and from the Midwest east	Soybean maturity groups range from 000 in the far north to VI in the far southeastern U.S.; harvested in early/early fall in the Midwest	National average yield was 44 bushels per acre in 2010; oil content is approximately 18-19 percent. Biodiesel yield is 1 gal. biodiesel per gal. of vegetable oil.	Soybeans are planted April to June in the Midwest; can be planted no-till; plant 75,000 to 200,000 plants per acre	Soybeans are legumes and produce their own nitrogen (in association with rhizobium bacteria); test soil for P and K; soil pH near neutral is optimal	Soybeans leave very little residue to prevent erosion; oil content is low compared to other oil seeds, so oil yield per acre is low	Insects and diseases are common but can be controled with pesticides and proper scouting	Biodiesel: 272 gal/ton Biopower**: 32.5 MMBTU/ton
Sunflower (A)	Native to the U.S.; sunflower seeds have high oil content	Predominantly grown in the Great Plains; South Dakota and Kansas are major producers; widely adapted and highly drought tolerant	Sunflower can be grown full season or double-cropped with wheat; double-cropping in the eastern Great Plains reduces insect pressure	Yield of 500 to 2500 lbs per acre; sunflower seeds contain ~ 45 to 50 percent oil; biodiesel yield is 1 gal biodiesel per gal. of vegetable oil	Plant 15,000 to 25,000 seeds per acre in wide rows (30 inch is common)	Responds to N fertilizer in high moisture ownhen irrigated, less so in dry conditions	In areas where sunflower is not a major crop, transportation for commercial delivery can be long; sunflower leaves little residue to prevent erosion.	Sunflower head moth and other insects can be a major issue in eastern Great Plains.	Biodiesel: 272 gal/ton Biopower**: 32.5 MMBTU/ton
Camelina (A)	Another member of the Brassicae, camelina seeds have high oil content; the oil has been converted to biodiesel and jet fuel	Mainly northern Midwest and Mountain States.	Planted in spring for spring varieties or fall for winter varieties; harvested in mid- to late summer; very small seeds require special handling during harvest	Camelina can yield between 1,800-2,000 lbs of seed/acre. Camelina seeds contain 25-41 percent oil; biodiesel yield is 1 gal biodiesel per gal of vegetable oil	Plant 2 to 3 lbs. of seed per acre	Moderate N requirement and growing season similar to wheat which makes camelina attractive in semi-arid areas.			Biodiesel: 272 gal/ton Biopower**: 32.5 MMBTU/ton

* based on 91 gallon ethanol per ton3

** Values are High Heating Values (HHV)

1. Bassam N. EI. 2010. Handbook of Bioenergy Crops. Earthscan, London UK.

2. Perrin R., Vogel K., Schmer M., Mitchell R., 2008. Farm-scale Production Cost of Switchgrass for Biomass. Bioenerg. Res. 1: 91-97.

3. Schmer M.R., Vogel K.P., Mitchell R.B., Perrin R.K., 2008. Net energy of cellulosic ethanol from switchgrass. Proceeding of the National Academy of Sciences of the United States of America.

4. NREL Theoretical Ethanol Yield Calculator online at http://www1.eere.energy.gov/biomass/ethanol_yield_calculator.html

5. NREL - (High Heating Value) Biomass Feedstock Composition and Property Database online at http://www.afdc.energy.gov/biomass/progs/search1.cgi

6. Shinners, K. J., and B. N. Binversie. 2007. Fractional yield and moisture of corn stover biomass produced in the northern U.S. corn belt. Biomass and Bioenergy 31(8): 576-584

7. Sorensen A., Teller P.J., Hilstrom T., Ahring B K., 2008. Hydrolysis of Miscanthus for bioethanol production using dilute acid presoaking combined with wet explosion pre-treatment and enzymatic treatment. Bioresource Technology 99, 6602-6607

8. Heaton E., 2010. Factsheet/Biomass: Miscanthus: Miscanthus. Iowa State University Department of Agronomy AG201

Soil erosion and organic matter concerns should determine the amount of stover left on surface after harvest; harvested amount needs to be limited on most soils

CORN GRAIN

Energy Potential	
Biofuel	462 gal/acre
Biopower	0.24 MMBTU/ton

Characteristics
Corn (*Zea Mays*) is a popular feedstock for ethanol production in the United States due to its abundance and relative ease of conversion of grain to ethanol. The conversion process for fuel includes grinding, cooking with enzymes, fermentation with yeast, distillation, and molecular sieve to remove any remaining water and adding denaturant to make it undrinkable.

Climate
Although best suited from growth in the upper Midwest with moderate temperatures and adequate rainfall, corn is grown on every continent and in much of the United States.

Soil
Highly fertile soil is preferable for best corn growth.

Seedbed preparation
No-till, reduced till, disking, and other tillage methods are common in corn production. Disease cycles can be broken with basic rotation between corn and other crops such as soybeans or wheat.

Fertility and lime requirements
Corn is a large user of nitrogen, as well as needing phosphorus and potassium. Nitrogen can be fertilized either inorganically or in the form of manure.

Variety selection
Many hybrids are available. Biotech hybrids for pest and weed resistance have gained popularity.

Water/irrigation requirements
Ninety percent of corn is grown under rain fed conditions. Much of the corn grown in the Western Cornbelt is grown under irrigation, predominately through center pivot systems.

Weed/insect control
Rotation of crops will reduce weed pressure. Additionally, biotechnological advances offer weed and insect control opportunities through insect resistance or herbicide resistant corn.

Disease control
Many prevention and treatment options are available. For more information, please visit the University of Missouri Extension Pest Management Guide for Corn Production.

Yield potential

Average national corn yield was 165 bu./acre in 2009. This translates into 462 gallons of ethanol produced per acre at the average of 165 bu./acre. Historical yield trends show an increase of about 2 bu./acre/year.

Economics

Corn is a well-established crop with much research and support. However, its high inputs and production costs make it a more intensive crop than most biofuel crops. Although other crops may be less intensive to produce, they require new machinery and have no other uses in case of limited demand for biofuel production, whereas corn is still potentially salable as a commodity.

Corn has also received criticism for the high level of fossil-fuel input required to produce ethanol. Studies show corn ethanol represents a 20–45% positive energy balance.

CORN STOVER

Energy Potential	
Biofuel	143 gal/acre
Biopower	6.68 MMBTU/ton

Characteristics

Corn stover (*Zea Mays*) is one of the largest potential annual crop-based biofuel feedstocks for several key reasons: the quantity of feedstock in the United States is large, the feedstock is relatively uniform, and large quantities are concentrated for biofuel production. Harvesting corn stover can have long-term impacts on soil quality and erosion unless it is done with a sustainable approach. Corn stover consists of the stalk, leaves, sheaths, husks, shanks, cobs, tassels, lower ears, and silks.

Production challenges

Challenges to production are directly related to the timing of harvest and how harvested material will be stored. Field drying is influenced by temperatures and precipitation and can allow for some leaching of nutrients, additionally forage harvesting with higher percent moisture allows for faster and sooner harvesting. Additionally, drawbacks could include contamination of the bales with dirt and plastic, a low labor efficiency of the process, bale breakage, and consistency of the product. Development of new economic models for stover harvest are needed that include the potential for less tillage, carbon and renewable energy credits, and rapid harvesting systems with minimal soil contamination.

Because corn stover acts to reduce wind and water erosion, increase soil moisture holding capacity and adds to long term soil carbon it is essential to consider environmental and long term impacts of corn stover harvest. This will be influenced greatly by the type of tillage and region of the country. There are several tropical temperate lines of *Zea mays* currently being evaluated for high total biomass, higher partitioning of stover versus grain, adaptation to perform at low nitrogen rates, and stalk sugar levels rivaling sugar cane. Multiple-use strategies addressed during the development of this "corn" crop may prove valuable to producers and end-user markets.

Corn cobs

Corn cobs are currently being used for heat in some parts of Europe; whereas in the United States, there is a notable level of cobs milled as base products for various industries (feed filler-additive, oil-drilling adsorbent, desiccant, and as a bio-abrasive). This feedstock is rapidly being developed as a feedstock for cellulosic ethanol, co-firing, and gasification projects. Corn cobs are dense and relatively uniform, and they have a high heat value, generally low N and S contents, and can be collected during corn grain harvest. Harvesting cobs has little potential impact on soil residue, soil carbon, or the nutrient requirements of subsequent crops. Corn cobs appear to be a relatively sustainable, but relatively low-yielding feedstock.

Yield potential

On a dry matter basis, cob yields average about 14% of grain yield and represent about 16% of the total stover biomass in a field.

Production challenges

Cob moisture is a logistical and storage challenge as cobs are typically wetter than the corn grain. At corn grain moisture levels of 20%, cobs may still be retaining 35% moisture.

SWITCHGRASS

Energy Potential	
Ethanol	400 gal/acre
Biopower	15.5 MMBTU/ton

Characteristics

Switchgrass (*Panicum virgatum*) is a native warm-season grass that is a leading biomass crop in the United States. Grown for over 70 years, it is native to the United States and grows from 3 to 10 ft tall. Of benefit is the potential use of haying equipment for harvest and the possibility of harvesting after frost kill. Challenges to production include the necessary storage to provide biomass to a bio-refiner consistently through the year, losses if left standing through winter, and necessary proximity of stands to bio-refinery.

Climate

Switchgrass is a warm-weather crop, but is capable of being grown in all of the United States (excluding Alaska).

Soil

Grows best in neutral pH soil (6–8), but slightly acidic soil is acceptable. Can grow in a variety of soil conditions from fertile soils to marginal land.

Seedbed preparation and planting

The two most common approaches are to plant into a tilled seedbed or into an herbicide-killed sod by using no-till planting equipment. Prepare soil and planter to achieve good seed to soil contact. Seeds should be planted at 5 lbs of pure live seed per acre, planting depth of 0.25–0.50 inches via drill, and with row width of 7–10 inches.

Fertility and lime requirements

Twelve to twenty pounds of nitrogen is required per acre for each ton of expected yield. Nitrogen fertilization is not recommended during establishment year be-cause it will encourage weed competition. Soil testing will reveal need for other macronutrients phosphorus and potassium.

Variety selection

Lowland varieties perform well in the south. Upland varieties are superior in the north. Selecting high yielding varieties is an important management decision. A hybrid between upland and lowland has been developed and shows great promise of 30–50% higher yields.

Growth

Primarily planted as a bunch-grass, switchgrass roots are up to 10 ft deep, but short rhizomes can form sod over time. While stands are recommended to last 10 years, there is potentially no re-plant necessary.

Weed control

Weed control during establishment is critical. Typically, little to no herbicide is needed after establishment.

Yield potential

An F1 hybrid of Kanlow and Summer yielded over 9 tons per acre. Yield is significantly reduced if a cultivar is planted over 300 miles north or south of its origins. Expect 4–5 tons per acre in the Great Plains, 2–7 in the Midwest, and approximately 10 tons per acre in the southeast.

Production challenges

Harvest after frost kill is ideal due to the fact that nutrients are translocated to the roots after frost. Cutting earlier in the season can improve yield, but will also remove additional nutrients that will need to be replaced. If crop is left standing over winter, up to 40% loss by weight can be expected. Economically, transportation is not feasible beyond 25 miles, so stands must be grown close to bio-refinery. Commonly used Midwestern varieties will establish in nearby uncultivated areas making it somewhat invasive but definitely opportunistic in growth habit and range.

MISCANTHUS

Energy Potential	
Ethanol	560 gal/acre
Biopower	21 MMBTU/ton

Characteristics

Giant Miscanthus (*Miscanthus gigantius*) is a promising bioenergy crop for some regions of the Midwest and eastern United States, with average biomass yields over two times that of switchgrass in Illinois trails. Challenges include overwinter survival, high cost of establishment, and availability of rhizomes for propagation.

Climate

Experience in Europe suggests that giant miscanthus will be productive over a wide geographic range in temperate regions, including marginal land, but is not appropriate for arid regions.

Soil

Grows best on well-drained soils but can tolerate heavy soils and periodic flooding. Yields are reduced on marginal land.

Seedbed preparation

To avoid weed competition, rhizomes should be planting on a clean seedbed. A cover crop of small grains may be used to protect soil overwinter and provide weed suppression in spring during establishment. Attention to soil moisture conditions at the rhizome placement zone and good rhizome to soil contact are important for establishment.

Fertility and lime requirements

Importance of fertilizer to increasing harvestable yield is not yet clear. In some research miscanthus responds to nitrogen fertilizer, but in others the response is negligible.

Variety selection

It is important to utilize the triploid hybrid sterile cross M. x *giganteus* rather than other miscanthus species; non-sterile varieties can become invasive.

Water/irrigation requirements

Yield usually increases with increased water available to crop, although irrigation should not be necessary during an average year in the Midwest. Miscanthus will not withstand continuously water-logged soils.

Weed control

Weed control during establishment is necessary. Typically, little to no herbicide is needed by the third year after planting. Higher plant population rates at establishment can provide some element of weed suppression.

Insect/pest control

A few new insects and pathogens have been identified in Miscanthus. Yield decreases have not been documented, so there are currently no controls recommended. New evidence indicates that corn borers may overwinter in Miscanthus residues.

Yield potential

Small plot trials of giant Miscanthus in Illinois suggest yield ranges of 10–14 tons per acre. However, it is critical to establish yields strongly the first year to increase average yields by being able to harvest on year two. Yield expectations, based on both European and U.S. data, range from 4 to 18 tons per acre with yields impacted by heat, moisture, and type of soil. Further research is needed to narrow these wide gaps. Energy potential was calculated at 7 tons DM/acre. Field trials in central Illinois (2008–2013), using a delayed harvest strategy (as late as May 1st), resulted in dry matter yields ranging from 7½ to 10 tons per acre with no mechanical raking to gather plant litter.

Economics

Currently, the difficulty in procuring rhizomes for propagation translates to a high cost of establishment, with costs of between $900 and $1800 per acre for planting material alone. Stands are productive for 15–20 years.

SORGHUM

Energy Potential	
Ethanol	1120 gal/acre

Characteristics

Sweet sorghum is similar to grain sorghum (same species). Sorghum bicolor has a higher concentration of sugar in the stalks. The specific advantage of this crop is its increased drought resistance over corn, and its limited need for nitrogen fertilization. Challenges include commercial harvesting equipment, rapid degradation of sugars, and transportation cost.

Climate

Similar to grain sorghum, sweet sorghum is drought resistant and hardier than corn. It can be grown throughout the continental United States.

Soil

Although sweet sorghum will grow on any land type, it best produces under loam or sandy silt loam. Deep soil is preferred with moderate drainage. Clay-like or shallow soils will produce lower yields with poorer quality.

Planting temperature

Direct seed planting requires a soil temperature of 65 °F.

Fertilization requirements

Fertilization requirements are primarily determined by nutrient quantity removed via biomass. If plant biomass byproduct is not spread on field, then P and K requirements will be higher. Due to high N efficiency of sweet sorghum, N fertilization will be low.

Variety selection

The four most prominent varieties are Dale, Keller, M81E, and Theis. A male sterile hybrid (KNMorris) was released in 2007 with other hybrids currently being developed.

Weed/disease control

Herbicides are available and labeled for use in sorghum. Seed treatment can be used to minimize crop injury. Disease is best controlled by rotation with non-grass crops and planting disease resistant varieties, such as Dale (most diseases), Keller and M81E (red stalk rot), and Theis (red stalk rot and maize dwarf virus).

Insect/predator control

Many corn worms and insects are also a problem with sorghum. In addition, chinch bugs are more prevalent in drought conditions in sweet sorghum, and webworms and greenbugs pose an extra consideration when watching for insects.

Yield potential

Sweet sorghum can grow to be 14 ft tall, yielding between 7 and 15 tons dry matter per acre under favorable conditions, and is four times more energy efficient than corn. To maximize ethanol yield, juice extraction must remove at least 50% of juice from the stalks. Concentrations of juice in the stalks can be monitored by sugar measurements.

Pre-harvest monitoring

Remove seed heads when seeds are in late milk stage to prevent allocation of sugar to seed formation. Approximately 2½ weeks later, sugar will reach concentrations for harvest. Refractometers can be used to estimate sugar content, with ideal brix percentages between 14 and 20%.

Economics

Availability of a cattle operation to feed plant biomass byproducts enhances profitability. On farm processing reduces transportation costs, but lack of commercial harvesting equipment still limits practical farm size, especially with rapid breakdown of sugars upon harvest.

POPLAR

Energy Potential	
Biopower	14.1 MMBTU/ton

Characteristics

Hybrid poplar is a fast growing tree, which has multiple benefits and uses from pulp wood products to bioenergy. This short rotation woody crop can be grown in the Midwest. Plantings can benefit soil and water resources in crop or non-cropland areas.

Climate
Native eastern and plains cottonwood trees range from the southern states to much of Canada. Western poplar species can grow in the western states.

Soil
Grows best on well-drained soils but can tolerate heavy soils and periodic flooding. Soil pH should be 5–7.5. Sandy and clay loam soils with organic matter from 3 to 8% are best.

Seedbed preparation
Perennial weeds need to be controlled the fall be-fore spring planting. A cover crop of small grains may be used to protect soil overwinter and provide weed suppression in spring during establishment. Control weeds in the spring and plant in late May or early June.

Variety selection
Many research studies have been done on hybrid poplar varieties or clones. Clones to consider: (old) DN-17, DN-34, DN-182, NM-6 (new) DN-2, DN-5, DN-70, NE-222, and I-45/51. Identify clones that have grown successfully in your area.

Planting
Softwood cuttings (sticks 8–10 inches long) or bareroot plants are usually planted in large plantings. Rows 12 feet apart with 8- to 12-foot spacings within the row are common. (450–300 trees per acre) Plant May to early June.

Fertility and lime requirements
Nitrogen is the limiting nutrient for poplars. Annual applications of 50 lbs of nitrogen per acre per year are recommended. Liming is needed if pH is below 5.

Water/irrigation requirements
Adequate annual rainfall of at least 18 inches per year is needed without stress or supplemental water.

Weed control
Weed control during establishment and first 3 years is necessary. Typically, little to no herbicide is needed by the third year after planting.

Insect/disease/predator control
The cottonwood leaf beetle, forest tent caterpillar, and the poplar borer are the main insects affecting poplars. Septoria canker disease is possible in the Midwest. Deer and rodents can damage plantings.

Yield potential
In the Midwest, yields from 3 to 6 dry tons per acre per year are realistic. Harvested in 12–15 years would yield from 36 to 90 tons per acre.

Economics
Economics have been variable with demand and supply. Pulp prices have been from $20 to $50 a cord and biomass prices have been low. Check the industries in your area for the current poplar market price.

WILLOW

Energy Potential	
Biopower	16.6 MMBTU/dry ton

Characteristics

Willow is a fast growing shrub, which has multiple environmental benefits and can be used as a bioenergy crop. This short rotation woody crop can be grown in the Midwest. Plantings can benefit soil and water resources in crop or non-cropland areas. Willows can be harvested every 3 or 4 years with modified forage harvesting equipment.

Climate

Willows are found in many climate zones but most prefer temperate regions. Midwest and northern states, and Canada are good growing regions for willows. Willows are not usually found in arid climates.

Soil

Grows best on well-drained soils but can be productive on soils with higher clay content and periodic flooding. Soil pH should be 5.5 to 8. Sandy, silt or clay loam soils preferred but some water-holding capacity is important.

Site preparation

Perennial weeds need to be controlled the fall before spring planting. A cover crop of small grains may be used to protect soil overwinter and provide weed suppression in spring during establishment. Control weeds in the spring with a combination of chemical and mechanical techniques.

Variety selection

Many research studies have been done on shrub willow varieties and clones. Identify clones that have grown successfully in your area. Planting material is available commercially for 15–20 varieties including hybrids that have been developed in the past decade.

Planting

Hardwood cuttings (sticks 8–10 inches long) are usually planted in large plantings. Rooted plants can also be used, but these are more expensive. Rows 5 feet apart between twin rows, 2.5 feet between twin rows with 2 foot spacings within the row, are best. (5800 + plants per acre). Plant April to mid June. After leaf fall in the first year, coppice (cut or mow off plants to 1 or 2 inches above the soil). This forces the plant to shoot multiple stems the following spring. The first harvest occurs 3–4 years later. Plants re-sprout after harvesting and subsequent harvests occur on 3–4 year cycles.

Fertility and lime requirements

Applications of 100 lbs of nitrogen (urea) per acre are recommended after the first year and after every harvest on 3–4 year cycles.

Water/irrigation requirements
Adequate annual rainfall of at least 18 inches per year is needed without stress or supplemental water.

Weed control
Weed control during establishment and first 2 years is necessary. Some weed control or cultivation may be needed after the first 2 years.

Insect/predator control
Generally minor insect problems, willow leaf, and Japanese beetles. There are some leaf defoliators, which can be controlled by Bt products. The main disease is willow melampsora (rust).

Yield potential
In the Midwest, yields from four to six dry tons per acre per year are realistic. Harvest every 3 years may produce 12 to 18 dry tons per acre. One planting could be harvested seven times on 3-year cycles in 21 years.

Economics
Economics have been variable with demand and supply. Check the biomass industries in your area for the current market price.

CANOLA (RAPESEED)

Energy Potential	
Biofuel	80 gal/Acre
Biopower	10.8 MMBTU/ton

Characteristics
Rapeseed is an old crop of family Brassica that has been grown since the twentieth century BC. However, its current form is more edible and more applicable in the biodiesel industry. Canola is a variety of rapeseed that was developed in the 1970s for high-quality oil characteristics that make it an excellent edible oil, livestock feed, and biofuel feedstock. Rapeseed is still the popular name in Europe for similar varieties as canola, both industrial and edible. Canola and rapeseed bio-diesel gels at a lower temperature than other oilseed crops, and thus provides an ideal colder temperature fuel.

Climate
In temperate climates (such as Pacific Northwest), canola/rapeseed can be planted in both spring and fall, although excessive cold and/or wet winter conditions can greatly hamper growth and yield for fall plantings. North Dakota produces most of American canola.

Soil
Grows well on most soil types, but must be well drained.

Seedbed preparation

To avoid cross-pollination, buffer zones between canola/rapeseed and other Brassicas are critical. Additionally, seedbed preparation must be good to prevent soil crusting on young plants. Canola/rapeseed must be planted in time to ensure adequate maturity prior to hot or cold weather.

Fertility and lime requirements

Canola/rapeseed has similar need to other small grain plants. However, it is a heavy user of sulfur. With a yield of 2000 lbs/acre (low-average), between 12 and 15 lbs of the straw and seed will be sulfur.

Variety selection

Rapeseed and canola have two main types: "Polish type" and "Argentine." Additionally, another type was developed in Canada, a variety of brown mustard.

Water/irrigation requirements

Fall planted canola/rapeseed is more drought-hardy due to more extensive root systems.

Disease control

Canola is susceptible to blackleg and Sclerotinia stem rot. If not rotated with resistant crops, seed treatment may be necessary.

Yield potential

Canola seed is about 40% oil and has an approximate yield of 75–80 gallons/acre compared with about 48 gallons/acre with soybeans. The energy box for this sheet is calculated for 80 gallons/acre.

Storage

Canola/rapeseed is handled and stored similarly to flax.

Production challenges

Canola/rapeseed can cross-pollinate with many other crops, such as rutabaga, Chinese cabbage, broccoli rabe, and turnip unless buffer distances are adequate. In addition, it is problematic to grow canola among infestations of mustard-family weeds for similar reasons. The crop should be windrowed once seed moisture reaches 35% to avoid seed shattering during harvest.

CROP HANDLING AND LOGISTICS
HARVEST

Generally, biofuels can be grown and harvested with traditional farm and logging equipment. However, new developments in biofuel production may call for alternative or modified production or harvesting techniques. Examples of recent equipment innovations include in-field cob collectors that follow behind the combine, and stalk-cutting attachments mounted on combine heads. Regardless, harvesting equipment will be unique to the crop type and production enterprise. Producers considering purchasing this equipment will need to evaluate the cost of new equipment versus the expected return from the investment. An alternative to this would be to hire a custom harvester, which would save the farmer the capital investment and maintenance costs.

DENSIFICATION

One of the main barriers to building a successful biomass energy industry is that most raw biomass is not immediately useful as an energy source. Corn stover cannot be poured directly into a fuel tank, and switchgrass cannot be easily fed into a woodstove. Typically, biofuel crops must be compacted or densified, making them easier to handle and transport. Pelletizing, briquetting, and chipping are the most common densification methods used. Densification increases the weight per cubic foot, which reduces transportation and handling costs.

STORAGE

Many variables affect the storage of bioenergy crops. Things like the type and condition of feedstock and how it was harvested are important factors in how it should be stored. For example, grains should be kept at the proper moisture content and in an approved storage bin. Cellulosic feedstock crops such as native grasses and forbs are usually baled (round or large square) and stored in a stack in the field or near the farm site. Stacked bales should be covered or put in a covered storage area; they can also be individually wrapped with a waterproof cover. Woody bioenergy crops like poplar and willow are usually cut to length (8–10 feet, depending on facility specifications) and hauled and piled near the field. They can then be stored as logs or chipped and hauled to the processing plant. Willow harvesters cut and bale or chip on-site. Woody bales and chip piles can maintain quality even if they are exposed to the weather. They can be stored in the field or in a covered storage area.

Storage policy and practices should be discussed with the biomass processer before contracts are finalized. Improperly stored feedstock can lose quality, which will decrease the price per ton that the farmer receives for the crop.

TRANSPORTATION

Hauling low bulk-density, unconsolidated material is not economical. Densified biomass increases the weight per cubic foot and allows it to be transported more cost-effectively. Engineers are working to develop equipment that can harvest and densify biomass all in one pass. It will be some time before this technology will be available to farmers. Equipment will also be needed for transporting the densified biomass. Most farms are equipped to handle free flowing grain. This equipment is not designed to handle bales, chips, or chopped material.

SUSTAINABLE MANAGEMENT FOR BIOMASS PRODUCTION

There are environmental benefits and challenges associated with the utilization of biomass for energy and other bio-based products. For example, although the use of biomass for energy can help moderate greenhouse gas emissions, protect soil, water and air quality, and enhance habitat for many species, improper practices can reduce these benefits or even place these resources at risk. Recognizing these concerns, various strategies, generally labeled as best management practices (BMPs) or guidelines, have been developed to mitigate these impacts. Here, we review environmental concerns and mitigation strategies associated with the management and harvesting of biomass on soil productivity, water quality and quantity, and biodiversity and wildlife.

Soil physical properties. Growing and harvesting biomass may alter soil structure, texture, porosity, density, drainage, and surface hydrology. Equipment traffic on unfrozen soil tends to compact the soil (increase its density), which reduces its porosity, aeration, and drainage, and can impact surface hydrology. Soil compaction also can reduce root survival and growth. Mechanical site preparation can change the soil structure and texture.

Soil chemical properties. The nutrient status and pH of soils may be altered by removing biomass, as it takes away material that would have decomposed and provided nutrients to the soil. When biomass is removed, the amount of organic matter in soil (the residues of dead plants and animals in various stages of decomposition) is also reduced, thereby affecting the soil pH and nutrient availability.

Soil biology. Soil contains many organisms, such as bacteria and fungi, which improve its quality by breaking down organic material to smaller particles and releasing nutrients. Soil organisms are also important links in many nutrient cycles. Mechanical site preparation techniques such as tillage, raking, windrowing, disking, and piling can reduce soil organic matter, which is important for soil microbes, soil structure, soil carbon storage, nutrient cycling, and regulation of soil hydrological processes. A reduction in soil organic matter and soil carbon storage will reduce the availability of nutrients, especially nitrogen (N), phosphorus (P), potassium (K), and calcium (Ca).

Residue management: Studies suggest that at least 2.3 tons per acre should be retained in a no-till continuous corn system and 3.5 tons per acre in a corn–soybean rotation system. For woody crops, recommendations are that 1/6 to 1/3 of the residue should be retained. It may be possible to increase the residue removal rate if other management practices are implemented, including establishment of cover crops; using no-tillage, narrow row spacing; growing higher plant populations; increasing fertilization rates; or applying bio-char or other organic amendments. Harvesting methods that leave enough crop residues to keep soil loss to a minimum can be used for perennials such as switchgrass or *Miscanthus* spp. Surface cover is especially important during the establishment of these perennial crops, which may take several years.

Water Quality and Quantity: The use of chemical fertilizers and pesticides in today's intensive row crop production has led to contaminated drinking water and pollution in lakes and streams. The construction of roads and trails for biomass harvest can cause erosion, resulting in sediment movement to streams. Biomass harvesting and site preparation operations also increase the risk of erosion and sediment flow to surface waters. Soil erosion increases in a non-linear, exponential manner with declining residue cover, which in turn reduces water infiltration and increases water runoff.

Production of short-rotation woody crops through monoculture plantations (e.g., willow, poplar) can deplete groundwater resources, especially if irrigation is required. Power plants processing this biomass usually require water for steam production and cooling. If water used in the conversion process is released untreated into lakes or rivers, it can negatively affect water quality due to the chemicals used during the conversion process.

Water Erosion and Soil Saturation: Removing vegetation and litter during biomass harvesting can increase the risk of higher rates of soil and water erosion. It also allows more water to fall directly on the soil surface and infiltrate to groundwater, resulting in a higher water table that could lead to soil saturation and a loss of productivity.

CONCLUSION

Starch- and sugar-based crops like corn and sugarcane have given rise to the fuel ethanol industry in the United States. Farmers have increased production in attempt to keep up with rising world demand for corn. Corn is one of the most researched and best understood crops grown in the United States, but its production also requires some of the highest inputs of fertilizer, pesticides and mechanical tillage.

Cellulose can be burned to generate electricity or converted into ethanol and other energy sources. Cellulosic crops have an advantage over corn because they have lower production inputs and in many cases are perennial crops that need to be established only once every 10–15 years. However, cellulosic crops have lower yield and less income potential per acre. Also, there are challenges in handling, storing, and transporting the quantity of cellulosic biomass needed to achieve federal production mandates. Recent research is making great progress in advancing these crops and addressing these challenges, but we are not there yet.

Each bioenergy crop has its own set of niches and challenges. Ultimately, if we could grow a high-yielding crop with little or no inputs, we would have an ideal situation. However, each crop responds differently to soil, climate, and management conditions. The highest-yielding crops tend to require the most inputs.

ACKNOWLEDGMENTS

This material is based upon work (Pennington, D., C. Gustafson, J. Hay, R. Perrin, G. Wyatt, and Zamora, D. 2011. Bioenergy Crop Production and Harvesting: Module two in S. Lezberg and J. Mullins (eds.) Bioenergy and Sustainability Course. On-line curriculum. Bioenergy Training Center) supported by the National Institute of Food and Agriculture, U.S. Department of Agriculture, under Agreement No. 2007-51130-03909. Any opinions, findings, conclusions, or recommendations expressed in this publication are those of the author(s) and do not necessarily reflect the view of the U.S. Department of Agriculture.

FURTHER READING

Angelstam, P., Mikusinski, G., Breuss, M., 2002. Biodiversity and forest habitats. In: Richardson, J., Bjorhe- den, R., Hakkila, P., Lowe, A.T., Smith, C.T. (Eds.), Bioenergy from Sustainable Forestry: Guiding Prin- Ciples and Practice. Kluwer Academic Publishers, Dordrecht, The Netherlands, pp. 216–243.

Brakenhielm, S., Liu, Q., 1998. Long-term effects of clearfelling on vegetation dynamics and species diversity in a boreal pine forest. Biodiversity and Conservation 7, 207–220.

Cruse, R., March 18, 2010. Life Cycle Analysis for Biofuels. Retrieved February 22, 2011, from http://www. extension.org/pages/Life_Cycle_Analysis_for_Biofuels. extension.org.

De Kam, M.J., Morey, R.V., Tiffany, D.G., 2009. Integrating biomass to produce heat and power at ethanol plants. Applied Engineering in Agriculture. ISSN 0883–8542, 25(2): 227–244.

Ecke, F., Lofgen, O., Sorlin, D., 2002. Population dynamics of small mammals in relation to forest age and structural habitat factors in northern Sweden. Journal of Applied Ecology 39, 781–792.

Ethanol Opportunities and Questions. By Mike Morris and Amanda Hill, NCAT Energy Specialists. © 2006 NCAT, Paul Driscoll, Editor Amy Smith, Production. This publication is available on the Web at: www.attra.ncat.org/attra-pub/ethanol.html.

F. John Hay, Extension Educator, University of Nebraska-Lincoln Extension. https://www.extension.org:443/pages/27536/corn-for-biofuel-production. Published by extension.org, Author Jennifer Rees, Extension Educator, University of Nebraska-Lincoln Extension Dennis Pennington, Bioenergy Educator, Michigan State University.

Gunnarsson, B., Nitterus, K., Wirdenas, P., 2004. Effects of logging residue removal on ground-active bee- tles in temperate forests. Forest Ecology and Management 201, 229–239.

McInnis, B.G., Roberts, M.R., 1994. The effects of full-tree and tree-length harvest on natural regeneration. Northern Journal of Applied Forestry 11, 131–137.

Merriam, W.J., 1994. Life Cycle Assessment. WLG 306, West Virginia University Extension Service. Publisher: West Virginia University Extension Service.

Minnesota Forest Resources Council (MFRC), 2007. Biomass Harvesting Guidelines for Forestlands, Brushlands and Open Lands. Minnesota Forest Resources Council, St. Paul, MN, 42p. Available at: http://www.frc.state.mn.us/documents/council/site-level/MFRC_forest_BHG_2001-12-01.pdf.

Morey, R.V., Kaliyan, N., et al., 2010. A corn stover supply logistics system. Applied Engineering in Agriculture 26 (3), 455–461.

Olson, Rich, 1992. Cover: It's Important to Wyoming's Wildlife. B-967. http://ces.uwyo.edu/PUBS/B967R.pdf. Source: University of Wyoming Cooperative Extension Service.

Tiffany, D.G., Morey, R.V., De Kam, M.J., 2009. Economics of biomass gasification/combustion at fuel ethanol plants. Applied Engineering in Agriculture. ISSN 0883–8542, 25(3): 391–400.

Webster, C.R., Flaspohler, D.J., Jackson, R.D., Meehan, T.D., Gratton, C., 2010. Diversity, productivity and landscape-level effects in North American grasslands managed for biomass production. Biofuels, 10.4155/BFS.10.18.2010, Future Science Ltd, ISSN 1759–7269. reprints@future-science.com.

Werling, B.P., Landis, D.A., 2011. Ecosystem Services and Biofuel Landscapes. Being reviewed for eXtension. http://www.extension.org/ag_energy. Michigan State University Extension.

APPENDIX A: ACRONYMS USED IN THIS CHAPTER

- BCAP: Biomass Crop Assistance Program
- BCF: Biomass conversion facility
- CCX: Chicago Climate Exchange
- CHST: Collection, harvest, storage, and transportation
- CRP: Conservation Reserve Program
- EISA: Energy Independence and Security Act
- EMO: Eligible material owners
- FSA: Farm Service Agency
- ISO: International Organization for Standardization
- LCA: Life cycle assessment
- MBTU: British thermal unit (stands for one million British thermal units per hour)
- NDFU: North Dakota Farmers Union
- SRC: Short-rotation crops (woody)

ON-FARM OIL-BASED BIODIESEL PRODUCTION

Heather M. Darby[1], Christopher W. Callahan[2]

[1] *University of Vermont, St. Albans, VT, USA;* [2] *University of Vermont, Rutland, VT, USA*

INTRODUCTION

Motivated by high diesel fuel prices and general volatility of energy costs, some farmers in the Northeast have explored the use of oilseed crops as feedstock for fuel production. Oilseeds are defined as a subset of grains that are valuable for the oil content they produce. Although the seeds are bought and sold as commodities throughout the nation and the world, their main value is derived from extrusion, when oil is separated from the rest of the seed. In this process, the seed is converted into two coproduct streams that include meal and oil. While the oil can be used for food and/or converted into liquid fuels, the meal can be used as an animal feed, fertilizer, and solid pellet fuel, among other things. These crops, including sunflower, canola, and soybeans, are not commonly grown or used this way in the region, so there are few central oilseed processing facilities. This necessitates adoption of on-farm handling, processing, and conversion technologies. This is particularly important if a farm is interested in producing fuel cost-effectively. General processing steps in on-farm oilseed processing, which will be covered in more detail in later sections, are outlined in Figure 8.1. Noteworthy is the flexibility afforded to farms by including these oilseed processing steps on the farm. For example, once a press exists on the farm, seeds from other enterprises can be processed to meal and oil for a fee. Additionally, once pressed from the seed, oil can be used to make fuel or sold as food oil, depending on the oil type, the need for fuel on the farm, and the market. Similarly, once the farm owns a biodiesel processor it can make fuel from various oil feedstocks, including pressed oils from the farm or used vegetable oil collected from restaurants.

Economic analysis of these on-farm enterprises shows promising cost per gallon outcomes and return on investment (Callahan and White, 2013), as well as strong return on energy investment and net carbon emission avoidance potential. Biodiesel production costs of between $0.60 and $2.52 per gallon have been estimated for farm-scale production models (Callahan and White, 2013). The main factors in the cost per gallon are crop production costs, crop yield, and biodiesel conversion equipment and chemicals. The energy return in Vermont on-farm biodiesel operations has been estimated at between 2.6 and 5.9 times the invested energy, demonstrating strong returns and potential for improvement with increased scale (Callahan). Furthermore, oilseed-based production of biodiesel has been estimated to result in a net reduction of carbon dioxide emissions of up to 1420 lbs per acre (Campbell, 2009).

Bioenergy. http://dx.doi.org/10.1016/B978-0-12-407909-0.00008-0

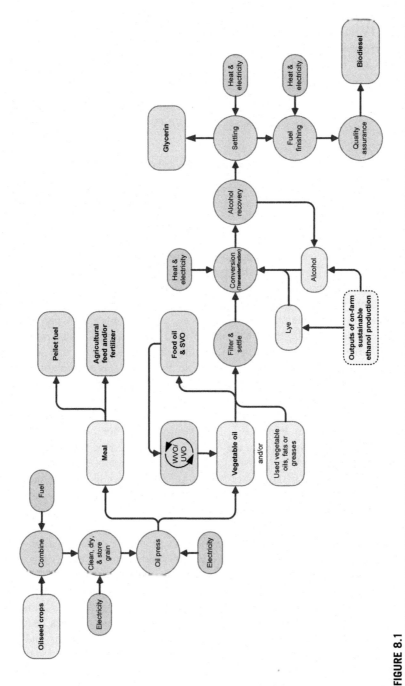

FIGURE 8.1

Overview of the farm-based biodiesel process using oilseeds as feedstock. In this model, several feedstock and coproduct options become possible.

SPECIES AND VARIETY SELECTION

On-farm oil-based fuel production begins with the selection and subsequent growth of an oilseed crop. Oilseed crops are grains that have been identified as having oil content usually above 20% of total seed weight. This is in comparison to other crops such as small grains that have only 1–2% oil content. Although there are many oilseed crops, the primary three include soybeans (*Glycine max*), sunflower (*Helianthus annuus*), and canola (*Brassica napus, brassica rapa, brassica juncea*). Sunflower and canola have oil content over 40% of their total seed weight. Soybeans have relatively low oil content, often below 25% of its total seed weight. Selecting the proper species and variety of oilseed crop to produce on the farm is critical to the overall success of this enterprise.

Species selection will be somewhat determined by the current crop production scheme of the farm. The pros and cons for each species need to be weighed properly prior to selection of a crop suitable for the farm. For example, if the farm currently raises livestock, then a farmer may decide to raise soybeans, as the meal is a well-known protein source fed to animals in the region. In addition, soybeans produce nitrogen, and hence have lower fertility needs than sunflower or canola. If a grower has a short growing season, they may choose to produce canola and not a full-season crop such as soybeans. Sunflowers might be grown to improve public relations. Regardless of the reasoning, these three crops can produce adequate yields in a variety of conditions. Once the oilseed crop species is selected, a variety or two will likely be selected for growth on a farm.

There are a number of varietal characteristics that must be considered to make the best choice, and the best variety for the situation often strikes a balance between yield potential, oil content, maturity, and resistance to disease. Growing oilseed crops for biodiesel production will only prove to be profitable for growers if they can regularly achieve their fullest yield potential; therefore, it is critical to pick a variety that has the highest yield potential within the limits of varieties that perform well in the given climate.

The following are guidelines for selecting varieties.

MATURITY

The most significant limiting factor for growing many oilseed crops is the shortness of the growing season. Canola is adapted to a relatively short and cool growing season. Soybeans require a relatively long growing season (100–140 days) and warm temperatures. Sunflowers will tolerate a somewhat shorter and cooler growing season. On average, sunflowers require about 2500 heat units (i.e., growing degree days) to reach physiological maturity. Once these crops are physiologically mature, it takes additional time for the crop to reach adequate moisture content that will allow for harvest with a combine. The time that it takes for the crop to dry down to adequate harvest moisture will partially depend on the variety as well as the climate. For most areas in the Northeast region, varieties that are considered "early" or "medium" maturing perform the best in terms of yield and quality. It is also important to select varieties that are reported to have excellent "dry down."

YIELD POTENTIAL

Seed yields in oilseed crops are determined by a large number of factors, the majority of which are environmental. However, plant genetics also play a major role. Some varieties exhibit a greater yield

FIGURE 8.2

University variety trials determine seed and oil yield potentials.

potential than others, which is important to know when determining which variety to plant. In general, shorter season crops have slightly less yield potential than longer season varieties, but this is not always the case. Seed company charts can be helpful in determining the yield potential, but are difficult to accurately compare between companies. Some universities conduct yearly variety trials that provide good insight into which varieties are appropriate for the climate and cropping systems common to the region (Figure 8.2). In general, the seed yields for canola, soybeans, and sunflowers range between 1000–3000 lbs/acre, 2000–4000 lbs/acre, and 1500–4500 lbs/acre, respectively.

OIL CONTENT AND QUALITY

Like yield potential, oil content of the seed is largely based on environmental factors, but genetic potential is also significant. For producers who grow for biofuel feedstock, oil content is as important as yield potential. In the Northeastregion, oil yields tend to be slightly lower than published values, likely because of the higher water availability in our soils, which can decrease oil content of the seed. The extraction potential of screw presses utilized in on-farm production also tend to be less efficient than most commercial presses. Even so, University of Vermont (UVM) Extension trials have regularly achieved oil content averages of approximately 33%, and often reaching 40% (Figure 8.3).

DISEASE RESISTANCE

Many varieties are bred for resistance to diseases. The resistance in most crops was developed through traditional breeding techniques, and the traits have been obtained either from wild types or natural mutations in these crops. Therefore, they are not genetically modified (GM) traits. Of the available disease resistance traits in canola, the most significant is resistance to black leg. Sunflower varieties resistant to downy mildew, rust, and wilt are also available. Soybean varieties can be purchased that have resistance to a host of pests, including brown stem rot (*Phialophora gregata*), *Phytopthora* root rot (*Phytophthora sojae*), and soybean cyst nematode (*Heterodera glycines*). The most common and economically problematic fungal disease for oilseeds in our region is *Sclerotinia sclerotiorum*, for which there is little resistance available among oilseed varieties.

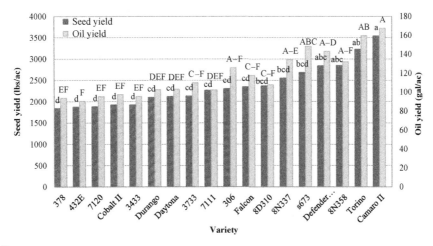

FIGURE 8.3

Seed and oil yields of 18 commercially available sunflower varieties, Alburgh, VT, 2013. Treatments that share a letter were not significantly different from one another ($p = 0.10$); compare lowercase letters for seed yield and capital letters for oil yield (Darby, unpublished).

HERBICIDE TOLERANCE

Like disease resistance, herbicide tolerances have been bred into most oilseed crops. Both sunflower and canola have herbicide resistances developed using traditional techniques, and are available as non-GM traits in sunflowers. There are currently two non-GM herbicide tolerance traits available for sunflower and canola hybrids (Clearfield® and ExpressSun®). Clearfield gives resistance to the herbicide Beyond® (imazamox), which can control several problem weeds in our area, including redroot pigweed, lambsquarters, and velvetleaf. The ExpressSun® trait provides tolerance to Express® herbicide (tribenuron methyl), which provides control of many broadleaf weeds in sunflowers. Other common herbicide tolerances available in soybean and canola crops include Roundup Ready® (glyphosate), Liberty Link® (glufosinate), and STS® technology (sulfonylureas). These hybrids tend to be much more expensive than conventional hybrids, and so producers should consider other methods of broadleaf weed control, such as well-planned rotations, secondary tillage, and cover cropping.

PLANTING CONSIDERATIONS

One of the most common production problems in crop fields across the country is planter error that leads to long skips or clusters in fields and subsequent increased weed pressure and yield losses (Figure 8.4). Many of these errors can be avoided with proper field selection, seedbed preparation, and planter calibration (Karsten, 2012; Kirkegaard et al., 1996).

SOILS AND FERTILITY

Oilseed crops are well adapted to a variety of soil types. However, they are best-suited for well-drained soils that have good water-holding capacity (i.e., high organic matter and good soil structure). They

FIGURE 8.4

Dense sunflower population due to planting error.

will tolerate heavy, wet soils, and light, excessively drained soils better than other crops, but yields will be lower compared to oilseeds grown under better soil conditions.

Sunflowers will grow in a wide range of soil types from sands to clays[10]. The demands of a sunflower crop for soil nutrients are less than corn. The optimal soil pH for growth is 6.0–7.2. Sunflowers in the Northeast often require far less added soil fertility than sunflowers grown in the Plains. Vermont soils can have high levels of nitrogen (N), phosphorus (P), and potassium (K) native in the soil, as well as high levels of organic matter (>6–8%). This high fertility is most often associated with our long history of using manure as a primary source of fertility. Sunflowers are exceptional scavengers of soil nutrients because of their extremely long taproots (Karsten, 2012). In deep soil, sunflowers are able to access nutrients from between three and four feet, far below the profile of corn and hay. Many nutrients in the lower soil profiles that are lost to other crops are still accessible to sunflowers. In the Great Plains, sunflowers generally require 5 lbs of N for every 100 lbs of desired seed yield. Sunflowers also require relatively low levels of P and K[10]. Standard soil tests will estimate available P and K in the soil. The testing laboratory will provide soil nutrient recommendations.

Soybeans can grow in a variety of soil types but the ideal is a loose, well-drained loam. Many fields have tight, high clay soil that becomes waterlogged when it rains. When the soil dries out, a hard crust surface may form, which is a barrier to emerging seedlings. Leguminous crops, such as soybean, that have been inoculated with nitrogen-fixing bacteria rarely give an economical yield response to N fertilization. The species of symbiotic bacteria that fix atmospheric nitrogen in the nodules of soybean plant roots is Bradyrhizobia japonicum. Inoculation of the seed with these bacteria is especially recommended for fields where soybean has not been grown recently. Maintaining a pH between 6.2 and 7.0 is optimum for soybean growth, as well as the growth of the nitrogen-fixing bacteria. Overall, soybeans require moderate amounts of fertility for high yields. Soil testing is the best way to determine the P and K needs of the crop. Manganese is a common micronutrient deficiency observed in soybeans, especially on light textured soils with low organic matter. The symptoms of manganese deficiency are upper leaves ranging from pale-green (slight deficiency) to almost white (severe deficiency) with green veins. Soil tests and plant analyses are useful in predicting where manganese deficiencies are likely to occur.

Canola does well in a wide variety of soil conditions, though it excels in loamy soils with good internal drainage and a pH in the range from 6.0 to 7.0 (Canada Canola Council, 2011; Kandel, 2013). Canola does not perform well in soils that are saturated, especially when they do not drain quickly. In excessively drained, dry soils, drought stress can be problematic, as canola is also intolerant of water stress. Canola has nutrient requirements that are similar to most small grains[7]. Nitrogen should be applied preplanting, due to the high sensitivity of canola seed to salt and ammonia injury. Nitrogen availability of 90–125 pounds per acre will allow a yield potential in the range of 2000–3000 lbs per acre, assuming other field conditions are favorable and production practices are done properly. Phosphorus and K recommendations should be determined through a standard soil test. Canola seed is especially susceptible to fertilizer salts and hence, high levels of N and K fertilizers can cause detriment to the emerging seeding. Canola has a much higher sulfur demand than most other crops, and significant yield boosts can occur from sulfur additions. Soils with test values lower than 10-ppm sulfur should receive additional sulfur at rates of 20–40 lbs per acre.

SEEDBED PREP AND SEEDING

Larger seeded oil crops such as sunflower and soybean should be planted 1.5–2.0 inch deep into a moist, even seedbed, once soil temperatures have reached 10 °C. Because of canola's small seed size, it requires a fairly shallow planting depth with good soil to seed contact, which in turn requires a smooth and level seedbed. To plant canola, grain drills should be set to plant at a depth of 0.5–1.0 inch, which helps to provide even emergence.

Like most crops, oilseeds benefit greatly from a well-prepared, even seedbed. Inattention to seedbed preparation will cause cascading problems throughout the season, beginning with uneven emergence, and ending with premature sunflower maturation and dry-down. Good seedbed preparation can reduce workload through the rest of the season and help ensure a good crop.

In the Plains, no-tillage oilseed production systems are common (Karsten, 2012). In the Northeast, some producers grow sunflowers and soybeans in no-till fields with good success as well, but it is not suitable for all fields. Because these crops need fairly warm soils to ensure good germination, they are generally not a good candidate for no-till systems on soils that are poorly drained or have a propensity to remain cold late into the spring. However, there are plenty of good opportunities on more moderate and light soils for reduced tillage. Reduced tillage systems can improve soil quality and reduce crop production costs.

PLANTING DATES

Because sunflowers and soybeans are a fairly long-season crop, it is important to plant them as soon as the soil conditions permit. In the Northeast, soil temperatures of 50 °F typically occur in the second or third week of May; it is a good rule of thumb to have sunflowers planted before June 1. If a short-season variety is planted at this time, it will generally be ready for harvest by the first or second week of October. Planting can extend to the first week in June, but planting later than that point introduces a significant risk that the crop will not reach maturity before first frost. Immature plants that are killed by a frost will have reduced seed-set, test weight, oil content, and oil quality.

Spring canola should be planted as early as the fields can support tillage and planting equipment. Because germination occurs once soils have reached 38 °F, canola could be planted as early as mid-April, but often early May is more feasible. Planting spring canola early also allows an earlier harvest, which is helpful to producers because it spreads out workload and equipment availability in the fall, and allows for the establishment of a cover crop or a winter cereal grain. Because heat and humidity during the flowering period can be detrimental for pollination and seed set, spring canola should be planted no later than mid-May. Winter canola should be planted early enough in the fall that it can establish a strong rosette before the first killing frost, when it will become dormant. Usually, this requires that it be planted before September (Figure 8.5). In the University of Vermont winter canola planting date trials, only plots planted before September 1 were harvestable.

SEEDING RATES AND POPULATIONS

Oilseed yields are influenced by population density, which in turn causes a number of other responses in lodging rates (the condition when the stems of a plant weaken at its base and are unable to support the plant weight) and oil contents. For example, in sunflower higher plant populations will decrease head size, seed size, and the number of seeds per head. However, increased plant populations can lead to increased oil content (Figure 8.6). Optimum plant populations for sunflowers vary depending on water availability and can range from 10,000 to 30,000 plants per acre. Because of the plasticity of many of these plant stand characteristics, seed and oil yields are roughly equivalent over a wide range of seeding rates. However, changes in head size and seed size could have an effect on bird damage, insect predation, oil content, and perhaps most notably, in-field drying rate, which in turn influences harvest date and seed drying time. As population increases, head width decreases, and drying time is shortened. At harvest time, moisture remains higher in sunflower stands with wider heads, or lower populations (Figure 8.7).

FIGURE 8.5

Winter canola planted on varying dates (left to right: 15-Aug, 23-Aug, 29-Aug, and 6-Sep).

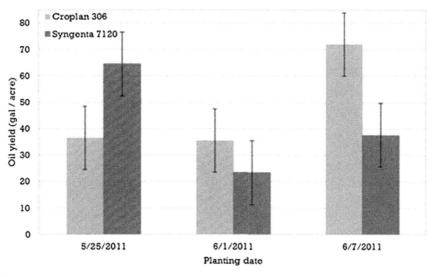

FIGURE 8.6

Effects of planting date on long and short season variety oil yields (Darby, unpublished).

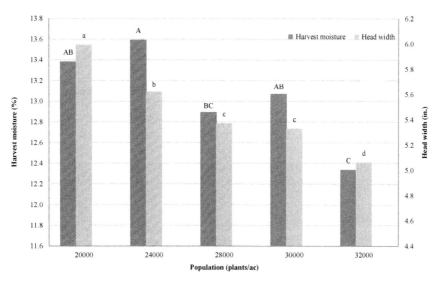

FIGURE 8.7

Effects of population on harvest moisture and head width, combined 2010–2012 (Darby, unpublished). Treatments that share a letter did not differ significantly from one another ($p = 0.10$, compare capital letters for harvest moisture and lowercase letters for head width).

Trials conducted in northern Vermont showed that the highest canola yields result from seeding rates between five and nine pounds per acre, though no significant yield increase occurs from seeding rates higher than five pounds per acre. Therefore, to save seed costs, lower rates should be used. In addition, lodging and disease transmission are more likely at higher plant densities, leading eventually to reduced yield.

Soybean seeding rates vary depending on row spacing and range from 210,000 to 130,000 seeds per acre. The narrower the rows, the higher the seeding rate for soybeans.

ROW SPACING

The standard row spacing for sunflower crops in the Northeast region is dictated by the available planting, cultivation, and harvest equipment, most of which operate on 30 inch rows. Maximum soybean yield is generally obtained with row spacing between 7.5 and 15 inch (Oplinger and Philbrook, 1992). Though planting in narrower rows confers some advantages, such as reducing competition between individuals and reducing weed pressure by achieving more canopy closure, narrower rows can also lead to higher incidence of fungal pathogens due to higher humidity beneath the canopy. Thirty-inch rows relieve some of the pressure of fungal diseases by allowing more air circulation while still achieving adequate yields.

Data from field trials in northwestern Vermont have also shown that row spacing from 7.5 to 18 inch produce similar yields. Individual canola plants respond to differences in plant density by adjusting the amount of branching and the number of flowers per branch. Depending on the planting and harvesting equipment available, UVM Extension recommends planting in 6.0–7.5 inch rows to help control weed pressure and promote even plant spacing across the field. Wider rows may be used, especially if white mold issues are prevalent. Wider rows may help increase airflow in the field, helping to eliminate the climate conducive for fungal growth.

PEST MANAGEMENT

Though oilseed crops are a fairly new field crop to the Northeast, a variety of pests already exist and show signs of economic significance. Oilseed crops grown in research trials have shown susceptibility to a variety of weeds, insects, and diseases. The various strategies for pest management can be grouped into four categories: cultural, mechanical, chemical, and biological. Several tools exist within each category, but not all options make economic or agronomic sense in every case. As a general rule of thumb, preventative pest management is far more effective and costs less than reactive pest management.

CULTURAL PEST CONTROLS

Simple modifications of a pest's environment often prove to be effective methods of pest control. These tactics are known as cultural control practices, because they frequently involve variations of standard agronomic practices. Simplicity and low cost are the primary advantages of cultural control tactics, and disadvantages are few as long as these tactics are compatible with a farmer's other management objectives (high yields, mechanization, etc.).

The most effective cultural control tool against pests of all kinds is a carefully planned rotation of oilseed crops with grass and other broadleaf crops. Virtually every production guide in the U.S.

recommends rotations that call for oilseed crops to be planted in a field once every three to five years, and this is appropriate for the Northeast as well. Where levels of serious and destructive pests are high, periods of five to seven years are recommended between successive crops until pressure subsides. In the Plains, where sunflowers are grown on a much larger scale, pests are not generally mobile enough to move between fields. In the Northeast, fields tend to be small, close together, and surrounded by hedgerows that can act as refuges for many pests. Therefore, insect and fungal pests are generally able to spread between fields, and it is important to provide enough space and time between successive sunflower crops that movement of those pests is limited. Eliminating white mold (*Sclerotinia sclerotiorum*), a devastating disease of oilseed crops, can be very difficult, but a few basic principles will help keep this disease under control. The best tool for managing white mold is a well-planned rotation that employs crops that are not susceptible (i.e., grasses) and long periods between successive oilseed crops. For disease-free fields, oilseed crops should be planted no more than every four years in the same field with other susceptible crops (essentially all broadleaf crops) making only rare appearances. If fields are heavily infested with white mold, more grass crops should be used in place of broadleaf crops to allow more time for the sclerotia fungal bodies to be consumed in the soil, and time between sunflower crops should be increased to six years. Sclerotia fungal bodies can linger in soils and also be difficult to remove from harvested seeds (Figure 8.8).

FIGURE 8.8

Sclerotia bodies, seen above, are often similar in size and shape to sunflower, and can be difficult to remove during cleaning.

In addition to good rotations, other cultural practices can help reduce pest pressure. Row spacing and stand density have a strong influence over the microclimate conditions under the plant canopy. Where plants are dense and rows narrow, humidity near the soil surface can become very high, which provides better conditions for fungal infections. As rows widen and plant spacing increases, however, conditions become better for weeds that can compete with the crop. Thirty-inch rows seem to be the best compromise for reducing fungal pressure in sunflowers and soybeans. Increasing row width of canola from 6 to 18 inch has been shown to reduce some fungal diseases. Planting and harvesting crops outside of normal ranges can also help avoid damage from certain pests. Banded sunflower moth is currently the most problematic insect pest in sunflower fields in Vermont. It is extremely widespread, and overwinters in our soils and field margins. The best management option is a good rotation where successive crops are located far enough from each other that the number of moths that can move between the fields is limited. Deep fall plowing after sunflower harvest has also been shown to reduce emergence of the adults by up to 80%, but that strategy can be costly in fuel and time, and is not practical for every field. Recent research suggests that delaying planting dates to early June may also reduce banded sunflower moth incidence and severity, as well as bird damage (Figure 8.9).

Mechanical pest controls
Mechanical pest control works well with weed populations. Mechanical control is a physical activity that inhibits unwanted plant growth. Mechanical weed control techniques manage weed populations through physical methods that remove, injure, kill, or make the growing conditions unfavorable. Techniques include but are not limited to mowing, pulling, tillage, steaming, and burning (flaming) the weeds.

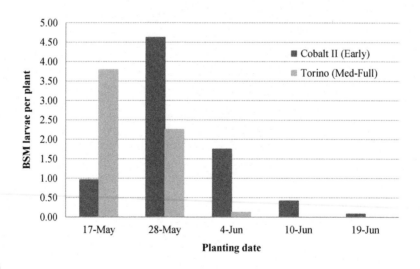

FIGURE 8.9

Effects of planting date on banded sunflower moth (BSM) larvae for two varieties (Darby, unpublished).

FIGURE 8.10

Rear-mounted tineweeder used to mechanically control weeds.

Standard cultivation equipment, such as row crop cultivators or spring-toothed harrows are highly successful at eliminating weeds between wider rows, but in-row weeds can remain problematic after using these implements. Generally, two cultivations before the sunflowers or soybeans canopy are sufficient to eliminate weeds for the rest of the season. Often cover crops are seeded at the time of final cultivation. Seeding a cover crop at this stage will not interrupt crop growth. Another mechanical weed control strategy is tineweeding. This tool is extremely effective at eliminating weeds when they are just germinating. In organic systems, mechanical weed control is popular and compliments strong rotations and cover crops as an integrated weed control program.

Sunflowers are especially well-adapted to mechanical weed control. Cultivation with a tineweeder can be a successful method for weed control in sunflowers. both pre- and postemergence (Figure 8.10). Trials conducted in Vermont indicate that cultivation with a rear-mounted tineweeder at both six (preemergence) and 12 (postemergence) days after planting can provide weed control similar in effectiveness to herbicide (Figure 8.11). Timing of tineweeding is very important to achieve maximum weed control. The best time to tineweed is when the weeds are in the white thread stage and have yet to emerge. Once the weeds have emerged and become established, removal with a tineweeder becomes more difficult. At this stage, the tineweeding must become more aggressive, and as a result, can also lead to crop loss. Higher seeding rates should be used if mechanical cultivation is used for weed control in sunflowers.

Chemical controls

Chemical controls are often used in oilseed crop production because they are readily available, rapid acting, and highly reliable. Fungicides and insecticides are common as seed treatments, but are usually not sprayed after the crop has emerged, though several fungicides are registered for use on oilseeds. In larger oilseed growing regions, insecticide and fungicide applications are used to protect the crop and ensure high yields. Resistant varieties will help reduce the need of fungicides to control such diseases as downy mildew. Because oilseeds are a broadleaf crop, herbicide options for broadleaf weed control are limited, and are essentially restricted to preplant incorporated herbicides that have relatively short periods of activity in the soil. As mentioned earlier, there are also herbicide-tolerant varieties that can also be used to assist with postemergence broadleaf weed control. Herbicide options for annual and perennial grass weeds are fairly extensive. However, care must be taken to ensure that these are registered for use in oilseeds and will not cause damage to the crop. Always closely follow the

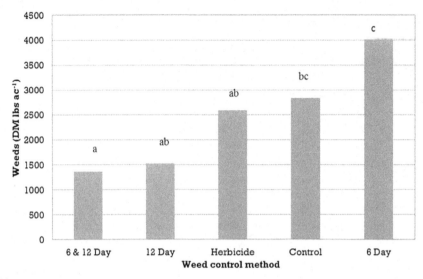

FIGURE 8.11

Effects of tineweeding and chemical controls on weed populations in sunflowers (Darby, unpublished).

directions found in the label, and contact your local Extension agency or herbicide dealer for more information.

Biological controls

Biological control uses other living organisms to reduce or eliminate pests. Biological control relies on predation, parasitism, herbivory, or other natural mechanisms. Natural enemies of insect pests, also known as biological control agents, include predators, parasitoid, and pathogen. Biological control agents of plant diseases are most often referred to as antagonists. Biological control agents of weeds include seed predators, herbivores, and plant pathogens. The only actively applied biological agent for pest management in oilseed production is Contans®, a fungus that parasitizes the dormant fungal bodies of white mold. Though it is not as effective as well-planned rotations, Contans® is organic-certified and does reduce the number of sclerotia fungal bodies in the soil. Contans® does not fully eliminate them or prevent white mold infestations in the field, however, so it should not be used as a substitute for other management tools, but it can be helpful where white mold is well-established.

HARVESTING, CLEANING, DRYING, AND STORAGE OF SEED

The part of the oilseed plant which is most critical for biofuel production is the seed. Oilseeds are distinct in their disproportionate storage of energy in the lipid form in the seed. These lipids exist as a combination of mono-, di-, and triglycerides, otherwise known as vegetable oil, which can be converted to methyl or ethyl-esters as biodiesel or used as "straight vegetable oil" (SVO) fuels. But first, the seed needs to be harvested and separated from the rest of the plant for efficient postharvest handling.

Oilseed production in the Northeast has focused on sunflower, canola, and soybean, although trials of camelina, winter canola/rapeseed, crambe, pennycress, and flax have been conducted. A combine is the implement used for the harvest of oilseeds. It combines the functions of reaping and threshing in one piece of equipment, resulting in a partially cleaned seed being sorted from the rest of the plant. Though some farms have harvested and used the straw or stover from oilseed crops for bedding or fuel, it is uncommon.

Early entrants into oilseed production in the Northeast were able to make use of older, smaller combines such as a Massey Harris SP35 (Figure 8.12). These combines are often sufficient for the small fields (one to five acres) that host early trials of the crops. Multiple heads are used to address the various needs of each crop; row crop head, bean head, small grain head, etc. Often customization of the implements, especially the head, is required to achieve a satisfactory harvest. In the case of sunflowers, plywood "pan" extensions have been added to a row crop combine to help direct the long, top-heavy crop that sometimes falls unpredictably when cut at the base. Similarly, attention should be paid to the sealing of the entire path of the crop through the combine when small seeds such as canola are being harvested. Along with confirming all fasteners are in place, flashing, aluminum tape, and duct tape are commonly used to ensure complete sealing.

Once the oilseed is in the combine bin, it has been partially cleaned. The first step to clean seed is to select the correct screens for the combine. If the seed is not clean enough after one pass through the combine, passing it through a high volume seed cleaner can remove the necessary amount of trash (Figure 8.13). If trash comprises more than about 3% of the harvested seed bulk, problems such as hot spots, mold, and insect activity can erupt in the storage bin. However, long-term storage and post-harvest pressing require secondary cleaning and drying with dedicated equipment. The secondary

FIGURE 8.12

John Williamson of State Line Farm Biofuels in Shaftsbury, VT started growing and harvesting oilseeds in 2009. Here he uses a 1967 Massey-Harris SP35 combine with a modified corn head to harvest sunflowers. John has since upgraded to a larger, more modern combine for the majority of his harvest, but the smaller combine is useful for smaller fields and smaller trial plots.

cleaning prevents weed sprouting or heat generation and mold from green material during storage. Cleaning can also be used to grade seeds into different sizes if seed saving for future planting is advantageous.

The machine of choice in the Northeast is either a fanning mill (new or old) or a rotary drum cleaner (Figure 8.13). Both make use of specifically sized screens that separate the desired seed

FIGURE 8.13

(a) A stationary fanning mill can be used to clean oilseeds, (b) A fanning mill used to clean oilseeds.

from chafe or weed seed. The rotary drum cleaner does so using a rotating, inclined drum with one or more screens of different sizes. The screens generally are successively finer in the outward radial direction, allowing for chafe to be held back in the early stages and smaller seed to pass in the latter stages. Similarly, a fanning mill makes use of vibratory inclined screens and winnowing air for successive cleaning. Both approaches allow tailoring of the process to the crop of interest by replacement of the screens with appropriately sized mesh and also by allowing the various isolated streams to be drawn off for collection or disposal.

Some modifications to normal grain dryers may be required to accommodate oilseeds such as canola because of its small size. A secured layer of burlap or muslin over a standard larger-grain floor can provide a simple, low-cost alternative to buying a canola-specific seed-drying floor, though the edges of the bin can still present leakage issues if the fabric is not secured well. Similar to other grains, concrete flooring in storage or drying bins is not recommended, because seed on the bottom will absorb moisture, and eventually mold and spoil. Additionally, as in the combine, any holes or seams in the seed bin will leak seed, and should be sealed before the seed is put into the bin, if possible.

Oilseeds are optimally stored in a dried state, with the optimal storage moisture depending on the crop. Generally, oilseeds are dried using ambient air (i.e., not heated) to avoid potential reduction of oil quality and/or combustion in the process. Specially built drying bins are used in this process, which includes a "false floor" that holds the grain up above an air plenum. A drying fan forces air into this plenum and up through the stacked seed, drying it by absorbing moisture into the air as it travels through the seed, eventually being exhausted out vents at the top of the bin. One grower has made use of a solar hot water system to augment his drying process, with a 50% reduction in drying time as a result (Callahan and Williamson, 2008). Small dryers that can be placed in bulk bags with the seed are also adequate for drying relatively small batches of seed (a ton or less). These auger-like aerators are often physically screwed into a one-ton tote of seed to aerate the batch (Figure 8.14).

Once the seed has reached 8–10% moisture, it is ready for long-term storage. Once the seed has gone into the grain bin for storage, producers should perform weekly checks for seed heating and for condensation on the bin walls and ceiling until the crop has cooled to below freezing (Figure 8.15). If any changes in humidity or temperature occur within the bin, the seed should be cooled with an aerator immediately until the bin has regained its normal condition. Once the bin and the seed have reached mid-winter temperatures, reducing checks to once per month is adequate. When properly cleaned and dried, oilseeds can be stored for months. Some have found storage of oil in the form of oilseeds to be the most stable approach, if markets allow. In other words, pressing oil from stored seed only as needed to provide for the freshest oil possible.

FIGURE 8.14

A screw-in auger used for drying small volumes of oilseeds.

FIGURE 8.15

Grain bin interior. Check on seed regularly during long-term storage.

OIL EXTRACTION AND BYPRODUCTS

On-farm oilseed pressing has generally taken the form of cold-pressing using screw presses which convert oilseeds to oil and meal (Figure 8.16).

There are a number of presses available commercially, and a recent evaluation was conducted among the presses in the Northeast. These presses separate the oil from the rest of the seed through a combination of heating, crushing, and mechanical pressure. They are essentially screw augers with an intentional restriction at the end (Figure 8.17). This restriction both crushes the seed structure and increases pressure locally, which results in separation of the oil from the seed meal. The oil flows backward through a sieve around the barrel that is too fine for the meal to pass through (Figure 8.18). The seed meal passes through the restricting die on the end of the screw press barrel. Calibrating the restriction, screw rotation speed, and operating temperature for optimal pressing can be challenging, counterintuitive, and often comes down to patient trial and error.

An evaluation of six commercially available, small-scale presses was conducted by researchers at the University of Vermont (Callahan et al., 2014). The price range of the presses was $6000 to $15,000 when purchased with claimed pressing capacity of 700–1800 pounds of seed in a 24 h period. These presses were assessed based on pressing capacity (pounds per 24 h) and oil separation performance

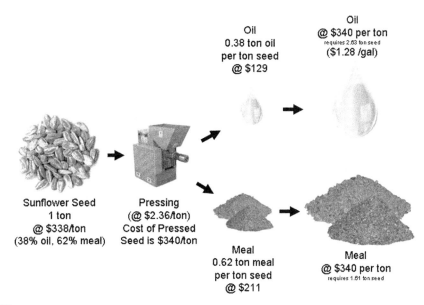

Oil
0.38 ton oil
per ton seed
@ $129

Oil
@ $340 per ton
requires 2.63 ton seed
($1.28 /gal)

Sunflower Seed
1 ton
@ $338/ton
(38% oil, 62% meal)

Pressing
(@ $2.36/ton)
Cost of Pressed
Seed is $340/ton

Meal
0.62 ton meal
per ton seed
@ $211

Meal
@ $340 per ton
requires 1.61 ton seed

FIGURE 8.16

Breaking down the cost to press one ton of sunflower seed into equivalent costs per ton of oil and meal.

(net oil % by weight) at various speeds using three crops: sunflower, canola, and soybeans. The owners and operators of the presses were also interviewed to collect anecdotal and qualitative information about the use of each press, common pitfalls or challenges, and how they are resolved.

This work documented actual through-put rates of three common oilseed crops at various speeds on each press. The results demonstrate how different presses have different capacities

KernKraft 40 at Borderview Research Farm in Alburgh, VT.

FIGURE 8.17

Small screw press used on farms to extrude oil from crops.

FIGURE 8.18

Operation of small screw press used for on-farm oil extraction.

"Screw "or "Worm" advances seed, pressing it against the backside of the die. Crushing it along the way. Oil and meal are separated by pressure.

based on the crop being pressed. Additionally, press rates were documented at different press speeds (RPM).

A key finding of this evaluation was that maximum net oil extraction tends to occur at lower than maximum press capacity. In other words, to extract the maximum amount of oil from a particular oilseed, one may have to run the press at a lower speed than the maximum it is capable of. This results in a trade-off between through-put (pounds per 24 h) and net oil yield (% by weight) from a particular crop. This can be an important consideration for a farm when just starting out with oilseed production. Often even the smallest scale presses are well above the required capacity for year-round pressing of typical harvests. This, combined with stability of oilseeds when stored as seed, suggests a slower pressing speed with an extended pressing schedule can result in higher net oil yields.

This work also noted decreasing phosphorus levels in oils with decreased pressing speed across all crops tested. Phosphorus content is used as an indicator for oxidative agents, which lead to oil instability and also to potential quality issues in oil converted to biodiesel.

The solids coming off the press are generally referred to as oilseed meal or meal cake. These solids are generally cellulosic and have caloric and nutritional value. Livestock meal often includes oilseed solids such as soy meal for its protein value. Farmers pressing oilseeds in the Northeast have experimented with feeding oilseed meal directly, and nutritional analyses have been conducted. Generally, the meal resulting from cold-pressing, solvent-free extraction is high in residual fat, which limits its overall proportion in common rations, especially for ruminant animals (Table 8.1).

A feeding trial was conducted on-farm in Vermont to determine the impact of homegrown canola meal on milk yield and quality. The study utilized 20 dairy cows in various stages of lactation. Overall, the feeding trial confirmed that farm-grown canola meal can serve as an adequate protein source on dairy farms in Vermont. However, the high fat content of the farm grown meal can potentially have some negative impacts on milk fat content (Table 8.2).

Table 8.1 Feed Analysis for Farm Grown and Purchased Canola Meal (Darby, Unpublished)

Canola Meal Source	Crude Protein	Crude Fat	Acid Detergent Fiber	Neutral Detergent Fiber	Phosphorus	Nonfibrous Carbohydrate	Net Energy Lactation
	% DM						Mcal/lb
Farm grown	33.1	21.7	14.8	20.7	0.98	20.9	1.15
Purchased	36.3	2.94	16.2	23.3	1.03	18.9	0.79

Table 8.2 Impact of Farm Grown Canola Meal on Milk Yield and Quantity (Darby, Unpublished)

Canola Source	Milk Yield	Fat Content	Protein Content
	lb	%	%
Farm	40.4	5.11	2.80
Purchased	39.1	5.25	2.80
LSD	NS	NS	NS

There is a potential market for the oilseed meal to be used for fertilizer and weed control. Because oilseeds are high in protein, they contain nitrogen (Table 8.3). In addition to their fertilizer value, meals from brassica crops (such as mustard and canola) have unique biochemical properties and have shown promise in other areas of the country as a soil amendment used for suppressing weeds (Table 8.4), plant parasitic nematodes, and fungal pathogens (Rahman and Somers, 2005; Haramoto and Gallandt, 2004).

Meal can also be densified by pelletizing and used as a combustion fuel in a standard pellet furnace or boiler, with similar heat output to wood pellets where allowed by state and local laws.

Table 8.3 Nutrient Analysis of On-farm Oilseed Meal (Darby, Unpublished)

Nutrient Content (Dry Matter Basis)	Sunflower	Canola	Mustard
% Dry matter	81.3	92.3	91
% Crude Protein	34.9	28.7	37.8
% Nitrogen	5.6	4.6	6.0
%Phosphorus	1.26	0.74	1.02
%Potassium	1.49	0.68	1.02
%Magnesium	0.64	0.30	0.42
%Calcium	0.76	0.48	0.52
%Sulfur	0.39	0.40	1.50

Table 8.4 Weed Counts in Oilseed Amended Plots in 2008 and 2009 (Darby, Unpublished)

Amendment	2008	2009
	Weed count*	Weed count
Sunflower meal	218a	33b
Canola meal	368a	38b
Mustard meal	269a	15a
Control (synthetic N)	272a	52c

*Within each column, numbers followed by the same letter are not significantly different ($P < 0.05$).

CHALLENGES AND OPPORTUNITIES WITH FARM-BASED FUEL PRODUCTION

A critical consideration for any farm exploring on-farm production of biofuels is safety. The production of biodiesel generally involves the use of two chemicals, caustic lye and alcohol, which present potential hazards.

A caustic is required as a catalyst in the typical transesterification process: most producers use either sodium or potassium hydroxide. These chemicals are highly basic (high pH) and can cause burns on contact. Care should be taken to always wear appropriate personal protective equipment such as goggles, dust mask, apron, and gloves, to avoid direct contact. Another practice that is often employed is to ensure a transport container for the catalyst that will hold the entire measure needed for a batch. This prevents pouring and dumping multiple times, and therefore limits the hazardous dust that is generated each time figure 8.19.

FIGURE 8.19

Jerrod LaValley demonstrates proper personal protective equipment while performing a water test on a batch of oil at Borderview Farm.

An alcohol is also part of the transesterification process. Typically this is methanol, although it is possible to use ethanol. Methanol poses several human health risks, since it is less easily processed in our bodies than more benign ethanol. Care should be taken to avoid direct contact and inhalation of fumes. One farm in the Northeast has used a vacuum system with rated tanks to allow for moving methanol directly from the shipping barrels to their biodiesel processor without ever pouring it or exposing personnel to it. This approach has the added benefit of ensuring any system leaks are of air into the system instead of methanol out of the system (since it is under vaccum).

Alcohols also present a combustibility risk, and it is recommended that all biodiesel processing areas include combustibility sensors that alarm when a combustible material is sensed in the space. These sensors can be relatively inexpensive and are useful for catching slow growing leaks in the system, before they become a larger issue.

Conducting a hazard review and a failure modes and effects analysis (FMEA) is a good practice for any process, but especially for one involving known risks. An example of one FMEA for a farm-based biodiesel process is available online from the Vermont Bioenergy Initiative (Callahan et al., 2009). A piping and instrumentation diagram (P&ID) of the processor and also a set of standard operating procedures (SOP's) as included in the referenced FMEA can be helpful to avoid mistakes. The diagram helps to quickly orient the operator to the process and to visually think about how changes may impact operation. A set of standard operating procedures in the form of a checklist helps to avoid missing a step or getting steps mixed up. This can prevent a failed batch of fuel or even avoid a hazardous situation.

Having a space on the farm dedicated to producing and handling oilseeds, biodiesel, and associated supplies will help to ensure safety and environmental protection. One part of this is spill prevention containment and countermeasures (SPCC). Although generally not required for smaller scale producers, SPCC is a best practice and can help prevent waste, cost, injury, loss of property, and environmental damage. The SPCC helps review your process to prevent a spill of a potentially hazardous material and identify where you may need supplemental containment and countermeasures to clean it up if it happens.

EXAMPLES OF ON-FARM FUEL SYSTEMS

The potential for farm-based biodiesel production is summarized in a breakeven economics report published by the Vermont Bioenergy Initiative (Callahan and White, 2013) using the Vermont Oilseed Cost and Profit Calculator (Callahan) (see Table 8.5). In this report, several of the Vermont farms adopting this approach are profiled. Additionally, two case studies are presented, which highlight a 13,000 gallon per year and a 100,000 gallon per year processing capacity. The smaller scale option is presented to demonstrate the feasibility of even small volume production to result in predictable fuel supply and costs, even when used at only partial capacity of 4000 gallons per year. This smaller scale operation is shown to produce fuel at a cost of $2.52 per gallon in the 4000 gallon per year start-up phase on 66 acres of sunflower production. Meal is generated at a cost of $389 per ton in this case. When at full capacity of 13,000 gallons per year (on 214 acres of sunflowers), the same equipment produces fuel for $2.29 per gallon and meal for $353 per ton. This case study demonstrates how volume of production can reduce the per unit cost of both products.

Table 8.5 Breakeven Pricing for Sunflower Products at Various Costs of Crop Production and Yield (Callahan and White, 2013)

Breakeven Pricing for Sunflower Products at Various Costs of Crop Production and Yield

Recurring costs (of crop production)	1,000 pounds per acre	1,500 pounds per acre	2,000 pounds per acre	Units
	295	204	158	Seed, $/ton
	298	206	161	Meal, $/ton
$100/acre	1.12	0.77	0.60	Oil, $/gallon
	1.98	1.64	1.47	Biodiesel, $/gallon
	1,974	1,316	987	Acres required
	395	271	208	Seed, $/ton
	398	273	211	Meal, $/ton
$150/acre	1.49	1.02	0.79	Oil, $/gallon
	2.36	1.89	1.66	Biodiesel, $/gallon
	1,974	1,316	987	Acres required
	495	337	258	Seed, $/ton
	498	340	261	Meal, $/ton
$200/acre	1.87	1.27	0.98	Oil, $/gallon
	2.73	2.14	1.84	Biodiesel, $/gallon
	1,974	1,316	987	Acres required

▢ = not profitable at assumed market prices

Note: Unless otherwise noted, all acreage above are net acres. These figures do not account for necessary rotation acres (4-5 x annual net oilseed acres), which will be necessary for sustained production as well as minimized pest and disease pressure. If the oilseed production is integrated with a diversified crop enterprise, this may not be a large factor in the operation.

A larger production model is also presented, which includes a detailed breakdown of the investment and operating cost assumptions as well as a parametric study using yield and cost of production as independent variables to assess product costs. In this case, a much higher investment is required to establish production on 1645 acres to produce 100,000 gallons per year. This model produces fuel at a nominal cost of $2.14 per gallon. One of the benefits of on-farm production of fuel for on-farm (off-road) use is that transportation costs and taxes are avoided. The fuel is available at a wholesale production cost instead of at a retail price. Price volatility can also be controlled in this way; the farm's fuel cost is divorced from the market price.

REFERENCES

Callahan, C., Harwood, H., Darby, H., Elias, R., Schaufler, D., March 3, 2014. Small-scale Oilseed Presses: An Evaluation of Six Commercially-available Designs. Pennsylvania State University. Available at: http://www.uvm.edu/extension/cropsoil/wp-content/uploads/OilseedPressEval_report.pdf.

Callahan, C.W., White, N., March 2013. Vermont On-Farm Oilseed Enterprises: Production Capacity and Breakeven Economics. Vermont Sustainable Jobs Fund. Available at: http://www.vsjf.org/assets/files/VBI/VT Oilseed Enterprises_March_2013.pdf.

Callahan, C.W., Williamson, J., October 2008. Feasibility Analysis: Solar Seed Dryer and Storage Bin. Vermont Sustainable Jobs Fund. Available at: http://www.vsjf.org/assets/files/VBI/Feasibility Study_Solar Seed Dryer_ October 2008.pdf.

Callahan, C.W., Williamson, J., January 15, 2009. Project Report: State Line Biofuels Safety Review and Engineering Study of an On-farm Small-scale Biodiesel Production Facility. Vermont Sustainable Jobs Fund. Available at: http://www.vsjf.org/assets/files/VBI/Oilseeds/State%20Line%20Safety%20Review_Engineering %20Study_Jan%202009.pdf.

Callahan, C.W. The Vermont Oilseed Cost and Profit Calculator. Vermont Sustainable Jobs Fund. Available from: http://www.vsjf.org/resources/reports-tools/oilseed-calculator.

Campbell, E., 2009. Master's Thesis: Greenhouse Gas Life Cycle Assessment of Canola and Sunflower Biofuel Crops Grown with Organic versus Conventional Methods in New England. University of Vermont (Natural Resources).

Canola Council of Canada, 2011. Canola Growers Manual [Online]. Published by Canola Council of Canada. Available at: http://www.canolacouncil.org/crop-production/canola-grower%27s-manual-contents/(verified 22 Jan. 2013).

Garza, E., April 24, 2011. The Energy Return on Invested of Biodiesel in Vermont. Vermont Sustainable Jobs Fund. Available at: http://www.vsjf.org/assets/files/VBI/Oilseeds/VSJF_EROI_Report_Final.pdf.

Haramoto, E.R., Gallandt, E.R., 2004. *Brassica* cover cropping for weed management: a review. Renewable Agriculture and Food Systems 19 (4), 187–198.

Kandel, H., 2013. 2012 national sunflower association survey. In: National Sunflower Association Research Forum, Fargo, ND. 9–10 Jan. 2013. National Sunflower Association.

Karsten, H., 2012. Penn state university. canola research update 2011. In: Annual Oilseed Producers Meeting, White River Jct, VT. 26 Mar. 2012. University of Vermont Extension.

Kirkegaard, J.A., Wong, P.T.W., Desmarchelier, J.M., 1996. *In vitro* suppression of fungal root pathogens of cereals by *Brassica* tissues. Plant Pathology 45 (3), 593–603.

National Sunflower Association, 2012. Growers [Online]. Published by the National Sunflower Association. Available at: http://www.sunflowernsa.com/growers/(verified 22 Jan. 2013).

Oplinger, E.S., Philbrook, B.D., 1992. Soybean planting date, row width, and seeding rate response in three tillage systems. Journal of Production Agriculture. 5, 94–99.

Rahman, L., Somers, T., 2005. Nemfix as green manure and seed meal in vineyards. Australian Plant Pathology 34 (1), 77–83.

LIFE-CYCLE ASSESSMENT: THE ENERGY RETURN ON INVESTED OF BIODIESEL

Eric L. Garza

Rubenstein School of Environment and Natural Resources,
University of Vermont, Burlington, VT, USA

INTRODUCTION

In the aftermath of the oil shocks of the 1970s, many governments, businesses, and entrepreneurs searched for sources of liquid fuels not derived from petroleum (Körbitz, 1999). Rising petroleum prices after 2000, particularly the 2008 spike that lifted oil prices to nearly $150 per barrel, and concerns of fossil fuel depletion (see, for example, Deffeyes, 2010) have again renewed interest in alternative fuels, particularly biologically-derived fuels (Demirbas, 2007; Pradhan et al., 2009). One biofuel that is of growing interest is biodiesel, which is made most commonly from plant oils and animal fats through the chemical process of transesterification (Pradhan et al., 2009; Meher et al., 2006). Biodiesel readily substitutes for diesel fuel in most engines, and can be used as a substitute for heating oil as well. Although current production of biodiesel is small relative to petroleum diesel, opportunities for expanding production exist in many regions.

A critical criterion that should be used to judge the potential of a biofuel—or any fuel—is the amount of energy it yields as a finished fuel relative to the energy required to produce it. This ratio of energy output relative to energy inputs is termed the energy return on invested ratio (EROI) (Mulder and Hagens, 2008). Calculated values of EROI range from zero to infinity, and the break-even point of 1:1 defines the point at which a fuel's energy yield exactly equals the energy required to produce it.

The economic and social ramifications of EROI are best explained within a societal context (Hall et al., 2009). A society's EROI is calculated as the weighted average EROI of all of the fuels that the society uses. If this weighted average is 1:1, it means that all of a society's energy output must be reinvested to generate tomorrow's energy. This "reinvested" energy constitutes the energy sector of the economy, and in a society with an EROI of 1:1, the energy sector of the economy is effectively the entire economy. When society's EROI rises above 1:1, surplus energy is generated that may be put to other uses besides further procurement of energy. With a societal EROI greater than 1:1, economic sectors outside of the energy sector may emerge such as art, information technologies, construction, etc. The higher society's EROI is above this 1:1 cutoff, the larger the nonenergy sector of their economy can be. Figure 9.1 illustrates this with two energy-flow diagrams, the one on the left showing the surplus versus reinvested energy for a society with high EROI, and the one on the right showing the surplus versus reinvested energy for a lower EROI society. In order for societies to have functioning

Bioenergy. http://dx.doi.org/10.1016/B978-0-12-407909-0.00009-2

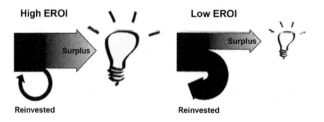

FIGURE 9.1

Two simplified energy-flow diagrams. The one on the left compares surplus energy to that which is reinvested for a high energy return on energy invested (EROI) fuel, showing the high surplus delivered to the economy that allows for a large and thriving economy outside of the energy sector. The image on the right compares surplus energy to that which is reinvested for a lower EROI fuel, showing the smaller relative surplus and thus the smaller economy outside of the energy sector. Only surplus energy can contribute to economic productivity outside of the energy sector.

nonenergy sector economies, the weighted average EROI for all of their fuels must be greater—and ideally substantially greater—than 1:1. Societies that cannot find a mix of fuels that yield surplus energy face economic and social decline. The higher the EROI of a society's fuel mix, the more substantial its nonenergy sector can be.

An additional societal impact of EROI deals with economic growth. For a society's economy to grow over time, the production of goods and services must expand. For the production of goods and services to expand, capital investments must also expand, and these capital investments require energy—to build the new machinery, transport the additional goods, etc. For a growing economy, the energy to power these new capital investments must come from existing energy surplus. Therefore, in order for a society's economy to grow, it must have a large enough energy surplus that it can afford to invest some of its energy surplus into building additional capital to fuel growth. Although a society can certainly persist at a steady-state with a low societal EROI, for it to grow its economy, the society needs a higher EROI.

Fossil fuels, particularly petroleum, have historically delivered fuels with very high EROI. Analysts estimate that crude oil yielded an EROI of 100:1 in the 1930s (Murphy and Hall, 2010). Given crude oil's dominance as an energy source, this suggests that an oil-powered economy at this time would readily have the capacity to expand its nonenergy sectors, given its high EROI fuel mix. Since the 1930s, however, the quality of crude oil resources has steadily declined and the infrastructural investments needed to extract a barrel of crude oil have increased, causing the EROI of crude oil to fall to approximately 20:1 globally. These 100:1 and 20:1 figures equate to EROI of crude oil, where crude oil flows from the well before it is delivered to a refinery and turned into finished products like gasoline, diesel, and kerosene. EROI estimates of finished fuels such as gasoline and diesel are typically much lower than these values, due to the additional energy costs of transport and refining. Data presented in Sheehan et al. (1998), for instance, suggest that the EROI of diesel fuel was about 5:1 in the late 1990s, and as EROI of crude oil at the wellhead has fallen since then, the EROI of diesel fuel has likely fallen as well.

Estimates of biodiesel EROI range widely (Table 9.1), from values below 1:1 as reported by Pimentel and Patzek (2005) for soybean- and sunflower-derived biodiesel, to the opposite extreme of

Table 9.1 Energy Return for Biodiesel by Feedstock and Reference

Feedstock, Reference	Energy Return
Reclaimed Vegetable Oil	
Elsayed et al. (2003)	4.85–5.88
Soybean Oil	
Pimentel and Patzek (2005)	0.78
Carraretto et al. (2004)	2.090
Ahmed et al. (1994)	2.5
Sheehan et al. (1998)	3.215
Hill et al. (2006)	3.67
Pradhan et al. (2009)	4.56
Sunflower Oil	
Pimentel and Patzek (2005)	0.76
Edwards et al. (2006)	0.85–1.08
Bona et al. (1999)	1.3–8.7
ADEME and DIREM (2002)	3.16
Kallivroussis et al. (2002)	4.5
Rapeseed Oil	
Edwards et al. (2006)	1.05–1.38
IEA (1999)	1.09–2.48
Elsayed et al. (2003)	2.17–2.42
ADEME and DIREM (2002)	2.99
Richards (2000)	3.71

Elsayed et al. (2003)'s estimated range of 4.85–5.88:1, based on the use of reclaimed vegetable oil as a feedstock and Bona et al. (1999)'s estimate of 8.7 from sunflower oil. The large range of EROI estimates stems from three main issues:

- No standard system boundaries are used in deciding which energy costs are to be included,
- No standard methods are applied to estimate energy costs,
- No standard methods are used to account for coproducts, which are other valuable products generated incidentally during the process of making biodiesel.

These three issues beg a more thorough discussion.

EROI APPROACHES AND CONTROVERSIES

Given the wide range of estimates listed, it comes as no surprise that the question of whether or not biodiesel is a viable fuel that yields an EROI greater than 1:1 remains controversial. One important reason for the variability seen in estimates is that there exists no consistent framework for assessing

EROI (Mulder and Hagens, 2008). EROI can be studied at three levels: first order, second order, and third order. First-order EROI analysis is represented as:

$$\text{EROI} = \frac{E_O}{E_D} \tag{25.1}$$

where E_O is the energy contained in a unit of fuel, and E_D is the direct energy costs required to deliver that unit of fuel. Direct energy costs include all fuels consumed in the production process, including liquid fuels such as gasoline, diesel, or biodiesel, heating fuels such as natural gas, propane, or fuel oil, and electricity to power machinery. Appropriate efficiency factors may be applied to the quantities of each fuel to account for energy requirements of their own processing or transport (Pradhan et al., 2009), although this is not always done. E_O and E_D are customarily measured in joules, the International System's energy unit, making EROI a unitless ratio. First-order EROI analysis is relatively straightforward, and direct energy costs can often be measured precisely or estimated with high precision, leading to EROI estimates with high certainty.

Second-order EROI analysis is represented as:

$$\text{EROI} = \frac{E_O}{E_D + E_I} \tag{25.2}$$

where E_I is the indirect energy costs required to deliver a unit of fuel, which are also called embodied energy costs. Indirect or embodied energy costs include energy required to design, build, and maintain structures, machinery, vehicles, and to produce other inputs such as chemical reagents, fertilizers, and pesticides, etc. The energy costs of structures, machinery, and vehicles are normally amortized over the expected amount of fuel they will produce over their expected lifetime. E_I is also measured in joules, so EROI remains unitless. Second-order EROI analysis is much more expansive in its scope than first-order EROI analysis, and there are no agreed upon methods stating which indirect energy costs must be included and how one should assess those costs. Furthermore, indirect energy costs are challenging and tedious to estimate, and although their inclusion makes the final EROI estimate more reflective of the real energy costs of a fuel, they also increase the uncertainty in the final EROI estimate. Most analysts ignore this added uncertainty, and unfortunately report estimates of indirect energy costs and second-order EROI as if the result was precisely known. A better approach would incorporate an analysis of uncertainty and report EROI estimates with associated standard deviations or confidence intervals.

Tertiary EROI analysis expands the scope of analysis even further, looking beyond merely energy inputs. For instance, a tertiary EROI analysis may inquire about nonenergy inputs such as water (Mulder et al., 2010), or other ecosystem services, or may account for environmental externalities per energy unit of fuel output. Although it is possible to attempt to translate these nonenergy inputs into energy units, this introduces uncharacterizable uncertainty into the analysis and should be avoided. Carrying out a tertiary EROI analysis while leaving nonenergy inputs in their native units requires a more expansive analytical process.

Perhaps one of the greatest challenges in assessing the EROI of individual fuels is that no two analysts include the same list of direct and indirect energy costs in their analysis. This yields the suite of biodiesel EROI estimates presented in Table 9.1 above, each with its own unique accounting framework and a final estimate that cannot readily be compared to other estimates due to differences in methodology. Beyond issues of methodological differences, analysts are not always clear on which direct and indirect energy costs they do include in their analyses. Some studies cited above, such as Elsayed et al. (2003) and Hill et al. (2006), take great pains to be transparent. Others, such as Bona et al. (1999), report results while giving virtually no information about their methodological approach. In an ideal world, analysts studying EROI would adopt a standardized way of delineating, or at the very least reporting on, their study boundaries (Mulder and Hagens, 2008). Such a standardized approach would make EROI analyses more useful, if only because the resulting numbers would be more believable and analysts would be better able to compare EROI estimates from study to study.

Other types of energy analyses can be expressed in ratios that appear similar to EROI, but they measure a very different set of processes or inputs and should not be compared to a standard first or second order EROI estimate. If boundaries are drawn strategically and embodied energy is estimated carelessly or left out entirely, even a dismal biofuel can be made to appear as though it yields a high EROI.

Another important factor in previous studies of biofuel EROI is how data is gathered. All studies cited in Table 9.1 gathered highly-aggregated data from regional or national datasets and by borrowing some assumptions on indirect energy costs from other, unrelated studies. This divorces the final estimate of EROI from any ecological context, thereby ignoring differences in oilseed yield, fertilizer and pesticide requirements, or other important inputs, as they vary as a function of physical geography and climate. A superior approach would be to study EROI on an individual farm, acknowledging the specific growing conditions and specific production process. Not only would this provide a more accurate appraisal of EROI, it would also allow the analyst—and the grower—to study how to improve the EROI of the biofuel produced. Kim and Dale (2005) focus on a more local area (Scott County, Iowa) in one study, although they do not narrow their study to a specific farm and its specific agricultural practices, nor do they publish an estimate of energy return for their biodiesel on its own (the study models the production of biodiesel and ethanol as dual biofuel products on land grown with corn and soybeans in rotation). It is necessary to study agricultural systems at a finer resolution to understand how the production process interacts with both production scale and geography to deliver, or fail to deliver, a positive energy balance for biodiesel. Producer-specific studies of energy return that actually measure direct energy inputs and that estimate indirect energy inputs specific to a supply chain represent an ideal to be strived for, as only these studies allow researchers and producers to observe the impacts of different elements of the production process on the fuel's energy balance. Producers can change their production process to make it more profitable, from an energy perspective, but only if they understand where the energy costs and potential for greater efficiencies lie.

PURPOSE

The purpose of this work is to gain a better understanding of the EROI for biodiesel produced in small-scale, distributed facilities in Vermont, USA. In particular, this work will calculate EROI for biodiesel produced from oilseed grown on five farms, and for one processor who uses reclaimed fryer oil to make biodiesel. A full accounting of direct and indirect energy costs for each producer will be presented using an expansive, second-order EROI analysis to assess the value of biodiesel as a fuel.

DATA AND METHODS
SYSTEM BOUNDARIES

Direct energy costs tabulated in the analysis include all liquid transport fuels used to move oilseed around or to power farm machinery, all fuels used to provide space or process heat for buildings and equipment, all electricity used to power machinery, and an estimate of the direct energy input associated with human labor. Transport fuels included diesel and biodiesel, whereas fuels used for space heat included biodiesel. Figure 9.2 illustrates the process of growing oilseed and processing this into biodiesel. Energy costs are apportioned between biodiesel and other coproducts by mass similar to Pradhan et al. (2009). I assume the energy content of diesel fuel is 136 MJ/gal, the energy content of biodiesel is 126 MJ/gal, and assume that 1 kWh of electricity equates to 3.6 MJ. The lower heating value for diesel fuel was adjusted by an efficiency factor of 0.84 to account for the life-cycle energy costs of production (Pradhan et al., 2009). An efficiency factor of 0.31 was applied to electricity estimates to account for generation and transmission losses (Pradhan et al., 2009). The producer who uses reclaimed fryer oil used some of their biodiesel as an input into their production process, so their EROI was converted to an efficiency factor of 0.75 to account for the life-cycle energy costs of producing this biodiesel. I assume that the direct energy input of human labor equates to a power output of 300 W (Smil, 2008). In reality, human power output varies as a function of the type of labor being done (heavy lifting versus driving a vehicle, for instance), but in practice, the direct energy value of human labor is negligible relative to other direct and indirect energy inputs so investing substantial effort further delineating types of labor is unnecessary. Surveys were used to collect data from biodiesel growers and producers on direct energy costs and labor inputs. All estimates of direct energy inputs are translated to energy input per gallon of biodiesel produced.

FIGURE 9.2

The process of growing oilseed and processing it into biodiesel. All inputs and machinery (at left) have both direct and indirect energy costs and labor costs associated with their use. All nonbiodiesel outputs that have value (at right) are apportioned some of energy costs as a function of mass.

(Source: (Pradhan et al., 2009.))

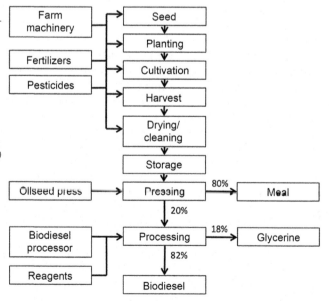

Resources and data are never available to carry out a complete accounting of indirect energy costs, forcing researchers to draw an inevitably arbitrary boundary line between that which is counted and that which is not. Indirect energy costs accounted for in this study included the embodied energy of all buildings, farm and processing machinery, fertilizer and other chemical agricultural inputs, seed, used vegetable oil, chemical reagents required for transesterification, and the indirect energy costs of human labor. Surveys collected data on the purchase price and purchase date of all farm machinery and implements, and these data were translated to embodied energy costs using the Carnegie-Mellon Economic Input–Output Life Cycle Assessment (EIOLCA) tool (Carnegie Mellon Green Design Institute, 2008). This model equates the purchaser price of a product to the amount of energy required to produce it by attempting to trace energy costs in different sectors as a function of monetary flows. This method is not as accurate or precise as assessing the physical makeup of all agricultural inputs and mechanical equipment, but offers a less resource- and time-intensive means of estimating embodied energy costs and was used, in this case, due to resource limitations. The consumer price index for years 1970–2010 was used to adjust purchase price estimates to 2002 dollars for input into the EIOLCA model.

The EIOLCA estimates embodied energy for products from particular economic sectors, acknowledging that products from some economic sectors are more energy intensive to produce than those of other sectors. Embodied energy of all farm machinery and implements were estimated using the EIOLCA sector "machinery and engines—farm machinery and equipment manufacturing". Embodied energy of biodiesel processing equipment was estimated using the EIOLCA sector "petroleum and basic chemical—petroleum refineries" if the processing unit was purchased, and using the sector "food, beverage and tobacco—breweries" if the processing unit was hand built. Embodied energy of vehicles was estimated using the sector "vehicles and other transportation equipment—light truck and utility vehicle manufacturing". Embodied energy of buildings was estimated using the sectors "construction—other residential structures" or "construction—nonresidential manufacturing structures". The EIOLCA documentation acknowledges that there is an unknown amount of uncertainty associated with estimates of embodied energy generated by its model. To attempt to account for at least some of this uncertainty, I assumed the EIOLCA model output represents a mean value and assumed a standard deviation of 25% of the mean for all estimates taken from the model.

Aside from using the EIOLCA model to estimate embodied energy of farm implements, buildings, and processing equipment, I use values from the USDA report to translate fertilizer, pesticide, seed, an reagent masses into energy values. These include 23.3 MJ/lb N, 4.1 MJ/lb P, and 2.7 MJ/lb K for fertilizer inputs; and 148 MJ/lb of pesticides (Pradhan et al., 2009). As with the EIOLCA model, there is unacknowledged uncertainty in these estimates, and to attempt to capture it, I assume that these values represent a mean and that the standard deviation around the mean is 25% of the mean. The embodied energy of manure was assumed to be 30% that of synthetic fertilizer per hectare (Wiens et al., 2008). The embodied energy of chemical reagents are assumed to be 11.2 MJ/gal of biodiesel for methanol, and 1.1 MJ/gal of biodiesel for other reagents (Pradhan et al., 2009). Other estimates of embodied energy for these components are available (see Elsayed et al., 2003; Nelson and Schrock, 2006 for examples), some higher and some lower, but estimates from Pradhan et al. (2009) were used to facilitate comparison of final EROI estimates to those put forward by the USDA. To estimate the embodied energy of human labor, I divided total energy use in Vermont in 2008 by the population of Vermont in 2008 estimate per capita energy consumption, divided this number by the number of hours per year to yield an estimate of per capita energy consumption per hour, and allotted this value as embodied energy as a function of the hours of labor embodied in each gallon of biodiesel (EIA, 2008; United States Census Bureau, 2008). All indirect energy inputs are translated to energy inputs per gallon biodiesel produced.

ACCOUNTING FOR UNCERTAINTY

Monte Carlo simulation was used to estimate final EROI for each producer while accounting for uncertainty. The Monte Carlo process begins with the generation of a random number with a normal distribution between 0 and 1 for each direct or indirect energy input that has uncertainty associated with it. This random number is used to generate a normal random value of that energy input based on its estimated mean and standard deviation. EROI is then calculated as in Eqn (25.2) by dividing the energy content of a gallon of biodiesel by the sum of the normal random values of all indirect and direct energy costs. This process is repeated 1000 times, generating 1000 unique estimates of EROI based on the means and standard deviations of all direct and indirect energy costs. The mean EROI estimate for each producer is calculated from these 1000 runs, along with the standard deviation of the mean.

PROJECTIONS

Because many of the producers included in this study are in the process of scaling up their operations, biodiesel production forecasts were used to project how their estimated EROI will change in the future. Table 9.2 presents biodiesel production forecasts for the processor who uses reclaimed fryer oil and the five growers. These forecasts are based on planned increases in oilseed planting by using current equipment and current or foreseeable land acquisition, and do not assume additional equipment purchase. Labor, energy, agricultural input, and reagent costs are assumed to rise proportionately to biodiesel production, whereas the embodied energy costs of existing machinery and buildings are assumed to remain constant.

Table 9.2 Processors and Production Forecasts, Specifying Years 2011 (Oilseed Grown in the 2010 Growing Season), 2014, and 2016. Projections Assume that Biodiesel Production Will Remain Constant from 2016 to 2020. Producer One Operates a Nonfarm Facility that Uses Reclaimed Fryer Oil as a Feedstock, whereas Producers 2–6 Use Vegetable Oil Pressed from Oilseed Grown On-farm

Producer	2011	2014	2016
1	16,000	36,000	50,000
2	10,000	20,000	26,000
3	3000	4000	5000
4	1000	2700	6700
5	1500	10,000	25,000
6	550	1000	1000

EROI OF VERMONT BIODIESEL

Table 9.3 shows the mean EROI for the production of biodiesel at each of the five farms and one reclaimed fryer oil processing center, along with estimates or forecasts of their total biodiesel production from oilseed grown in 2010. Mean estimates of EROI for Vermont-produced biodiesel range from a low of 2.63:1 to a high of 5.89:1. The weighted average for Vermont is 4.04:1, which is unambiguously above 1:1. For the small-scale growers studied here, direct energy costs represented 3–23% of total energy costs, whereas indirect energy costs represented 77–97% of total energy costs. Energy surplus ranges from 61 to 83% for all Vermont producers.

Figure 9.3 shows the energy costs associated with biodiesel production for the five growers and one used oil processor studied, as well as equivalent energy costs from the USDA study (Pradhan et al., 2009). These energy costs are normalized to each gallon of biodiesel produced. Figure 9.4 shows these same energy costs and energy surplus as a percentage of the lower heating value of a gallon of biodiesel, clearly delineating energy costs from the energy surplus delivered by produced biodiesel. The numbers assigned to each grower correspond to those in Tables 9.2 and 9.3. The embodied energy of reagents, particularly methanol, is consistently a large component of all growers' and producers' energy costs, including that of industrial growers studied by Pradhan et al. (2009). For the relatively small-scale growers studied here, the embodied energy costs of their processor can also be substantial, particularly in the case of grower four who purchased a sophisticated processor and who produces only 1000 gal of biodiesel, preventing him from apportioning the embodied energy costs of this processor over a larger amount of finished product. The direct energy costs of liquid fuels, primarily diesel or biodiesel, are important for some Vermont growers and producers, as is the embodied energy costs of fertilizers and pesticides. Tables showing itemized energy costs for all growers and producers studied in Vermont are provided in the appendix.

The impacts of projected biodiesel production increases on the EROI of the six producers studied are shown in Figure 9.5. Producer 4, who has the lowest EROI and currently produces 1000 gal of biodiesel per year, is projected to see the largest increase in EROI as his production scale increases.

Table 9.3 Feedstock, Mean Second Order Energy Return on Energy Invested (EROI) ± One Standard Deviation, and 2010 Biodiesel Production for Six Producers in Vermont

Processor	Feedstock	EROI	Energy Surplus (%)	Production (gal/year)
1	Reclaimed oil	3.60 ± 0.38	72	16,000
2	Soybean oil	4.24 ± 0.44	76	10,000
3	Sunflower oil	3.61 ± 0.39	72	3000
4	Sunflower oil	2.63 ± 0.41	61	1000
5	Sunflower oil	5.89 ± 0.73	83	1500
6	Sunflower oil	5.12 ± 0.56	81	550

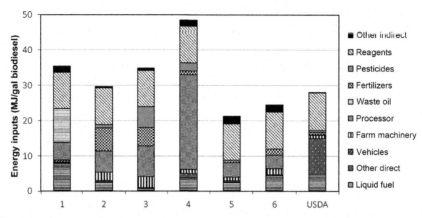

FIGURE 9.3

Energy costs for producing biodiesel for all producers in Vermont presented next to energy costs reported by USDA (Pradhan et al., 2009). "Liquid fuel" includes gasoline, diesel, and biodiesel fuel, "Other Direct" includes electricity, natural gas, propane, and human labor. "Other Indirect" includes the embodied energy of buildings, vehicles, oilseed presses, seed, and human labor, except for the USDA analysis, which does not include the embodied energy of human labor. "Reagents" is dominated by the embodied energy of methanol, but includes all reagents involved in transesterification.

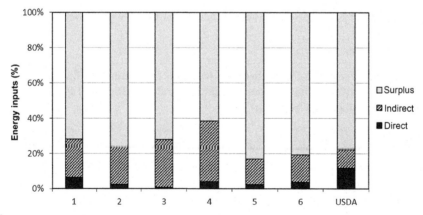

FIGURE 9.4

Energy costs and energy surplus from six Vermont biodiesel producers normalized to the lower heating value of biodiesel. "Indirect" includes all indirect energy costs, "Direct" includes all direct energy costs.

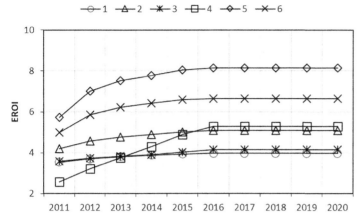

FIGURE 9.5

Projected EROIs for the six biodiesel producers based on expected increases of biodiesel processing, as noted in Table 9.2.

This is largely due to his ability to apportion the energy costs of his biodiesel processor over a larger number of gallons of finished biodiesel product. Producer 5, an organic grower with the highest EROI for the 2010 growing season, is projected to reach an EROI of roughly 8:1 if he reaches his production targets without requiring additional equipment.

CONSIDERATIONS

EROI estimates from this study are comparable to those noted in Table 9.1, although one must be cautious comparing EROI numbers from different studies due to methodological differences. Notably, even the smallest-scale grower studied in Vermont, who produces 550 gal of biodiesel per year, shows an EROI well above those calculated by Pimentel and Patzek (2005).

One important benefit of studying biofuel production at the level of individual grower and individual supply chain is that it affords the researcher and other stakeholders the opportunity to compare the influences of alternative growing practices and alternative production practices on final estimates of EROI. Some of the energy costs cataloged are effectively fixed energy costs and cannot be reduced relative to the energy benefit of a gallon of finished biodiesel. The energy costs of chemical reagents are the perfect example of this, as methanol and other reagents must be used at a standard ratio when reacting vegetable oil to produce biodiesel. These fixed energy costs impose an upper limit of roughly 20:1 on the EROI that can be achieved when transesterification is used to convert vegetable oil to biodiesel. In practice, other variable energy costs, such as those associated with transporting feedstock, powering machinery, and preparing, fertilizing, and cultivating land, can be reduced but can never be reduced to zero. Other chemical processes are being investigated for their ability to yield quality biodiesel products (Körbitz, 1999; Meher et al., 2006), but they will certainly also require reagents, so care must be taken to assess the energy costs of alternatives before producers make substantial investments to switch production processes.

Although acknowledging this maximum is important, it is also important to realize that all of the producers outlined in this report can improve their operations and deliver a fuel with a higher second-order EROI than that measured by this study. Direct energy costs are made up primarily of liquid fuels used for transporting feedstocks and to power agricultural machinery. Although it may not be cost-effective to purchase new machinery that is more energy efficient to reduce liquid fuel use, a mixture of behavioral shifts and more strategic decision-making when scheduling on-farm or off-farm tasks will likely allow for the reduction of liquid fuel use. The degree to which these direct energy costs can be reduced will vary from farm to farm, but the potential is certainly there, and in many cases may allow for modest EROI gains.

Direct energy requirements represent a small proportion of the total energy requirements to produce a gallon of finished biodiesel in Vermont. Besides reagents, which account for roughly 20% of the energy value of the finished fuel, all other indirect energy costs are variable and subject to change based on the grower and processor's strategic decision-making. The embodied energy of farm machinery, seed presses, and the biodiesel reaction apparatus generally represent a small component of the total indirect energy, but this energy input can still be reduced. By reducing the use of machinery, sharing machinery among growers and producers, or more thoroughly using existing machinery for a wider variety of purposes, the proportion of the embodied energy attributable to a single gallon of finished biodiesel can be reduced. Scaling up biodiesel production without adding new equipment, as investigated in the projections shown in Figure 9.4, is one means of doing this and allows producers and growers to spread the embodied energy costs of equipment over a larger amount of finished product. Additionally, if other production practices allowed for higher yields of oilseed and thus biodiesel, the embodied energy of machinery could be spread among more finished gallons of fuel, reducing it further. Among the growers, there are farmers who use less fertilizer than others and one that replaces synthetic fertilizer with manure, lowering energy costs. Some growers also do not use pesticides, lowering their energy costs. More experimentation with the level of fertilization and pesticide application, combined with programs of crop rotation to reduce pest infestations, could allow for modest increases in biodiesel EROI.

No other state in the United States or the world is currently studying its biofuel producers with the rigor that the state of Vermont currently is, creating the opportunity for Vermont to emerge as a global leader in acknowledging and using EROI as a criteria in judging the viability of its biofuel enterprises. But EROI should not be the single criteria used to judge biofuel viability, which prompts a return to third-order EROI analysis, noted above. In addition to requiring energy inputs to generate a gallon of finished biodiesel, a variety of other inputs are required. One vitally important input is cash flow. Other work is underway to estimate the monetary costs associated with producing a gallon of biodiesel, and although monetary cost accounting is not typically thought of as EROI analysis, it fits in perfectly with Mulder and Hagens' (2008) notion of third-order EROI analysis. Second-order EROI analysis asks what we get back in terms of energy relative to what we invest, or E_O/E_I based on the notation used above. A third-order EROI that studies monetary cost inputs might estimate the energy return on money invested (EROMI) as $E_O/\$_I$. Knowing that E_O generally represents the energy content of a gallon of fuel (or other set volume), EROMI is effectively the reciprocal of cost per gallon of fuel. Other inputs or impacts can be accommodated in EROI analysis by using a generalized ratio of E_O/X as well, where X might be the mass of carbon released or sequestered by the production process, the amount of water required to produce a gallon of fuel, the amount of nitrogen or phosphorus lost from farm fields due to runoff, the mass of soil lost from fields due to erosion, or any number of relevant social, economic, or environmental variables. Although the basic second-order EROI analysis that accounts for energy inputs and outputs is a vital element of biofuel viability over the short to moderate term, third-order EROI analysis that broadens the scope of assessment is necessary to determine biofuel sustainability over the long term.

CONCLUSIONS

The totality of evidence suggests that biodiesel can be produced in Vermont with a positive EROI, and that modest changes in production practices would improve EROI on most farms. When compared, using similar methods, to studies that survey national oilseed datasets, Vermont farmers appear to be better able to efficiently grow and turn oilseed into finished biodiesel, despite their small production capacity.

ACKNOWLEDGMENTS

This study was paid for by Vermont Sustainable Jobs Fund with funds secured by U.S. Senator Patrick Leahy from the U.S. Department of Energy.

REFERENCES

Direction of Agriculture and Bio-Energies of the French Environment and Energy Management Agency (ADEME) & French Division of the Energy and Mineral Resources (DIREM), 2002. Energy and Greenhouse Gas Balances of Biofuels Production Chains in France (a report by ADEME and DIREM).

Ahmed, I., Decker, J., Morris, D., 1994. How Much Energy Does it Take to Make a Gallon of Soydiesel? Institute for Local Self-Reliance.

Bona, S., Mosca, G., Vamerali, T., 1999. Oil crops for biodiesel production in Italy. Renewable Energy 16, 1053–1056.

Carnegie Mellon University Green Design Institute, 2008. Economic Input–Output Life Cycle Assessment (EIO-lca), US 2002 Industry Benchmark Model [Internet]. Available from: http://www.eiolca.net (accessed 12–28 January, 2011).

Carraretto, C., Macor, A., Mirandola, A., Stoppato, A., Tonon, S., 2004. Biodiesel as alternative fuel: experimental analysis and energetic evaluations. Energy 29 (12–15), 2195–2211.

Deffeyes, K., 2010. When Oil Peaked. Hill and Wang.

Demirbas, A., 2007. Importance of biodiesel as transportation fuel. Energy Policy 35, 4661–4670.

Dewulf, J., van Langenhove, H., van de Velde, B., 2005. Exergy-based efficiency and renewability assessment of biofuel production. Environmental Science and Technology 39 (10), 3878–3882.

Edwards, R., Larivé, J.–F., Mahieu, V., Rouveirolles, P., 2006. Well-to-Wheels Analysis of Future Automotive Fuels and Powertrains in the European Context. Joint Research Center of the European Commission.

Elsayed, M.A., Matthews, R., Mortimer, N.D., 2003. Carbon and Energy Balances for a Range of Biofuels. Resources Research Unit, Sheffield Hallam University.

Hall, C.A.S., Balogh, S., Murphy, D.J., 2009. What is the minimum EROI that a sustainable society must have? Energies 2 (1), 25–47.

Hill, J., Nelson, E., Tilman, D., Polasky, S., Tiffany, D., 2006. Environmental, economic, and energetic costs and benefits of biodiesel and ethanol biofuels. Proceedings of the National Academy of Science 103 (30), 11206–11210.

International Energy Agency (IEA), 1999. Automotive Fuels for the Future: The Search for Alternatives (a report by the IEA).

Kallivroussis, L., Natsis, A., Papadakis, G., 2002. The energy balance of sunflower production for biodiesel in Greece. Biosystems Engineering 81 (3), 347–354.

Kim, S., Dale, B.E., 2005. Life cycle assessment of various cropping systems utilized for producing biofuels: bioethanol and biodiesel. Biomass and Bioenergy 29, 426–439.

Körbitz, W., 1999. Biodiesel production in Europe and North America, an encouraging prospect. Renewable Energy 16, 1078–1083.

Meher, L.C., Sagar, D.V., Naik, S.N., 2006. Technical aspects of biodiesel production by transesterification—a review. Renewable and Sustainable Energy Reviews 10 (3), 248–268.

Mulder, K., Hagens, N.J., 2008. Energy return on investment: toward a consistent framework. Ambio 37, 74–79.

Mulder, K., Hagens, N.J., Fisher, B., 2010. Burning water: a comparative analysis of the energy return on water invested. Ambio 39 (1), 30–39.

Murphy, D.J., Hall, C.A.S., 2010. Year in review—EROI or energy return on (energy) invested. Annals of the New York Academy of Sciences 1185, 102–118.

Nelson, R.G., Schrock, M.D., 2006. Energetic and economic feasibility associated with the production, processing and conversion of beef tallow to a substitute diesel fuel. Biomass and Bioenergy 30, 584–591.

Pimentel, D., Patzek, T.W., 2005. Ethanol production using corn, switchgrass, and wood; biodiesel production using soybean and sunflower. Natural Resources Research 14, 65–76.

Pradhan, A., Shrestha, D.S., McAloon, A., Yee, W., Haas, M., Duffield, J.A., Shapouri, H., 2009. Energy Life-Cycle Assessment of Soybean Biodiesel. United States Department of Agriculture.

Richards, I.R., 2000. Energy Balances in the Growth of Oilseed Rape for Biodiesel and of Wheat for Bioethanol (a report by Levington Agriculture Ltd).

Sheehan, J., Camobreco, V., Duffield, J., Graboski, M., Shapouri, H., 1998. An Overview of Biodiesel and Petroleum Diesel Life Cycles. National Renewable Energy Laboratory.

Smil, V., 2008. Energy in Nature and Society: General Energetics of Complex Systems. MIT Press.

United States Census Bureau, 2008. At: http://quickfacts.census.gov/qfd/states/50000.html (accessed on 14.01.11.).

United States Energy Information Administration, 2008. At: http://www.eia.gov/state/state_energy_profiles.cfm?sid=VT (accessed on 14.01.11.)

Wiens, M.J., Entz, M.H., Wilson, C., Ominski, K.H., 2008. Energy requirements for transport and surface application of liquid pig manure in Manitoba, Canada. Agricultural Systems 98 (2), 74–81.

APPENDICES

The following tables show energy costs estimated at all growers and producers included in this study, including their current production rate and feedstock type. All energy costs are normalized per gallon of finished biodiesel (Tables A1–A6).

Table A1 Energy Costs of Producer 1. Numbers after the ± are 1 Standard Deviation of the Mean

Feedstock	Reclaimed Oil
Production (gal/year)	16,000
Energy cost	MJ/gal
Biodiesel	7.87 ± 0.47
Electricity	0.22 ± 0.02
Labor, direct	0.06 ± 0.01
Vehicles	0.75 ± 0.19
Processor	4.90 ± 1.23
Building	0.02 ± 0.00
Used vegetable oil	9.58 ± 2.39
Methanol	9.13 ± 2.28
Other reagents	1.14 ± 0.28
Labor, indirect	1.74 ± 0.44

Table A2 Energy Costs of Producer 2. Numbers after the ± are 1 Standard Deviation of the Mean

Feedstock	Soybean Oil
Production (gal/year)	10,000
Energy cost	MJ/gal
Diesel	2.61 ± 0.26
Electricity	0.35 ± 0.04
Labor, direct	0.01 ± 0.00
Tractor	0.46 ± 0.11
Tractor	0.09 ± 0.02
Tractor	0.20 ± 0.05
Tractor	0.36 ± 0.09
Combine	0.65 ± 0.16
Plow	0.14 ± 0.03
Land finisher	0.16 ± 0.04
Planter	0.30 ± 0.07
Roller	0.06 ± 0.01
Seed cleaner	<0.01
Press	0.06 ± 0.02
Processor	6.01 ± 1.50
Building	0.17 ± 0.04
Seed	0.43 ± 0.11
Nitrogen	4.99 ± 1.25
Phosphorus	0.89 ± 0.22
Potassium	0.57 ± 0.14
Pesticides	0.96 ± 0.24
Methanol	9.13 ± 2.28
Other reagents	1.14 ± 0.28
Labor, indirect	0.18 ± 0.04

Table A3 Energy Costs of Producer 3. Numbers after the ± are 1 Standard Deviation of the Mean

Feedstock	Sunflower Oil
Production (gal/year)	3000
Energy cost	MJ/gal
Diesel	0.93 ± 0.23
Electricity	0.07 ± 0.02
Labor, direct	<0.01
Tractor	0.06 ± 0.02

Continued

Table A3 Energy Costs of Producer 3. Numbers after the ± are 1 Standard Deviation of the Mean—cont'd

Feedstock	Sunflower Oil
Production (gal/year)	3000
Energy cost	MJ/gal
Combine	2.81 ± 0.70
Harrows	0.22 ± 0.05
Planter	0.04 ± 0.01
Press	0.09 ± 0.02
Processor	8.68 ± 2.17
Building	0.24 ± 0.06
Seed	0.44 ± 0.11
Nitrogen	4.38 ± 1.09
Phosphorus	0.31 ± 0.08
Potassium	0.61 ± 0.15
Pesticides	5.83 ± 1.46
Methanol	9.14 ± 2.28
Other reagents	1.15 ± 0.29
Labor, indirect	0.36 ± 0.09

Table A4 Energy Costs of Producer 4. Numbers after the ± are 1 Standard Deviation of the Mean

Feedstock	Sunflower Oil
Production (gal/year)	1000
Energy cost	MJ/gal
Diesel	3.82 ± 0.38
Electricity	1.19 ± 0.12
Labor, direct	0.01 ± 0.00
Tractor	0.36 ± 0.09
Tractor	0.04 ± 0.01
Combine	0.12 ± 0.03
Planter	0.04 ± 0.01
Sprayer	0.01 ± 0.00
Dump wagon	0.03 ± 0.01
Dryer	0.63 ± 0.16
Press	0.35 ± 0.09
Processor	26.73 ± 6.68

Table A4 Energy Costs of Producer 4. Numbers after the ± are 1 Standard Deviation of the Mean—cont'd	
Feedstock	Sunflower Oil
Production (gal/year)	1000
Energy cost	MJ/gal
Building	1.12 ± 0.28
Seed	0.44 ± 0.11
Nitrogen	0.66 ± 0.16
Phosphorus	0.23 ± 0.06
Potassium	0.15 ± 0.04
Pesticides	2.32 ± 0.58
Methanol	9.27 ± 2.32
Other reagents	1.15 ± 0.29
Labor, indirect	0.32 ± 0.08

Table A5 Energy Costs of Producer 5. Numbers after the ± are 1 Standard Deviation of the Mean	
Feedstock	Soybean Oil
Production (gal/year)	1500
Energy cost	MJ/gal
Diesel	2.62 ± 0.26
Electricity	0.36 ± 0.04
Labor, direct	0.01 ± 0.00
Tractor	0.48 ± 0.12
Combine	0.03 ± 0.01
Cultivator	0.07 ± 0.02
Planter	0.03 ± 0.01
Seed cleaner	0.21 ± 0.05
Seed dryer	0.29 ± 0.07
Press	0.36 ± 0.09
Processor	4.13 ± 1.03
Building	1.63 ± 0.41
Seed	0.44 ± 0.11
Manure	0.57 ± 0.14
Methanol	9.17 ± 2.29
Other reagents	1.13 ± 0.28
Labor, indirect	0.21 ± 0.05

Table A6 Energy Costs of Producer 6. Numbers after the ± are 1 Standard Deviation of the Mean

Feedstock	Sunflower Oil
Production (gal/year)	550
Energy cost	MJ/gal
Diesel	4.54 ± 0.45
Electricity	0.15 ± 0.01
Labor, direct	0.01 ± 0.00
Tractor	0.14 ± 0.04
Combine	0.39 ± 0.10
Cultivator	0.08 ± 0.02
Tine weeder	0.01 ± 0.00
Planter	0.36 ± 0.09
Plow	0.01 ± 0.00
Dryer	0.79 ± 0.20
Press	0.32 ± 0.08
Processor	3.79 ± 0.95
Building	1.47 ± 0.37
Seed	0.44 ± 0.11
Nitrogen	1.52 ± 0.38
Phosphorus	0.22 ± 0.05
Potassium	0.11 ± 0.03
Methanol	9.25 ± 2.31
Other reagents	1.15 ± 0.29
Labor, indirect	0.16 ± 0.04

ENERGY MANAGEMENT DURING FIELD PRODUCTION PRACTICES 10

Mark Hanna[1], Scott Sanford[2]

[1] *Agricultural and Biosystems Engineering Department, Iowa State University, Iowa, USA;*
[2] *Rural Energy Program, Biological Systems Engineering, University of Wisconsin, Madison, USA*

INTRODUCTION AND OVERVIEW

Almost all direct energy used during field operations is consumed by an engine during operation of tractors or self-propelled equipment such as combines, forage harvesters, or sprayers. Transmitting engine power as efficiently as possible for the task (pulling implements through soil, cutting plants, pumping, etc.) has significant effects on the amount of energy being used.

Before investigating specific ways to increase efficiency during these tasks, one should first ask if the field operation is necessary. Increasing fuel efficiency may gain 5%, 10%, 20%, or even more, but omitting the operation and leaving the tractor parked saves 100% of fuel. Some type of seeding, harvest, and weed or pest control operation is nearly always necessary. The type and frequency of tillage operations to prepare or weed a seedbed are often variable, however. Row crops, such as corn or soybeans, may be produced with one or no tillage passes prior to planting. Establishment of perennial alfalfa and small grains traditionally used primary and secondary tillage operations. New no-till seeders with better seedbed preparation and seed placement have resulted in yields equal to stands established using conventional tillage. Factors in choosing tillage operations include comfort with a specific management style along with local soil, crop, and weather conditions. Successful reduced and no-till operations are often found in the same neighborhood as fields with more aggressive tillage schemes, suggesting that options to reduce tillage frequently exist. For example, although surface cornstalks from the previous year can appear daunting, no-till soybean yields are frequently equal to those of full-width tillage systems in yield trials.

If tillage is required, consider using only a single-pass tillage system prior to planting. Strip-till systems till only a part of the field in the row zone for the subsequent crop to be planted. Ridge-till systems use row-crop cultivation for weed control to build ridges, then plant into the ridge the following year. Even when tilling the entire field area, consider why the tillage is being done and do not till any deeper than necessary. For example, a chisel plow operation at a 6- or 8-inch depth requires less drawbar pull and tractor energy than operation of a subsoiler or ripper at depths of a foot or more. Drawbar pull is directly related to tillage depth for many specific tillage implements. Aggressive primary tillage operations, such as moldboard plow or subsoiler, often require around 1.5 gal of diesel fuel or more per acre, whereas chisel plowing may require about 1 gal per acre depending on depth, soil conditions, and speed.

Other cultural or production schemes that generally increase efficiency also potentially reduce energy use for the amount of crop harvested. Narrow corn rows can help stimulate vegetative growth

Bioenergy. http://dx.doi.org/10.1016/B978-0-12-407909-0.00010-9

and increase potential harvested yield, particularly in northern areas of the Corn Belt. A leguminous cover crop can supply nitrogen to a subsequent crop, reducing the need for fertilizer nitrogen as well as field trips for transport and application.

TRACTOR USE

Because so many field operations are tractor-powered, special attention must be given to optimizing how tractor engine power is generated and transmitted for work. For higher horsepower tractors, many operations are drawbar work that involves pulling tillage and seeding implements through the soil. Efficient transfer of engine power through the tractor's transmission, along with proper attention to ballasting and tire inflation, are important issues. Other tasks require transfer of engine power through the power-take-off (PTO) shaft (e.g., baling) or hydraulic or electrical systems (e.g., some spray pumps or planter seed metering drives). Taking time to assess how tractor engine power is being transferred and used for field operations allows one to focus on management strategies that can make a difference. Major areas affecting tractor fuel consumption include ballasting/slip/tire inflation, maintenance, transmission, and tractor selection.

BALLASTING, SLIP, AND TIRE INFLATION

Excessive wheel slippage during drawbar pull operations creates an obvious waste of labor, fuel, and tractor hours. Conversely, a tractor ballasted so heavily that there is little or no wheel slip sinks too far into the soil, causing rolling resistance as the wheel tries to climb out of the track and extra energy use as tire sidewalls flex. Optimum wheel slip range for maximum tractive efficiency (equal to the ratio of drawbar power to power available at the drive axle) depends on surface conditions (Figure 10.1). Higher-horsepower tractors often have sensors allowing drive wheel slip to be monitored from the cab. Slip can be conveniently checked during fieldwork with significant drawbar loads. On tractors without

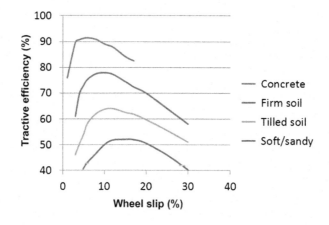

FIGURE 10.1

Tractive efficiency of transferring axle power to the drawbar as affected by wheel slip for various surface conditions.

slip measurement, slip can be approximated by measuring the distance a tractor covers during 10 wheel revolutions under drawbar load and comparing this with the distance traveled during 10 wheel revolutions without drawbar load. For example, if loaded wheel distance is 180 ft and unloaded wheel distance is 200 ft, the tractor under load is covering only 90% of the unloaded distance, or experiencing a 10% wheel slip. As a quick visual check, optimal wheel slip on soil usually occurs when wheel lug marks near the tire centerline are obliterated but lug marks near the outer edge are reasonably distinct.

If wheel slip is outside the optimal range of approximately 9–15% (depending on soil conditions) or if there are questions regarding whether the tractor is ballasted properly to use power available from the engine, the tractor operation manual or various references can be checked for advice on ballasting (e.g., Table 10.1). Specific amounts of total tractor weight per tractor horsepower are generally suggested depending on tractor style (two-wheel drive, front-wheel-assist, four-wheel drive) and operational speed. Tractors using faster field speeds (e.g., 6–7 mph instead of 4–5 mph) have optimal fuel efficiency using slightly less weight because they do not need to pull quite as much load to accomplish an equivalent amount of fieldwork in a given time. Because power is efficiently transferred from engine to drawbar over a range of slip, some variation in weight is allowed. Because most tractors spend a significant amount of time requiring only 70–90% of rated power available, weight values in Table 10.1 are near the low side of the appropriate range. Carrying extra ballast for unused horsepower during operations with light drawbar loads (e.g., pull-behind sprayer, mower/conditioner, or baler) results in small amounts of slip. If the tractor is used for long periods of time for light drawbar loads but has been optimally ballasted for full drawbar horsepower, consider removing ballast to avoid burning fuel to carry dead weight.

Just as important as total tractor weight is splitting weight appropriately between front and rear axles. The correct percentage of total weight on each axle depends on tractor style (two-wheel drive, front-wheel-assist, four-wheel drive) and whether any rear implement weight is transferred to wheels on the rear axle (pull-type or mounted implement, significant tongue weight, etc.; Table 10.2).

Table 10.1 Gross Tractor Weight

Speed	<4.5 mph	5 mph	>5.5 mph
Tractor Type			
Two-wheel drive and MFD, mechanical front-wheel drive (lb/hp)	130	120	110
Four-wheel drive (lb/hp)	110	100	90

Table 10.2 Front-to-Rear Axle Weight Ratio as Percentage of Total Weight (%Front/%Rear)

Tractor Type	Towed/Drawbar	Semimounted	Fully Mounted
Two-wheel drive	25/75	30/70	35/65
MFD	35/65	35/65	40/60
Four-wheel drive	55/45	55/45	60/40

Tires should be correctly inflated for the load they carry to maximize the ability for lugs to engage soil and develop pull. Contrary to ensuring automotive tires are well inflated to minimize fuel consumption, off-road tires operating on soft soil surfaces increase pull by exposing more of the lug surface at lower tire pressure. Overinflated tires can create excessive slip as lug surfaces near the tire sidewall do not penetrate the soil surface. Knowing the weight carried by the front and rear axles when making ballasting decisions allows the weight carried by each tire to be known. Correct pressure can be determined from tire load and inflation tables of the tire manufacturer or in the tractor operation manual. Maintaining correct rather than low pressure is important because under inflation causes premature tire failure.

MAINTENANCE

Following a prescribed schedule for tractor maintenance is often a source of pride for agricultural tractor owners. Earlier studies with owner operators indicate that, on average, many operators are timely with maintenance and filter replacement. Still, one study indicates that following scrupulous maintenance results in measurable savings.

In a University of Missouri study (Schumacher et al., 1991), 99 tractors were brought in to be tested "as is" at six locations in the state. Tractor horsepower was measured on a PTO dynamometer. Primary and secondary air and fuel filters were then replaced on each tractor before retesting tractor power. Average engine horsepower increased 3.5% after filter replacement. Factory tractor specialists indicated such an increase was normal and expected.

In a check of maintenance records on the tractors, most operators were current on filter replacement. Although some were near the end of the service interval, others were near the beginning. Such results suggest that an average increase of 3.5% power, or alternatively setting the throttle/fuel supply back 3.5% to obtain the same engine power, can be easily obtained by being scrupulously vigilant on air and fuel filter replacement.

Researchers at the time (1988–1989) estimated fuel savings to be 105 gal of diesel annually per tractor tested. Increases in average engine horsepower since then suggest that vigilant tractor maintenance may save more than this, depending on annual hours of tractor use.

TRANSMISSION

If the tractor is using only part of its power when pulling a lighter drawbar load, significant fuel savings are possible by shifting the transmission up to a higher gear and pulling the throttle back (reducing engine RPM). Pulling a sprayer or a smaller field cultivator, disk, or planter that is not well matched to the total tractor power available are common examples. Unless the implement requires PTO operation at a specific engine speed, shifting up and throttling back to reduce engine speed saves fuel. Avoid lugging the engine by only reducing speed to a point somewhat above where the engine starts to lug.

Some newer, higher-horsepower tractors manufactured in recent years offer infinitely or continuously variable transmissions using electronic control to automatically set the transmission at the most fuel efficient point for a given speed and drawbar load. Taking advantage of this new technology as a tractor is replaced saves fuel.

An example of actual fuel savings can be found from an Organization of Economic Cooperation and Development (OECD) tractor test done at the Nebraska Tractor Test Laboratory on a Case IH

Magnum 275 rated at 227 hp. Fuel use at 75% of maximum drawbar power was reduced by 8% when the transmission was shifted from 9th to 11th gear and engine speed was reduced from 2091 to 1589 rpm. In similar conditions, fuel use was reduced 21% when only 50% of drawbar power was used. Average fuel savings as indicated by tractor tests from 1979 to 2002 indicate fuel savings of 13% at 75% load and 21% at 50% load are possible by reducing engine speed and operating in a higher gear (Grisso et al., 2004).

TRACTOR SELECTION

Although matching available tractor power to the task at hand is desirable, moving a smaller tractor several miles to perform a limited task usually does not save fuel. Diesel tractors are generally efficient in fuel use for partial loads of 75% or even 50% if the throttle is reduced and a higher gear selected. For example, the tractor test data on the Case IH Magnum 275 show that fuel use efficiency is not reduced (from 100% loading) at 75% drawbar load and reduced only 14% at 50% drawbar load. The fuel efficiency of a smaller tractor properly sized to handle the reduced load is often not great enough to justify moving unless the new application is only 10–30% of tractor power and involves significant hours of use.

If a new or used tractor will be acquired, obtain and read OECD tractor tests for tractors being considered. Fuel use efficiency as measured in tests is listed as hp-h/gal under fuel consumption. Greater numbers indicate better fuel efficiency. Fuel efficiency values are listed for several levels of PTO and drawbar loading. Because average tractor use is often at a partial load, using fuel efficiency from a 50% pull and reduced engine speed may serve as a good overall estimate. When comparing tractors, make sure to compare fuel efficiency values with similar loading conditions. As with automotive fuel efficiency estimates, fuel use will vary and depend on actual operation, but test values give an indication of relative efficiency between tractors.

OTHER ISSUES

Adding new technology, such as auto-steering or auto-swath control for seed, pesticide, and fertilizer inputs, can help to avoid wasting time and materials in the field. Auto-steering allows global positioning system (GPS) information to steer the tractor and avoid excessive overlap of swaths, which wastes field time and energy. Auto-swath control allows sections across the implement swath to be turned off when previously treated areas would be overlapped. These technologies can be added to existing tractors and equipment, but it may be more cost-effective to purchase them as options when equipment is upgraded and replaced. Manufacturers are starting to embed this type of technology into new equipment, further decreasing prices. Cost for auto-steering can range from $5000 to $50,000, depending on the accuracy desired.

A study at Auburn University (Troesch et al., 2010) indicated input savings ranged from 1% to 12% for each pass across a field when using automatic section control. This study indicated that on average a 4.3% savings on seed cost could be observed for a farm, while some operations could see as high as a 7% savings. Savings are dependent upon field shape and size, with the highest benefits occurring in small, irregular shaped fields or fields containing conservation management structures such as grass waterways and terraces. Generally, automatic section control technology can pay for itself within 2 years.

Modern diesel engines require less idling time to cool the engine. Recommendations for specific equipment can usually be found in the operation manual or through the dealer. Do not let newer engines idle for periods of many minutes and waste fuel. Check with state regulatory officials regarding proper fuel storage. Vacuum/pressure relief valves protect fuel from water condensation. Shade and/or reflective white or aluminum paint on the fuel supply barrel can help avoid evaporative fuel losses.

If an engine block heater is used to assist starting during cold weather, use a timer to avoid heating for many hours before startup. A typical engine block heater can warm the engine up in about 2 h. A low-cost timer used to control swimming pool pumps can be used for most 120 V heaters; it will pay for itself in about 2 months or less, depending on the heater size and the amount of time currently being used.

Diesel fuel mixtures are different for summer and winter. Do not purchase fuel ahead in late summer if it cannot be used up before cold weather sets in. Use a fuel conditioner or fuel-line anti-freeze in equipment that is not used much during the winter.

HARVEST OPERATIONS

Significant amounts of energy are required to cut, chop, and at times pack forage material. Roughly 40% of the energy needed by a forage chopper is used for cutting. Keeping knives sharp and in close tolerance to a shear bar, along with avoiding length-of-cut any shorter than is required for good storage, has significant impact on required energy. Kernel processor rolls are used to crush and crack corn kernels and cobs to make them more digestible, but they provide no benefit for hay forage crops; therefore, they should be removed from the crop stream for chopping hay crop silage.

Sharp knives also reduce the energy required for mower/conditioners. On cutterbar mowers, knives should be in close proximity and register with guards. Conditioning roll clearance should be adjusted for adequate but not excessive pressure to wilt stems.

Many advances have been made with merging or inverting equipment for hay crop forage. Rakes and mergers can be used to promote drying by turning windrows and exposing new surfaces to air movement and sunlight. Merging windrows saves fuel by better matching harvester or baler capacity with crop yield. Harvesters and balers work more efficiently, and take less fuel per ton of crop, if they are operated close to capacity. If the harvester is picking up a single swath and not loaded near machine capacity, it will not only take longer but it also wastes fuel.

Forage blowers require significant amounts of energy to pneumatically move crop. Energy-saving points are the clearance between blower fan tip blades and the fan housing, as well as maintaining generally straight dent-free conduit pipe. Check the operation manual for proper fan tip clearance. A common recommendation is that blades should "move a nickel but not move a dime."

Following a lubrication schedule as outlined in the operator's manual for any machine affects energy use and avoids premature wear. A sharp knife and maintenance of recommended plunger clearance reduces the energy required for square balers.

Grain harvest combines use a governed engine operating gathering, threshing, separating, cleaning, and conveying components. Because the engine is loaded (with some reserve capacity) for these operations plus movement through the field and grain unloading, significant ways to conserve energy are generally limited to good engine maintenance and attention to certain functional areas with larger power requirements. A chopper attachment used near the rear of the combine requires significant power. Replacing worn chopper knives and checking belt tension to maintain shaft speed is frequently overlooked. Closely following air and fuel filter replacement guidelines along with other engine maintenance has a direct effect on fuel efficiency.

OTHER INDIVIDUAL EQUIPMENT OPERATIONS

Because a tractor powers most tillage, seeding, and many application operations and also because the total energy available from the engine rapidly dissipates if there are significant losses in transmission, drives, and at the tire/soil interface, the primary attention for energy saving should be paid to the tractor. Look for ways to combine field operations into a single pass, such as tilling and applying fertilizer with a strip-till implement or using one-pass tillage.

Many individual points regarding saving energy with tillage, seeding, application, and other types of field equipment involve good management and maintenance practices to ensure a good field job is accomplished and avoid the need for another field pass. If the objectives of the desired amount of seed, fertilizer, or pesticide being applied or soil tilled to a certain condition are not met, fuel and perhaps additional crop inputs are needed for a second pass.

On tillage equipment, worn bearings, scrapers, or cutting edges affect soil manipulation and potentially draft (drawbar pull). Good planter operation involves a prefield check of seed and fertilizer metering components, along with in-field checks of seed placement, proper operation of soil-engaging components, and periodic lubrication.

FERTILIZER AND OTHER CULTURAL AND TECHNOLOGY ISSUES AFFECTING ENERGY USE

Although energy is not purchased directly, large amounts of natural gas are used to manufacture ammonia and other nitrogen-based fertilizer. Net energy used for nitrogen fertilizer manufacture is several times greater than that used to produce phosphorous or potassium fertilizer or for pesticide manufacture. For crops requiring nitrogen fertilizer, particularly corn, the energy used to supply commercial nitrogen fertilizer can be equal to or greater than diesel fuel used in all field operations. Use no more than the recommended nitrogen rates for maximum economic return and energy efficiency.

Consider supplying corn, wheat, or other crops requiring nitrogen by means other than commercial fertilizer. Growing alfalfa, or to a lesser extent soybeans or other leguminous crops, in rotation prior to corn reduces the need for nitrogen fertilizer. Nitrogen and other nutrients in manure can be used in place of commercial nitrogen fertilizer. Manure nutrient availability varies with livestock species, manure storage, and livestock production practices, as well as separation of liquid and dry components. Although general guidelines for manure nutrient content can be found in state extension bulletins, periodic testing for nutrient analysis of applied manure provides more confidence in application.

Precision agricultural technology allows the opportunity for better, more precise management of field production. Applying the correct amount by more precise application of fertilizers and pesticides and variable seeding can help to avoid the need for subsequent field operations. GPS and geographic information system technologies allow mapping for rotations, weed and insect management, crop yield, and other management items that may reduce the need for extra field operations.

REFERENCE

Grisso, R.D., Kocher, M.F., Vaughn, D.H., 2004. Predicting tractor fuel consumption. Applied Engineering in Agriculture 20 (5), 553–561.

FURTHER READING

Field operations—general

Hanna, M., Harmon, J., Flammang, J., 2010. Limiting Field Operations–Farm Energy. Iowa State University Extension Publication. PM 2089D. Available at: http://www.extension.iastate.edu/Publications/PM2089D.pdf.

Svejkovsky, C., 2007. Conserving Fuel on the Farm. National Sustainable Agriculture Information Service/ National Center for Appropriate Technology. Available at: http://attra.ncat.org/attra-pub/PDF/consfuelfarm.pdf.

Tractor—general

Schumacher, L.G., Frisby, J.C., Hires, W.G., 1991. Tractor PTO horsepower, filter maintenance, and tractor engine oil analysis. Applied Engineering in Agriculture 7 (5), 625–629.

Staton, M., Harrigan, T., Turner, R., 2010. Improving Tractor Performance and Fuel Efficiency. Michigan State University Extension Publication.

Tractor ballasting/slip/tire inflation

Hanna, M., Harmon, J., Petersen, D., 2010. Ballasting Tractors for Fuel Efficiency–Farm Energy. Iowa State University Extension Publication. PM 2089G. Available at: http://www.extension.iastate.edu/Publications/PM2089G.pdf.

Tractor transmission

Grisso, R., Pitman, R. Gear up and Throttle Down—Saving Fuel, Virginia Tech Cooperative Extension publication, pp. 442–450. Available at: http://pubs.ext.vt.edu/442/442-450/442-450.pdf.

Tractor selection

Grisso, R.D., Vaughn, D.H., Perumpral, J.V., Roberson, G.T., Pittman, R., Hoy, R.M., 2009. Using Tractor Test Data for Selecting Farm Tractors. Virginia Tech Cooperative Extension Publication, 442-072. Available at: http://pubs.ext.vt.edu/442/442-072/442-072.pdf.

Nebraska Tractor Test Laboratory Reports. Available at: http://tractortestlab.unl.edu/index.htm.

Other tractor issues

Troesch, A., Mullenix, D.K., Fulton, J.P., Winstead, A.T., Norwood, S.H., Sharda, A., 2010. Economic analysis of auto-swath control for Alabama crop production. In Proceedings of the 10th International Conference on Precision Agriculture. Denver, CO, July, 18–21.

Fertilizer issues

Sawyer, J.E., Hanna, M., Petersen, D., 2010. Energy Conservation in Corn Nitrogen Fertilization–Farm Energy. Iowa State University Extension Publication. PM 2089D. Available at: http://www.extension.iastate.edu/Publications/PM2089I.pdf.

No-till seeding

Duiker, S.W., Myers, J.C., 2006. Steps Towards a Successful Transition to No-till. Pennsylvania State University. Bulletin No. UC192. Available at: http://pubs.cas.psu.edu/FreePubs/pdfs/uc192.pdf.

Leep, R., Undersander, D., Peterson, P., Min, D., Harrigan, T., Grigar, J., 2003. Steps to Successful No-till Establishment of Forages. Michigan State University Extension. Bulletin E-2880. Available at: http://fieldcrop.msu.edu/documents/E2880.pdf.

Schneider, Nick, 2006. No-till Planting of Alfalfa with Italian Ryegrass, Field Research Study Report. University of Wisconsin. Available at: http://winnebago.uwex.edu/ag/documents/No-TillPlantingofAlfalfawithItalianRyegrass.pdf.

Wolkowski, R., Cox, T., Leverich, J., 2009. Strip-tillage: A Conservation Option for Wisconsin Farmers. University of Wisconsin Extension. Bulletin No. A3883. Available at: http://learningstore.uwex.edu/Assets/pdfs/A3883.pdf.

Corn production

Staggenborg, S.A., et al., 2001. Narrow Row Corn Production in Kansas. Kansas State University Extension. Bulletin No. MF-2516. Available at: http://www.ksre.ksu.edu/library/crpsl2/mf2516.pdf.

Stahl, L., Coulter, J., Bau, D., 2009. Narrow-row Corn Production in Minnesota. University of Minnesota Extension. Bulletin No. M1266. Available at. http://www.extension.umn.edu/distribution/cropsystems/M1266.html.

Thomison, P., 2010. Twin-row Corn Production: 2009 Research Update in C.O.R.N. Newsletter 2010-07. Ohio State University Extension. Available at: http://corn.osu.edu/newsletters/2010/2010-07/twin-row-corn-production-2009-research-update.

STRAIGHT VEGETABLE OIL AS A DIESEL FUEL?

11

Department of Energy, Office of Energy Efficiency and Renewable Energy, U.S.

Biodiesel, a renewable fuel produced from animal fats or vegetable oils, is popular among many vehicle owners and fleet managers seeking to reduce emissions and support US energy security. Questions sometimes arise about the viability of fueling vehicles with straight vegetable oil (SVO), or waste oils from cooking and other processes, without intermediate processing. But SVO and waste oils differ from biodiesel (and conventional diesel) in some important ways and are generally not considered acceptable vehicle fuels for large-scale or long-term use.

PERFORMANCE OF SVO

Research has shown that there are several technical barriers to widespread use of SVO as a vehicle fuel.

The published engineering literature strongly indicates that the use of SVO leads to reduced engine life,[1] caused by the buildup of carbon deposits inside the engine and the buildup of SVO in the engine lubricant. These issues are attributable to SVO's high viscosity and high boiling point relative to the required boiling range for diesel fuel. The carbon buildup doesn't necessarily happen immediately upon use of SVO; it typically takes place over the long term. These conclusions are consistent across a substantial body of technical literature, including an SAE technical paper that reviewed the published data on the use of SVO in diesel engines.[2] The SAE paper states:

- Compared to No. 2 diesel fuel, all of the vegetable oils are much more viscous, are much more reactive to oxygen, and have higher cloud point and pour point temperatures.
- Diesel engines with vegetable oils offer acceptable engine performance and emissions for short-term operation. Long-term operation results in operational and durability problems.

Some investigators have explored modifying vehicles to preheat SVO prior to injection into the engine. Others have examined blends of vegetable oil with conventional diesel. These techniques may mitigate the problems to some degree but don't eliminate them entirely. Studies show that carbon buildup (coking) continues over time, resulting in higher engine maintenance costs and/or shorter

---

[1]Sidibé, S.S., Blin, J., Vaitilingom, G., Azoumah, Y., 2010. Use of crude filtered vegetable oil as a fuel in diesel engines state of the art: literature review. Renewable and Sustainable Energy Reviews 14(9):2748–2759.
[2]Babu, A.K., Devaradjane, G. Vegetable oils and their derivatives as fuels for CI engines: an overview. SAE Technical Paper No. 2003-01-0767.

Bioenergy. http://dx.doi.org/10.1016/B978-0-12-407909-0.00011-0

FIGURE 11.1

Buildup of carbon deposits in the engine as a function of the proportion of oil in the fuel.

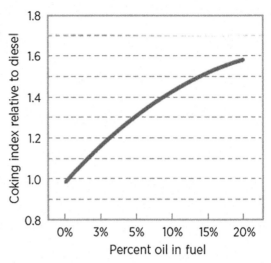

engine life. Figure 11.1 shows that the tendency to form carbon deposits increases with increasing proportions of vegetable oil blended into the fuel.[3]

Viscosity (the measure of a fuel's resistance to flow) is another important consideration related to the use of SVO. The viscosity of SVO is much higher than that of diesel fuel at normal operating temperatures.[1] Figure 11.2 illustrates the viscosities of diesel fuel and of 100% sunflower oil over a

FIGURE 11.2

Viscosities of sunflower oil and conventional diesel fuel as a function of temperature.

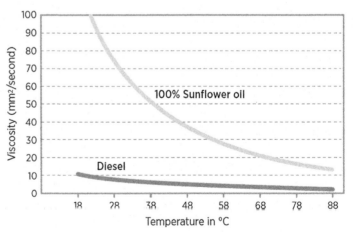

---

[3]Figure 1 adapted from Jones, Samuel T.; Peterson, Charles L.; Thompson, Joseph C. Biological and Agricultural Engineering Department, University of Idaho, Moscow, Idaho. "Used Vegetable Oil Fuel Blend Comparisons Using Injector Coking in a DI Diesel Engine." Presented at 2001 ASAE Annual International Meeting, Sacramento, Calif., July 30–August 1, 2001. SAE Paper No. 01–6051.

range of temperatures. High fuel viscosity can cause premature wear of the fuel pumps and injectors. It can also dramatically alter the structure of the fuel spray coming out of the injectors: increasing droplet size, decreasing spray angle, and increasing spray penetration. These effects tend to increase wetting of the engine's internal surfaces, thereby diluting the engine lubricant and increasing the tendency for coking.

The long-term effect of using SVO in diesel engines equipped with modern emission control systems is also a matter of concern. Buildup of fuel in the lubricant is more significant in these engines—even for petroleum diesel—and would likely be severe with SVO. In general, these systems were not originally designed to accommodate the properties of SVO, and they can be seriously damaged or poisoned by out-of-spec or contaminated fuel.

BIODIESEL: FUEL MADE FROM SVO

Biodiesel is an alternative fuel that can be made from SVO or other fats in a chemical process called transesterification, which involves a reaction with methanol, using caustic soda (sodium hydroxide) as a catalyst. Biodiesel has substantially different properties than SVO and results in better engine performance. In particular, biodiesel has a lower boiling point and viscosity than SVO.

Biodiesel is most commonly used as a blend with petroleum diesel fuel. All manufacturers of diesel vehicles and engines have approved the use of B5 (a blend containing 5% biodiesel and 95% petroleum diesel), and some approve the use of blends up to B20 (20% biodiesel and 80% petroleum diesel) or higher.

To ensure good performance in engines, biodiesel must meet quality specifications developed by ASTM International. ASTM Specification D6751 is for pure biodiesel (B100), used for blending with petroleum diesel fuel. Biodiesel that meets ASTM D6751 is legally registered with the U.S. Environmental Protection Agency. Blends up to B5 may be found in conventional diesel fuel without additional labeling at the pump. Properties of these low-level blends are covered by the diesel fuel ASTM Specification D975. For blends of biodiesel ranging from B6 to B20, a separate specification exists, ASTM D7467, and pump labeling is required to inform the consumer that a biodiesel blend is being sold.

For a complete list of ASTM biodiesel requirements, see the Biodiesel Handling and Use Guidelines at www.nrel.gov/vehiclesandfuels/pdfs/43672.pdf. In addition, the biodiesel industry has instituted a quality assurance program for biodiesel producers and marketers. To learn more about the BQ-9000 program, visit bq-9000.org.

WHERE CAN I GET MORE INFORMATION?

- The U.S. Department of Energy's Alternative Fuels Data Center, at *afdc.energy.gov*, contains a vast collection of information on alternative fuels and alternative fuel vehicles.
- The National Biodiesel Board is the national trade association representing the biodiesel industry. Its Website, biodiesel.org, serves as a clearinghouse of biodiesel-related information.

Credit:
Straight Vegetable Oil as a Diesel Fuel?
January 2014.
U.S. Department of Energy.
Office of Energy Efficiency and Renewable Energy.
DOE/GO-102,014-3449 www.cleancities.energy.gov/publications<http://www.cleancities.energy

CELLULOSIC ETHANOL—BIOFUEL BEYOND CORN

Nathan S. Mosier

Department of Agricultural and Biological Engineering, Purdue University,
West Lafayette, IN, USA

INTRODUCTION

Fuel ethanol production in the U.S. is expected to exceed 7.5 billion gallons before 2012. This represents a *doubling* of ethanol production from 2004, which consumed approximately 10% of the corn produced in the U.S. in that year. Increased demands for domestically produced liquid fuel is increasing competition between animal feed and fuel production uses of corn. Cellulosic feedstocks (wheat straw, corn stover, switch grass, etc.) can also be converted to ethanol. Overcoming the technological and economic hurdles for using cellulose to produce liquid fuel will allow the U.S. to meet both food and fuel needs.

CELLULOSE AS ETHANOL FEEDSTOCK

Cellulose is a polymer of sugar. Polymers are large molecules made up of simpler molecules bound together much like links in a chain. Common, everyday biological polymers include cellulose (in paper, cotton, and wood) and starch (in food). Cellulose is a polymer of glucose, a simple sugar that is easily consumed by yeast to produce ethanol (Mosier and Illeleji, 2006).

Cellulose is produced by every living plant on the earth, from single-celled algae in the oceans to giant redwood trees. This means that cellulose is the most abundant biological molecule in the world.

A study completed by the U.S. Department of Agriculture and the U.S. Department of Energy concluded that at least one billion tons of cellulose in the form of straw, corn stover, other forages and residues, and wood wastes could be sustainably collected and processed in the U.S. each year. This resource represents an equivalent of 67 billion gallons of ethanol, replacing 30% of U.S. gasoline consumption (U.S. Department of Energy Biofuels: 30% by 2030 Website).

Bioenergy. http://dx.doi.org/10.1016/B978-0-12-407909-0.00012-2

Plants use cellulose as a strengthening material, much like a skeleton that allows plants to stand upright and grow toward the sun, withstand environmental stresses, and block pests. People have used cellulose for centuries in paper, wood, and textiles (cotton and linen). If cellulose chains are broken down into the individual links, the released sugar can be used to make ethanol. This ethanol can then be purified using the same technology as corn-based ethanol production (Mosier and Illeleji, 2006). A number of technological advances are currently under development to make this approach to biofuels economical.

CHALLENGES IN CELLULOSIC ETHANOL

Using technology available today, cellulose can be converted into ethanol. The major difference between cellulosic ethanol and grain ethanol is the technology at the front end of the process. The technology for fermentation, distillation, and recovery of the ethanol are the same (Mosier and Illeleji, 2006).

The major challenges (Figure 12.1) are linked to reducing the costs associated with the production, harvest, transportation, and up-front processing in order to make cellulosic ethanol competitive with grain-based fuel ethanol and gasoline (Eggeman and Elander, 2005). The major processing challenges are linked to the biology and chemistry of the processing steps.

Advances in biotechnology and engineering will likely make significant impacts toward achieving the goals of improving efficiency and yields in processing plant material to ethanol.

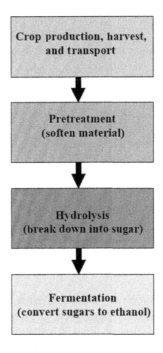

FIGURE 12.1

Major challenges in producing cellulosic ethanol.

PLANT BIOTECHNOLOGY

Most of the recent biotechnological advances in crop genetics have focused on grain production. Bioenergy crops will be grown for the inedible portion of the plant (cellulose-rich cell wall material in leaves and stems) and not primarily for grain.

The tools of biotechnology are beginning to unlock the secrets of how plant cell walls are synthesized (Yong et al., 2005; Humphreys and Chapple, 2002). This knowledge is being used to alter genes in plant that make them more productive in cellulose production and make the cellulose more easily converted to biofuels (Vermerris et al., 2007). In summary, plant genetics research and biotechnology are giving researchers the tools to increase the agricultural yield of cellulosic plant material that is tailor-made for conversion to biofuels (Ragauskas et al., 2006).

PRETREATMENT

Pretreatment is the name given to processing done to the plant material before it is broken down into simple sugars (hydrolysis). This is done to soften the cellulosic material to make the cellulose more susceptible to being broken down. Thus, the subsequent hydrolysis step is more efficient because the breakdown of cellulose into sugar is faster, higher in yield, and requires fewer inputs (enzymes and energy).

The leading pretreatment technologies under development use a combination of chemicals (water, acid, caustics, and/or ammonia) and heat to partially break down the cellulose or convert it into a more reactive form (Mosier et al., 2005). Better understanding of the chemistry of plant cell walls and the chemical reactions that occur during pretreatment is leading to improvements in these technologies, which lower the cost for producing ethanol (Eggeman and Elander, 2005).

HYDROLYSIS

In the hydrolysis step, the cellulose and other sugar polymers are broken down into simple sugars through the action of biological catalysts called enzymes. These enzymes are produced by fungi that feed upon dead plant matter in the natural world.

Our understanding of how these enzymes work is aiding efforts in determining what collection of enzymes working together can best hydrolyze cellulose in industrial applications (Mosier et al., 1999). Biotechnology has allowed these enzymes to be produced more cheaply and with better properties for use in biofuel applications (Knauf and Moniruzzaman, 2004).

FERMENTATION

The equipment and processing technology for producing ethanol from cellulose is the same as for producing ethanol from grain. In addition, yeast used in grain-based ethanol production can use glucose obtained from cellulose. However, only about 50–60% of the sugar derived from cellulose-rich plant materials is glucose. The remaining 40–50% is largely a sugar called "xylose," which naturally occurring yeast cannot ferment to ethanol.

Biotechnology has been used to genetically modify yeast (Sedlak and Ho, 2004) and some bacteria (Ohta et al., 1991) to allow them to produce ethanol from both glucose and xylose.

These advances increase the amount of ethanol that can be produced from a ton of cellulosic material by as much as 50%. Additional improvements, based upon understanding the basic metabolism and genetics of microorganisms, are underway to increase the efficiency and rates that the microorganisms convert xylose to ethanol (Bro et al., 2006).

CONCLUSION

Fuel ethanol produced from nonfood plant materials can contribute significantly to reducing the use of petroleum for automobile transportation in the U.S. Advances in basic science and technology across a number of disciplines are required to make the process of converting this abundant raw material into fuel economically competitive with grain-based ethanol and gasoline. Applications of biotechnology to crops, industrial microorganisms, and industrial biocatalysts (enzymes), together with bioprocess engineering to integrate and optimize production technologies, are likely to make production of cellulose-based ethanol a reality.

ACKNOWLEDGMENT

This paper was prepared for Purdue University Extension BioEnergy Series (technical ID# ID-335).

REFERENCES

Bro, C., Regenberg, B., Forster, J., Nielsen, J., 2006. *In silico* aided metabolic engineering of *Saccharomyces cerevisiae* for improved bioethanol production. Metabolic Engineering 8 (2), 102–111.

Eggeman, T., Elander, R.T., 2005. Process and economic analysis of pretreatment technologies. Bioresource Technology 96 (18), 2019–2025.

Humphreys, J.M., Chapple, C., 2002. Rewriting the lignin roadmap. Current Opinion in Plant Biology 5 (3), 224–229.

Knauf, M., Moniruzzaman, M., 2004. Lignocellulosic biomass processing: a perspective. International Sugar Journal 106 (1263), 147–150.

Mosier, N.S., Hall, P., Ladisch, C.M., Ladisch, M.R., 1999. Reaction kinetics, molecular action, and mechanisms of cellulolytic proteins. Advances in Biochemical Engineering/Biotechnology 65, 24–40.

Mosier, N., Wyman, C., Dale, B., Elander, R., Lee, Y.Y., Holtzapple, M., Ladisch, M.R., 2005. Features of promising technologies for pretreatment of lignocellulosic biomass. Bioresource Technology 96 (6), 673–686.

Mosier, N., Illeleji, K., 2006. How Fuel Ethanol Is Made from Corn. ID-328. Purdue University Cooperative Extension Service.

Ohta, K., Beall, D.S., Mejia, J.P., Shanmugam, K.T., Ingram, L.O., 1991. Genetic-improvement of *Escherichia coli* for ethanol-production – chromosomal integration of *Zymomonas mobilis* genes encoding pyruvate decarboxylase and alcohol dehydrogenase-II. Applied and Environmental Microbiology 57 (4), 893–900.

Ragauskas, A.J., Williams, C.K., Davison, B.H., Britovsek, G., Cairney, J., Eckert, C.A., Frederick, W.J., Hallett, J.P., Leak, D.J., Liotta, C.L., Mielenz, J.R., Murphy, R., Templer, R., Tschaplinski, T., 2006. The path forward for biofuels and biomaterials. Science 311 (5760), 484–489.

Sedlak, M., Ho, N.W.Y., 2004. Production of ethanol from cellulosic biomass hydrolysates using genetically engineered *Saccharomyces* yeast capable of cofermenting glucose and xylose. Applied Biochemistry and Biotechnology 113–116, 403–405.

U.S. Department of Energy Biofuels: 30% by 2030. http://www.doegenomestolife.org/biofuels/.

Vermerris, W., Saballos, A., Ejeta, G., Mosier, N.S., Ladisch, M.R., Carpita, N.C., 2007. "Molecular breeding to enhance ethanol production from corn and sorghum stover". Crop Science 47, S142–S153.

Yong, W.D., Link, B., O'Malley, R., Tewari, J., Hunter, C.T., Lu, C.A., Li, X.M., Bleecker, A.B., Koch, K.E., McCann, M.C., McCarty, D.R., Patterson, S.E., Reiter, W.D., Staiger, C., Thomas, S.R., Vermerris, W., Carpita, N.C., 2005. Genomics of plant cell wall biogenesis. Planta 221 (6), 747–751.

BIOHEAT

13

Robert G. Hedden

*President- Hedden Company, a full service Efficient Heating consulting and training firm;
Executive Director-Oilheat Manufacturers Association; Director of Education-National Oilheat Research Alliance.*

BIODIESEL

Biodiesel is a nontoxic, biodegradable, combustible liquid fuel. It is a domestic, renewable fuel derived from natural vegetable and animal oils. It has an American Society for Testing and Materials (ASTM) specification D6751 for pure biodiesel (B100). Biodiesel is comprised of mono-alkyl esters of long-chain fatty acids. Current feedstocks (Figure 13.1) for the creation of biodiesel are soybean, canola, sunflower, mustard, and rapeseed oils, as well as waste cooking grease, and trap grease, tallow, and animal fats such as fish oil. The protocols for handling biodiesel are similar to handling the oils and fats from which it is made.

Biodiesel is the most diverse fuel available. It is made from regionally available renewable bio-resources that are abundant in the United States. The U.S. Environmental Protection Agency (EPA) recently confirmed biodiesel as an "advanced biofuel" that meets the demands of the EPA's Renewable Fuel Standard. Future production from algae and bacteria feedstocks promise to make this fuel even better.

Biodiesel is manufactured by reacting 100 lb vegetable oil, animal fat, and/or waste grease with 10 lb alcohol and sodium as a catalyst to yield 100 lb biodiesel and 10 lb glycerin. This process, called *transesterification* (Figure 13.2), enjoys the highest energy balance of any renewable fuel. For every

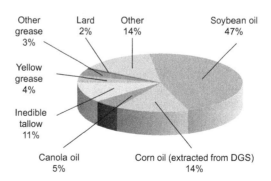

Biodiesel feedstocks
in marketing:
2012–2013

Other grease 3%
Lard 2%
Other 14%
Soybean oil 47%
Yellow grease 4%
Inedible tallow 11%
Canola oil 5%
Corn oil (extracted from DGS) 14%

FIGURE 13.1

Biodiesel feedstocks.

Courtesy of the National Biodiesel Board.

FIGURE 13.2

Transesterification.

Courtesy of the National Biodiesel Board.

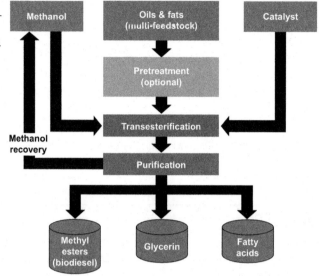

unit of energy used to produce biodiesel you get 5.54 units of energy back. The vegetable oils and animal fats used for biodiesel feedstocks are not produced specifically for biodiesel; they are a minor by-product of food production. So, they do not suffer from the fuel-from-food stigma the ethanol does.

BIOHEAT

Bioheat fuel is a blend of 95%–98% heating oil (#2 oil, ASTM D396) and 5%–2% B100 biodiesel (ASTM D6751) that creates a heating fuel equivalent to and virtually indistinguishable from #2 oil. A 5% blend is called *B5*. Bioheat can be used in oil burners with little or no modifications to the equipment or operating practices and protocols. Although the flashpoint is higher, ignition with blends of 5% or less is no problem. The viscosity is higher yet still within ASTM limits for heating oil, but flow rate and atomization are similar. Bioheat creates slightly less deposits on the heat exchanger thanks to the reduced sulfur levels. B100 has 10%–12% oxygen, so it increases excess air to the flame. B100's heating value is 118,170 BTU/gal. It has a slightly greater density and greater cloud and pour points than #2 oil.

The National Oilheat Research Alliance is working with Penn State, the National Biodiesel Board, and Brookhaven National Laboratory on tests to determine how quickly they can recommend to UL and ASTM that biofuel concentrations be increased. Currently, we are introducing data and rationale to ASTM to move to B20. We hope to have ASTM and UL approval by June 2015. The oil heat industry goal is to be at B100 and a zero carbon footprint by 2040. There is B100 heating equipment available on the market today. The issues to be addressed before going to a greater concentration are cold-weather performance of the fuel, and determining retrofit strategies to bring the legacy heating equipment installed in the field up to the point at which it can burn B100 safely and reliably.

Bioheat has strong public appeal as a renewable fuel. It has good lubricity, which helps with low-sulfur fuels. It increases our fuel source diversity, reducing our dependence on foreign petroleum, and is a potentially huge market for American agriculture.

FIGURE 13.3

BQ-9000.

Raw or refined plant oil, or recycled greases that have not been processed into biodiesel, are not biodiesel and should be avoided. Research shows that plant oils or greases used in concentrations as low as 10–20% can cause serious operations and maintenance problems. These problems are caused mostly by the greater viscosity, or thickness, of the raw oils (~ 40 mm^2/s) compared with that of the heating oil. Through the process of converting plant oils or greases to biodiesel by transesterification, the viscosity of the fuel is reduced to values similar to conventional diesel fuel (biodiesel values are typically 4–5 mm^2/s).

Biodiesel is a legally registered fuel and fuel additive with the U.S. EPA. The EPA registration includes all biodiesel that meets the ASTM biodiesel specification, ASTM D6751.

BQ-9000

The National Biodiesel Accreditation Commission has created a voluntary quality assurance program that accredits producers, marketers, and laboratories. BQ-9000 (Figure 13.3) promotes the success of biodiesel, warrants that the quality is maintained at ASTM D6751 specifications, and helps monitor quality throughout the distribution system.

BIOHEAT AND ITS PROPERTIES

Bioheat is made from blending #2 heating oil and up to 5% biodiesel. The resulting fuel is virtually indistinguishable from regular #2 oil. Bioheat operation characteristics and maintenance requirements are the same as petroleum #2 oil.

PETROLEUM

Heating oil is a fossil fuel, as is natural gas, propane, and coal. We call them *fossil fuels* because they are all made from the prehistoric plants and animals that form fossils. Fossil fuels are hydrocarbons.

Biodiesel is also a hydrocarbon; the only real difference between it and petroleum is it does not require 10,000 years of pressure and heat to turn organic material into a fuel, as petroleum does.

Hydrocarbon molecules are the building blocks of life. Everything that is or was ever alive is made of molecules composed of hydrogen and carbon atoms. Carbon is normally a solid that, if not totally burned, becomes smoke and soot. Hydrogen is the lightest gas and the smallest atom. Bonded together, hydrocarbons can be a gas such as propane, a liquid such as heating oil, or a solid such as candle wax. The hydrocarbon gases contain more hydrogen; the liquids and solids contain more carbon.

Petroleum comes out of the ground in the form of crude oil and wet gas. They are a complex mix of compounds consisting mostly of the elements carbon and hydrogen. Sulfur and nitrogen are also bound to some of these hydrocarbon compounds. This mixture of molecules is separated at the oil refinery (Figure 13.4) by distillation into their various boiling ranges. Heating oil, diesel fuel, jet fuel, and kerosene are classified as middle distillates because their boiling range is in the middle of the sweep of petroleum products separated in the distillation process. Heating oil produced directly by the distillation process is called a *straight-run product*. Heating oil is also produced by cracking, catalytically and thermally, heavier, more complex molecules into the small heating oil hydrocarbon molecules. This is called *cracked product*. Blending a mixture of various middle distillate products together also creates heating oil. Blending biodiesel into this mix makes it Bioheat.

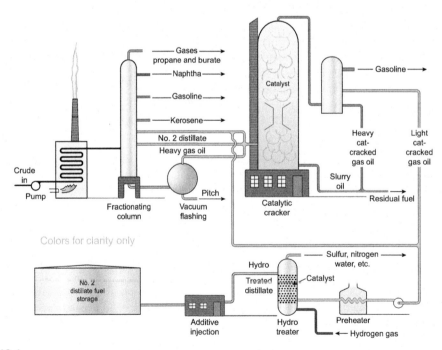

FIGURE 13.4

Oil refinery.

Courtesy of the National Oilheat Research Alliance.

PROPERTIES OF HEATING OIL AND BIOHEAT

Bioheat should be treated with the same care and technical skill required for handling heating oil (Figures 13.5 and 13.6). There are technical and service aspects of Bioheat, such as material contamination factors and cold-flow properties, that should be considered and are outlined in this section.

AMERICAN SOCIETY FOR TESTING AND MATERIALS

ASTM publishes industry specifications for many different materials, including petroleum products. The specification for fuel oils is ASTM D396. This standard sets the minimum specifications for the fuel. ASTM has defined D396 as containing up to 5% ASTM D6751 biodiesel.

FLASHPOINT

The flashpoint of fuel oil is the maximum temperature at which it can be stored and handled safely without serious fire hazard. The ASTM-specified flashpoint for #1 and #2 oil is 100 °F (minimum). When oil is heated to its flashpoint, some of the hydrogen flashes off but the fuel does not continue to burn. The flashpoint for B100 is slightly greater, at 150 °F (Figure 13.7).

IGNITION POINT

The ignition or fire point is the lowest temperature at which rapid combustion of a fuel takes place in air. It is the temperature at which all the fuel has been heated and vaporized sufficiently to continue to burn for at least 5 seconds. For #2 oil and Bioheat, the ignition point is more than 500 °F.

Heating oil's physical properties (no. 2 oil)

ASTM Specification:	D 396
Flashpoint:	100°F minimum (37.8 °C)
Ignition Point:	>500°F (260 °C)
Pour Point:	17°F (8.3 °C)
Cloud Point:	Pour point temp. plus 10–20° (F)
Viscosity:	Varies: increases as temp. drops
Water/Sediment:	ASTM allowable amount of H_2O: 0.1% (Water content is usually much lower in practice)
Sulfur Content:	Ranges from 0.5% to 0.05% (5000 to 500 parts per million); ASTM maximum allowable amount is 0.5%.
Color:	Colorless, but heating oil is dyed red for tax compliance reasons. Color resembles cranberry juice.
BTU Content:	139,000 (approx.)

FIGURE 13.5

Properties of heating oil.

Courtesy of National Oilheat Research Alliance.

SPECIFICATION FOR
BIODIESEL (B100) – ASTM 6751-11a

#Biodiesel (B100) and the petroleum diesel must meet their respective ASTM specifications before blending.

Property	ASTM Method	Limits	Units
Calcium & Magnesium, combined	EN 14538	5 maximum	ppm (g/g)
Flash Point (closed cup)	**D 93**	**93 minimum**	**°C**
Alcohol Control (one to be met)			
1. Methanol Content	EN 14110	0.2 maximum	mass %
2. Flash Point	D93	130 minimum	°C
Water & Sediment	**D 2709**	**0.05 maximum**	**% vol.**
Kinematic Viscosity, 40 C	D 445	1.9 – 6.0	mm^2/sec.
Sulfated Ash	D 874	0.02 maximum	% mass
Sulfur			
S 15 Grade	**D 5453**	**0.0015 max. (15)**	**% mass (ppm)**
S 500 Grade	**D 5453**	**0.05 max. (500)**	**% mass (ppm)**
Copper Strip Corrosion	D 130	No. 3 maximum	
Cetane	D 613	47 minimum	
Cloud Point	**D 2500**	**report**	**°C**
Carbon Residue 100% sample	D 4530*	0.05 maximum	% mass
Acid Number	**D 664**	**0.5 maximum**	**mg KOH/g**
Free Glycerin	**D 6584**	**0.020 maximum**	**% mass**
Total Glycerin	**D 6584**	**0.240 maximum**	**% mass**
Phosphorus Content	D 4951	0.001 maximum	% mass
Distillation	D 1160	360 maximum	°C
Sodium/Potassium, combined	EN 14538	5 maximum	ppm (g/g)
Oxidation Stability	**EN 15751**	**3 minimum**	**hours**
Cold Soak Filtration	**D7501**	**360 maximum**	**seconds**
For use in temperatures below -12 °C	**D7501**	**200 maximum**	**seconds**

BOLD = BQ-9000 <u>Critical Specification Testing</u> Once <u>Production Process Under Control</u>

*   The carbon residue shall be run on the 100% sample.
#   A considerable amount of experience exists in the US with a 20% blend of biodiesel with 80% diesel fuel (B20).
      Although biodiesel (B100) can be used, blends of over 20% biodiesel with diesel fuel should be evaluated on a
      case-by-case basis until further experience is available.

FIGURE 13.6

Specifications for B100.

Courtesy of the National Biodiesel Board.

POUR POINT

Pour point (Figure 13.8) is the lowest temperature at which fuel flows. Below this point it turns to a waxy gel. The ASTM standard for untreated #2 oil is 17 °F. Additives or kerosene are added to heating oil during the winter to ensure it flows.

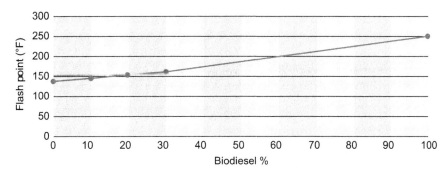

FIGURE 13.7

Bioheat flash point.

Source: Brookhaven National Labs.

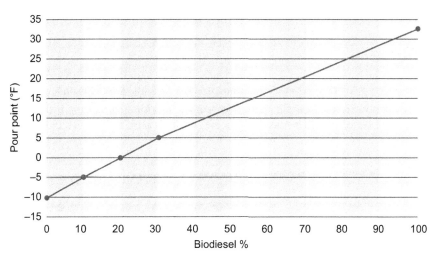

FIGURE 13.8

Bioheat pour point.

Source: Brookhaven Nation Labs.

CLOUD POINT

Cloud point is the temperature at which wax crystals begin to form in the fuel, typically 10–20 °F greater than the pour point. These crystals can clog filters and strainers, restricting fuel flow. Increasing the temperature causes the crystals to go back into solution. ASTM does not list a specification for cloud point for heating oil. Both pour point and cloud point affect winter performance and could cause problems if the fuel is not treated properly.

VISCOSITY

Viscosity is the thickness of the fuel—its resistance to flow. Grease has a high viscosity. Gasoline has a low viscosity; it flows easily. Heating oil's viscosity changes dramatically with temperature. As the temperature decreases, viscosity increases. Normally, the temperature of oil in a basement tank is 60 °F. In the winter, you might get a delivery of 5 °F oil. The colder oil has a higher viscosity, and burner performance is affected until the fuel warms. Cold oil causes poor atomization, delayed ignition, noisy flames, pulsation, and possible sooting. Bioheat has a slightly higher viscosity than heating oil (Figure 13.9).

WATER AND SEDIMENT

Accumulation of water in tank bottoms is undesirable because it leads to the formation of sludge and ice. Sludge is largely oil and water. Water and oil usually do not mix, but if organic sediment is present in the fuel, it acts as a binder to stabilize the mix of fuel and water, and forms a white milky substance that does not burn. The ASTM D396 limit for water is 0.1%, but most fuel sold has much less water. Unfortunately, water can get in the system from condensation, leaks in lines, or missing vent and fill caps.

Bioheat is able to hold more water in suspension than #2 oil. So, avoiding condensation, checking for water leaks into the tank, and removing water in the tank are even more important for Bioheat.

SULFUR CONTENT

Sulfur exists in varying degrees in all fossil fuels. The sulfur content of heating oil ranges from 0.5% to 0.05%. Biodiesel has virtually no sulfur in it. When the sulfur burns it mixes with oxygen and forms sulfur dioxide. It also creates a small amount of sulfur trioxide. The sulfur trioxide reacts with the water vapor in the combustion gases to create sulfuric acid aerosol. If the acid condenses

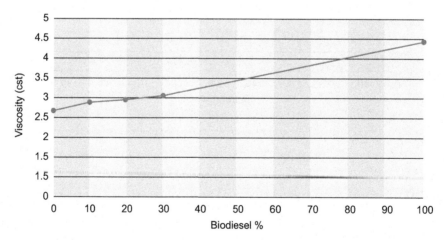

FIGURE 13.9

Bioheat viscosity.

Source: Brookhaven National Labs.

(at 150–200 °F), it adheres to the heat exchanger, flue pipe, and inside of the chimney. It creates a scaly yellow-to-red crust. Scale typically makes up 50% of deposits on the heat exchanger. It downgrades efficiency by 1–4% during the year. It also blocks flue passages, restricting airflow and increasing smoke and soot. Using ultralow-sulfur fuel and blending with biodiesel all but eliminates scale and soot formation on heat exchanger surfaces. The efficiency does not degrade during the heating season, which saves energy. It also results in decreased appliance service.

COLOR

Heating oil and Bioheat are dyed red to differentiate them from on-road diesel fuel, for tax compliance reasons. Problems with the fuel are not indicated by the darkness of the color. A murky appearance, however, may indicate a fuel quality problem.

FUEL-RELATED SERVICE CALLS

The oil heat industry's top four service priorities are safety improved reliability, maximized efficiency, and reduced heating equipment service costs. A significant number of unscheduled no-heat service calls are caused by inconsistent fuel quality, fuel degradation, and contamination.

BACKGROUND

Heating oil varies during the season. Petroleum comes from all around the world—from Malaysia to North Dakota. Biodiesel feedstocks vary widely. Every refinery is slightly different, and each of the products they make is slightly different. A great deal of our product is created by blending various fuels together to meet the rather loose definition for #2 heating oil laid out in the ASTM D396 specification. As a result, the fuel in the customer tank may be a mixture of a variety of fuels. In addition, over time, fuel degrades, water may enter the system, and bacteria have an opportunity to grow. Good house-keeping, installation of filters on all burners, and an aggressive problem-tank replacement program can cut fuel-related service calls dramatically.

POTENTIAL PROBLEMS

As oil tanks age, rust and sediments build up in the tank. The age of the fuel in the tank can also be a problem. Oil and Bioheat have a finite shelf-life and they break down over time. Another problem is the size and speed of delivery. Filling a tank kicks up all the sediments and rust at the bottom of the tank, which leads to plugged lines, filters, and nozzles. The solutions here are not to let the level of fuel in the tank get too low, to slow down the pumping speed of the truck, and to use diverters on the "blow or whistle pipes" (underground fill pipes) when filling underground tanks.

MAJOR FACTORS IN FUEL DEGRADATION

1. Chemistry of the fuel
   a. Heat, which causes the oxidation of organics
   b. The presence of sulfur and nitrogen, which hasten degradation
   c. Corrosion, creating iron oxides (rust)

    d. Gels caused by mercaptan sulfur
    e. Incompatible fuels
2. Microbiological effects
3. The tank and its environment—moisture, fuel circulation resulting from temperature differences, and so on
4. Lack of tank maintenance, and poor design and installation that prevent adequate tank inspection; withdrawal of water and sediment; improper or no filtration; and lack of corrosion protection.

FUEL STABILITY

As mentioned, fuels degrade over time. If the fuels are contaminated, they degrade even more quickly. The stability of heating oil depends a great deal on the crude oil sources from which it was made; the severity of the refinery process; the use of additives, including biodiesel; and any additional refinery treatments. Fuels that are stored for long periods of time and subjected to temperature extremes may form excessive amounts of sediments and gums that can plug filters, strainers, and nozzles.

WATER PROBLEMS

The worst fuel problem is water in the oil tank. Water enters the tank in the following ways:

1. Condensation
2. Broken tank gauge (outside tank)
3. Loose fill or vent fittings and missing caps
4. Directly from delivery trucks
5. Leaking vent, fill pipes, or tank
6. Pumping old oil into a new tank

SLUDGE HAPPENS

Sludge (Figure 13.10) is a combination of water, colonies of bacteria, degraded fuel, and other contaminants such as sand, grit, and rust. The ability of bacteria to grow almost anywhere and reproduce amazingly fast makes it an all-too-common problem. The bacteria live in the water and eat the fuel. They break down the fuel into hydrogen, carbon dioxide, and a carbon-rich residue. The bacteria also create sticky slime or gum to protect themselves. Scientists call this slime *biofilm*. This deterioration of fuel is a natural occurrence that appears in all tanks unless proper maintenance is performed. The sludge grows at the oil–water interface and, when stirred, can lead to serious and recurring service problems, most notably plugged fuel lines, filters, strainers, and nozzles. Sludge is acidic and destroys the tank from the inside.

To reduce sludge formation,

- Never pump oil from one tank to another. You may be transferring tank-killing sludge.
- Slow down delivery rates. High-pressure filling can stir up existing sludge, causing it to be drawn into the oil line.
- Check the tank routinely for water. After you have removed the water, clean the sludge from the tank, if possible, and treat the tank with a fuel-conditioning additive.
- Draw the fuel from the bottom of the tank. Water condenses and collect in all tanks, so it is best to draw off the water as it forms. Small amounts burn off during the combustion process. Allowing water to accumulate creates conditions favorable for the formation of sludge.

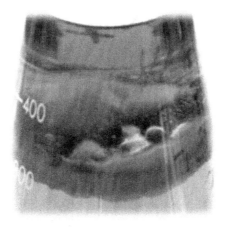

FIGURE 13.10

Sludge.

Courtesy National Biodiesel Board.

The exception to this rule is outdoor aboveground tanks. In cold weather, the water in the bottom suction line may freeze, causing a blockage and no heat. It appears that the best solution to this problem is to run the suction line into one of the top tapings on the aboveground tank, use a floating suction line device, and, occasionally, remove the water that condenses in the bottom of the tank.

LOW-TEMPERATURE PERFORMANCE

As fuel gets cold, several bad things happen:

1. Any water in the fuel freezes, plugging lines and filters.
2. The viscosity of the fuel begins to increase, causing burner operation problems.
3. Wax crystals begin to form in the fuel. This wax, or paraffin, is a natural component of heating oil. It plugs nozzles and filters.

Oil temperature is the main factor in changing fuel viscosity (Figure 13.11). As the temperature of the fuel decreases, the viscosity increases. The fuel gets thicker, which can cause a smoky fire.

HOW TO DEAL WITH "FROZEN" TANKS AND OIL LINES

Cold-flow additives called *pour point depressants* can help avoid "frozen" lines, but after the tank or lines have frozen or waxed, other solutions are needed. The best solution is to top off the tank with kerosene. The agitation of the fuel in the tank caused by the kerosene delivery and the solvency of kerosene break up and dissolve the wax crystals. You may also have to remove the filter, temporarily convert to a one-pipe system, and heat the suction line with a hair dryer or heat lamp. If you are unable to arrange for a delivery, some technicians report that adding as little as 5 gal kerosene can help. Others report having success "shocking" the tank with a pour point depressant.

FIGURE 13.11

Viscosity changes with temperature.

Courtesy National Oilheat Research Alliance.

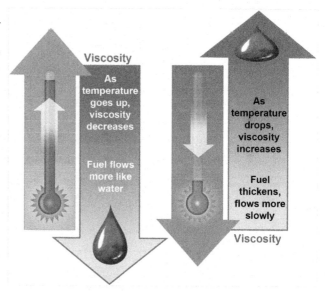

Be very careful with heat tapes. If you wrap a heat tape over itself, it can burn through its own insulation, causing a short that can result in a fire. The insulation on the wires can also crack with age and exposure to the elements, creating the potential for a fire.

BIOHEAT AND COLD WEATHER

Like regular heating oil, Bioheat can gel at low temperatures, and water in fuel can freeze. However, with good fuel management and use of ASTM D6751 and ASTM D396 blends with cold-flow additives and/or kerosene, you can have a very reliable cold weather fuel.

TO ENSURE A PROBLEM-FREE WINTER

- Buy only quality ASTM specification fuels.
- Cold-flow properties can be improved by blending Bioheat with kerosene (#1 oil), which has excellent cold-flow properties. For outdoor aboveground tanks in colder climates, at least a 25% kerosene blend is recommended.
- A number of additives are available for improving the low temperature of Bioheat. These additives include pour point depressants, flow improvers, and wax antisettling additives. All additives must be blended according to manufacturers' recommendations.

THREE STEPS TO SUCCESSFUL BIOHEAT MANAGEMENT
CHECK QUALITY

First, look for BQ-9000 suppliers when ordering Bioheat, and accept ASTM D6751 and ASTM D396 fuel only. Doing so avoids poor fuel quality and resulting plugged nozzles, filters, and strainers. It is okay to request documentation from your fuel distributor.

MAINTAIN AS USUAL

Using a Bioheat blend (B5) requires the same maintenance protocols as heating oil. The fuel–air mixtures and pump pressure settings are standard. Make sure the tanks you are filling are not contaminated with water, old fuel, or bacterial growth (sludge). Stored fuel should be used within 6 months. Check fuel filters on each service call, and change as needed.

KNOW LIMITATIONS

Burner and component manufacturers' warranties only cover up to and including B5 concentrations. However, blend levels up to B20 are becoming increasingly popular. With a clean fuel system and good service procedures, this usually does not present a problem. Simply monitor fuel lines, O-rings, pump seals, and gaskets like normal. Any blend greater than B20 can suffer from fuel system clogging, leakage, cold-weather problems, and false-flame failure. These higher blends are compatible with most oil heating equipment, but will affect gaskets, O-rings, and seals over time.

QUICK TESTS FOR FUEL QUALITY
CLEAR-AND-BRIGHT TEST

The purpose of the clear-and-bright test is to detect possible water or solid contaminants in the fuel by visual inspection. Using a clean glass container, take the sample at the bleed port of the fuel unit. Be sure the fuel sample tap (the bleed valve) is clean and free of loose contaminants by flushing it out at maximum flow before drawing a sample. Let the sample settle for a minute to remove the air bubbles. Observe the sample against a light background for a clear, bright condition. The sample should look more like cranberry juice than red wine (Figure 13.12). Swirl the container to create a whirlpool. Free

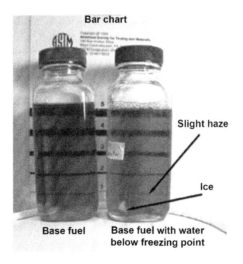

FIGURE 13.12

Clear-and-bright test.

Courtesy National Oilheat Research Alliance.

water and solids tend to collect at the bottom of the whirlpool. The term phrase *clear and bright* does not refer to color. Clear-and-bright fuel has no floating or suspended matter, and no free water. Bright fuel tends to sparkle.

WHITE-BUCKET TEST

The white-bucket test is a good, quick test for drivers to be sure they are filling their trucks with good fuel. The purpose of this test is to determine visually the possible presence of contaminants and water in the fuel. Fill a clean white bucket halfway with fuel and let the sample stand for a minute to remove the air bubbles. Place the bucket on a level surface with good light in the bucket. Inspect the fuel; it should be clear and bright with no water or solids. It should not be hazy or cloudy, and there should be no brown or black slime. Drop a shiny coin into the bucket. If you can read the date easily, the fuel is probably okay. The fuel should also smell "normal." Strange odors can indicate problems. With either the clear-and-bright or white-bucket tests, a haze caused by wax crystals may appear in the fuel if it is too cold. A haze in fuel that is not too cold may be the result of contamination with water.

VISUAL DETECTION OF BACTERIAL CONTAMINATION

The clear-and-bright and white-bucket tests can also be used for testing tank bottoms, filter cans, and fuel pump drainings for the presence of microorganisms and sludge. There will be evidence that can be seen and smelled. Put the fuel into a clean white bucket or clear glass jar. Allow the sample to settle for 2 min. Tip or swirl the container from side to side, looking for any evidence of dark solids, dark water, substances that cling to the side of the container, or a scummy mucuslike material. Hold the sample in front of a light.

To determine whether the solids are rust, move a small magnet along the outside of the container. Rust particles collect and follow the magnet. If the sample is a dark sludgelike material and it does not respond to the magnet, then it is probably bacterial contamination. Other indicators of these microorganisms are a "matty," lumpy, or stringy consistency and a rank, moldy odor.

WATER DETECTION PASTE

Use water detection pastes to determine the depth of water at the bottom of the storage tank. Apply the paste in a thin coating on a gauge stick from zero up to a couple inches above the suspected fuel–water interface. Lower the stick carefully into the tank until it touches the bottom lightly. Hold it in this position for 30 seconds to 1 min. Remove the stick. The water level will be indicated clearly by a definite color change where the water contacts the paste (Figure 13.13). Water paste does not detect a fuel–water emulsion. Check customers' tanks for water during every tune up, and drain off or pump out the water if detected.

OIL FILTRATION

The installation of filters in burner fuel suction lines is strongly recommended. Filters protect the pump and nozzle by trapping contaminants before they reach these components. There are passages in the oil burner nozzle that are smaller than the diameter of a human hair. It takes very little contamination to

FIGURE 13.13

Water detection paste.

Courtesy National Oilheat Research Alliance.

plug these passages in the nozzle. This is why it is critical to do everything to be sure clean fuel is delivered to the burner.

FILTERS AND SLUDGE

Filters may fail because they have become blanketed with biologically active slime or sludge (Figure 13.14). The resulting black or gray "ball of grease" is a tough service problem. This sludge has not been sucked from the tank; it is alive and actually growing in the filter. When small particles of sludge are in the tank and are drawn into the oil line, the bacteria in these particles look for moist places to reproduce. If there is any water in the bottom of the filter canister, or if there is water emulsified in the fuel, bacteria can grow their biofilm. Contrary to popular belief, there does not have to be a layer of free water to support the growth of biologically active sludge. There is always some water dissolved in the fuel. This is why sludge can sometimes grow faster on filters and strainers than it can in the tank. The rate at which sludge grows depends on the temperature and the availability of moisture and nutrients.

Filters may plug, even with new tanks and lines. The "seed" sludge particles can arrive with the fuel from a contaminated tank upstream in the distribution system. They can be drawn directly into the suction line before they have a chance to settle to the bottom of the tank. If the conditions are right, a filter can plug within weeks of installation, even with an immaculately clean tank. Also, sludge is corrosive. Untreated sludge can attack the steel filter housing, causing pinhole leaks.

FUEL-ADDITIVE TREATMENT

Additives are designed to prevent or retard fuel deterioration. Numerous types of additives are available on the market. A successful fuel treatment program requires knowledge of the quality of the fuel in the tank and the specific service problems. Using an additive off the shelf without testing may be more harmful than doing nothing at all.

FIGURE 13.14

Sludge.

Courtesy of National Oilheat Research Alliance.

SELECTION OF ADDITIVES

The multifunctional aftermarket additives available for heating oil and Bioheat are proprietary products that offer a range of properties. Guidelines to select additives are as follows:

- Define the problem and the additive needed.
- Make sure the fuel sample being tested represents the fuel being treated.
- Will the additive be used once or is continuous treatment required?
- Does the additive perform more than one function?
- Does the additive supplier have technical support if you have questions or problems?
- Can the supplier provide a way to determine effectiveness in specific cases?
- Follow all safety and handling instructions on the labels and Material Safety Data Sheets that should accompany the package.
- Follow the recommended treatment rates.
- Dispose of the additive containers properly. Know and follow the local laws concerning disposal of sludge and water bottoms.

TYPES OF ADDITIVES

Cold-flow improvers

Cold-flow improvers are designed to lower the cold-temperature operability limit for the fuel and avoid wax plugging of the filters. Pour point reducers or antigels lower the temperature fuel gels or cause them to solidify. Cold-filter plug point reducers lower the temperature at which wax plugs the filter.

After wax has formed in the fuel, an additive does not change the waxes present. To dissolve wax, a solvent such as kerosene must be used.

Dispersants or detergents
Keep the little chunks of junk floating in fuel so they can slip through the fuel system and be burned, rather than letting them settle to the bottom of the tank. Initial use of a dispersant may cause filter plugging as existing deposits, sludge, and dirt are broken up, suspended in the fuel, and picked up by the pump.

Antioxidants and metal deactivators
Fuel degradation caused by oxidation or aging leads to gum deposits. Antioxidant additives can slow this process. Dissolved metals, such as copper, can speed aging and degradation, and produce mercaptide (sulfur-containing) gels. To minimize these effects, metal deactivators combine with the metals and render them inactive. Periodic monitoring of fuel stability is recommended if these additives are being used.

Biocides
Serious problems can arise from microbial proliferation, including sludge formation, and acid and surfactant formation, leading to operational problems. (Translation: Critters can grow in the tank. They create a sludgy mess that causes lots of no-heat calls.) Biocides kill or prevent the growth of bacteria and other microorganisms. They must be fuel soluble and must be able to sink to the water in the bottom of the tank, where the microbes live. Microbiological organisms in fuel are bacteria, molds, and yeast. Because biocides are poisons, you have to very careful with them. Read the label to determine product use, treatment rate, and human exposure hazard warnings.

PREVENTATIVE MAINTENANCE
Good housekeeping means doing everything you can to minimize dirt and water from entering tanks. Water promotes the growth of microbes, which use the fuel as a food source, and accelerate the growth of sludge and internal corrosion of the tank. Water can enter the tank through cracked or leaking fill pipes and vents, which should be checked periodically and when water contamination is suspected. Varying air temperature and humidity can cause condensation in the tank. In general, dirt and debris are introduced into the fuel through careless handling.

STEPS TO BETTER FUEL PERFORMANCE
- Before removing the fill cap for a buried tank, drivers must be sure water, dirt, snow, or ice cannot fall into the tank. After delivery, drivers should check gaskets and O-rings, if needed, on the fill cap to be sure they are in good shape; reinstall the fill cap; and make certain it is sealed tight.
- While making a delivery, drivers should check to be sure the vent cap is in place, there is no water around the fill, the vent pipe is solid, and there is no water in the tank.
- On aboveground outdoor tanks, drivers should determine whether the tank legs are stable, on a solid foundation; whether there are there signs of rust, weeps, wet spots, deep scratches, or dents on the tank surface; whether there are oil leaks or signs of spills; and whether the tank needs painting.

- Sampling of tank bottoms should be done routinely (during the tune-up) for cleanliness and lack of water.
- If excessive sludge and water are found, they should be removed as soon as possible.
- Hold up on deliveries to problem tanks until the sludge and water problem is solved.
- After the sludge and water are removed from the tank, fill the tank with kerosene or specially treated fuel, tune up the burner, hand-pump the oil lines thoroughly, and replace the filter, strainer, and nozzle. Schedule a follow-up call a month later to verify that the tank and lines remain clean.
- The tank's fill box, fill pipes, vent cap and pipes, and remote fills should be checked for cracks and leaks on every delivery and tune-up. Often, the problem is a hole in the vent pipe just below ground level. Dig a few inches of soil away from the vent to check for rusting. If the fill box is in a driveway, it should have a "mushroom-type" fill box with a watertight gasket rather than a metal-to-metal fit.
- When additives are used, they should be added before filling the tank, if possible, to facilitate proper mixing.

TANK CLEANING

With massive accumulations at the bottom and on the sides of the tank, mechanical cleaning, fuel filtration, the use of additives, and a preventative maintenance program are the only way to remove the sludge effectively. Portable tank cleaning/filtration machines are available. Their effectiveness depends on the condition of the tank, access to the interior, and the operator's skill. Before attempting to clean the tank, let the burner draw the oil down as low as possible to minimize the amount of fuel you have to dispose. There are companies that offer a tank-cleaning service; however, cleaning a residential heating oil or Bioheat tank is usually expensive and difficult, and a tank replacement may be more economical and effective.

TANK REPLACEMENT

If tank erosion has gone too far, tank and fuel treatment remedies only buy you some time. A tank's surface contains microscopic pits and craters where bacteria can "hide." When fresh fuel is added and a bit of water condenses, the bacteria can reproduce at an astounding rate, and sludge formation begins. Often, the only solution is to replace the tank and oil lines. Never pump the fuel from the old tank into the new one. You will transfer the contaminants that caused the problem in the first place. It takes surprisingly little time to make the nice new tank as dirty as the old one.

If you are installing (or maintaining) an outdoor aboveground tank, it is recommended that you paint it a light color to reflect the light. This helps keep the tank cooler and minimizes moisture condensation inside the tank. Also, there are several types of tank sheds available. They minimize water buildup and frozen lines because the tank temperature is not subject to intense fluctuations.

KEEP THE TANK FULL

Brookhaven Lab tests have found that topping off fuel tanks, especially outdoor aboveground tanks in the spring, helps prevent condensation. The less air in tank, the less condensation.

SOURCES

National Oilheat Research Alliance's *Oilheat Technicians Manual*, 2008 edition.
National Biodiesel Board.
National Renewable Energy Laboratory's *Biodiesel Handling and Use Guide*, 4th edition, 2008.
Brookhaven National Laboratory.

FURTHER READING

A number of resources provide additional details and technical information in support of these guidelines. They include the following.
National Oilheat Research Alliance's websites: noraed.org and nora-oilheat.org.
The National Biodiesel Board's online documents library and information: www.nbb.org; www.biodiesel.org/resources/reportsdatabase/.
1-800-841-5849; The National Renewable Energy Laboratory's technical publications on biodiesel: www.nrel.gov/vehiclesandfuels/npbf/pubs_biodiesel.html.
The U.S. Department of Energy's Alternative Fuels and Advanced Vehicles Data Center: http://www.eere.energy.gov/afdc/.
The U.S. Department of Energy's technical publications: www.eere.energy.gov/biomass/document_database.html.
The U.S. EPAs Renewable Fuel Program: www.epa.gov/otaq/renewablefuels/index.html.

ALGAE BIOMASS CULTIVATION FOR ADVANCED BIOFUEL PRODUCTION

Anju Dahiya[1,2]

[1] *University of Vermont, USA;* [2] *GSR Solutions, USA*

INTRODUCTION

Imagine a tiny living cell—so small that its size may range from less than a micrometer (micron) to about a millimeter. For instance, a *Chlorella* cell measures between 2 and 10 μm (the diameter of a typical human hair is about 40–50 μm wide). This cell and others of its kind are known as the fastest growing plants on the planet earth, belonging to a large group of phytoplankton called algae (singular *alga*; plural *algae*). Algae is bound in the equation of the life support system on earth—as a food base in the food chains and a major producer of oxygen (over 70%). Can these microscopic algae cells be potentially scaled up to solve major global problems? Such as:

- *Energy crisis*
- *Global freshwater demand* (by 2025 the global water demand will exceed the supply by 56 percent as estimated by Clarke and Barlow (2005))
- *Wastewater treatment* (currently an expensive treatment, and subject to Environmental Protection Agency and state regulations on what and how much can be discharged in natural water bodies)
- *Atmospheric pollution* from flue gases

Research and development in the algae area over the past many decades has proven that algae could be a potential solution to these problems.

ALGAE AS A SUSTAINABLE FEEDSTOCK FOR MULTIPLE USES

Algae as a biofuel feedstock provides beneficial options (Dahiya, 2012):

- Algal productivity can offer high biomass yields per acre of cultivation (DoE, 2010) as algae grow significantly faster than land crops used for biodiesel and are reported to produce **15–300 times more oil for biodiesel than traditional crops** on an area basis.
- Algae as biomass feedstock for fuel does not compete with food and water resources.
- Algae biomass production can be integrated with the treatment of industrial, municipal, and agricultural wastewaters.

Bioenergy. http://dx.doi.org/10.1016/B978-0-12-407909-0.00014-6

- Algae can capture carbon from flue gases from wood-burning power plants and industrial operations besides atmospheric carbon dioxide, for cost-effective carbon-neutral cultivation.
- Low-temperature fuel properties and the energy density of algae-derived fuel makes it suitable as jet fuel.
- Algae biomass production ensures a continuous renewable supply for biofuel production.
- Algae production can provide valuable byproducts like lubricants, bio plastics, animal feed, nutraceuticals, and pharmaceuticals.

HISTORICAL PERSPECTIVE OF ALGAE BIOMASS AND JOURNEY TOWARD ALGAL BIOFUEL

In order to meet the global challenges and to harness the benefits of using algae as a bioenergy feedstock, algae would have to be grown in enormous amounts. That is unequivocally possible because the algal blooms occurring in natural water bodies are a well-known menace.

HOW DO ALGAE GET THEIR MASS AND FROM WHERE?

This question was evidently raised in the context of higher plants in 1450 (Nicolaus of Cusa's 1450 book *De Staticus Experimentis*) and the answer was demonstrated in the late 1640s by a Belgian scientist, Von Helmont, whose willow experiment is considered a classic in the history of botany (Hershey, 2003). Helmont filled an earthen container with 164 pounds of soil and planted a willow tree branch and recorded the weight of the branch, soil, and water added. He found that after five years the plant gained about 164 pounds, however, the amount of soil remained the same. Hence, he deduced the weight gain resulted from water (published in *Ortus Medicinae* in 1648). With the scientific advancement during the following centuries it became clear that the majority of mass in plants comes from *carbon dioxide* fixed through a process called photosynthesis in plants (including algae), as the well-known equation was worked out by Julius Sachs in early 1860s:

$$\underset{\text{Carbon dioxide}}{6CO_2} + \underset{\text{Water}}{6H_2O} + \text{Solar energy} \rightarrow \underset{\text{Glucose(carbohydrate)}}{C_6H_{12}O_6} + \underset{\text{Oxygen}}{6O_2} \qquad (6.1)$$

By the time this equation surfaced, microscopic algae (microalgae) were very well known after they were first observed by a Dutch scientist, Antonie van Leeuwenhoek (the father of microbiology) around 1670s.

IMPORTANCE OF LIGHT IN PHOTOSYNTHESIS

The modern science techniques and concepts developed during the twentieth century helped in learning about the photosynthesis equation in great detail. In 1905, investigations by Blackman and Matthaei showed the photosynthesis equation followed a two-step mechanism: *"light reaction"* and *"dark reaction"* dependent on solar energy. That was the time when a whole new era of quantum mechanics was emerging—the Scottish physicist, James Clerk Maxwell had already described light as an electromagnetic wave which culminated in 1900 in German physicist Max Planck's description of the energy of waves consisting of small packets termed "quanta." At that time, Albert Einstein's ideas about mass to energy were also shaping up, which led to the calculations of mass conversion to energy and to the understanding of how solar matter was converted to the energy radiating in the space and how a fraction of the electromagnetic energy is reaching out to the earth (making the photosynthesis possible). In 1926, an American physical chemist, Gilbert Newton Lewis with his colleague introduced the term **photon** for light quanta. All these developments helped the twentieth century photo-biologists to further investigate the photochemical reactions in the "*light reaction*" phase of photosynthesis that depends on the absorption of photons by the cellular chlorophyll molecules triggering the onset of photosynthesis reaction (see equation above), thereby making it possible to convert the light energy into chemical energy and storing it in the form of carbohydrate molecules. Likewise, it was found the light-independent "dark reaction" process facilitated carbon fixation from atmospheric CO_2, and that is when the conversion to carbohydrates and lipids takes place. Light and dark reactions were further studied in great detail (refer Hall and Rao, 1999).

USE OF ALGAE FOR BIOFUEL

For the first time in 1942, Harder and Witch suggested that algae (specifically microalgae called di-atoms) could be a useful source of lipids (oil) as both a food and a fuel source. A decade later, an "algae mass culture symposium" held at Stanford University, California, brought together algae experts of that time who were engaged in the massive cultivation of algae from around the world (Burlew, 1953). A more focused approach to using algae for energy started as an application as a throughput feedstock for methane production with the use of anaerobic digestion (Meier, 1955; Oswald and Golueke, 1960). The 1970s energy crisis led researchers to seriously consider algae as a bioenergy feedstock that actually resulted in a $25 million investment for the 18 year long Algae Species Program (ASP) (1978 until 1996) supported by the U.S. Department of Energy (Sheehan, 1998). This program was shut down due to the fall in gasoline prices back to $1 per gallon, and there was no competition with the 1996 petroleum costs. ASP had started with a focus to produce hydrogen, but in the early 1980s switched to liquid fuel production. Major advances were made through algal strain isolation and characterization, studies of algal physiology and biochemistry, genetic engineering, process development, and demonstration-scale algal mass culture (DoE, 2010), and most importantly, the ASP demonstrated biofuel production from algae. However, a cost-effective technology is yet to be found. One of the recommendations of ASP was the integration of algal biofuels with waste treatments to offset the cost of production since photosynthetic algae (and cyanobacteria, like other higher plants), besides capturing carbon from atmospheric CO_2, requires other nutrients from the growth media such as nitrogen to make proteins, phosphorus for nucleic acid, and other supplementary ions (e.g., sodium, calcium, potassium, and iron) for several processes that are readily present in the waste streams.

POSSIBILITIES OF BIOFUEL PRODUCTION FROM ALGAE BIOMASS

Over the decades, different approaches of algae biomass processing have emerged that can produce different types of biofuels depending on the pathway used as follows (Figure 14.1):

- **Biogas** produced by the anaerobic digestion of algae biomass (see chapters 17 and 18 on Anaerobic digestion in this book for more info about the process).
- **Liquid Hydrogen Fuels** by gasification using syngas (synthetic gas mixture containing hydrogen, carbon monoxide, and carbon dioxide) (see chapters 17 and 26 on "Gasification" and

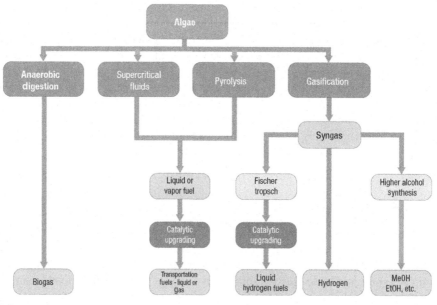

FIGURE 14.1

Algae-based biofuels.

Source: Bioenergy Technologies Office, DoE, 2010.

"Pyrolysis" respectively in this book for more info about the process, including the liquid fuel formation process used in conventional fuel extraction).

- **Hydrogen** also produced by gasification using syngas (see "Gasification" chapter 16).
- **Alcohols** also produced by gasification using syngas.
- **Transportation fuels (liquid or gas)** by using *supercritical fluids* and *pyrolysis*: Supercritical fluids exist in multiple phases (solid, liquid, and gas) at a critical temperature and pressure. The supercritical form of carbon dioxide is a good option that can be used for the selective extraction of oil from algal biomass via a diffusion-based process. Pyrolysis requiring thermochemical decomposition of biomass (in the absence of oxygen) is another option for producing high-quality drop-in fuel (as a blend or substitute for conventional fuel such as gasoline) from algae.

Figure 14.2 shows the start to finish process of liquid biofuel production from algae biomass in the following four steps: (1) *algae strain selection* (some strains are capable of accumulating oil), (2) *algae cultivation*, (3) *biomass harvesting*, and (4) *algae oil extraction*.

FIGURE 14.2

Algae biomass to biofuel production steps.

Courtesy: GSR Solutions LLC.

The next four sections describe the algae biomass to biofuel production steps (Figure 14.2.) followed by other topics such as economics and life cycle analyses in the later sections.

ALGAE STRAIN SELECTION: ALGAE TYPES, STRAINS, AND USE IN BIOFUEL PRODUCTION

Depending on the size, algae are classified as microalgae or macroalgae (seaweed). Both the algae types have been explored for biofuel production as described below.

MICROALGAE

Based on the type of nutrition, algae growth can be photoautotrophic, mixotrophic, or heterotrophic as described below.

PHOTOAUTOTROPHIC ALGAE

Photosynthetic algae absorb light and carbon dioxide to produce glucose and oxygen (Eqn (6.1)). Figure 14.3 shows different types of algae cells.

There are over 40,000 algae species already identified that are classified in multiple major groupings as follows: cyanobacteria (Cyanophyceae), green algae (Chlorophyceae), diatoms (Bacillariophyceae), yellow-green algae (Xanthophyceae), golden algae (Chrysophyceae), red algae (Rhodophyceae), brown algae (Phaeophyceae), dinoflagellates (Dinophyceae) and "pico-plankton" (Prasinophyceae and Eustigmatophyceae). All these algae types vary in the lipid (oil) content as shown in Figure 14.4: (a) green microalgae, (b) diatoms, (c) oleaginous species/strains from other eukaryotic algal taxa, and (d) cyanobacteria (Hu et al., 2008).

To grow algae biomass for fuel at massive scales, the algae species used should contain high lipid content (at least 35%) (Dahiya, 2012). Table 14.1 presents algae lipid contents in different strains.

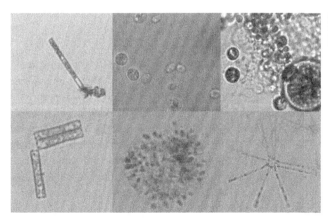

FIGURE 14.3

Different microscopic algae cells.

Courtesy: GSR Solutions.

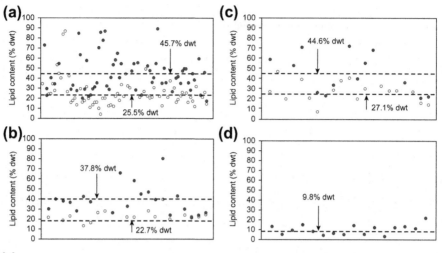

FIGURE 14.4

Cellular lipid content in various classes of microalgae and cyanobacteria under normal growth (open circles) and stress known to enhance lipid (closed circles) conditions.

Adapted from Hu et al., 2008 with permission from the publisher.

HETEROTROPHIC AND MIXOTROPHIC ALGAE

Unlike the photoautotrophic algae, heterotrophic algae cannot use sunlight and inorganic carbon from CO_2, but requires a source of organic carbon such as sugars. Some strains of algae can take advantage of both phototrophic and heterotrophic nutrition modes and can utilize carbon from CO_2 as well as sugars, these are called mixotrophic.

Although heterotrophic algae are not limited by sunlight exposure and can produce high amounts of biomass, the major limitation of the heterotrophic mode of algal biomass production is associated with the availability of cheap sources of organic carbon and economical investment and operation costs. Its use in biofuel and byproduct production has been demonstrated by one of the leading algal companies, Solazyme Inc.

Genetically modified algae and its use in biofuel production

Microbes including bacteria, fungi, etc., have been successfully genetically modified for biotechnological applications. Similarly, it is possible to leverage the metabolic pathways of algae via genetic engineering methods to produce biofuel. Cyanobacteria, the blue-green algae which like bacteria is a prokaryote (cells do not contain a true nucleus), are thought to be better candidates for genetically engineered biofuel production. As noted in the algal biofuel roadmap (DoE, 2010), "cyanobacteria generally do not accumulate storage lipids (see Figure 14.4), but they are prolific carbohydrate and secondary metabolite producers. Some strains can double quickly (less than 10 h), and some strains can fix atmospheric nitrogen and produce hydrogen. Moreover, many can be genetically manipulated, making them attractive organisms for biofuel production." Among eukaryotic algae (cells containing true nucleus), the alga *Chlamydomonas reinhardtii* has been extensively studied as an important model

Table 14.1 High End Oleaginous Algae (Single Species) Showing Lipid Content (Percentage Dry Weight) Grown in Different Media Types (Dahiya, 2012)

Algae	Lipid %	Media	Literature
Botryococcus braunii	86	AM (mod. Chu)	Brown et al. (1969), Wolf (1983)
	80[a]		Brown et al. (1969)
	>75		Banerjee et al. (2002)
	63		Metzger and Largeau (2005)
	25−75		Chisti, 2007
Schizochytrium sp.	50−77		Chisti, 2007
Nitzschia species			
Nitzschia dissipata	66	AM	Sheehan et al. 1998
Nitzschia palea	40	AM	Shifrin and Chisholm (1981)
Boekelovia hooglandii	59	Urea-enriched AM	Sheehan et al. (1998)
Monallantus salina	41−72	AM	Shifrin and Chisholm (1981)
Navicula species			
Navicula saprophila	58	AM	Sheehan et al. (1998)
Navicula acceptata	47	AM	Shifrin and Chisholm (1981)
Navicula pelliculosa	45	AM	Sheehan et al. (1998)
Navicula pseudotenelloides	42	AM	Sheehan et al. (1998)
Chlorella species			
Chlorella minutissima	57	AFW	Shifrin and Chisholm (1981)
Chlorella vulgaris	41	AFW	Shifrin and Chisholm (1981)
Chlorella pyrenoidosa	36	AFW	Shifrin and Chisholm (1981)
Dunaliella sp.	45−55	AM	Sheehan et al. (1998)
Neochloris oleoabundans	35−54	AM	Sheehan et al. (1998)
Monoraphidium sp.	52	AM	Sheehan et al. (1998)
Amphora	51	AM	Sheehan et al. (1998)
Ourococcus	50	AM	Shifrin and Chisholm (1981)
Nannochloris sp.	48	ASW	Shifrin and Chisholm (1981)
	35	ASW**45	Sheehan et al. (1998), Rodolfi et al. (2009)
Nannochloropsis salina	46	ASW	Sheehan et al. (1998)
Scenedesmus	45	AFW	Sheehan et al. (1998)
Scenedesmus obliquus	41	AFW	Shifrin and Chisholm (1981)
Ankitodesmus	40	AM	Sheehan et al. (1998)
Chaetoceros species			
Chaetoceros calcitrans	40	ASW	Rodolfi et al. (2009)
Chaetoceros muelleri	39	ASW	Sheehan et al. (1998)
Cyclotella cryptica	37	AM	Shifrin and Chisholm (1981)
Amphiprora hyalina	37	AM	Sheehan et al. (1998)
Cylindrotheca sp.	16−37		Chisti (2007)
Pavlova lutheri	36	ASW	Rodolfi et al. (2009)

Artificial media (AM), artificial freshwater (AFW), and artificial seawater f2 (ASW).
*unsaponifiable lipids
**also a higher lipid amount

for biological research, with its genome available.[1] And the *Chlorella* sp. have been well explored for biofuel production. Solazyme Inc. has produced many thousands of gallons of algae biofuel using genetically altered algal strains and has signed contracts with the US Navy to supply algae biofuel. In 2009, ExxonMobil backed the Celera Genomics scientist, Craig Venter, with a five-year $600 million deal to produce biofuel from genetically modified algae at commercial scale.

Biofuel production from genetically modified algae sounds promising, however, many concerns have been expressed by various groups worldwide from the stand point of the unknown long-term impacts on ecosystems. Therefore, using genetically modified algae for biofuels, due to the expression of the foreign genes when inserted inside algal cells, is still questionable.

MACROALGAE (SEAWEED)

Macroalgae is multicellular macroscopic algae growing largely in a marine environment. Macroalgae can be of red, green, or brown algae types (example sushi-porphyria). It has been explored for bio-energy options via anaerobic digestion (biogas) and fermentation (ethanol, etc.) since the 1960s. It is a well-established aquaculture-based product, largely in Asia, and has been tested elsewhere.

ALGAE CULTIVATION: GROWTH SYSTEMS

Three major types of algae growth systems are in use at different scales: photobioreactors, open ponds, and closed fermenters (Figure 14.5).

PHOTOBIOREACTORS

Photobioreactors are the closed systems (Figures 14.5 and 14.6) that provide a controlled environment for growing photosynthetic algae under sterile conditions. As the name suggests, these reactors are designed to provide adequate light exposure for photosynthesis by means of either artificial or natural light. In this type of reactor the algae growth parameters (pH, temperature, mixing, etc.) can be regulated to maximize algae biomass production under sterile conditions. The energy and materials required for the controlled environment raise the capital costs making this system less cost efficient for algae cultivation for biofuel. Common photobioreactor designs are as follows (Figure 14.6):

- Flat plate (Figures 14.6(a)-left, and (b))
- Tubular (Figure 14.6(d)-right)
- Hanging bags (Figure 14.6(c))
- Bubble columns (Figures 14.6(a)-right, 14.6(c)-left, and 14.6(d)-right)
- Closed tanks (Figure 14.6(d)-left)

OPEN SYSTEM

Compared to photobioreactors, open systems (Figure 14.7) popularly known as raceways are relatively easier to build and maintain and, therefore, economical for mass cultivation of algae. In a

---

[1]http://phytozome.jgi.doe.gov/pz/portal.html#!info?alias=Org_Creinhardtii.

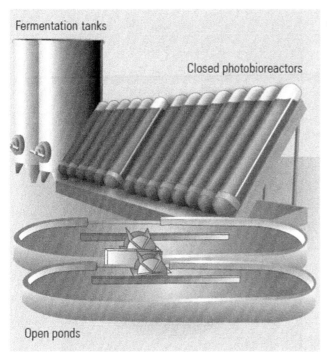

Fermentation tanks

Closed photobioreactors

Open ponds

FIGURE 14.5

Algae growth systems.

*Source: Bioenergy Technologies Office, DoE,
2010.*

typical raceway system, the nutrient-rich algae growth media is mixed through a paddle wheel (Figure 14.7(a)-middle). These systems have been in operation and can be scaled on many acres of land (Figures 14.7(b) and 14.8).

FERMENTERS

Heterotrophic algae is grown in fermenters (Figure 14.5). Fermenters are available in different sizes and can be vertically installed to great heights. For example, one of the largest fermenters in use for algae production in the world is 12 stories high in Winchester, Kentucky run by Alltech Inc.

STERILE PHOTOBIOREACTORS VERSUS OPEN PONDS VERSUS HETEROTROPHIC CULTIVATION

Sterile photobioreactors have shown to produce significantly higher volumetric cell densities compared to open raceway ponds. However, commercial-scale photobioreactors are not yet cost-effective due to the high costs required for investments in infrastructure development and mainte-nance compared to open ponds (Chisti, 2007; Schenk et al., 2008). Open pond systems are limited by contamination issues such as the grazing of algae by zooplanktons and being outcompeted by other microbes (e.g., bacteria, fungi, and wild algae). Similar to photobioreactors, the heterotrophic algae

FIGURE 14.6

Closed photobioreactor systems. Pictures (a), (b), (c), and (d) are showing different types of photobioreactor systems.

Courtesy: Robert Henrikson.

biomass cultivation in fermenters can maintain a sterile environment, but for larger scale systems those are also not cost-effective due to high capital and operation costs. Table 14.2 summarizes the advantages and challenges of photoautotrophic, closed photobioreactors and open pond systems and the heterotrophic systems.

FIGURE 14.7

Open pond systems note the paddle wheel-operated raceway in the top picture (a) and circular pond along with aerial views (b) of outdoor pond systems.

Courtesy: Robert Henrikson.

FIGURE 14.8

An aerial overview of the algae grown in open raceway ponds.

Courtesy: Earthrise Inc.

Table 14.2 Advantages and Challenges of Different Systems

		Advantages	Challenges
Photoautotrophic cultivation	Closed photobioreactors	• Less loss of water than open ponds • Superior long-term culture maintenance • Higher surface to volume ratio can support higher volumetric cell densities	• Scalability problems • Require temperature maintenance as they do not have evaporative cooling • May require periodic cleaning due to biofilm formation • Need maximum light exposure
	Open ponds	• Evaporative cooling maintains temperature • Lower capital costs	• Subject to daily and seasonal changes in temperature and humidity • Inherently difficult to maintain monocultures • Need maximum light exposure
Heterotrophic cultivation		• Easier to maintain optimal conditions for production and contamination prevention • Opportunity to utilize inexpensive lignocellulosic sugars for growth • Achieves high biomass concentrations	• Cost and availability of suitable feedstocks such as lignocellulosic sugars • Competes for feedstocks with other biofuel technologies

Source: Table credit: *Bioenergy Technologies Office, DoE, 2010.*

The heterotrophic mode of algal production requires less space compared to autophototrophic production in photobioreactors or open ponds as it can be grown in a more controlled environment in fermenters or like containers. However, it has many other downsides (Perez-Garcia et al., 2011): (1) there is a limited number of microalgal species that can grow heterotrophically, (2) increasing energy expenses and costs by adding an organic substrate or sugars compared to waste-grown phototrophic strains, (3) contamination and competition with other microorganisms, (4) inhibition of growth by excess organic substrate, and (5) inability to produce light-induced metabolites.

ALGAE HARVESTING

As shown in Figure 14.2, algae biomass cultivation is followed by the next step—algae harvesting. Algae harvesting methods are selected based on mainly algae size, algae strain type, algae density, and the type of media it is grown in. It is difficult to harvest algae due to its small size and its tendency to form stable suspensions due to the negative charge on its surface, posing difficulty in concentrating and harvesting. Some examples of the harvesting methods are:

Flocculation and Sedimentation: Chemical additives that bind algae or otherwise affect the physiochemical interaction between algae are known to promote flocculation. Alum, lime, cellulose, salts, polyacrylamide polymers, surfactants, chitosan, and other man-made fibers are some examples of chemical additives.

Filtration: Solid/liquid filtration technologies are simple and especially applicable for large-sized algae. Small-sized algae need to be first flocculated before filtration.

Centrifugation: A centrifuge is used to separate algae from the medium. This causes the algae to settle to the bottom of the vessel. Due to its high efficiency, it is a widely used method, but the current level of centrifugation technology makes this approach cost-prohibitive for the large-scale algae biorefineries.

Drying: Because drying generally requires heat, methane drum dryers and other oven-type dryers have been used. It is a very energy intensive option.

OIL EXTRACTION FROM HARVESTED ALGAE BIOMASS

Multiple pathways of oil extraction are available for downstream processing of algae to extract oil as described below. Figure 14.9 shows some of these pathways.

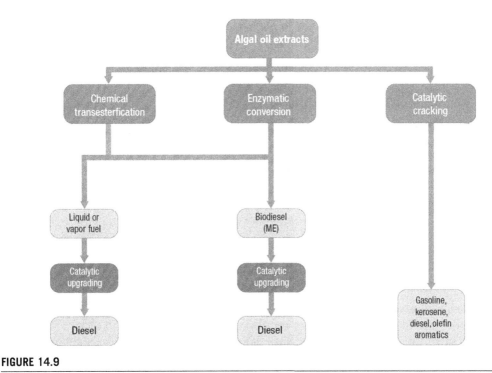

FIGURE 14.9

Algae oil extraction pathways.

Figure credit: *Bioenergy Technologies Office, DoE, 2010.*

Mechanical methods: An oil press is the simplest available method to press the dry biomass to extract oil. Oil presses are successfully used for the extraction of oil from crop seeds (e.g., sunflower, canola, and olives). In the case of algae, the biomass requires drying prior to pressing it through the press, which is a significant cost making this option less cost-effective.

Enzymatic conversion: Natural or synthetic enzymes can be used for the degradation of algae cell walls, and then water can be used as a solvent for fractionation of the oil.

Catalytic cracking: this process can be used for breaking down the long-chain hydrocarbon molecules into shorter chains which can be further refined to be blended into gasoline or other fuels.

Supercritical fluid method: As described earlier, supercritical CO_2 (liquefied under pressure and heated to the point that it has the properties of both a liquid and a gas), could be used as a solvent for oil extraction from algae.

Chemical methods: The hexane solvent method is one of the most popular methods based on the principle of transesterification (see chapter 21 in this book on this topic). Prior to oil extraction from algae, the algae cells require pretreatment with a cell disruption method (e.g., sonication) for releasing oils out of the algae cells. Up to 95% of oil released can be extracted in a solvent (e.g., hexane). Transesterification is a well-known process of the conversion of vegetable oil to biodiesel in which triglycerides are reacted with an alcohol (methanol or ethanol) in the presence of an alkaline catalyst (potassium hydroxide) and the reaction forms a mixture of fatty acids including fatty acids and glycerol as byproducts. During this process, an exchange of an organic group of an ester with the organic group of an alcohol takes place that can be speeded up with a catalyst. Biodiesel primarily contains fatty acid methyl esters (FAME) molecules (note: FAME and related extraction processes are described in detail in two respective chapters 22 and 20 in this book by Laurens and by Pruszko).

In order to use biomass-derived fuel for transportation fuels, they require chemical transformation to increase *volatility* and *thermal stability* and to reduce *viscosity* (Elliott, 2007), and FAME as such causes problems due to the oxygen content. The fuel performance can be improved by using additives. Algae oil can be transformed into *renewable diesel* via *hydrogenation* (treatment with the addition of hydrogen). See chapter 28 titled, *"Cutting Edge Biofuel Conversion Technologies to Integrate into Petroleum-based Infrastructure and Integrated Biorefineries,"* in this book by Dahiya for details of the renewable diesel manufacture process.

Figure 14.10 summarizes the different steps of algae biofuel production as described so far. It outlines the algae species selection based on the mode of growth, the cultivation systems, the intermediate constituents produced including hydrocarbons and alcohols, lipids, carbohydrates, proteins and biomass conversion processes required for end use fuel products.

CHALLENGES IN UPSCALING OF ALGAL BIOFUEL OPERATIONS FROM BENCH TO COMMERCIAL SCALES

The big economic barrier in the production of algal-based drop-in biofuel is the cost-efficiency involved in the development and production of algae biomass for biofuel (Sheehan, 1998; DoE, 2010; Dahiya et al., 2012). According to the algal biofuel roadmap (DoE, 2010), the following four broad cultivation challenges have emerged that are important to address for economically viable,

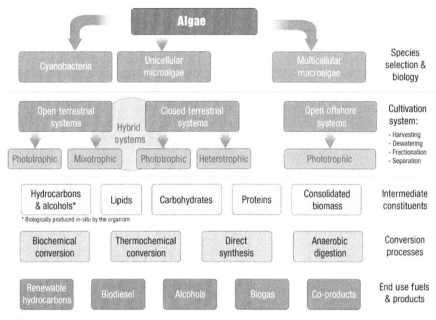

FIGURE 14.10

Algae biofuel start to finish processes.

Figure credit: *Bioenergy Technologies Office, DoE, 2010.*

commercial-scale algal cultivation followed by three important questions concerning the stability of large-scale systems:

Cultivation challenges:

- Culture stability
- Standardized metrics for system-level productivity analysis
- Nutrient source scaling, sustainability and management
- Water conservation, management, and recycling

Questions concerning large scale systems stability:

- Are agricultural or municipal waste streams—a potentially significant source of nutrients for algal cultivation—actually a liability because of significant reservoirs of algal pathogens and predators?
- To what extent will local "weedy" algae invade and take over bioreactors and open ponds?
- What prevention or treatment measures might limit such takeovers?

INTEGRATED ALGAL BIOFUEL PRODUCTION

The integrated approach for algae biofuel production involving wastewater treatment has been estimated to have enormous potential. The possibility of off-setting costs through coupling with waste treatment, as well as the possibilities of creating new products through production of biomass byproducts (fertilizers, animal feed, new food/agricultural food webs, pharmaceuticals etc.), offers additional cost advantages of integrated systems over conventional algae production approaches. Figure 14.11 shows synergistic algae biomass integration with cheap sources of carbon such as flue gases from power plants. The algae production systems can be integrated with waste management to utilize nutrients including nitrogen and phosphorus from waste streams such as municipal, industrial or dairy effluents (Dahiya, 2012; Dahiya et al., 2012). The successful role of algae in wastewater treatment has been very well documented since the early 1950s in the studies by Prof. William Oswald and his group (1960, 1990, and 2003). Many different types of integrated systems have been explored for capturing nutrients from wastewaters as follows:

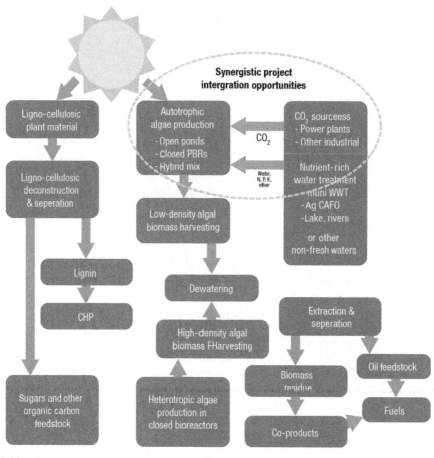

FIGURE 14.11

Integrated algae biofuel production.

Figure credit: *Bioenergy Technologies Office, DoE, 2010.*

- The "Algal Turf Scrubber" system has proven that algae fed on a diet of dairy manure can recover over 95% of the nitrogen and phosphorous in agricultural manure wastewater (Mulbry et al., 2005).
- Algal-based "Advanced Integrated Treatment Pond" systems used for municipal sewage treatment have been recorded removing over 90% of total nitrogen in the wastewater stream (Oswald, 1990).
- Algae grown in municipal wastewater was reported to successfully remove over 99% of both the ammonium and orthophosphate (Woertz et al., 2009).

Integrated systems hold great potential for municipalities, industries and dairies producing effluents that are required to meet the standards for the handling and recycling of manure (nutrients) per guidelines from the state and the U.S. Environmental Protection Agency (EPA). Algae is known to utilize the extra nutrients including nitrogen, phosphorus, potassium, heavy metals and other organic compounds from the wastewaters (Oswald, 1990; Mulbry et al., 2008). Lignocellulosic plant material can be broken down into sugars to produce heterotrophic algae biomass (Figure 14.11). CO_2 from power plants is a significant source of low cost carbon to offset the costs of algae production.

LIFE CYCLE ANALYSIS, ECONOMICS, AND ENVIRONMENTAL IMPACTS

This section based on Dahiya et al. (2012) provides an overview of the life cycle analysis of algae biofuel production, related economics, and environmental impacts as follows.

As of now the cost of producing algae for oil and the related cost-effectiveness of an integrated system is difficult to estimate because no optimized commercial-scale operation is yet in existence. Considering the widely quoted estimates by Chisti (2007), the cost of algae biomass produced for providing a liter of oil would be around US$1.40/L for photobioreactors and around US $1.81/L for open ponds. Algae biofuel production is not yet comparable to fossil-based petroleum. One of the ways algae biofuel could be made cost-effective is by integrating it with wastewater treatment to offset the investment and maintenance costs, which are the main drivers of algae biomass production costs (Dahiya et al., 2012). The calculations from studies (Lundquist, 2008) show costs of approximately US $6 per gallon of oil (or US $1.57/L of oil) produced by algae integrated wastewater treatment (saving 15 kWh per gallon oil produced), which represents a net savings of US $0.24/L for oil produced by open ponds in Lundquist's open pond system as compared to US $1.81/L in Chisti's traditional open pond system.

According to a life cycle analysis (LCA) performed by Levin et al. (2009), over 1.552×10^6 L (410,000 gallons) of oil could be produced from algae per year in 100 ha of open ponds treating wastewater. However, since no commercial scale facility is producing oil from algae (DoE, 2010), no thorough LCA following the production chain, from algae culturing through to biodiesel, is available. Most of the partial LCA performed (Beal et al., 2012; Sills et al., 2012) have been based on extrapolations of lab-scale or pilot studies and combined with known processes developed for first generation biofuels (Lardon et al., 2009; Clarens et al., 2010). Clarens et al.'s LCA has been highly criticized by the algal biofuel community due to their over estimation of the negative environmental effects of algae production. The water demand for algae production has not been directly addressed in any of the preliminary algae LCA analyses so far (Aresta et al., 2005). However, based on their

LCA model, Clarens et al. (2010) demonstrated that when algae was grown to treat wastewater, the algae was found to have lower environmental impacts (water, energy use, emissions) than the conventional crops of switchgrass, canola and corn, as well as compared to algae production fed with clean water and fertilizer, which all showed higher upstream burdens. According to the recent DoE report (2010), one of the major benefits of growing algae is that, unlike terrestrial agriculture, algal culture can utilize water with few competing uses, such as saline and brackish water. This observation is crucial when we consider the water resources needed for the future development and expansion of algal biofuel production.

The LCA done by one of the leading algae-oil companies, Sapphire Energy, revealed that algae-based fuels emit approximately two-thirds less CO_2 than petroleum-based fuels at scale, and when compared with conventional biofuels, such as corn ethanol and soy biodiesel, algae "green crude" oil has significantly less than half their carbon impact. A report by Sánchez Mirón et al. (1999) states that algal biomass contains approximately 50% carbon by dry weight; and thus, production of 100 tons of algal biomass fixes approximately 183 tons of CO_2 (Chisti, 2007). These numbers make it clear that algae provide a tremendous potential to capture free-of-cost CO_2 emissions from power plant flue gases and other fixed sources. When taken together, all of this suggests that biodiesel from algae can be carbon neutral because all the power needed for producing and processing the algae could potentially come from biodiesel itself and from methane produced by the anaerobic digestion of the biomass residue (or bioethanol from fermentation) left behind after the oil has been extracted (Chisti, 2007).

Clarens et al.'s LCA model (2010) demonstrated that, in terms of land use impact, algae offers clear and appreciable improvement over corn, canola, and switchgrass. Their land use estimates indicate that algae cultivation on roughly 13% of the US land area using current technologies could meet the nation's total annual energy consumption. In contrast, use of corn would require 41% of the total land area, while switchgrass and canola would require 56% and 66%, respectively. Land use changes implicit in large-scale bioenergy deployment are expected to have important implications for climate change and other impacts, and these so-called "indirect" changes are associated with conversion of arable land into biofuel production. The potential annual oil yields from oil-rich algae are projected to be at least 60 times higher than from soybeans per acre of land, around 15 times more productive than jatropha, and around 5 times more than oil palm (DoE, 2010).

CONCLUSION

Algae biofuel is a promising emerging fuel. As of 2014, the leading algae biofuel companies have made tremendous progress. Solazyme Inc. that supplied many thousands of gallons of algae biofuel to the US Navy, partnered with Brazil based Bunge to manufacture 100,000 metric tons of fuel per year. Sapphire Energy Inc. in New Mexico is operating 2200 acre algae farms and has a target of 10,000 barrels of crude oil per day production by 2018. However, many hurdles need to be overcome before algae biofuel is ready to compete with fossil fuel costs and is commercially available. These hurdles include the use of improved oil-rich algae strains, cost efficiency of growth systems and related operation costs, efficient oil extraction technologies, and a complete supply chain based on the availability of a continuous supply of low-cost throughput nutrient sources for algae cultivation.

REFERENCES

Aresta, M., Dibenedetto, A., Barberio, G., 2005. Utilization of macro-algae for enhanced CO_2 fixation and biofuels production: development of a computing software for an LCA study. Fuel Processing Technology 86, 1679–1693.

Banerjee, A., Sharma, R., Chisty, Y., Banerjee, U.C., September 2002. *Botryococcus braunii*: a renewable source of hydrocarbons and other chemicals. Critical Reviews in Biotechnology 22 (3), 245–279.

Beal, C.M., Stillwell, A.S., King, C.W., Cohen, S.M., Berberoglu, H., Bhattarai, R.P., et al., 2012. Energy return on investment for algal biofuel production coupled with wastewater treatment. Water Environment Research 84 (9), 692–710.

Brown, A.C., Knights, B.A., Conway, E., 1969. Hydrocarbon content and its relationship to physiological state in the green alga *Botryococcus braunii*. Phytochemistry 8, 543–547.

Burlew, J.S. (Ed.), 1953. Algal Culture from Laboratory to Pilot Plant. Carnegie Institution of Washington, Washington DC, p. 357.

Chisti, Y., 2007. Biodiesel from microalgae. Biotechnology Advances 25 (3), 294–306.

Clarens, A.F., Resurreccion, E.P., White, M.A., Colosi, L.M., 2010. Environmental life cycle comparison of algae to other bioenergy feedstocks. Environmental Science and Technology 44, 1813–1819.

Clarke, T., Barlow, M., March 1, 2005. The battle for water. YES! Dec 2004. YES! Magazine. http://www.yesmagazine.org/article.asp?id=669.

Dahiya, A., 2012. Integrated approach to algae production for biofuel utilizing robust algae species. In: Gordon, R., Seckbach, J. (Eds.), The Science of Algal Fuels: Cellular Origin, Life in Extreme Habitats and Astrobiology, vol. 25. Springer, Dordrecht, pp. 83–100.

Dahiya, A., Todd, J., McInnis, A., 2012. Wastewater treatment integrated with algae production for biofuel. In: Gordon, R., Seckbach, J. (Eds.), The Science of Algal Fuels: Cellular Origin, Life in Extreme Habitats and Astrobiology, vol. 25. Springer, Dordrecht, pp. 447–446.

DoE, 2010. National Algal Biofuels Technology Roadmap. US Department of Energy, Office of Energy Efficiency and Renewable Energy, Biomass Program (accessed 30.06.10.). http://www1.eere.energy.gov/biomass/pdfs/algal_biofuels_roadmap.pdf.

Elliott, D.C., 2007. Historical developments in hydroprocessing bio-oils. Energy and Fuels 21 (3), 1792–1815.

Hall, D.O., Rao, K.K., 1999. Photosynthesis, sixth ed. Cambridge University Press, Cambridge.

Hershey, D., 2003. Misconceptions about van Helmont's willow experiment. Plant Science Bulletin 49, 78.

Hu, Q., Sommerfeld, M., Jarvis, E., Ghirardi, M., Posewitz, M., Seibert, M., Darzins, A., 2008. Microalgal triacylglycerols as feedstocks for biofuel production: perspectives and advances. Plant Journal 54, 621–639.

Lardon, L., Hélias, A., Sialve, B., Steyer, J.-P., Bernard, O., 2009. Life-cycle assessment of biodiesel production from microalgae. Environmental Science and Technology 43, 6475–6481.

Levine, R., Oberlin, A., Adriaen, P., 2009. A Value Chain and Life Cycle Assessment Approach to Identify Technological Innovation Opportunities in Algae Biodiesels. Paper presented at the Nanotech Conference & Expo, Houston, TX.

Lundquist, T.J., February 19–21, 2008. Production of algae in conjunction with wastewater treatment. Paper presented at the National Renewable Energy Laboratory-Air Force Office of Scientific Research Joint Workshop on Algal Oil for Jet Fuel Production. Arlington, VA.

Meier, R.L., 1955. Biological cycles in the transformation of solar energy into useful fuels. In: Daniels, F., Duffie, A. (Eds.), Solar Energy Research. Univ. Wisconsin Press, Madison,Wisconsin, USA, p. 179.

Metzger, P., Largeau, C., 2005. *Botryococcus braunii*: a rich source for hydrocarbons and related ether lipids. Applied Microbiology and Biotechnology 66, 486–496.

Mulbry, W., Kebede-Westhead, E., Pizarro, C., Sikora, L.J., 2005. Recycling of manure nutrients: use of algal biomass from dairy manure treatment as a slow release fertilizer. Bioresource Technology 96, 451–458.

Mulbry, W., Kondrad, S., Buyer, J., 2008. Treatment of dairy and swine manure effluents using freshwater algae: fatty acid content and composition of algal biomass at different manure loading rates. Journal of Applied Phycology 20, 1079–1085.

Oswald, W.J., 1990. Advanced integrated wastewater pond systems. In: Paper Presented at the Supplying Water and Saving the Environment for Six Billion People 1990 ASCE Convention EE Div/ASCE. San Fransisco, CA, November 5–8, 1990.

Oswald, W.J., 2003. My sixty years in applied algology. Journal of Applied Phycology 15, 99–106.

Oswald, W.J., Golueke, C.G., 1960. Biological transformation of solar energy. In: Umbreit, W.W. (Ed.), Advances in Applied Microbiology, vol. 2. Academic, New York, pp. 223–262.

Perez-Garcia, O., Escalante, F.M.E., De-Bashan, L.E., Bashan, Y., 2011. Heterotrophic cultures of microalgae: metabolism and potential products. Water Research 45, 11–36.

Rodolfi, L., Zittelli, G.C., Bassi, N., Padovani, G., Biondi, N., Bonini, G., Tredici, M.R., 2009. Microalgae for oil: strain selection, induction of lipid synthesis and outdoor mass cultivation in a low-cost photobioreactor. Biotechnology and Bioengineering 102 (1), 100–112.

Sánchez Mirón, A., Contreras Gómez, A., García Camacho, F., Molina Grima, E., Chisti, Y., 1999. Comparative evaluation of compact photobioreactors for large-scale monoculture of microalgae. J Biotechnol 70, 249–270.

Schenk, P., Thomas-Hall, S., Stephens, E., Marx, U., Mussgnug, J., Posten, C., Kruse, O., Hankamer, B., 2008. Second generation biofuels: high-efficiency microalgae for biodiesel production. BioEnergy Research 1, 20–43.

Sheehan, J., Dunahay, T., Benemann, J., Roessler, P., 1998. A Look Back at the U.S. Department of Energy's Aquatic Species Program: Biodiesel from Algae. National Renewable Energy Laboratory, Golden, Colorado.

Shifrin, N.S., Chisholm, S.W., 1981. Phytoplankton lipids: interspecific differences and effects of nitrate, silicate and light–dark cycles. Journal of Phycology 17, 374–384.

Sills, D.L., Paramita, V., Franke, M.J., Johnson, M.C., Akabas, T.M., Greene, C.H., Tester, J.W., 2012. Quantitative uncertainty analysis of life cycle assessment for algal biofuel production. Environmental Science & Technology 47 (2), 687–694.

Wolf, F.R., 1983. *Botryococcus braunii*: an unusual hydrocarbon-producing alga. Applied Biochemistry and Biotechnology 8 (3), 249–260.

Woertz, I., Feffer, A., Lundquist, T., Nelson, Y., 2009. Algae grown on dairy and municipal wastewater for simultaneous nutrient removal and lipid production for biofuel feedstock. Journal of Environmental Engineering 135, 1115–1122.

BIOMASS TO LIQUID BIOFUELS SERVICE LEARNING PROJECTS AND CASE STUDIES

15

Anju Dahiya[1,2]

[1] *University of Vermont, USA;* [2] *GSR Solutions, USA*

The Michigan State University Extension Senior Bioenergy Expert Dennis Pennington recommends before planting a biofuel crop, the first step should be to start with questions worth asking (see Chapter 7): "Do I have access to a reliable market? Do I need to invest in additional equipment or labor to plant, harvest or handle biomass? What is my cost of production? Which energy crop species fit my situation? What are the potential yield and prices I need for a biofuel crop to be at least as profitable as my current crop? Do I need to develop an exit strategy or timeline in consideration of the above listed items?"

All these questions indicate the importance of a strong business model for farmers to employ toward energy crop production. The service learning students need to first understand these questions and look at the success stories before diving deeper into this topic. In Chapter 8, the University of Vermont Extension experts Dr Heather Darby and Chris Callahan present two case studies, which highlight farm-based biodiesel production at a 13,000 gallon per year and a 100,000 gallon per year processing capacity. The smaller scale option is presented to demonstrate the feasibility of small volume production to result in a predictable fuel supply and costs, even when used at only partial capacity of 4000 gallons per year. This smaller scale operation is shown to produce fuel at a cost of $2.52 per gallon in the 4000 gallon per year start-up phase on 66 acres of sunflower production that Dr. Darby describes in her lectures.

The in-class lectures, followed by actual field trips, give students an opportunity to have face-to-face discussions with farmers and the farm team (Figure 15.1), and directly impacts the service learning experience. For instance, during the 2012 version of the Biomass to Biofuels course,[1] one of the students reflected in the class forum on the energy crops topic, *"They (dairy farmers) hate to convert acres from forage production in fear of not having enough feed. The farmers can feed the meal from these crops, but they still have to accept it. What it will take is a respected, innovative farmer to start saying how much he saved on fuel, how well his tractors run, and how well his cow's milk on the meal, then this may take off…"* Toward that end, a couple of students, Chuck Custeau, a loan officer, and Meghan Seifert, a junior in environmental studies, undertook a service learning project, *Cost Analysis of Oilseed Production for Biodiesel* in partnership with a local bioenergy farm, and worked with Dr Darby to identify the tall barriers to break down in getting farmers to grow the oilseed crops analysis—to see if dairy farmers could realize an economic benefit from dedicating 20% of their acreage to growing an oilseed for biofuel conversion. They found the dairy had an adequate land base,

---

[1] http://go.uvm.edu/7y1rr.

Bioenergy. http://dx.doi.org/10.1016/B978-0-12-407909-0.00015-8

FIGURE 15.1

Oilseed crop-based biodiesel process demonstration at the farm.

using acres that would have rotated in to corn or hay, and raised an oilseed crop for one year, and their analysis compared sunflowers, canola, and soybeans. The value of the by-product meal was also analyzed for its value as a substitute for purchased grain. The results of the analysis were favorable to biofuel production. This report is presented in this chapter. Note the tables in this report provide the analysis from different crops. Chuck and Meghan's effort was picked up by a 2013 Biomass to Biofuels course student, James MacLeish, who took their study to the next level and explored *The Volatility in Corn and Crude Oil Prices*. Chuck comprehensively combined the two service learning studies to present in this chapter (see below).

Earlier in 2011, a Biomass to Biofuels student, originally from Cameroon, Africa, an engineer employed in the Canadian oil industry, had also partnered with the same local farm that Chuck and Meghan collaborated with. For his service learning project, his goal was to learn from the established oilseed crops-based biodiesel farm, and bring the knowledge to Cameroon to establish a local energy crop-based biofuel facility. His long-term plan was to use the local feedstock supply, either oilseed from plants grown on land by local farmers or from the sap of palm trees grown on land by local farmers and/or palm oil producing businesses. After completing his service learning project over the next few years, he established a business and has been working with the local farmers.

The heating oil market is another consumer of biodiesel blends, and the blended fuel is termed Bioheat, as described by the National Oilheat Research Alliance (NORA) Director of Education and Efficient Heating Consultant and Educator Bob Hedden (see Chapter 13). He states, "The Oilheat Industry's top three service priorities are improved reliability, maximizing efficiency, and reduced heating equipment service costs. A significant number of unscheduled no heat service calls are caused by inconsistent fuel quality, fuel degradation, and contamination… As oil tanks age rust and sediments build up in the tank. The age of the fuel in the tank can also be a problem. Oil and Bioheat have a finite shelf life and break down over time. The third problem is the size and speed of delivery. Filling a tank kicks-up all the sediments and rust in the bottom of the tank that leads to plugged lines, filters and nozzles." Going through a similar situation at his family farm, a 2013 Biomass to Biofuels student, Engineering Energy Consultant Ethan Bellavance undertook a service learning project, and partnered with his own medium-sized commercial dairy farm to quantify the economics of reducing the farm's thermal heating costs by using harvested biomass from the farm. His farm has been using fuel oil and kerosene to heat its existing building infrastructure and hot water (spending upwards of $20,000 a year in heating fuel) with two 100 MBH[2] kerosene furnaces in onsite trailers and one 350 MBH fuel oil boiler system (see the case study *Thermal Heating Fuel Switch—a feasibility study for a farm*, presented in this chapter).

Oilseed-based crops are used for biodiesel production, whereas corn has been heavily used for ethanol production. As Dennis Pennington stated (see Chapter 7), "Starch and sugar-based crops like corn and sugarcane have given a rise to the fuel ethanol industry in the U.S. Farmers have increased production in an attempt to keep up with the rising world demand for corn. Cellulosic crops have an advantage over corn because they have lower production inputs and in many cases are perennial crops that need to be established only once every 10–15 years. However, cellulosic crops have lower yield and less income potential per acre." This thought was earlier reflected in the service learning project of another 2012 Biomass to Biofuel student, Tracey McCowen, also a doctoral student focused on the ethics and economics of ethanol and the role of ethanol policy in the 2007/2008 food crisis. In her service learning project *Mobile Ethanol Distillery Unit: a feasibility study*, she found that although ethanol is a costly product to produce, if the feed value of sorghum distillers' grains is included it may become viable. This may be of great assistance to dairy farmers faced with high feed input costs due to the 300% price increase of corn since 2004. See details in this study as presented in this chapter.

---

[2]1 MBH = 1000 BTU/hr. Thousand BTUs per Hour (MBH)

COST ANALYSIS OF OILSEED PRODUCTION FOR BIODIESEL AND THE VOLATILITY IN CORN AND CRUDE

15A

Chuck Custeau

2012 Biomass to Biofuels Student, University of Vermont, Burlington, VT, USA

Projects: (1) *Cost Analysis of Oil Seed Production for Biodiesel* project by Chuck Custaeu and Meghan Seifert (2012) and (2) *The Volatility in Corn and Crude Oil Prices* (project by James MacLeish (2013)—built his project on the 2012 project report, community partner Yankee Credit Union).

Community partners: Yankee Farm Credit, ACA (YFC)—a member borrower owned cooperative, which is a part of the farm credit system. Borderview farm in Alburgh, Vermont (worked closely with Dr Heather Darby, Associate Professor University of Vermont Extension).

Background: In 2012, an analysis was conducted to measure the economic feasibility of a dairy farm to take acreage from forage production to raise oilseed crops for biodiesel production. The study assumed a farmer with a 125 cow dairy farm with 300 tillable acres approached Yankee Farm Credit with a loan request to fund the purchase of an oilseed press, biodiesel reactor, storage bin, and used combine. The total amount of the loan request was $69,000 on a five-year term. The cost of the equipment was based on actual amounts for the various pieces of equipment. This hypothetical farm was raising 125 acres of corn silage and 175 acres of hay and hay silage. At the time of the study, corn was approaching $7.00/bu[3]. on the Chicago Board of Trade (CBOT). Diesel fuel was $4.00/gallon. Feed costs were increasing due to the demand for corn for ethanol production. As the price of corn increased, the demand for land to grow it on increased, displacing soybeans and consequently increasing their price as well. Vermont dairy farmers were faced with increasing energy and feed costs. The analysis used Dr Heather Darby's logic of using 20% of tillable acres available for biodiesel production.

Methods: The analysis compared the value of corn silage to soybeans, canola, and sunflowers. All of the crops made use of manure to supplement commercial fertilizer. The assumed rate of manure was 5000 gallons/acre, providing 60 pounds of nitrogen (N), 20 pounds of phosphorus (P), and 80 pounds of potassium (K) per acre. Commercial fertilizer was used to meet the nutrient requirements to grow 20 tons per acre of corn silage, 2400 pounds of soybeans per acre, 1500 pounds of canola per acre, and 1500 pounds of sunflowers per acre. Fertilizer prices were an average of local dealers in Northern Vermont. As with energy and grain prices, fertilizer was at historic price levels with urea at $800/ton, Phosphorus (DAP) at $600/ton, and potash at $600/ton. All of the crops used in the analysis were glyphosate-resistant to keep the cost of weed control equal. Seed prices were from Northern Vermont

---

[3]bu stands for bushel.

dealers. Debt service requirements (yearly principal and interest payments) were based on a five-year term with monthly payments at 5.5% interest, the standard rate charged by Yankee Farm Credit for a $69,000 loan.

Yields used in the analysis were from various trials conducted by the University of Vermont Extension Service (Darby, 2011a,b,c,d,e,f; White, 2007). Yields of meal and oil from the oilseed crops were from the same trials. Yields of biodiesel were based on one gallon of oil yielding one gallon of biodiesel. The amount of diesel fuel required for a 125 cow dairy was from the 2010 Northeast Dairy Farm Summary conducted by the Northeast Farm Credit Associations (2012). Planting and harvest cost were assumed to be equal for all crops.

Results: The value of the total yield (oil and meal) was compared to corn silage. None of the oilseed crops could produce enough oil to meet the entire needs of the farm. This is not to say that it was not economically feasible to grow and produce biodiesel. The value of corn silage was determined by using corn silage yielding 20 ton per acre at $35 ton per acre. After deducting the growing cost and the purchase of fuel for the farm, corn silage contributed $954 of net profit to the farm. The same calculation was made for soybeans yielding 2040 pounds of meal and 48 gallons of oil per acre. Total revenue from meal and biodiesel was $648 per acre, or $38,889 for 60 acres. After factoring debt service and purchasing the balance of fuel needed, soybeans had a net profit of $12,021. The same analysis was used for canola. After factoring in the value of canola meal and biodiesel and deducting the cost of diesel fuel that was not supplied from biodiesel, canola contributed $7528 of net profit. Sunflowers were subjected to the same analysis, and they had a loss of $1701.

Based on spring 2012 prices of fuel, corn silage, soybean, canola, and sunflower meal, a dairy farm could profitably grow some of its own diesel fuel requirements. The prices of all the items in the study are volatile. For example, corn silage in the fall of 2012 was selling for $50–70/ton due to increasing grain prices. The term of the loan was for five years. Before borrowing or investing in oilseed/biodiesel production, a farmer would need to know if this investment would be profitable for the life of the loan. A subsequent analysis was performed comparing the price of corn and soybeans to that of crude oil. Three scenarios were compared: a Rising Price Scenario, a Business as Usual Scenario, and a Dropping Price Scenario. In the Rising Price Scenario, the study determined that for every dollar increase in the barrel price of crude oil, corn increased $3 per metric ton, and soybeans increased $5.60 per metric ton. In the Business as Usual Scenario, crude oil increased $15 per barrel. Corn and soybeans increased $3.33 and $5.60 per metric ton. In the Dropping Price Scenario, crude oil falls by $10 per barrel. Corn price stays the same, and soybeans would decrease $4.50 per metric ton. This based on historic data.

This analysis demonstrates that a farmer's investment in capital for the equipment to produce biodiesel and the opportunity cost of taking land out of forage production to be used to grow soybeans has a good chance of being profitable (Tables 15A.1–15A.9).

Benefits to community partners: Through this research and report we were able to provide benefits to our community partners. Yankee Farm Credit will benefit from this project by using this report as a template to help analyze loan requests for biofuel projects. This report will give credit analysts and loan officers a basis for determining whether a request will be beneficial to the borrower and if the endeavor will generate enough revenue and profit to pay for itself. Yankee Farm Credit is not our only community partner who will benefit. This report demonstrates that Borderview Farms is generating a profit by raising and producing oilseed crops for biofuels. This cost analysis will show the farm owner which of the oil feed stock crops generated the most profit for him. Overall, we believe that

Table 15A.1 Equipment Total

Item	Amount
Used gleaner N6 combine with six row corn head and 20' flex head	$20,000.00
Ayang press	$3500.00
Spring board biodiesel reactor	$17,000.00
Used 20,000 bu. bin/drier	$15,000.00
20 ft. auger to load bin	$5000.00
Misc tanks, pumps, filters, etc.	$7500.00
Equipment total	**$68,000.00**

Table 15A.2 Loan Total

Item	Amount
UCC recording fee	$25.00
YFC stock purchase	$1000.00
Monthly payment 84 months, 5.5%	$992.00
Loan total	$69,025.00
Yearly debt service	$11,904.00

Table 15A.3 Crop Cost

Crop	Seed Cost	Seed Rate/ Acre	Seed Cost/ Acre	Fertilizer Cost/ Acre	Pesticide Cost/ Acre	Total Input
Corn	$300.00	30,000	$112.00	$157.39	$50.00	$319.89
Sunflower	$250.00	30,000	$50.00	$ -	$ -	$ -
Canola	$650.00	50	$65.00	$ -	$ -	$ -

this report will show not only our community partners, but the community as a whole, which oilseed crop has the most profit for biofuel production.

Future Directions: After reviewing and analyzing all of the costs of inputs and the net return, it appears that a dairy farmer could benefit from producing their own biodiesel. With the research on oilseed biodiesel production throughout Vermont, there are many ways for farmers to obtain the resources needed for starting oilseed production. Despite our conclusion, the potential benefits from producing biodiesel will vary greatly on each individual farm. Based on the farmer's management

Table 15A.4 Fertilizer Requirements

| Crop | Requirements | | |
	Nitrogen (N)/acre	Phosphorus (P)/acre	Potassium (K)/acre
Corn	150	80	160
Manure contribution	60	20	80
Fertilizer requirement	90	60	80
Soybeans	0	30	45
Manure contribution	60	20	80
Fertilizer requirement	0	10	0
Sunflowers	125	25	30
Manure contribution	60	20	80
Fertilizer requirement	65	5	0
Canola	112.5	26.25	45
Manure contribution	60	20	80
Fertilizer requirement	52.5	6.25	0

Table 15A.5 Fertilizer Costs

| Crop | Cost/Acre | | | |
	$/lb N	$/lb P	$/lb K	Fertilizer Cost/Acre
Corn	$0.87	$0.65	$0.50	$157.39
Soybean	$ -	$6.52	$ -	$6.52
Sunflower	$56.52	$3.26	$ -	$59.78
Canola	$45.65	$4.08	$ -	$49.73

Table 15A.6 Cost Comparison

Crop	Fertilizer Cost/Acre	Seed Cost/Acre	Herbicide Cost	Total
Corn	$157.39	$112.50	$35.00	$304.89
Soybean	$6.52	$ -	$35.00	$41.52
Sunflower	$59.78	$50.00	$35.00	$144.78
Canola	$49.73	$65.00	$35.00	$149.73

Table 15A.7 Input Cost Comparison

Crop	Input Cost/Acre	Yield Tons/Acre	Meal/ Acre	Biodiesel	Cost/ Unit	Total
Corn	$304.89	20	—	0	$40.00	$800.00
Soybean	$41.52	—	2040	48	$0.22	$448.80
Sunflower	$144.78	—	930	74	$0.11	$102.03
Canola	$149.73	—	1630	74	$0.16	$260.80

Table 15A.8 Revenue and Cost Comparison

Crop	Gross Revenue	Input Cost	Debt Service	Total Cost
Corn	$48,000.00	$18,293.40	$ -	$18,293.40
Soybean	$38,448.00	$2491.20	$11,904.00	$14,395.20
Sunflower	$23.881.80	$8686.80	$11,904.00	$20,590.80
Canola	$33,408.00	$8983.80	$11,904.00	$20,887.80

Table 15A.9 Net Comparison

Crop	Net	Fuel Purchase: 5688 Gallon at $/gal	Net after Fuel Costs
Corn	$29,706.60	$22,752.00	$6954.60
Soybean	$24,052.80	$12,032.00	$12,020.80
Sunflower	$3291.00	$4992.00	$(1701.00)
Canola	$12,520.20	$4992.00	$7528.20

techniques and environmental conditions of the farm, the yields and start-up costs will vary widely. Therefore, research and a complete analysis would be required for each farm. As we have seen through this report, farmers can certainly benefit from producing their own biodiesel.

REFERENCES

Custeau, C., Seifert, M., 2012. Cost Analysis of Oil Seed Crop Production for Biodiesel. UVM, Burlington, VT. Unpublished raw data.

Darby, H., 2011a. Soybean Tineweed Trail. University of Vermont Extension. http://www.uvm.edu/extension/cropsoil/wp-content/uploads/Soybean_tineweed_report_20111.pdf.

Darby, H., 2011b. Winter Canola Variety Trial. University of Vermont Extension. http://www.uvm.edu/extension/cropsoil/wp-content/uploads/2011WinterCanolaVarietyTrialReportfinal.pdf.

Darby, H., 2011c. Sunflower Tineweeding Trail. University of Vermont Extension. http://mysare.sare.org/MySare/assocfiles/9432902011%20Sunflower%20Tineweed%20report.pdf.

Darby, H., 2011d. Sunflower Variety Trail. University of Vermont Extension. Retrieved from: http://www.uvm.edu/extension/cropsoil/wp-content/uploads/2011_Sunflower_VT_report_final.pdf.

Darby, H., 2011e. Sunflower Seeding Rate × Nitrogen Rate Trial. University of Vermont Extension. http://www.uvm.edu/extension/cropsoil/wp-content/uploads/Sunflower_SR×NR_report_final.pdf.

Darby, H., 2011f. Vermont Sunflower Planting Date Study. University of Vermont Extension. http://www.uvm.edu/extension/cropsoil/wp-content/uploads/2011_Sunflower_Planting_Date_Study-final.pdf.

MacLeish, James, 2013. Volatility in Crude Oil and Corn Prices. UVM, Burlington, VT. Unpublished raw data.

Northeast Farm Credit Associations, May 2012. 2011 Northeast Dairy Farm Summary. Prepared by Lidback, J., Laughton, C., Farm Credit East, ACA, Farm Credit of Maine, ACA, Yankee Farm Credit, ACA. https://www.farmcrediteast.com/Knowledge-Exchange/~/media/Files/Knowledge%20Exchange/2011%20DFS%20Report.ashx.

White, N., August 2007. Alternatives for On-farm Energy Enhancement in Vermont: Oilseeds for Feed and Fuel. Vermont Biofuels Association. http://www.uvm.edu/~susagctr/Documents/SAC%20Oilseed%20ExecSumm_Final.090107.pdf.

THERMAL HEATING FUEL SWITCH

15B

Ethan Bellavance

2013 Biomass to Biofuels student, University of Vermont, Burlington, VT, USA

Summary: The community partner, Sunset Lake Farms is a medium-sized commercial dairy farm located on the peninsula of Alburgh, Vermont. This farm, like nearly every other in state, relies on outsourced resources to operate efficiently and produce its core product, dairy milk. In recent history the cost of sourcing these resources has increased. Sunset Lake currently uses #2 fuel oil and kerosene to heat its existing building infrastructure and hot water. The current heat load for the farm is distributed between two trailers, a farm office, the milking parlor complex, and hot water used for sanitization. In the future, there are plans to expand the heating demand by constructing a calf barn and another administration building. Thermal heating at Sunset Lake Farms consumes approximately 6000 gallons of fuel oil equivalent in its annual operation.

Project Objectives: The goal of this report is to quantify the economics of reducing Sunset Lake Farms' thermal heating costs by utilizing woody biomass. This ensures Sunset a constantly priced energy supply to meet its growing needs. In analyzing this goal, the costs of retrofitting the existing heating systems to a biomass alternative will be examined. The main options to consider in this scenario are whether to create a centralized boiler system that pipes hot water to different load centers, or continue to provide decentralized heating, utilizing existing furnace and boiler systems. These recommendations will create a road map for Sunset to follow, and give them information they need to take the next step in contacting biomass engineering firms. This road map will look at and weigh differing biomass systems from harvest to combustion stages. This will provide Sunset with the best information available to receive the most cost-effective system possible. The biomass feedstocks to be analyzed are cord wood, wood pellets, and wood chips.

Background: Currently, Sunset Lake Farms spends upwards of $20,000 a year in heating fuel. There are two 100 MBH kerosene furnaces in on-site trailers, and one 350 MBH fuel oil boiler system. This system heats the business office, milking parlor, and all the hot water used for the dairy. It is important to note that the existing boiler and furnace systems are antiquated and will need to be replaced in the near term if they are to provide the full heating load of the facility. Due to this required infrastructure investment, costs required for a new biomass heating system are reduced due to the need to replace the existing fossil fuel heating. Sunset Lake Farms views converting this system to biomass as a win-win situation where they can reduce capital expenditures, better utilize their land, and keep energy dollars within the Vermont economy.

Agricultural facilities in the Northeast are uniquely tailored to reap the benefits of biomass heating; farms have an intimate relationship with the land and are often located near wood resources. In addition, farms that crop large acreage commonly have to trim their field boundaries to keep them free of woody debris. This "cleaning" process in itself generates significant quantities of biomass. This biomass can be chipped, cut, or ground into pellets to produce a product that can be utilized for heating. While heating requirements on many farms, specifically dairy, are not extensive when compared to industrial or large commercial installations, the proximity of the farms to local wood resources and the relative ease of processing woody biomass make them ideal candidates to utilize this resource. Farmers want to reduce their reliance on fossil fuels and their economic costs of thermal heating. They also want to enhance their native forest landscape. If managed correctly, biomass heating can accomplish all of these goals.

Community Partner: The specific community partner that I'm working with at Sunset Lake Farms is Tom Bellavance, my father, and the owner of the facility. Tom believes if you can control the input costs of your agricultural enterprise, you have additional flexibility to diversify your core product. Tom and I have developed an energy plan to reduce Sunset's input costs, increase profitability, and reduce dependence on outside resources. The first step in this plan was to install energy-efficient equipment for dairy processing. Efficiency measures are traditionally the most cost-effective to implement in an operation. From there, we have begun to look at fuel switching for the farms' thermal energy demand. Sunset would like to reduce the quantity of fuel oil and kerosene used for heating its operation.

Plan of Work: The plan for this facility-wide analysis is as follows. Sunset Lake Farms will be looked at through a thermal efficiency lens. Recommendations will be recorded and provided as additional feedback to the farms' manager. These recommendations will be broad-based and will not include additional insulation measures, exclusively air sealing and best management practices. The reason for this is multifaceted. Due to process refrigeration that occurs in the facility, there is an extensive amount of heat rejection that is currently captured in the conditioned space. This heat has to be expended regardless, so insulation can only provide limited benefit. Sunset also has little interest in looking into extensive retrofits of the facility because the majority of the thermal heating required in the facility is for hot water use and not space heating.

- After the farm has been tested through the thermal efficiency lens and recommendations have been provided, an analysis of biomass feedstock availability will be looked at. This feedstock analysis will take into account best management logging practices, which allow harvesting of only the quantity of biomass that is annually created.
- Once this feedstock analysis has been completed, we will look into whether it is more economical to follow a centralized or distributed heating system for the facilities' footprint. The pros and cons of each type of system will be looked into, and a recommendation will be drawn up.
- Once the recommendation for following a distributed or centralized heating plan has been established, the pros and cons of each woody biomass feedstock will be looked at. The feedstocks options are cord wood, wood pellets, and wood chips. The economics will be analyzed, which will allow Sunset to make decisions that are applicable to their business. Once these areas have been analyzed, information will be presented.

Results/Expected Outcome: Efficiency testing was conducted at Sunset Lake Farms on March 9, 2013. The results of this testing were a list of recommendations for air leakage reduction. These recommendations are below.

- Input gasketed penetrations to allow for cables to be strung through the sealed wall cavity. This alleviates the problem of having to crack windows open to string cables through them.

- Fix broken windows that will not shut; these open windows, although in an area heated by condenser fan exhaust, cause condensation and mold when not properly closed.
- Doors need to have auto closing mechanisms on them, or training needs to be provided to staff to ensure doors are closed during the winter.
- Doors that close the area between the milking parlor and the free stall barn should be automated to open and close via a time clock; currently doors are left open except for on the coldest nights.

While these recommendations are basic in regards to thermal efficiency, they will reduce the amount of space heating required for the office and milk parlor facility. It is important to note that space heating is only one part of the equation; approximately 50% of thermal heating demand is due to process hot water. Process hot water use is critical for a dairy operation to ensure that milking equipment and infrastructure are sanitized. This keeps both cows and cow's milk clean for human consumption. Sunset Lake Farms consumes 870 gallons of 180° water and 360 gallons of 120° water per day to ensure safe operating conditions. This hot water heating alone uses 3500 gallons of fuel oil each year to heat. This is more than 50% of the total heating bill for the agricultural facility, and it is important to keep in mind when sizing biomass heating systems.

It is critical to analyze biomass feedstock availability to ensure adequate and sustainable wood supply for a biomass heating system. Based on conversations with foresters, it is safe to assume that Sunset's property can achieve approximately one ton of biomass production per acre. As shown in Figure 15B.1 below, Sunset Lake owns approximately 120 acres of woodlands. This correlates to 120 tons of woody biomass, which can be harvested each year sustainably. Pictures of the varying forested areas are shown below; all pictures were taken to the same scale. It is believed that with proper forest management practices, an adequate supply of wood can be harvested from these forests to cover the heating needs of the farm. The farm uses 780,000,000 BTU/year for heating. It's assumed that 7,600,000 BTU of energy can be gathered per ton of wood chips. Therefore, 102 tons of wood chips will be required each year to heat the agricultural facility. This value is less than the 120 tons expected

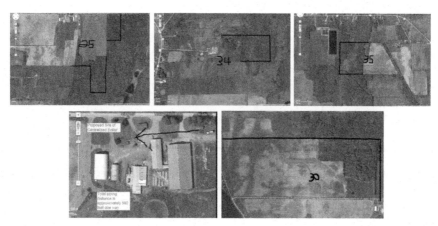

FIGURE 15B.1

Sunset farms' woodland acreage. Top right: proposed underground piping runs.

from the forest and gives Sunset Lake Farms a nice buffer of material. If efficiency measures are used, then this quantity of wood can be reduced further.

Sunset Lake Farms has enough biomass to sustainably heat their operations at current conditions. The next decision to make is whether Sunset should look into a centralized boiler system or a distributed heating system. The benefit of a centralized boiler system is that there are fewer pieces of equipment, which can lead to cost savings. In addition, maintenance can be reduced and system efficiency can be increased by having fewer centrally located mechanical systems. Detriments of a centralized boiler system are that heat must be transferred from the generation source to the load. This requires underground hot water piping that quickly adds up in cost and has the potential to lose heat. In addition, water to air heat exchangers will be required to transmit high temperature water to air, which can then be moved through the existing trailers' ductwork system. It is estimated that approximately 1000 ft of underground piping will be required to transmit hot water to and from all load centers. Piping costs are estimated to be around $15 per foot. Due to this cost, the longest run of the system (which is to the second trailer) may be eliminated due to piping distance. This elimination would reduce piping lengths by approximately 400 ft. Figure 15B.1 (bottom right) highlights the proposed underground piping runs and estimates total piping distance.

The hot water to air heat exchangers will be sized to match the existing load of the oil furnace in the trailers. These furnaces have a capacity of approximately 100,000 BTUS/h. Heat exchangers of this capacity can be readily bought at approximately $300 per system. In addition, the trailers' furnace fan will have to be reconfigured to ensure that proper system operation is obtained. Decentralized systems have their benefits and detriments as well. It would cost approximately $2000 to install an additional 100,000 BTU pellet stove, hearth, and venting system to match the output of the existing trailers. This cost would increase another $1000 if a cord wood stove were to be installed instead. This investment of $6000 has the potential to dramatically cut costs while ensuring redundancy; the existing oil furnace would not be removed. As you can see, there are some compelling arguments to keep a decentralized heating system established, especially for the second trailer that has a 300 foot pipe run.

Based on conversations with Sunset Lake Farms, if financially viable, it would be preferable to support a centralized heating system compared to a decentralized system. These discussions revealed

Table 15B.1 The Three Options Are Cord Wood, Wood Pellets, and Wood Chips

End Use	Benefits	Detriments
Cord wood	Least altered end use, high EROI/mile traveled, cheapest processing costs. Comparable infrastructure investment. When coupled with water storage, systems become quite efficient.	If incorrectly managed, fires can be polluting, low automation = high human involvement, stoves could have to be potentially stoked at least once every day.
Pellets	Automation allows Sunset greater freedom. Biomass can be stored outside, low emissions, high automation.	High cost, has to be processed, lower EROI. Wood chipper, hammer mill, and pelletizer would have to be purchased.
Wood chips	Cheapest when bought on the open market, has potential to be utilized in a smaller application like this. Easy processing, can be burned with high moisture content.	Chips have to be stored indoors or at least under cover to avoid freezing, mechanization and costs are involved with system. Freezing could be avoided by keeping wood in log form.

Table 15B.2 Cost of Systems

Sunset Lake Farms Information Biomass Installation Costs

Type of Systems	Proposed System Size (MBh)	Estimated Boiler Cost Delivered ($/MBh)	Installation Cost of Turnkey System ($/MBh)	1000 Gallon Hot Water Storage Tank ($)	Trench Cost at 400 ft of Pipe	Processing Equipment Cost ($)	Estimated Installation cost ($)
Cord wood	400	100	86	10,000	6000	1,500	91,786
Wood pellet	350	86	171	10,000	6,000	15,000	121,000
Wood chip	350	129	171	10,000	6,000	10,000	131,000

Biomass System Economics

Type of System	Estimated Installation Cost ($)	Estimated Thermal Heating Fuel Consumption (Gallons)	Estimated Thermal Heating Fuel Offset (Gallons)	% Fossil Fuel Reduction	Estimated Yearly Fossil Fuel Savings ($)	Estimated Rate of Return	Simple Payback
Cord wood	91,786	6000	5300	88%	21,200	23%	4.3
Wood pellet	121,000	6000	5300	88%	21,200	18%	5.7
Wood chip	131,000	6000	5300	88%	21,200	16%	6.2

that there is a high initial investment required when installing underground piping systems. Therefore, piping should be limited to as short a distance as possible. With this information in hand, Sunset Lake believes the most economical route forward is to install a centralized heating system that supplies hot water to all load centers except for the farthest trailer. For that trailer, a cord wood or wood pellet stove makes more financial sense, unless the cost of piping can come down. We have now determined we are going to follow a largely centralized boiler based wood system. The question now lies with which type of woody biomass feedstock to utilize. The three options are cord wood, wood pellets, and wood chips. Each of these system types has unique advantages and disadvantages—best displayed in Table 15B.1.

Sunset Lake Farms is determined to ensure their systems achieve the most efficient and environmentally friendly operation. In doing so, it is believed that operating with thermal storage is important. It is expected that approximately 1000 gallons of thermal storage are required at this location. This number was chosen because it matches the daily hot water load required for the farm. Hot water storage is planned to be implemented in all types of centralized boiler-based systems.

Costs of systems are extremely important; they need to fall into an area where it can become financially viable for systems to operate. A calculator has been created to quantify these costs; a readout is presented in Table 15B.2. As you can see, all systems fall within a relatively close range of paybacks. Based on the information provided above, Sunset Lake Farms plans to make their decision on which wood system to follow in the upcoming months. This baseline assessment has given them the necessary information to make an informed decision.

Future Directions: Sunset Lake has a lot to think about in regards to which woody biomass combustion system to pursue. They plan to contact a woody biomass engineering firm in the upcoming months to hammer down cost estimates, system benefits, and system design. With each of these systems, there are pros and cons, but in the end, the message is clear. Efficient thermal generation can be created from woody biomass. Wood heat systems not only stimulate local economies, but also reduce farm expenditures and emissions. If you have any questions in regards to how this project is moving along, do not hesitate to contact me.

Benefits to Community Partner: Sunset Lake Farms has another step in their road map towards sustainability completed. With the recommendations provided, they have the potential to reduce their annual operating costs by approximately $20,000 and achieve a return on their investment of around 16%. In addition, Sunset now has an avenue in which they can more effectively utilize the land they currently own. (Figure 15B.1)

MOBILE ETHANOL DISTILLERY UNIT: A FEASIBILITY STUDY

15C

Tracey McCowen

2012 Biomass to Biofuels student University of Vermont, Burlington, VT, USA

SUMMARY

Vermont dairy farmers have been economically harmed by the high price of corn and fuel inputs. Distillers Grains (DG), a coproduct of ethanol production, are an alternative feed source to corn. However, it is not feasible to transport wet DG over long distances. Therefore, the benefit of ethanol production coproducts are not realized by farmers in a state like Vermont. An increasing number of farmers in Vermont are producing on-farm biodiesel in an attempt to reduce costs. Ethanol could be a substitute for methanol, which must be added to oil at a concentration of 20% in order to create biodiesel. Since methanol as an additive is derived from fossil fuel, it is therefore priced accordingly. This makes methanol susceptible to global supply issues that drive up costs from which Vermont farmers have no buffer. However, the infrastructure necessary to produce ethanol is prohibitively expensive. This research seeks to discover if start-up costs can be lowered and if shared infrastructure by way of a mobile unit might be feasible when the coproduct feed source is added.

BACKGROUND
BIODIESEL PRODUCTION

Biodiesel is primarily tallow or vegetable oil, and it has been used as a fuel for millennia. However, ever since the Bush administration's biofuels mandate in 2005, there has been an explosion in their research and development. The demand for oilseed crops, from which oil to make biodiesel is derived, has jumped dramatically. In conjunction with these events, the global demand for fossil fuels has risen exponentially as other countries, once poor, have industrialized rapidly. As demand increases, so do prices, and this has been especially hard on farmers whose means of agricultural production are intrinsically linked to fossil fuels through the bonds of fuel, fertilizers, and other agricultural chemicals.

As a result, the incentive for farmers to disengage their ties from the fossil fuel industry has never been so great. Since farmers already have land, fuel self-sufficiency is only a few steps away. Roger Rainville of Borderview Farms in Alburgh, Vermont has demonstrated that he can produce biodiesel from any one of a number of oil seed crops for $1.70/gallon, and this includes the amortization on his equipment. That is a pretty big incentive at a time when "off-road" diesel is nearly $4/gallon. As previously mentioned, biodiesel is primarily vegetable oil. However, it has two other key ingredients that transform the vegetable oil into premium quality biodiesel that can run any conventional diesel engine. These other two ingredients are lye (potassium hydroxide or hydrated potash 99%) and methanol 99%, or ethanol. The lye is used in trace amounts and is therefore not a significant cost. However, methanol (ethanol) must make up 20% of the biodiesel mix, and when prices are high, this is a significant expense. For every 100 gallons of vegetable oil, 20 gallons of ethanol are needed to make biodiesel. The seed meal left over from pressing the oil seeds is a valuable fertilizer. Sunflower meal has an NPK ratio of 5.6:1.2:1.5, and canola has an NPK ratio of 4.6:0.74:0.68 Darby et al., 2014. A fertilizer seed meal provides organic matter to the soils, something synthetic fertilizers derived from fossil fuels do not. The meal can also be pelletized and used as a fuel for stoves or furnaces, or the pelletized meal can be used as a high protein feed for livestock. This means there are a number of coproducts associated with biodiesel beyond fuel use.

ETHANOL PRODUCTION

Ethanol is basically vodka, gin, or whiskey. It is made in much the same way, except that one does not need to worry about flavor, and one has to achieve 199 proof for ethanol. A brewer's mash is made from a product with a high sugar content grain. Enzymes or yeasts are added to this slurry that is heated, and fermentation is allowed to take place. Once the mash has fermented, it is distilled. According to Chris Cogburn of the National Sorghum Producers, sorghum in some environments can yield up to 1410 gallons of ethanol per acre, compared with corn's 499 gallons per acre (Cogburn, 2009). One of the coproducts of ethanol production is the bagasse, or DG, which is a valuable feed more easily digestible to livestock than straight grain. DG comes in various qualities, but dry DG now sells for a similar price as straight pure grains. Wet DG have not been further processed via energy intensive dehydration; therefore, they offer a cheap nutritious feed source for farms near an ethanol plant (Dooey and Martens, 2009). Sorghum DG is a direct substitute for corn DG and offers nearly three times the yield of corn per acre. The 2010 Callahan report for on-farm ethanol production indicates a breakeven cost of $5.50 per gallon, if bagasse is used to fuel the distillation process. However, the Callahan report failed to calculate the feed value of the sorghum DG. With a surplus DG yield of about three tons per acre, assuming the DG is worth about the same price as corn, currently 6.55/bushel. With 37.9 bushels to the ton, the Callahan report misses an additional value of $747 per acre.

FERMENTATION

The fermentation process for ethanol would be completed by the farmer on the individual farm premises. It takes 100 gallons of brewer's mash to produce 10 gallons of ethanol. So the space requirements are considerably larger than for a still. Also, hygiene is of importance in the fermentation process, but most farmers familiar with operating a dairy would understand the hygiene standards. The equipment necessary for this would be similar to a microbrewery.

DISTILLATION

The equipment necessary to distill brewer's mash into ethanol is one of the limiting factors to ethanol production due to its expense. Consequently, it was this expense that inspired the concept of a mobile distillery that could be shared among farmers wanting to make their own ethanol to make biodiesel. The idea is that the mobile unit would be a cooperative with a designated operator. The mobile unit would only contain the still. It takes about 2.7 MBTUs to distill 148 gallons of ethanol, which is the average yield per acre on Stateline Farms in Vermont (Callahan, 2010). The still would need to have an extra high stack since the goal is to get a nearly pure alcohol liquid with barely any water in it.

FRACTIONAL DISTILLATION

In winter, a farmer could use the freezing temperatures for fractional distillation. This is when the fermented mash at 10% alcohol by volume is allowed to freeze. Because alcohol has a lower freezing point than water, the water freezes at the top of the tank. The liquid can then be drained from the bottom, leaving the ice crystals behind and thereby increasing the alcohol concentration without additional energy expenditures.

REGULATIONS

Before setting up an ethanol distillery operation, the operator needs to obtain a license from the Bureau of Alcohol, Tobacco, Firearms, and Explosives (ATF). It is not clear if a mobile unit has ever sought a license before; therefore, other constraints may inhibit the feasibility of a mobile ethanol unit. It is beyond the scope of this short report to pursue an individual request from the ATF.

SURVEY RESULTS

The following survey results are based upon state distillers who are distilling a variety of different products, from apples to corn. It demonstrates that there is a wide cost range for still setup.

Operation	Still Cost	Production Rate	Capacity or Batch Size
Caledonia Spirits	Refused	—	—
Green Mountain Distillers	50,000 (self-made)	Variable ~ vodka different than ethanol	200 gallons
Stateline Farms	200,000	70 g/h	—
Vermont Spirits	Did not respond	—	—
Shelburne Orchards	7000	?	500 gallons
Chris Davis	120,000 (1980s)	65 g/h	1400 gallons
Internet	6178	>3g/batch	28 gallons

Could ethanol be viable in Vermont? Grain prices have fallen significantly in 2014, but Vermont dairy farmers would still be vulnerable to another spike in grain prices. That being said, there could still be a potential for ethanol made from milk whey, a waste product from the cheese industry. Although this would not provide livestock feed as DG do, it could still provide a methanol alternative that in times of fossil fuel price volatility could buffer Vermont farmers from global markets. In addition, it would solve nutrient pollution associated with whey disposal.

CONCLUSIONS

It cannot be determined whether a mobile ethanol unit would be financially viable or even legal due to licensing restrictions. However, the value of coproducts in ethanol production may change previously determined costs. In most situations, corn is processed for ethanol with fuel as the primary product in mind. In Vermont, the feed could be the most valuable product and the ethanol the secondary coproduct. Although ethanol is a costly product to produce, if the feed value of sorghum DG is included, it may become viable. This may be of great assistance to dairy farmers faced with high feed input costs due to the 300% price increase of corn since 2004 (2012).

This report demonstrates that ethanol production in Vermont sourced from grain is only viable when feed prices or fossil fuel prices are high. Ethanol can only be a substitute for methanol when it is of extremely high proof (nearly 100% alcohol). Corn-based ethanol is not as cost-effective as ethanol from sorghum, but sorghum is not easily stored. A shared mobile ethanol unit would reduce start-up costs. Finally, ethanol could be a means of reducing waste whey pollution from large cheese factories. This could have significant environmental ecological benefits for Vermont's waterways.

It is important to recognize the importance biofuels play in acting as a buffer against uncontrollable costs: the greatest risk to any manufacturing operation. As research into biofuel production increases, no doubt costs will change and new manufacturing opportunities will arise.

ACKNOWLEDGMENTS

Thank you to John Williamson of Stateline Farms, Duncan and Todd of Green Mountain Distillery, Nick Crowles of Shelbure Orchards, and Chris Davis of Meach Cove Trust. All of whom provided valuable time and information.

REFERENCES

Callahan, C., 2010. Producing Ethanol for Biodiesel in Vermont. Prepared for VT Agency of Agriculture, Food and Markets REAP Grant Report (Award #REAP070004.).

Cogburn, C., 2009. Sorghum as a Biofuels Feedstock. National Sorghum Producers. http://client-ross.com/lifecycle-workshop/docs/4.2_Cogburn_National_Sorghum_Producers_6-10-09.pdf.

Dooley, F., Martens, Bobby J., 2009. Using Distillers Grains in the US & International Livestock and Poultry Industries, p. 202. Available at: http://www.card.iastate.edu/books/distillers_grains/ (retrieved 02.05.12.).

Heather Darby, Karen Hills, Erica Cummings, and Rosalie Madden, Assessing the value of oilseed meals for soil fertility and weed suppression. University of Vermont Agricultural Extension report. Available at: http://www.uvm.edu/extension/cropsoil/wp-content/uploads/finalereportmeals10.pdf. September 17, 2014.

GASEOUS FUELS AND BIOELECTRICITY 4

ANJU DAHIYA[1,2]

[1] *University of Vermont, USA,* [2] *GSR Solutions, USA*

Gaseous fuels can be produced as a by-product of the breakdown of organic material by anaerobic digestion (e.g., biogas—mixture of methane and carbon dioxide) or gasification (e.g., syngas) or pyrolysis. Additionally, both of the processes can be used for generating electricity. Syngas could be converted to methanol, butanol, etc. The global biogas market is broken down by end use: *municipal power generation, on-site heat and power production,* and *transportation applications*[1]. This technology is well established in Europe and Asia. In the United States, it is emerging, as of January 2014, there are approximately 239 anaerobic digester systems operating at commercial livestock farms in the United States that generated approximately 840.6 million kWh equivalent of energy in 2013[2]. The Compressed Natural Gas (CNG) is an emerging solution for transportation.

Part IV contains one chapter on gasification and two on anaerobic digestion. In addition, the detailed description of the pyrolysis process is available in the respective chapter under part V.

Chapter 16 (Thermal Gasification of Biomass—A Primer) provides an introduction to the fundamental chemical and physical bases for gasification to show how various gas mixtures

---

[1]http://www.environmentalleader.com/2014/02/21/anaerobic-digestor-market-to-reach-7-bn/.
[2]http://www.epa.gov/agstar/news-events/digest/2014spring.html.

FIGURE 1

Gaseous fuels and bio-electricity.

Picture courtesy: A. Dahiya.

can be produced and how heat is internally generated and managed. It describes the ranges of the heat contents of gas mixtures that are produced in different gasification schemes. The importance of understanding the role of pyrolysis is also discussed. Several generic types of industrial gasifiers are described along with the applicability and limitations of each. It concludes with a description of the needs for gaseous product stream purification prior to end use. For additional information about pyrolysis see chapter 26.

Chapter 17 (Basics of Energy Production through Anaerobic Digestion of Livestock Manure) gives an overview of the anaerobic digestion process using livestock manure. The benefits of anaerobic digestion and what makes an operation appropriate for anaerobic digestion are described.

Chapter 18 (Bioenergy and Anaerobic Digestion) describes the detailed anaerobic digestion process steps (hydrolysis, fermentation or acidogenesis, acetogenesis, methanogenesis), the fermenting bacteria and methanogen relationship, feedstock, carbon to nitrogen ratio (C/N), volatile solids, the process of starting a digester, starting the digester during cold weather, the loading rate and its calculation, hydraulic loading, the operation and control of a digester, the role of mixing in biogas production, and types of anaerobic digesters.

Chapter 19 (Gaseous Fuels Service Learning Projects and Case Studies) includes service learning projects on these two topics i.e. biogas and gasification: effect of daily variation in food waste on biogas production during anaerobic digestion, potential for anaerobic digestion in meeting statewide energy needs, and biomass gasification as a strategy for rural electrification lessons from the field.

THERMAL GASIFICATION OF BIOMASS – A PRIMER

16

Robert G. Jenkins

School of Engineering, College of Engineering & Mathematical Sciences,
The University of Vermont, Burlington, VT, USA

INTRODUCTION

This chapter focuses on the principles involved in the production of gaseous fuel mixtures from carbon-rich solid[1] materials, such as solid biomass, municipal solid wastes, coals, cokes, etc., by chemical reactions and processes involving elevated temperatures (generally $>700\,°C$).[2] Almost all these processes involve, to varying degrees, partial oxidation of the carbonaceous material. That is, the solid fuel is subjected to oxygen-lean conditions and, thus, it is not combusted completely to carbon dioxide (CO_2) and water (H_2O). Since the solids also encounter elevated temperatures, there is accompanying thermal decomposition (pyrolysis, discussed later) producing hydrocarbon gases/liquids and highly carbon-rich solids.

As just stated, the principal goal of gasification is to convert a primary carbon-containing fuel into a secondary, or derived, gaseous fuel mixture. The reasons for wanting to achieve this goal are numerous but, in summary, the most important are as follows:

- Gaseous fuels can be burned more easily and completely than a solid, because they are much more amenable to mixing with oxygen (O_2), since these are homogeneous reactions. Thus, it is easy to ensure that they produce very little or no smoke/soot when combusted.
- Gas combustion produces minimal or no "ash," reducing concerns about fouling heat transfer surfaces and contamination of materials being heated by direct contact with the products of combustion.
- Liquid, solid, and gaseous impurities are easier to remove from a gas stream than from solids and liquids. Hence, it is possible to more easily dry and remove tars, dust, and undesirable gases, such as hydrogen sulfide (H_2S), carbonyl sulfide (COS), and ammonia (NH_3), from gases than from a solid fuel prior to combustion.
- Combustion of gaseous fuels generally allows for more flexible and controllable heat release rates (i.e., large turndown ratios) than for liquid and solid fuels.
- Gases can be stored quite easily in tanks and pipelines over a range of pressures.

---

[1]Many of the processes/chemical schemes described herein will, of course, apply to the conversion of carbonaceous liquid fuels (e.g., oils/fats/waxes) to gaseous fuels.
[2]Production of "biogas" by anaerobic digestion methods is outside the chapter's scope.

Bioenergy. http://dx.doi.org/10.1016/B978-0-12-407909-0.00016-X

- By the use of appropriate pressure regulation, gases can be delivered directly to the end user through pipelines, although the distances they can be delivered is limited by energy considerations for the compression processes involved.
- Hydrocarbon-rich gases have a higher energy concentration/unit mass than solid or liquid fuels. However, their densities are much lower.

Gas mixtures generated by thermal gasification comprise varying amounts of flammable components such as carbon monoxide (CO), hydrogen (H_2), methane (CH_4) and other low molecular weight hydrocarbons (lhc's, e.g., propane [C_3H_8]). In addition, mixtures can contain noncombustible gases like nitrogen (N_2), H_2O vapor, and CO_2. In a later section, the actual heating value (HV) of a particular gas mixture is determined by the concentration and HV of each component gas. The precise HVs of gas mixtures generated by a particular process are also influenced by such factors as equilibrium and kinetic considerations, the input materials and the design of the gasifier. Since gasification processes described in this chapter all involve biomass being exposed either concurrently or sequentially to thermal decomposition and partial oxidation chemical reactions, the principles of both of these processes are outlined and discussed in terms of actual gasification schemes.

CLASSIFICATION OF GAS MIXTURES

In any discussion of thermal gasification, it is important to consider the HVs (or calorific value) of the gas mixtures being produced. Generally, the gas mixtures are divided into three broad categories, based on the HV of the gas mixtures. The categories are summarized in Table 16.1.

Low HV Gas Mixtures: Production of these mixtures requires implementation of relatively simple technology at the lowest cost of any of the other categories. Low HV mixtures are primarily used as industrial gases that are burned for a very wide range of applications such as steam raising and generation of process heat. Common industrial names for these gas mixtures are *producer gas* and *water gas*. Because of the low level of energy concentration in these fuels, it is necessary they be used on or very close to the gasifier site. This arises from the fact that compression costs[3] for pressuring the gas into a pipeline outweighs the economic value of the gas. Combustion of these fuels by use of appropriately designed burners produces stable flame; however, they cannot utilize conventional natural gas burners.

Table 16.1 Classification of Gas Mixtures by Heating Value

Category	Higher Heating Value Range (Btu/scf)[a]	HH Value Range (MJ/m³)	Main Components
Low	90–190	3.35–7.1	N_2, CO & H_2
Medium	250–550	9.3–20.5	CO & H_2 + lhc's
High	950–1150	33.4–43.0	CH_4 + lhc's

[a]*British thermal units/standard cubic foot (60 °F & 14.73 psi).*

---

[3]From fundamental thermodynamic considerations, it can be shown that the amount of work required to compress a fluid is inversely proportional to the substance's density; thus, the work required to compress any gas is considerably larger than that for pumping liquids. Likewise, less dense gases (e.g., H_2) require more compression work than do denser gases, such as CH_4.

One special application of low HV gas mixtures has been for vehicles carrying on-board gasification units. While there have been a number of recent demonstrations of this technology, its most widespread use was in World War II, when a large number of mainly civilian cars and light vehicles driven in the Axis countries were powered by wood gasifiers because gasoline was in very short supply. The main reason for widespread nonacceptance of this technology is a lack of overall vehicle performance, since the HV of a N_2, CO and H_2 mixture is about 12.5% that of gasoline, and that the gasifier carried on the vehicle has substantial mass that adds to the total mass of the vehicle.

Medium HV Gas Mixtures: For medium HV mixtures, the applications are similar to those of the lower HV range. As a consequence of the relatively low density of these gas mixtures, the economic constraints of compression and pumping have to be considered. While some consumption is likely to be used on site, there can be some degree of off-site distribution, depending on the exact HV of the gases produced. Similar to low HV gases, these fuels have to be burned in specifically designed gas burners. As will be discussed later, an important additional use of these gas mixtures is in the production and conversion of synthesis gas (syngas). This term refers to mixtures of H_2 and CO that are used subsequently in the catalytic production of hydrocarbon fuels and chemicals, and ammonia.

High HV Gas Mixtures: In the context of thermal gasification, high HV gas refers to the production of CH_4-rich mixtures that have essentially the same properties of conventional natural gas; thus, they are often called *substitute* or *synthetic* natural gas (SNG). This being the case, SNG is indistinguishable from natural gas in terms of energy content, combustion characteristics, and compression considerations. However, production of SNG involves the use of sophisticated technologies and is, by far, the most expensive of the three broad categories of gas mixtures. Processes that are specifically designed for direct SNG production from biomass will not be discussed further, but detailed information is available in many of the suggested Further Readings.

CHEMICAL CONCEPTS AND BACKGROUND

The basic concept of gasifying carbonaceous fuels is rather simple and has been used industrially for over 200 years. If carbon is considered to be the primary source of gas production, then the following endothermic chemical reactions can be used to illustrate the basic principles of thermal gasification.

The Bouduard reaction:

$$C(s) + CO_2(g) \rightleftharpoons 2CO(g); \quad \Delta H = 172 \text{ MJ/kmol} \qquad \text{(Rxn 16.1)}$$

and the carbon steam reaction:

$$C(s) + H_2O(g) \rightleftharpoons H_2(g) + CO(g); \quad \Delta H = 131 \text{ MJ/kmol} \qquad \text{(Rxn 16.2)}$$

In both these reactions, carbon reacts with a reactive gas (either carbon dioxide or steam) at some elevated temperature. These oxidation/reduction reactions then produce gases (CO and H_2) that are themselves combustible. The overall net outcome is that the solid carbonaceous fuel has been converted into useful fuel gases—the principal goal of gasification. Strictly speaking, it is important to recognize that the origins of the fuel gases are not solely from the solid fuel but are equally derived from the reactants CO_2 and H_2O.

As indicated, both Rxns (16.1) and (16.2) are endothermic and, therefore, require heat and elevated temperatures for them to occur at reasonable rates. For example, the C/H_2O reaction is only significant above 750 °C. The question then becomes, how is the required heat supplied to these reactions? There are two different sources to be considered. One involves autothermal methods in which the heats of reaction are provided by some exothermic reactions taking place in conjunction (in situ) with the gasification reactions; and the other source involves allothermal (or indirect) methods, in which the heats of reaction are provided by some external heating scheme (such as an external combustion process). In this chapter, only the more common autothermal strategies are described in detail.

In autothermal gasification schemes, the primary source of heat is usually the strongly exothermic reaction of carbon with oxygen.

$$C(s) + O_2(g) \rightleftharpoons CO_2(g); \quad \Delta H = -394 \text{ MJ/kmol} \quad \text{(Rxn 16.3)}$$

Reaction (16.3) not only provides much of the energy for the endothermic reactions (Rxns (16.1) and (16.2)), but also produces the CO_2 for the Bouduard reaction (Rxn (16.1)), which in turn generates the fuel gas CO. While this oxidation reaction is written for pure O_2, a cheaper and most readily available source of O_2 is air. If air is used in the oxidation reaction, the equivalent stoichiometry is:

$$C(s) + O_2(g) + 3.76 \text{ N}_2 \rightleftharpoons CO_2(g) + 3.76 \text{ N}_2 \quad \text{(Rxn 16.4)}$$

In this case, N_2, the dominant component of air, is considered to be inert. Its influence is to reduce the overall exothermicity of the oxidation reaction, as the heat of reaction of the C/O_2 reaction is required to heat an additional 3.76 molecules of N_2.

It is interesting to note that an advantage of allothermal gasification processes is that they provide the opportunity for driving Rxns (16.1) and (16.2) without the need for dilution by N_2 or by the use of pure O_2. However, the costs of external heating can add substantially to the overall economics of the processes.

Another consequence of using air as the primary gasification medium is that in the subsequent Bouduard reactions, the N_2 acts as a diluent. For this situation, the Bouduard reaction is written as:

$$C(s) + CO_2(g) + 3.76 \text{ N}_2(g) \rightleftharpoons 2CO(g) + 3.76 \text{ N}_2(g) \quad \text{(Rxn 16.5)}$$

Again, the combined HV of the products of Rxn (16.5) is lower than that of Rxn (16.3) due to the presence of a large quantity of inert diluent N_2. This illustrates the fact that gasification of carbon in air, compared to that in pure O_2, produces gas mixtures of lower HV (Section Heating values of gases and gas mixtures).

There are other exothermic reactions that occur during gasification that contribute some energy required for the endothermic reactions. Considering first the partial oxidation of carbon:

$$C(s) + \frac{1}{2}O_2(g) \rightleftharpoons CO(g); \quad \Delta H = -111 \text{ MJ/kmol} \quad \text{(Rxn 16.6)}$$

In this reaction, heat and CO are evolved, contributing to both the thermal requirements and the production of a gaseous fuel.

Another reaction that has to be considered is the water-gas shift (WGS) reaction:

$$CO(g) + H_2O(g) \rightleftharpoons CO_2(g) + H_2(g); \quad \Delta H = -41 \text{ MJ/kmol} \quad \text{(Rxn 16.7)}$$

This gas-phase reaction is most important in modifying the gas composition, since it converts toxic CO to CO_2 and increases the mixture's H_2 content. Its exothermicity is relatively low compared to Rxns (16.3) and (16.6). This reaction can occur in a gasifier both in the gas phase and catalytically on the solid surface of carbon present. The WGS reaction is also used in postgasification treatment to adjust a mixture's H_2/CO ratio.

For completeness, there are a number of other reactions that can be considered, including in situ oxidation reactions of CO and H_2—both of which are strongly exothermic:

$$CO(g) + \frac{1}{2}O_2(g) \rightleftharpoons CO_2(g); \quad \Delta H = -283 \text{ MJ/kmol} \qquad \text{(Rxn 16.8)}$$

$$H_2(g) + \frac{1}{2}O_2(g) \rightleftharpoons H_2O(g); \quad \Delta H = -242 \text{ MJ/kmol} \qquad \text{(Rxn 16.9)}$$

In addition, there are two other important reactions, the direct hydrogenation of carbon to methane:

$$C(s) + 2O_2(g) \rightleftharpoons CH_4(g); \quad \Delta H = -75 \text{ MJ/kmol} \qquad \text{(Rxn 16.10)}$$

and the direct production of CH_4 from CO and H_2.

$$CO(g) + 3H_2(g) \rightleftharpoons CH_4(g) + H_2O(g); \quad \Delta H = -206 \text{ MJ/kmol} \qquad \text{(Rxn 16.11)}$$

The kinetics of all of the above reactions is strongly influenced by the presence of inherent catalytically active inorganic species in the carbonaceous material. For example, very small quantities of finely dispersed sodium or potassium compounds in the feed material will substantially increase reaction rates for almost all these reactions. Presence and concentration of the inorganic species associated with any biomass is a reflection of the environment of cultivation. If the soil and/or ground water are relatively rich in a given inorganic species, the inorganic constituents of the biomass derived from that location are extremely likely to be rich in that species. Since a number of species are very active catalytically in gasification, only small concentrations can markedly affect kinetics. An important outcome of these considerations is that various sources of apparently very similar biomasses can behave quite differently in a given gasification process if the compositions of the inorganic species are different.

HEATING VALUES OF GASES AND GAS MIXTURES

Typically quoted HVs[4] of a number of pure gases and natural gas (predominantly methane) are listed in Table 16.2.

As with any hydrogen-containing fuel, there are two distinct HVs, one the HHV[5] and the other the LHV. The difference between them is dependent on the phase of H_2O in the combustion products, which in turn is governed by temperature and pressure. During combustion, hydrogen in the fuel will be oxidized to H_2O vapor. This requires that some amount of energy be used to vaporize and maintain the H_2O as vapor. The amount of energy required to achieve this is the latent heat, or enthalpy, of vaporization at that condition. However, if the temperature of the products is reduced to below the saturation temperature of H_2O, then the latent heat of vaporization is recovered.

[4]Heating value is the quantity of heat produced per unit volume or mass when the fuel is combusted stoichiometrically (i.e., neither fuel rich nor fuel lean) from room temperature. When considering gaseous fuels, it is usual to report these values on a standard volumetric basis.
[5]HHV = Gross HV and Upper HV; LHV = Net HV.

Table 16.2 Higher and Lower Heating Values of a Range of Gases

Gas	Higher Heating Value (HHV)		Lower Heating Value (LHV)		HHV/LHV %
	(Btu/scf)	*(Btu/lb)*	*(Btu/scf)*	*(Btu/lb)*	
Carbon monoxide (CO)	323	4368	323	4368	0
Hydrogen (H_2)	325	61,084	275	51,628	18.2
Methane (CH_4)	1011	23,811	910	21,433	11.1
Ethane (C_2H_6)	1783	22,198	1630	20,295	9.4
Propane (C_3H_8)	2572	21,564	2371	19,834	8.5
Butane (C_4H_{10})	3225	21,640	2977	19,976	8.3
Natural Gas	950 to 1150	19,500 to 22,500	850 to 1050	17,500 to 22,000	9 to 11
N_2, CO_2, H_2O	Noncombustible				

Source of Heating Values: www.engineeringtoolbox.com (accessed March 2013).

The HHV is a fuel's HV when H_2O in the products has been returned to liquid (at some standard state, usually room temperature, 1 atm). In contrast, the LHV is that when the H_2O in the products remains in the vapor phase. The difference between HHV and LHV for any fuel is calculated from the amount of H_2O produced by its combustion and the latent heat of vaporization of H_2O. The LHV can be calculated from the HHV from:

$$LHV = HHV - mh_{fg} \left(Btu/lb \text{ or } kJ/kg \right) \qquad (16.1)$$

where m is the mass of H_2O in the products/unit mass of fuel and h_{fg} is the latent heat of vaporization at the specified temperature. If, within a particular process, the gaseous products of combustion are cooled to a temperature below the saturation temperature, then the HHV is applicable. However, in many processes the products are exhausted directly to the atmosphere at some elevated temperature; in this case, the LHV is germane.

Examination of Table 16.2 shows that for pure H_2, the HHV is just over 18% higher than its LHV. As would be anticipated, for the low molecular weight hydrocarbons, as H/C ratio decreases, the percentage difference between HHV and LHV also decreases. Since CO contains no hydrogen, its HHV and LHV values are identical.

Returning to Rxn (16.1)(C/O_2 reaction), the HV of the product (pure CO) of this stoichiometric reaction is 323 Btu/scf. By comparison, for Rxn (16.4) (carbon gasification in air), the HV of the product gas (2CO + 3.76 N_2) is [2 × 323/5.76 =] 112 Btu/scf. It can be seen that, in this case, the use of air as the primary gasification medium reduces the HV of the product gas by a factor of almost three.

To further illustrate the influence of air versus O_2 in gasification, a realistic product gas from an air-blown gasifier would be:

$H_2 = 4.1$ vol%, CO 23.9%, $CO_2 = 12.8\%$, $CH_4 = 3.1\%$, $N_2 = 56\%$.

Note that the large amount of N_2 confirms that the gasification process used air as the oxidant. It is easy to estimate that the HHV[6] of this mixture is approximately 122 Btu/scf. On the other hand, if this gas mixture were generated by O_2 alone (i.e., zero N_2 in the products), the calculated HHV_{mix} would be about 277 Btu/scf (an increase of about 127%).

Another important factor that governs HVs of these mixtures is the influence of light hydrocarbon gases, since their HVs are substantially larger than those of CO and H_2. As a consequence, relatively small variations in the hydrocarbon content of a gas mixture have substantial influences on HVs. Using the previous example, doubling the percentage of CH_4 to 6.2 vol% and reducing the N_2 by 3.1% increases the HHV_{mix} to 153 Btu/scf (a 26% increase).

MEASURES OF GAS PRODUCTION PERFORMANCE

Two general measures of performance are often quoted as measures of how effective a given gasification scheme may be. In the carbon conversion efficiency (CCE), the effectiveness is determined by the percentage of carbon converted to gas from the feedstock. The solid residual material (ash) is analyzed for carbon content and compared to the carbon content of the feedstock. Ideally, all the input carbon should be consumed in the gasifier and there should be no unreacted carbon in the ash. Clearly, the higher the CCE, the greater amount of carbon gasified.

$$CCE = \left(1 - \frac{\text{mass of unconverted carbon [in ash]}}{\text{mass of carbon entering gasifier [in feedstock]}}\right) \times 100\% \qquad (16.2)$$

The other widely reported efficiency is the cold gas efficiency (CGE), a comparison of the HV of the gas mixture produced to that of the feedstock.

$$CGE = \eta_{CG} = \frac{\text{Heating Value of Product Gas}}{\text{Heating Value of Feed Material}} \times 100\% \qquad (16.3)$$

In commercial gasifiers, values generally fall in the range of 60–80%. The higher the CGE, the more effective the process is in terms of converting the potential energy in the feedstock to useful energy in the gas.

PYROLYSIS

As thermal gasification processes take place at elevated temperatures (600–1500 °C), they all involve thermal decomposition of the carbonaceous feed material in addition to the chemical processes described in Section Chemical concepts and background. These thermal decomposition phenomena are usually referred to as *pyrolysis*.[7]

Pyrolysis is a dominant process when high molecular weight organic/carbonaceous materials are heated to temperatures greater than 275–350 °C in very low reactive (or in truly inert O_2-free)

---

[6]$HHV_{mix} = 0.041 \times 325$ (H_2) $+ 0.239 \times 323$ (CO) $+ 0.128 \times 0$ (CO_2) $+ 0.031 \times 1011$(CH_4) $+ 0.56 \times 0$ (N_2) Btu/scf.
[7]Pyrolysis is also referred to as *carbonization* or *destructive distillation*.

FIGURE 16.1

Major products of pyrolysis of carbonaceous solids.

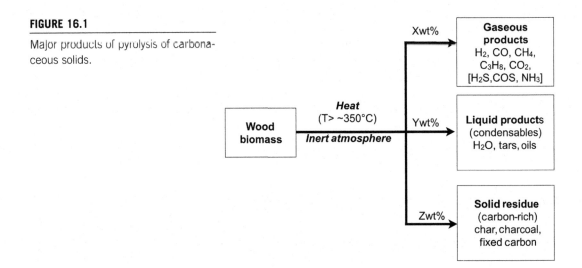

atmospheres. Under these conditions, the organic molecules primarily decompose (or "crack") to form complex mixtures of "volatiles" (gases, vapors, liquids) and high carbon-content solids.[8] Overall, pyrolysis reactions are somewhat endothermic or thermally neutral; they do not involve any significant degree of combustion since oxygen is excluded or in very low concentrations.

Figure 16.1 summarizes these product groups for pyrolysis of a solid such as wood, or other carbonaceous biomass heated to 900 °C in an inert atmosphere (it would apply equally for coals and heavy petroleum fractions).

The amount produced of each product group will vary significantly for different carbonaceous solids; some will produce substantially more or less quantities of solid, liquids, and gases. The product distribution will also be influenced by factors such as heating rate, ultimate heat treatment temperature, residence time, bed depth, particle/lump size, composition and reactivity of the atmosphere surrounding pyrolysis, and pressure.

We first discuss the carbon-rich solid residue, most commonly referred to as *char* (or charcoal) for biomass pyrolysis. This material has a very high carbon content, but it is not pure elemental carbon; it will contain small but measurable amounts of hydrogen, oxygen, and possibly sulfur. It is this carbon, however, that represents the carbon in chemical reactions discussed in Section Chemical concepts and background. The char will be porous, particularly on the molecular

---

[8]In its simplest form, cracking can be illustrated by:

$$C_nH_m \rightarrow \frac{m}{4}CH_4 + \left(n - \frac{m}{4}\right)C$$

where a large hydrocarbon molecule, C_nH_m, is cracked completely to CH_4 and pure C. However, the real situation is more complex, since this large molecule will contain a range of similar and differing strength carbon–carbon bonds. When cracking occurs, –C–C– bonds break more or less randomly, giving rise to a distribution of products ranging from CH_4 through numerous molecules with molecular weights lower than C_nH_m to pure C. In reality, the situation gets even more complex when rearrangements and the presence of heteroatom moieties are considered.

or microscale,[9] and its strength and friability will be determined by the extent of the porosity and the structure of the carbon. When biomass is pyrolyzed, it nearly always exhibits thermosetting properties, i.e., the decomposing solid does not exhibit any plastic or flow properties while being heated. Thus, the physical appearance of the char will be very similar to that of the original source material[10]; if the raw material is fibrous, then its char will appear fibrous. As an aside, some carbonaceous solids, such as a group of particular coals, do become plastic on pyrolysis and in this case, the solid residue, *coke*, does not have the appearance of its precursor material. The issue of thermoplasticity could be a consideration if a process to cogasify biomass with coking coals or certain polymers is envisaged because, during the plastic phase, particles will agglomerate, thus dramatically impeding gas and particle flow.

Char also contains nonvolatile thermal decomposition products of any inorganic constituents in the original material. For a number of biomass materials, the amount of inorganic constituents is relatively small; however, some of them are important in catalyzing some gasification reactions (as discussed in Section Chemical concepts and background). These inorganic components are most commonly described as *ash*. However, this term should be used with some caution, as it will be recalled that ash is the oxidized product of the inorganic components after combustion, which is not the case in pyrolysis and gasification. Chemical analysis of ash produced by combustion will not properly reflect the chemical nature of the inorganics in the pyrolyzed solid.

In regard to the volatile components of pyrolysis, they are subdivided into gases, light liquids, and heavy liquids referred to as *tars* (these heavy liquids are usually defined as the condensable volatile products, excluding light vapors). The gases are composed of light hydrocarbon gases and vapors (usually C_1 through C_6 organic compounds) and permanent gases. The distribution of gaseous products is dependent on the composition of the source material, e.g., high sulfur-containing biomass will produce larger quantities of H_2S than that of low sulfur content. The presence of light hydrocarbon gases generated during pyrolysis can enhance the HV of a gasifier output (see Section Heating values of gases and gas mixtures). Figure 16.1 shows that there are noxious gaseous components, such as H_2S, COS, and NH_3 that need to be removed via some gas cleanup process prior to utilization.

As will be seen, for some gasification processes, handling heavy tarlike volatile components is a very important consideration. In other schemes, all liquids are deliberately exposed to severe thermal conditions to further crack them into much lighter gaseous species.

Figure 16.2 further illustrates the important stages involved in pyrolysis at low to moderate heating rates. This figure represents a thermogravimetric analysis (TGA) plot of a quantity of a typical carbonaceous material heated in a flowing inert atmosphere. In this type of analysis, a small sample of material is heated at some constant rate while its mass is continuously recorded. TGA plots provide an excellent tool for characterizing pyrolysis phenomena, since the shape of the curves and magnitudes of the changes are unique for a given material under the proscribed, controlled conditions.

The solid line in Figure 16.2 represents the mass of the material as it is being heated at the low heating rate (say, 20 °C/min) from about room temperature to 900 °C in an inert atmosphere (usually

---

[9]The microporous nature of chars derived from many wood and other biomasses (e.g., coconut shell) has made them a primary source for activated carbon filters used for purifying liquids and gases.

[10]It will be recalled that the shape and texture of lump charcoal particles reflect the features of the wood from which it was derived.

FIGURE 16.2

Thermogravimetric depiction of pyrolysis.

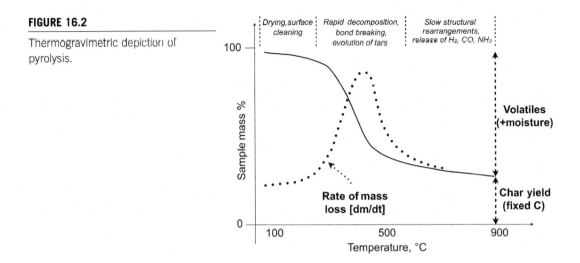

N_2 or helium [He]). The dotted line is the first derivative of sample mass with respect to time, indicating mass loss rates.

At low temperatures, the mass of the material shows a slow degree of loss as it is being heated. These low temperature losses are mainly due to drying and to a lesser extent the volatilization/evaporation of absorbed and adsorbed species associated with the material. Once temperatures reach between 250 and 350 °C, mass loss becomes more pronounced, passing through some maximum rate usually in the 450–550 °C range. Loss of mass in this region is a result of the decomposition of large organic biopolymers that exist within a particular biomass of other carbonaceous material. The volatiles in this part of the curve comprise heavy and light hydrocarbon liquids, oxides of carbon, and H_2. In the final stages of pyrolysis, once the primary large molecule decomposition reactions are completed, evolution of gases continues, typically, CO_2, CO, and H_2. These gases are generated by decomposition of peripheral groups and chemical rearrangements. It should be noted that even at temperatures above 900 °C, a small mass loss rate will be continue to be observed (from the evolution of H_2).

TGA can also be used to generate characteristic ignition data for the material if the experiment is performed in flowing air. In this type of plot, mass loss at the lower temperatures is almost identical to that observed for pyrolysis. However, at some temperature the mass loss rate becomes extremely rapid until only inorganic ash remains. The temperature defined by the onset of the rapid mass loss is characteristic of the material's ignition point.

In common practice, the gross thermal behavior of a particular material is characterized by the proximate analysis that consists of four parameters: moisture, volatile matter (VM), ash, and fixed carbon contents. Determination of these parameters is performed under appropriate standard ASTM procedures.[11]

---

[11]For example: ASTM E870–82(2006) Standard Test Methods for Analysis of Wood Fuels.

Table 16.3 Example Values of Properties of Various Biomass/Wastes	
Property	**Typical for Biomass**
Heating value (dry basis)	5,000−11,000 Btu/lb
Hydrogen/Carbon ratio	1.2−1.7
Moisture (AR[a], wt%)	2−75
Volatiles (wt%)	50−75
Ash (wt%)	0.3−2% (wood), 6% (corn stover), >20% (rice hulls), 25% MSW[b]
Sulfur[c] (wt%)	$\leq 0.3\%$
Nitrogen (wt%)	0.2−1.2%
Oxygen (wt%, by diff.)	30−45%

[a]*As Received.*
[b]*Municipal solid waste.*
[c]*Tires >> 0.3 wt%*
Partial Source: Sustainable Energy, J.W. Tester, E.M. Drake, M.J. Driscoll, M.W. Golay, W.A. Peters., MIT Press, 2005

- *Moisture Content*: Measures the percentage mass of moisture found in a sample by heating it to a temperature slightly above 100 °C.
- *Volatile Matter*: Method determines the percentage of volatile products in the dry sample, which are released when it is heated to 950 °C.
- *Ash Content*: This measurement determines the percent mass of the initial total sample remaining after it is burned in air under closely controlled conditions. The ash thus produced consists of the oxidation products of nonvolatile inorganics in the material being characterized.
- *Fixed Carbon:* The carbon-rich residue that remains after drying and devolatilization, on an ash-free basis is calculated for the sample by subtracting the percentages of moisture, ash, and volatile matter from 100%.

While the proximate analysis provides broad information on how a material will behave thermally, another set of procedures characterizes the material of interest in terms of elemental analysis (the ultimate analysis) and its HVs (higher and lower heating values). The ultimate analysis determines the percentages of the elements (usually on a dry-ash-free basis) carbon, hydrogen, nitrogen, and sulfur directly and oxygen content by difference. The methodologies for these analyses and determination of HVs are detailed in ASTM standards (see ASTM E870 - 82(2006) Standard Test Methods for Analysis of Wood Fuels).

Table 16.3 is included to depict the wide range of properties encountered in biomass and wastes to show that the ranges are large and can be very variable. These data clearly indicate that for any given biomass of interest, it is very important to have detailed characterizations made of its properties[12] and behavior before proceeding with any processing scheme.

[12]Extensive data are available from the USDoE Bioenergy Technologies Office and Bushnell, D., Biomass Fuel Characterization: Testing and Evaluating the Combustion Characteristics of Selected Biomass Fuels, BPA report, 1989 (available via http://cta.ornl.gov/bedb/pdf/BEDB4_Appendices.pdf).

There are a number of very important properties not included in either the proximate or ultimate analyses that must be considered. Of particular importance for gasification is the chemical composition of the VM. As determined, the VM is the weight percentage of all gases and condensable liquids produced by pyrolysis. It does not provide any information regarding the HVs of the products, nor does it distinguish between the relative quantities of individual gas species and the composition of the condensables. If a particular VM is rich in H_2/CO/CH_4/light hydrocarbon gases, then it is likely to produce a gas mixture with a high HV. On the other hand, if the VM is very rich in CO_2, the HVs of the gaseous components are going to be lower. In a similar vein, as there is no knowledge of the composition of its condensable components, it is impossible to determine its potential utility, caloric value, or its ease or difficulty in processing.

Four other examples of important properties of biomass feedstock to be used in gasification are measurements of the material's bulk density, chemical reactivity, behavior on comminution, and ash properties. The reasons for considering these parameters are as follows:

- *Density*: Knowledge of crop and bulk densities is important for biomass gathering, storage, and handling. It is well known that the densities of various biomasses are quite different.
- *Reactivity*: The chemical reactivities of many biomass chars to O_2, H_2O, and CO_2 are higher than those of most coals and petroleum-based carbons, possibly necessitating gasifier redesign in terms of feed rates and O_2 and H_2O demands.
- *Comminution*: Essentially, all processes require input solids to be of some prescribed size or size distribution to ensure proper mixing with reactant gases and controlled flow of both solids and gases. To achieve this condition involves some degree of material comminution. Since biomass tends to be fibrous in nature, and not brittle like coals, effective size diminution has to be addressed.
- *Ash Behavior*: All autothermal gasifiers require some degree of internal combustion to provide the heat to drive the endothermic gasification reactions. Under these conditions, it is extremely important that the high-temperature behavior of the ash generated from the biomass be well characterized and understood. Biomass ashes, especially those that contain amounts of alkali metals, can soften, or even become molten, at relatively low temperatures. Under circumstances where inorganic components soften, ash particles will at least sinter to form masses of clinker that inhibit flow of both gases and solids; reduce the ability of the carbon to gasify; and possibly interact with reactor linings and heat transfer surfaces. In the worst case, if ash becomes molten, slagging behavior will adversely interact with internal surfaces, the solid reactant, and the ability to remove the ash. There are two basic options for gasifier design regarding ash handling in the reactor. They are either designed to operate at nonslagging conditions or as slagging gasifiers.

In a nonslagging (or dry) gasifier, the ash cannot melt. This is achieved by specific selection of the feed materials and judicious temperature control. On the other hand, slagging gasifiers are designed specifically to ensure that the ash forms and remains liquid slag in the gasifier. In these types, slag is only allowed to solidify once it has been removed from the reactor. Clearly, the design and operation of a slagging gasifier is substantially more complex and expensive than one that operates in a nonslagging mode. These ash behavior considerations are most critical if the gasifier operates with O_2 as the primary gasification medium.

THERMOCHEMICAL PATHWAYS

There are a large number of pathways and schemes available for conversion of biomass by thermochemical means, most of which are summarized in Figure 16.3.

The number and diversity of processes available for converting biomass to produce useful combustible gases, liquids, and chemical products from biomaterials is remarkable. These schemes give rise to concepts of biomass "refineries" that can generate gaseous liquid and gaseous fuels, and a very wide range of chemicals such as alcohols, NH_3, tars, and waxes.

While the description of each pathway is beyond the scope of this chapter, several of the suggested Further Readings deal with them in detail. However, there are a number of important comments to be made about this figure. First, there is always a pathway of using the biomass as a fuel directly, the desirability of which is dependent on particular need and economics. Second, the other pathways are broadly divided into those that utilize autothermal or allothermal heat sources. But, as can be seen, there are some common pathways between these two schemes.

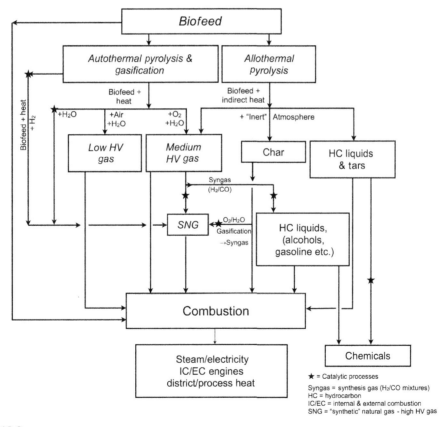

FIGURE 16.3

Thermochemical biomass conversion processes.

Figure 16.3 does not contain any information as to technical ease or even relative economics. For example, the technological complexity and expense of hydrogenating a biomass feedstock to produce substitute gas or SNG are orders of magnitude greater than for, say, generation of a low HV industrial gas mixture. Also to be noted are the number of processes that require the use of catalysts, requiring a relatively high degree of technological sophistication.

One pathway of historical interest is that of allothermal pyrolysis of carbonaceous feedstocks to yield medium HV gas mixtures, char, and liquids. This was the basis, using coal as the feed, for generation and subsequent distribution of coal gas (also known as *town* or *illumination* gas) for domestic and industrial use starting in the late nineteenth century. Prior to the widespread use of petroleum and natural gas, coal carbonization was the principal source of not only lighting and heating gases but also "smokeless" solid fuels (coke) and a very wide range of chemicals in Europe and the United States. The last main application for this type of process is the production of metallurgical cokes for the production of iron and steel. In addition to producing this specialized carbon, coke batteries generate medium HV gases used for process heat in the steel mill.

Finally, Figure 16.3 refers to syngas which, as noted previously, comprises mixtures of H_2 and CO that are used to synthesize a very wide range of hydrocarbon fuel and chemical products (ranging from CH_4 through gasoline and diesel fuels to waxes).[13] In the context of this chapter, it is important to recognize that biomass materials can be used for the production of synthesis gas and, thus, hydrocarbon fuels and chemicals.

Figure 16.4 focuses on the most widely used schemes for gasifying biomass through the use of autothermal pyrolysis and gasification to yield low and medium HV gas mixtures. As noted before (Rxns (16.3) and (16.4)), the difference between the production of these two categories of gas mixtures is the use of air versus oxygen as the main gasifying medium.

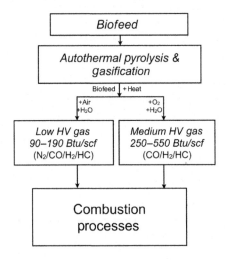

FIGURE 16.4

Gasification of biomass to low and medium heating value gas mixtures.

---

[13]*Chemistry of Fossil Fuels and Biofuels*, Cambridge University Press, 2013, provides extensive current details of the chemistry and processes involved.

When air is used to provide the O_2 for the exothermic carbon combustion reaction (Rxn (16.4)), the resulting gas mixture will be of low HV (typically 80–190 Btu/scf), composed mainly of N_2/CO. The CO_2 generated from the combustion reaction is then reduced to CO by the Bouduard reaction (Rxn (16.1)). The process consists of blowing air through a hot bed of carbon; once ignited, the bed is self heated. The actual HV of the resulting gas is determined by the extent of dilution by the N_2 and possible presence of light hydrocarbons generated by in situ pyrolysis. Classically, this gas mixture is generally referred to as producer gas.

If steam is concurrently (or intermittently) blown into an air-blown reactor, Rxn (16.2) occurs. This C/steam reaction adds H_2 to the CO/N_2 gas mixture. Since the addition of steam to the input gas mixture does not increase the total amount of N_2 added to the system, the volumetric HV is higher than in the case of producer gas. The extent of the increase of the HV is dependent on the amount of steam added. That quantity is limited, however, by the thermal balance between the exothermic combustion reactions and the heat and temperatures required for Rxns (16.1) and (16.2) to be sustained. When steam is added, the resulting gas mixture is commonly called water gas (less commonly, blue gas).

The sequence of the reactions involved in gasification is illustrated in Figure 16.5. This portrays the reactions and process zones of a simple counter flow gasifier using air and steam as the gasification media.

In this type of reactor, biomass is fed in the top of the gasifier, and air and steam are introduced through the reactor's base. The solids flow downward and the gases and vapors flow upward. The

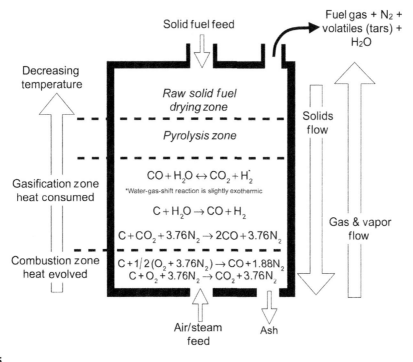

FIGURE 16.5

Reaction zones in an updraft moving-bed counter flow gasifier.

bottommost zone is where the C/O_2 combustion reactions take place (Rxn (16.4) and air-equivalent of Rxn (16.6)) and where, in essence, all the required heat is generated. In this manner, all the CO_2 is generated for subsequent Bouduard reactions, and some wanted CO results from partial oxidation of carbon. The fact that air is used results in a large presence of the diluent N_2 in the gasifier, leading to the reduced overall HV, as described in Section Chemical concepts and background.

Ash is removed from the bottom of gasifier; it will encounter maximum temperatures close to its exit in the most intense combustion region. In reality, the depth of the combustion zone is small, much smaller than depicted. Above this first zone, primary gasification occurs, mainly through the C/CO_2 and C/H_2O reactions to yield most of the desired CO and H_2. Some modification of the CO/H_2 ratio can result from the water–gas shift reaction that takes place on the surface of the solid char. As the height increases, the temperature falls. At some point the temperature of the bed is too low for further gasification.

When the down-flowing fuel enters the gasifier, it comes into contact with the rising hot gases/ vapors and begins to dry. The water/steam evolved from drying leaves via the exit, intimately mixed with the gases and heavy volatiles. On its continued travel downward, the dried fuel starts to pyrolyze. The generated volatiles exit the gasifier in the gas stream. In this type of gasifier, the volatile hydrocarbons and tars never experience very high temperatures that would cause them to crack to more desirable light hydrocarbon gases. As a consequence, these vapors contain high molecular weight species that are somewhat akin to complex mixtures like creosotes. Separation of water from these compounds is very difficult and can be an environmental challenge.

At some point in its downward path, all major stages of biomass pyrolysis are complete. The solid is now truly a char, composed dominantly of elemental carbon in association with its inherent inorganic components. Char then enters the gasification zone with reactions CO_2 and H_2O, forming the desired CO and H_2. Remaining ungasified carbon ultimately travels into the combustion zone, where it reacts with O_2 to produce the required heat and CO_2.

In Section Classification of gas mixtures, it was noted that the consumption of low HV gas mixtures is limited to on-site applications because gas compression costs make widespread distribution off site uneconomical. In the past, a remedy for overcoming this limitation was to spray oil into an internal hot heat exchange surface within the system. By this means, the added oil would crack to low molecular weight hydrocarbon gases, thus markedly increase to HV, to a the level of medium HV gas mixtures; this type of gas was referred to as *carbureted water gas*.

Generation of medium HV gas mixtures through autothermal gasification is accomplished by use of pure O_2 (or highly O_2-enriched air) rather than air as the principal gasification medium (see Section Chemical concepts and background). As a consequence, the N_2 dilution factor is removed, resulting in increased HV of the product gas to the medium HV level. Figure 16.5 for O_2 gasification would be identical to that for air except that N_2 is not present. The same issues would apply to the tars and liquids because they never encounter severe thermal conditions. Another consequence of the absence of N_2 is that the mixtures of CO and H_2 are free of the diluent, thus once purified are true synthesis gas mixtures.

Of course it will be appreciated that such a gasifier has to have a source of O_2 close by to operate. As this will require cryogenically derived O_2, the cost of the operation and its overhead must be considered. Temperatures in the O_2 gasifier will be enhanced due to combustion taking place in O_2 rather than in air. While this can influence the precise distribution of volatile products, its most important effect is likely to be on the behavior of the ash. Temperatures must be controlled to operate the gasifier at nonslagging conditions or the gasifier has to be designed specifically to cope with slagging ash behavior (see Section Pyrolysis).

GASIFIER TYPES

Gasification reactors must perform a number of vital tasks for successful operation. Among other factors, they have to provide:

- Housing to maintain the elevated temperatures and pressures required for gasification.
- Safe containment for the reactants and products: It is most important to recognize that the two main products are H_2 and CO. Hydrogen has a wide range of flammability in air and is prone to detonate, and carbon monoxide is a poisonous, odorless, tasteless, colorless gas.
- Manageable flows of solids (fuel and ash) that can be managed and maintained.
- Adequate mixing of the reactants and sufficient time to ensure high levels of conversion: Any ungasified carbon represents a loss of CCE (Eqn (16.2)). Effective mixing is a critical factor in gasification because the primary chemical processes involve heterogeneous reactions between solids and gases.
- Heat management to control the heat distribution, reactivity, and ash behavior.

While the economics of any proposed gasification project play an overarching role, there are a number of factors that must be considered in the selection of a specific gasifier for a given application. Among them are:

- The postgasification application: Determines the desired HV of the gas produced.
- Size and scale of operation: The requirements for a small, standalone application producing a low HV gas mixture are very different from those of a large-scale operation intended to generate syngas for manufacturing premium liquid fuels.
- Availability and nature of the biomass feedstock: In addition to those properties discussed in Section Pyrolysis (ash content and properties, moisture content, etc.), it is important to know, for example, the availability of the biomass, how homogeneous are its properties from batch to batch, and whether different or mixed feeds are to be processed on regular or seasonal time frames.

Actual gasification reactors come in many different designs. The following are generic descriptions of the most common autothermal reactors.

MOVING OR FIXED-BED GASIFIERS

For biomass, types of this group of gasifiers are usually used to produce fuel gas mixtures; however, they can be used to produce syngas.[14]

UPDRAFT GASIFIER

The reaction zones for this type of gasifier were shown in Figure 16.5, and its overall features were described in the previous section. Figure 16.6 again illustrates this type of reactor.

The bed of solids is supported on a grate that also acts as a gas distributor. In this counter flow reactor, solids are introduced into the reactor at its top via a solids distributor, which acts to deliver the

---

[14]Note for coals: In the large-scale SASOL Coal-to-Liquids technology, pressurized updraft gasifiers are the basis for syngas production.

FIGURE 16.6

Moving-bed, updraft gasifier.

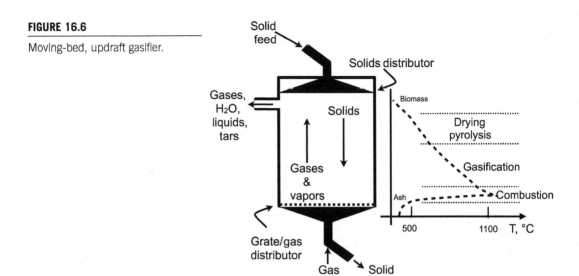

solids uniformly onto the bed below. For some designs, a stirring device submerged in the upper bed ensures uniformity and helps break up any agglomerating solid masses. Typically, the solid feed material is sized in the range of 0.5 to 10 cm (0.2–4″). These gasifiers do not tolerate fine particles well; the particles tend to cause large pressure drops across the bed. Throughout the reactor, its solids are always in contact with each other. This means that if there are any tendencies for the feed materials (or their inorganic constituents) to soften on heating, agglomerates will form and cause serious flow problems. Reiterating what was discussed in Section Pyrolysis, the issue of thermal plasticity/ softening is a very important concern for the gasification of caking or coking coals, but it should not come into play for biomass gasification. However, if cogasification is envisaged with some potentially thermoplastic cofeed, it must be taken into account.

In the representation of the gasifier's temperature profile,[15] the solids heating rates prior to the combustion zone are quite modest. As would be anticipated, temperature increases dramatically in the relatively narrow combustion zone. It would be in this region that issues could arise regarding the softening/slagging behavior of the ash.

One of the major drawbacks of this configuration is that the yields of tar, compared to other gasifiers, are relatively large; the reason for the large yield is that, one the tars have been evolved during pyrolysis, they quickly exit the gasifier without encountering any elevated temperatures. Therefore, they do not undergo further chemical cracking and remain essentially unchanged. The large tar yield is of some concern for two main reasons. First, as was described in the discussion of

---

[15]The temperature profiles presented herein are after those shown in EPRI; Technical Report 102,034, 1993.

Figure 16.5, the useful product gas stream is laden with a mixture of steam and tars. While separation of the gases from the condensables is relatively straightforward, separating the heavy hydrocarbons from the water is difficult—the tars contain hydrophilic functional groups, making complete removal from water technically challenging. The consequence of this is that, before process water can be returned to the environment, it has to undergo significant chemical treatment. Second, relatively large quantities of tar represent a loss of both carbon that could have been converted to CO and light hydrocarbon gases that could have been produced by tar cracking. In this manner, the CCE and CGE (Section Measures of gas production performance) are reduced.

MOVING-BED, DOWNDRAFT GASIFIER

If the heavy volatile products of pyrolysis can be passed through the combustion zone, then it is possible to significantly reduce the tar content of the product stream. Such a scheme is the basis for the moving-bed, downdraft gasifier. While there are many different designs for this type of gasifier, its salient points are depicted in Figure 16.7.

This is a concurrent flow reactor with both solids, gases, vapors, and tars traveling downward. Induced draft fans ensure that the reactant gases flow downward. Solids are fed into the gasifier at its top and removed at the bottom. Reactant gases are introduced in a midsection of the gasifier. This arrangement establishes a combustion zone in the "throat" that is defined by the horizontally opposed burners. As the solids travel down from the inlet, they are dried and undergo pyrolysis. Some of the resulting char is then burned in the combustion zone to CO_2 and CO. The volatile products of pyrolysis are exposed to elevated temperatures as they approach the combustion and will undergo a substantial degree of cracking to low molecular species. These cracked products pass though the combustion zone where most will be burned, or partially burned, to CO_2, CO, and H_2O. These gases, along with those

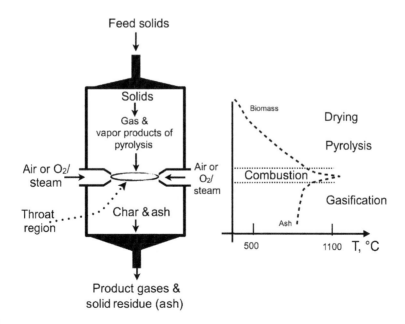

FIGURE 16.7

Moving-bed, downdraft gasifier.

generated from char combustion, gasify remaining char in the gasification zone. The ash and gaseous products exit the gasifier at its bottom. The net effect of employing this design on tar yields is to reduce the concentration in the product stream by as much as 50–100-fold. Under comparable conditions, gas compositions of both the moving-bed gasifiers are quite similar. The comments made about thermoplasticity for the updraft moving-bed gasifier are the same for this type, as the solids are always in contact.

Although the idealized temperature profile for this gasifier is shown for the solids, the temperature profile for gases will be similar. Thus, it can be seen that the exit temperature of the gases is substantial, resulting in a significant loss of sensible heat. As a consequence, for this type of gasification scheme, a heat exchange system is located in the product stream. This system recovers a substantial amount of the sensible heat loss and cools the gas stream. Addition of a heat recovery operation increases the overall thermal efficiency of the gasifier but, of course, adds to the total cost of investment. Applications of these gasifiers are limited to smaller scale operations.

Another operational factor that has to be considered in this gasifier is that ash, and small particles of partially converted char, are entrained in the gaseous product stream. As a result, entrained dust must be removed by from the product gases by devices such as cyclones. This issue is not serious for counter current moving-bed gasifiers, since in those, product gases pass through upper strata of pyrolyzing and drying solids, entrapping most of the entrained particles.

FLUIDIZED-BED GASIFIERS

In a fluidized bed, a gas flows through a bed of particles with sufficient momentum to lift individual particles but not enough to cause elutriation. Ideally, every particle is just separated from its near neighbors and the bed appears to be very similar to a boiling liquid. As a result, a fluidized bed behaves much more like a liquid than does a fixed or moving bed. Mixing between reacting gases and solids is extremely good, which in turn leads to uniformity of both temperature[16] and reaction distribution (zonal behavior, as observed in packed beds, does exist). Another advantage of these beds is that the pseudoliquid nature allows for heat transfer surfaces to be placed inside the bed, further enabling heat control and recovery.

The bed material is usually composed of partially reacted solid fuel (including its ash) in a bed of inert media such as sand. It is possible to add other solids that can contribute to in situ removal of unwanted gaseous species such as SO_x (e.g., limestone) generated from sulfur-containing groups in the solid fuel. It should be noted that, while the particles are essentially separate, they do come into contact with each other. As a consequence, any degree of feedstock thermoplasticity leading to particle agglomeration can cause serious operational difficulties.[17]

In these reactors, incoming reactant gas acts as the fluidization medium (see Figure 16.8) and the concentration of the oxidant is only sufficient to ensure partial combustion of the fuel. Fluidized-bed gasifiers operate at relatively low temperatures (800–900 °C) due partly to ensure dry-ash conditions in the bed and as a consequence of the absence of a true high-temperature combustion zone. If the ash

---

[16]A measure of the uniformity is that well-controlled inert particle fluidized beds are commercially available as constant temperature baths for calibrations at elevated temperature.

[17]The presence of inert solid material in the bed does reduce the influence of thermoplasticity.

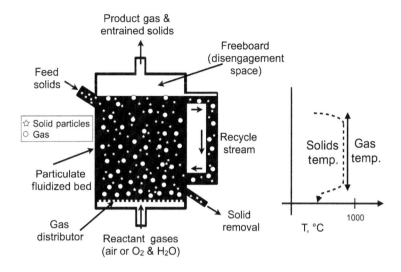

FIGURE 16.8

Circulating fluidized-bed gasifier.

does soften, it will form agglomerates with other ash particles and char. Once the agglomerates reach some critical size, they sink and ultimately cause the bed not to function in a fluidized manner.

Fluidized beds are designed to be operated at either atmospheric or elevated pressures. One major feature of all fluidized-bed gasifiers is that, due to high gas velocities, the concentration of small particles in the product gas stream is high. This requires that entrained particles be removed prior to further gas processing by devices such as cyclones. To enhance overall carbon conversion, these fine particles will be returned directly into the bed from the cyclones. Another consequence of the high gas throughput is that the product flow contains a large amount of sensible heat. Thus, to minimize sensible heat loss, a sizeable thermal recovery system must be incorporated into the exiting gas stream.

There are a number of specific designs but most are either bubbling beds or circulating beds. In the former, and simpler, design, there is no deliberate recycling of materials other than return of a portion of the entrained materials from the exit product gas flow stream. Bubbling beds operate at relatively low gas velocities only sufficient to agitate the bed composed of biomass particles sized in the 5–15 cm (2–6 inch) range. This type of gasification unit is generally used for smaller scale gasification operations; however, larger throughput can be achieved in pressurized units.

As the name suggests, the circulating-bed gasifier incorporates deliberate recycling of both solids and gases. This type of gasifier is generally employed in large-scale, pressurized, and atmospheric pressure gasification schemes because of higher carbon conversions, resulting from extended residence times and mixing. Circulation beds operate at gas velocities much higher than those used in bubbling beds, as the particulate solids are in a suspended state in the gasifier. In this case, biomass particles are nominally sized less than 2 cm (0.8 inch).

Figure 16.8 shows the main features of a circulating fluidized-bed gasifier.

Particles are fed directly into the hot bed. The incoming solids are subjected to high heating rates, nominally of the order of 10^2–10^3 °C/s. To enhance residence times, mixing, and carbon conversion, a portion of the main bed is deliberately recycled back into itself. As is the case for all fluid beds, its temperature profile for the bed is remarkably uniform. In some circulating fluidized-bed gasification schemes, more than one bed is employed. For example, the entrained partially reacted solids from an initial reactor are fed directly into a second bed where gasification is completed.

Once the incoming particles are subjected to the elevated bed temperature, rapid pyrolysis occurs with volatiles evolving immediately into a hot environment, which enhances cracking. However, the bed's relatively low operating temperatures and relatively short gas residence times ensure the presence of some light hydrocarbon gases and tars in the product gas stream. The amounts produced do depend on the actual bed temperature—the lower the bed temperature, the greater the amount of tar. The net effect of these considerations results in tar production levels somewhat between those of the updraft and downdraft moving-bed gasifiers discussed previously.

A major reason that fluidized beds are attractive candidates for biomass gasification is that they are relatively flexible in terms of choice of biomass selection. However, it is imperative not only to understand ash slagging properties to be well understood for a given biomass, but also to understand how they could be modified by the presence of ash components from some cofed material.

ENTRAINED-FLOW GASIFIERS

Entrained-flow gasifiers were originally developed for high-temperature (1200–1600 °C); high-throughput coal gasification processes operated at conditions that ensure that the ash was liquid—that is, a *slagging* gasifier. They have also been used to gasify (and cogasify) petroleum heavy oils, cokes, industrial and toxic wastes, and biomass.

Very small particles of the solid material are entrained in a flow of O_2 and steam into the reactor. In some operations, the solid feed is suspended as slurry prior to injection into the gasifier. Figure 16.9

FIGURE 16.9

Entrained-flow gasifier.

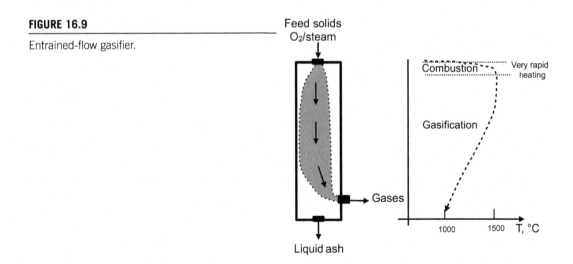

shows a down-fired vertical entrained-flow reactor. There are alternative designs in which the flow is upward. Entrained-flow gasifier operations are not affected by any feedstock thermoplastic behavior because, once entrained, the particles rarely come into physical contact with each other.

As soon as the solid particles enter the very high temperatures close to the burner ($>1200\,°C$), they are heated extremely rapidly (at nominal rates of $>10^4\,°C/s$). Under these conditions of rapid heating and high temperatures, essentially all hydrocarbons (C_1 through heavy tars) are cracked. Thus, the resulting gas mixtures are free of tar, and the concentration of light hydrocarbon gases is low compared to other gasification schemes. While the pyrolysis phase is extremely rapid, the partial oxidation of the char is much slower due to the heterogeneous nature of those reactions. The reactor has to be designed such that there is sufficient residence time for the char/H_2O/O_2 reactions to be completed—typical residence times are of the order of a second or so. Overall, the composition of the product gases will be almost exclusively CO and H_2, and as such, are syngas mixtures (limited to a medium HV of ~ 320 Btu/scf, because of the lack of hydrocarbons).

In a slagging gasifier, ash is always molten inside the reactor, otherwise it will lead to fouling, causing disruptions in heat transfer and material flow. Some of the liquid ash is incorporated in the gas stream flowing downward; the rest will impinge on the walls and flow down to the reactor base. It is possible that slagging requirements could be problematic for some biomass feeds because they produce ash rich in components that do not, or only partially, melt under the temperature regimes encountered in these gasifiers.[18]

The most important aspects of entrained-flow gasifier temperature profiles are that following the high-intensity combustion zone, gas and particle temperatures remain very high (sufficient to keep the ash molten). As a direct consequence, gases exit the gasifier at very high temperatures, requiring that a large amount of the products' sensible heat must be recovered through an extensive heat exchange system.

When coal is used as a feedstock in these gasifiers, it is pulverized into a very fine powder ($<100\,\mu m$ [4×10^{-3} inch] and typically $\sim 75\,\mu m$ diameter) prior to entrainment. This probably poses the biggest drawback for biomass use in entrained-flow gasifiers. Since biomass tends to be very fibrous, its reduction to these small sizes is extremely energy intensive and so costly that it is uneconomical.

One method for lowering pulverization costs is to pretreat the biomass by *torrefaction,* a low-temperature pyrolysis process. In this pretreatment, biomass is heated to temperatures in the range of 200–320 °C in the absence of air for about 30 min. Recalling Section Pyrolysis, under these conditions the biomass will dry completely and then begin to pyrolyze. The extent of pyrolysis will not be great, but some of the volatiles will be tarlike. Combustion of these tars is used to provide heat for the torrefaction process. The solid torrefied residue is a semichar, not a char, since it still contains relatively large quantities of oxygen and hydrogen. Torrefied biomass is easier and less costly to pulverize than its parent material, but it is still probably uneconomical to pulverize it to $<100\,\mu m$.

On the other hand, it may not be necessary for biomass to be ground that finely to be an effective feed for entrained-flow gasifiers. It will be recalled that biomass is more reactive than coals in gasification, thus requires less residence time for complete carbon conversion. Therefore, it is possible that

---

[18]For entrained-flow gasification, slagging is generally preferred over nonslagging gasification for two major reasons. First, a slagging entrained-flow gasifier is more fuel flexible; and, second, even in nominally dry-ash gasifiers, the elevated gas temperatures result in some degree of ash melting, leading to reactor wall fouling.

in a given design, the particle size of the feed could be more of the order of <1 mm ($< \sim 0.04$ inch). The only issue that then arises is related to the ability of the incoming gas streams to entrain particles of that size range.

POST-GASIFICATION PROCESSING

In all biomass gasification processes, ash must be collected and disposed of in an acceptable manner. Depending on the specific process and biomass ash properties, some will generate particulate residues and others solidified slag. Many uses have been proposed for ash ranging from manufacturing of construction materials (bricks, concrete, and asphalt agglomerates) to agricultural products (fertilizer supplements). Any potential value of a given ash is related to the quantity produced and its physical and chemical properties. In terms of quantity, Table 16.3, for example, shows that for some feedstocks the amount is quite minimal and for others substantial. Physical properties govern the ease or difficulty of handling and further processing. On one extreme, low-density dry ash might need to be consolidated; on the other extreme, slag is glasslike and, almost certainly, will have to be ground into smaller fragments prior to use. Ash chemical properties will determine its inertness or activity for an application. Likewise, presence of particular trace elements can either be advantageous or detrimental for a particular use. Thus, it is very important that the ash from a given biomass feed be very well characterized before seeking applications, and its real economic value realized.

Hot gases exiting any gasifier must undergo a number of treatments to make them acceptable for subsequent use, regardless of the desired application. Details of these processes are very well described in a number of the Further Readings listed later on. But in outline, they include removal of:

- Particulate Matter: Entrained solids composed of ash and partially converted carbonaceous material. Removal is achieved either by mechanical devices such as cyclones and bag filters or via liquid scrubbing.
- Sulfur-Containing Compounds: H_2S and COS removed by being dissolved in some solvent; sulfur is recovered by some subsequent integrated chemical process.
- Nitrogen-Containing Gases: NH_3 and HCN are removed by solvent treatment.
- Tars: These are generally removed by condensation somewhere in the postgasification processes (depending on the specific cleanup scheme). The challenge is to adequately separate tars from any condensed water before the water is recycled back into the process or is discharged back to the environment.
- Other Chemical Compounds/Species: Depending on the biomass, it might be necessary to remove (or substantially reduce) components such as chlorine compounds, mercury, and alkali metals.

Some applications will employ processes that take advantage of the water–gas shift reaction (Rxn (16.7)) to adjust the product H_2/CO ratio. This is critical for production of specific hydrocarbon liquids and ammonia from syngas. If it is important to increase the gas H_2 content, then dissolving CO_2 from the products of Rxn (16.7) will achieve this goal and increase the HV of the gas mixture. Removal of H_2O from Rxn (16.7) will drive the equilibrium in the opposite direction if pure CO is required.

In all of the processes outlined in Section Gasifier types, there has to be heat management of the product gas stream. In its simplest form, this could be gas temperature reduction by a water quench prior to further treatment, or it could involve more complex heat recovery systems by heat exchangers. All the high-throughput gasifiers described must have some integrated heat recovery units to reduce what would be substantial sensible heat losses. The recovered heat is in the form of generated steam that is used internally in the process and/or externally to provide process steam or heat.

INTEGRATED GASIFICATION COMBINED CYCLES (IGCC)

Of particular interest is biomass gasification in an integrated gasification combined cycle (IGCC) to produce electricity at high overall thermal efficiencies.

An IGCC operation consists of the integrated processes shown in Figure 16.10. Following gas purification, the syngas is fed to a stationary gas turbine, where it is burned in air in the combustion chamber. The gas turbine operates at high temperature ($T_{max} \sim 1500\,^\circ C$) on a thermodynamic Brayton cycle generating mechanical power ($\dot{W}_{GT}$) and very hot waste gases ($>500\,^\circ C$). The mechanical power is used to generate electricity through an electrical generator. Heat is recovered from the hot waste gases by a heat exchange system, a heat recovery steam generator (HRSG). The steam raised in the HRSG is then fed to a multistage steam turbine generator set producing additional power ($\dot{W}_{ST}$-through a Rankine cycle), thus, more electricity. If the gas turbine does not produce sufficient heat for the HRSG, additional heat can be added by the combustion of a supplemental fuel as needed. To close the cycle, some quantity of heat is rejected via flue gases and heat exchange in condensers. In thermodynamic terms, the combined cycle comprises a Brayton cycle on top of a Rankine cycle. The thermal efficiency of the combined cycle is described by:

$$\eta_{th,cc} = \frac{\dot{W}_{GT} + \dot{W}_{ST}}{\dot{Q}_{in}}$$

where $\dot{Q}_{in}$ = total heat input rate from all sources.

By combining both power outputs from a single heat input, the realized overall thermal efficiency falls between 45 and 60%, much higher than the individual thermal efficiencies of either Brayton or Rankine cycles.

An IGCC scheme provides environmental advantages because the integrated gas purification processes remove almost all sulfur and nitrogen-containing compounds, thus reducing

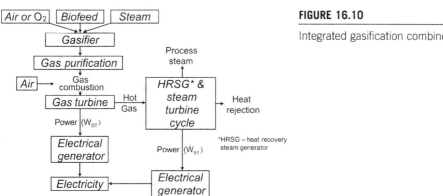

FIGURE 16.10

Integrated gasification combined cycles (IGCC).

fuel-derived SO_x and NO_x from the gas turbine combustion. Similarly, particulate emissions are reduced compared to electricity generated from conventional coal-fired electrical power plant. The downside of IGCC is that it is a more complex process with a higher capital cost than competing technologies.

FURTHER READING

Cheng, J. (Ed.), 2010. Biomass to Renewable Energy Processes. CRC Press.

Republished by Synthetic Fuels, Probstein & Hicks, 2006. Dover Books. Originally published by McGraw-Hill, (1982).

Higman, van der Burgt, 2008. Gasification, second ed. Gulf Professional Publishing.

Kishore (Ed.), 2009. Renewable Energy Engineering & Technology. Earthscan Publishing.

Kreith, Kreider, 2011. Principles of Sustainable Energy. CRC Press.

Miller, Tillman (Eds.), 2008. Combustion Engineering Issues. Academic Press.

Pandey, A., Larroche, C., Ricke, S.C., Dussap, C.-G., Gnansounou, E. (Eds.), 2011. Biofuels: Alternative Feedstocks & Conversion Processes. Academic Press.

deRosa, A., 2012. Fundamentals of Renewable Energy Processes, third ed. Academic Press.

Schobert, H., 2013. Chemistry of Fossil Fuels and Biofuels. Cambridge University Press.

Siedlecki, M., De Jong, W., Verkooijen, A.H., 2011. Fluidized bed gasification as a mature and reliable technology for the production of bio-syngas and applied in the production of liquid transportation fuels—a review. Energies 4, 389–434. Open access available online from: http://www.mdpi.com.

Tester, J.W., Drake, E.M., Driscoll, M.J., Golay, M.W., Peters, W.A., 2012. Sustainable Energy, second ed. MIT Press.

BASICS OF ENERGY PRODUCTION THROUGH ANAEROBIC DIGESTION OF LIVESTOCK MANURE

Klein E. Ileleji, Chad Martin, Don Jones

Department of Agricultural and Biological Engineering, Purdue University, West Lafayette, IN, USA

INTRODUCTION

Bioenergy generated from diversified sources provides local and emerging opportunities to reduce our dependence on foreign oil and petroleum-based fuels. Livestock manure from concentrated livestock operations can be a source of energy production that not only provides an alternative energy source for on-farm use, but also mitigates the negative consequences of odor from livestock operations. Biogas generated from manure can be used directly in a gas-fired combustion engine or a microturbine to create electricity. Additional energy in the form of waste heat from turbine operations can be used to provide heat or hot water for on-farm use, as well as maintain the temperature of a digester during a cold winter.

THE ANAEROBIC DIGESTION PROCESS

Anaerobic Digestion (AD) occurs when organic material decomposes biologically in the absence of oxygen. This process releases biogas while converting an unstable, pathogen- and nutrient-rich organic substrate, like manure, into a more stable and nutrient-rich material with a reduced pathogen load (Figure 17.1). Biogas is composed of approximately 65% methane, while the remaining content is mostly carbon dioxide and other trace gases (Jones et al., 1980). The leftover, more stable substrate can be a good source of fertilizer, or in some cases, further composted and reused as a bedding material.

An anaerobic digester is the unit of operation used to produce methane from manure. Figure 17.2 is a schematic that shows the anaerobic digestion process. In an anaerobic digester, the organic substrate is first liquefied by bacteria. This is followed by a two-step process involving acid production by acid-forming bacteria (acidogenesis) and methane production from the acids with methane-forming bacteria (methanogenesis). In most cases, after digestion the effluent can be relatively easily separated into solid and liquid fractions. In the case of dairy cows, the solid fraction may be used as recycled bedding, and the rest of the digested material may be land-applied at the agronomic rate to meet the soil and crop needs. The biochemical methane potential (BMP) of manure varies by livestock

Bioenergy. http://dx.doi.org/10.1016/B978-0-12-407909-0.00017-1

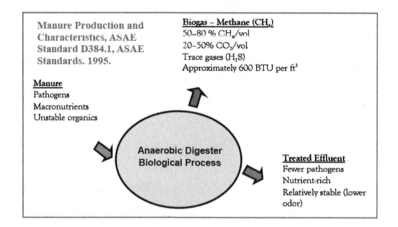

FIGURE 17.1

Schematic of basic process of anaerobic digestion.

Source: After Robert T. Burns, Iowa State University.

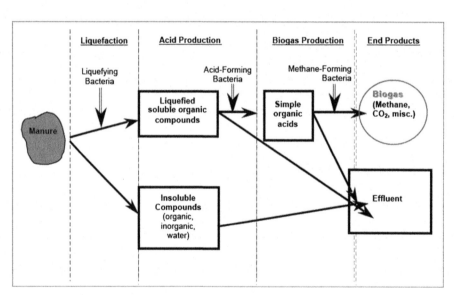

FIGURE 17.2

Schematic of the anaerobic digestion process.

species and is a measure of the methane production potential of the manure. Production is measured as cubic feet of methane gas per animal unit (AU). An animal unit, as defined by the American Society of Agricultural and Biological Engineers (ASABE) Standard, is based on 1000 live weight of the livestock (ASABE Standards, 1995).

Table 17.1 shows the BMP of some common livestock waste streams. In addition to livestock waste, residue from nearby food processing has been effectively utilized within the anaerobic digester system to boost methane production. This included high-starch or high-fat content materials.

Anaerobic digesters can be categorized based on (1) the operating temperature of the AD unit and (2) the AD unit process design. The latter allows either separate acidogenesis and methanogenesis reactions or mixed acidogenesis and methanogenesis reactions. These temperature ranges are identified as psychrophilic (68 °F, 20 °C), mesophilic (95–105 °F, 35–41 °C), and thermophilic (125–135 °F, 52–57 °C). The pH levels of the digester environment should be maintained as close to neutral (pH 7.0) as possible (Jones et al., 1980). There are a number of process designs currently used to digest livestock manure. The technologies listed below range from the very simple (covered lagoon) to more complicated in an upflow anaerobic sludge bed (UASB). However, most on-farm unit digesters use the simpler systems listed below (usually 1, 2, or 4) (Figure 17.3):

1. Covered lagoons
2. Plug-flow digesters
3. Mixed plug-flow digesters
4. Complete-mixed digesters
5. Fixed-film digesters
6. Temperature-phased anaerobic digesters
7. Anaerobic sequencing-batch reactor (ASBR)
8. Upflow anaerobic sludge bed (UASB)

The components of the typical anaerobic digestion system include: manure collection, anaerobic digesters, effluent storage, gas handling, and gas use/electricity generating equipment. In some cases, the gases produced are simply flared off into the atmosphere. The advantages of each system are dependent upon several variables within the livestock operation. Table 17.2 summarizes the characteristics of four common AD technologies used in livestock operations.

BENEFITS OF ANAEROBIC DIGESTION

In a study commissioned by the Great Lakes Regional Biomass Energy Program, the following benefits were documented for dairy operations (Kramer, 2004):

- Revenue from annual electricity sales or cost offsets generated $32–$78 per head.[1]
- Annual bedding costs were reduced by using digested manure instead of other bedding materials.
- After digestion, manure has improved nutrient availability, reduced acidity, and reduced odor. By avoiding fertilizer purchases, producers saved $41–$60 per head (from dairy cattle).
- Odor control is a key benefit in being a better neighbor. It increased quality of life on and off the farm, helped producers avoid complaints and lawsuits, allowed continuation of the operation or the ability to site new facilities, and increased operational flexibility.
- Anaerobic digestion reduced pathogens associated with manure discharges (Mosier, 1998).

[1]Resale of electricity depends on state and utility policies.

Table 17.1 Characteristics and Operational Parameters of the Most Important Agricultural Feedstocks for Anaerobic Digestion

Feedstock	Total Solids TS (%)	Volatile Solids (% if TS)	C:N Ratio	Biogas Yield[c] (m³/kg VS)	Retention Time (days)	CH₄ Content (%)	Unwanted Substances	Inhibiting Substances	Frequent Problems	References
Pig slurry	3–8[d]	70–80	3–10	0.25–0.50	20–40	70–80	Wood shavings, bristles, H_2O, sand, cords, straw	Antibiotics, disinfectants	Scum layers, sediments	Brachtl (1998), Braun (1982), Thomé-Kozmiensky (1995), Wellinger (1984)
Cow slurry	5–12[d]	75–85	6–20[a]	0.20–0.30	20–30	55–75	Bristles, soil, H_2O, NH_4^+, straw, wood	Antibiotics, disinfectants	Scum layers, poor biogas yield	Brachtl (1998), Braun (1982), Thomé-Kozmiensky (1995), Wellinger (1984)
Chicken slurry	10–30[d]	70–80	3–10	0.35–0.60	>30	60–80	NH_4^+, grit, sand, feathers	Antibiotics, disinfectants	NH_4^+-inhibition, scum layers	Brachtl (1998), Kuhn (1995)
Whey	1–5	80–95	n.a.	0.80–0.95	3–10	60–80	Transpiration impurities		pH reduction	Brachtl (1998), Thomé-Kozmiensky (1995)
Fermented slops	1–5	80–95	4–10	0.35–0.55	3–10	55–75	Undegradable fruit remains		High acid concentration, VFA-inhibition	Brachtl (1998), Thomé-Kozmiensky (1995)
Leaves	80	90	30–80	0.10–0.30[b]	8–20	n.a.	Soil	Pesticides		Brachtl (1998), Thomé-Kozmiensky (1995)

Material							Unwanted material		Problems	Reference
Wood shavings	80	95	511	n.a.	n.a.	n.a.	Unwanted material		Mechanical problems	Brachtl (1998), Thomé-Kozmiensky (1995)
Straw	70	90	90	0.35–0.45[e]	10–50[e]	n.a.	Sand, grit		Scum layers, poor digestion	Brachtl (1998), Thomé-Kozmiensky (1995)
Wood wastes	60–70	99.6	723	n.a.	n.a.	n.a.	Unwanted material		Poor anaerobic biodegradation	Brachtl (1998), Thomé-Kozmiensky (1995)
Garden wastes	60–70	90	100–150	0.20–0.50	8–30	n.a.	Soil, cellulosic components	Pesticides	Poor degradation of cellulosic components	Brachtl (1998), Thomé-Kozmiensky (1995)
Grass	20–25	90	12–25	0.55	10	n.a.	Grit	Pesticides	pH reduction	Brachtl (1998), Thomé-Kozmiensky (1995)
Grass silage	15–25	90	10–25	0.56	10	n.a.	Grit		pH reduction	Brachtl (1998), Thomé-Kozmiensky (1995)
Fruit wastes	15–20	75	35	0.25–0.50	8–20	n.a.	Undegradable fruit remains, grit	Pesticides	pH reduction	Brachtl (1998)
Food remains	10	80	n.a.	0.50–0.60	10–20	70–80	Bones, plastic material	Disinfectants	Sediments, mechanical problems	Nordberg and Edström (1997)

[a]Depending on straw addition
[b]depending on drying rate
[c]depending on retention time
[d]depending on dilution
[e]depending on particle size
n.a. = not available

FIGURE 17.3

(a) Two anaerobic reactors (foreground) showing flexible covers to collect gas as part of a community AD system in Juehnde in Southern Lower Saxony, Germany. Also shown: housing for an electric generation set and an underground mixing tank, (b) an anaerobic digestion reactor on a dairy farm in Oregon, and (c) piping for transporting digested liquid manure effluent from the AD reactor to an open lagoon on a farm in Oregon.

Table 17.2 Summary of the Characteristics of Digester Technologies (U.S. EPA AgSTAR Handbook, 2007)

Characteristics	Covered Lagoon	Complete Mix	Plug Flow	Fixed Film
Digestion vessel	Deep lagoon	Round/square in-/above-ground tank	Rectangular in-ground tank	Above-ground tank
Level of technology	Low	Medium	Low	Medium
Supplemental heat	No	Yes	Yes	No
Total solids	0.5–3%	3–10%	11–13%	3%
Solids characteristics	Fine	Coarse	Coarse	Very fine
HRT[a] days	40–60	15+	15+	2–3
Farm type	Dairy, hog	Dairy, hog	Dairy only	Dairy, hog
Optimum location	Temperate and warm climates	All climates	All climates	Temperate and warm climates

[a]*Hydraulic Retention Time (HRT) is the average number of days a volume of manure remains in the digester.*

WHAT MAKES AN OPERATION APPROPRIATE FOR ANAEROBIC DIGESTION?

When considering an on-farm anaerobic digestion facility, careful planning must include an understanding of key variables. The following checklist should benefit the initial process of identifying the feasibility of installing this system on-farm (U.S. EPA AgSTAR Handbook, 2007):

- Is the operation a confined feeding operation with at least 500 head of dairy/beef cattle or 2000 sows/feeder pigs?[2]
- Is 90% of the manure collected regularly?
- Is manure production and collection stable year-round?
- To be compatible with biogas energy production, is manure managed as liquid, slurry, or semisolid?

FIGURE 17.4

Electric generation unit showing piping for recovering heal that is used to heat up the AD reactor.

---

[2]Note: This will be dependent on state and utility policies on purchasing energy produced on-farm. In some states, the rate is at or near the retail value of electricity, while in other states, there is no state-level policy. At the time of this publication, most states (Indiana included) do not have a requirement that utilities purchase farm-generated electricity.

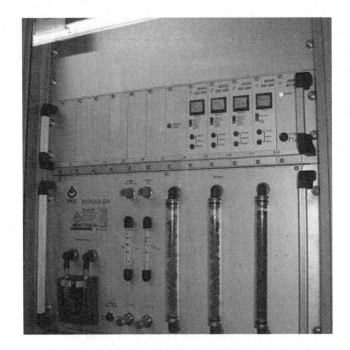

FIGURE 17.5

Control panel for biogas monitoring.

- Is manure free of bedding in the form of sand or other materials, such as rocks?
- Is there a ready use for energy recovered on-farm (heat, ventilation fans, etc.)?
- Can the operator manage the system with regular attention, repairs, and maintenance, and do they have the desire to see that the system runs successfully?
- Can the livestock production system be modified to add relatively fresh manure to the digester and store the digested manure?
- Safety is always an important concern when selling excess electricity to the utility grid. Specific wiring considerations are necessary to prevent electrocution during times when power lines are down. Biogas should also be handled with extreme caution (Figures 17.4 and 17.5).

CONCLUSION

Anaerobic digestion of livestock manure is an alternative pathway for managing large organic waste loads and its associated problems encountered in large feeding lots and confined animal feeding operations. When planned correctly, AD can result in revenue from energy sales or savings in on-farm energy generation. Even though AD is not a new technology, its practice on U.S. farms is not common, and requires careful planning and implementation in order to reap its benefits. Overall, AD technologies can help preserve and integrate livestock production within communities, and creates renewable energy resources to serve a growing bioeconomy within rural communities.

ACKNOWLEDGMENTS

The authors would like to thank the Indiana Office of Energy and Defense Development for their financial support for making this publication possible (contract number 8-BM-002).This paper was prepared for Purdue University Extension BioEnergy series (technical ID# ID-406).

REFERENCES

ASABE Standard, 1995. Manure Production and Characteristics. ASAE Standard D384.1, ASAE Standards.

Brachtl, E., 1998. Pilotversuche zur Cofermentation von pharmazeutischen Abfällen mit Rindergülle. Diplomarbeit. Interuniversitäres Forschungsinstitut für Agrarbiotechnologie, Abt. Umweltbiotechnologie, 3430-Tulln, Austria. (in Arbeit).

Braun, R., 1982. Biogas – Methangärung organischer Abfallstoffe, Grundlagen und Anwendungsbeispiele. Springer Verlag, Wien, New York.

Jones, D., Nye, J., Dale, A., 1980. Methane Generation from Livestock Waste. AE-105. Purdue University Cooperative Extension Service.

Kramer, J., 2004. Agricultural Biogas Casebook – 2004 Update. Great Lakes Regional Biomass Energy Program, Council of Great Lakes Governors.

Kuhn, E. (Ed.), 1995. Kofermentation. Kuratorium für Technik und Bauwesen in der Landwirtschaft e.V. (KTBL), Arbeitspapier 219, Darmstadt.

Mosier, M., 1998. Anaerobic Digesters Control Odors, Reduce, Pathogens, Improve Nutrient Manageability, Can Be Cost Competitive with Lagoons, and Provide Energy Too!. Resource Conservation Management, Inc. Presentation at Iowa State University.

Nordberg, Å., Edström, M., March 1997. Co-digestion of ley crop silage, source-sorted municipal solid waste and municipal sewage sludge. In: Proceedings from 5th FAO/SREN Workshop, Anaerobic Conversion for Environmental Protection, Sanitation and Re-Use of Residues; Gent, Belgium, pp. 24–27.

Thomé-Kozmiensky, K.J. (Ed.), 1995. Biologische Abfallbehandlung. EFVerlag für Energieund Umwelttechnik, Berlin, D.

U.S. EPA, September 2007. AgSTAR Handbook: A Manual for Developing Biogas Systems at Commercial Farms in the United States, second ed. http://www.epa.gov/agstar/resources/handbook.html.

Wellinger, A., 1984. Anaerobic digestion: a review comparison with two types of aeration systems for manure treatment and energy production on the small farm. Agricultural Wastes 10, 117–133.

BIOENERGY AND ANAEROBIC DIGESTION

18

M. Charles Gould

Extension educator, Agriculture and Agribusiness Institute,
Michigan State University, West Olive, MI, USA

INTRODUCTION

Anaerobic digestion, as a renewable energy technology, harnesses a natural biological process by using available biomass (e.g., food wastes, animal manures, and bioenergy crops) to produce biogas (renewable methane). Biogas can be used to produce electricity and heat, or upgraded for use as a vehicle fuel or injection into the natural gas grid. Biogas is made up of approximately 45–65% methane and 30–40% carbon dioxide, along with trace gases and moisture (Table 18.1). According to the USEPA (2013), methane (CH_4) is the second most prevalent greenhouse gas emitted in the United States from human activities. Methane's lifetime in the atmosphere is much shorter than carbon dioxide (CO_2), but CH_4 is more efficient at trapping radiation than CO_2. Pound for pound, the comparative impact of CH_4 on climate change is over 20 times greater than CO_2 over a 100-year period. Combusting biogas is a strategy to reduce greenhouse gas emissions.

THE ANAEROBIC DIGESTION PROCESS

Anaerobic digestion is a complex biochemical reaction carried out in a number of steps by several types of microorganisms that require no oxygen to live. This reaction produces biogas, which is primarily composed of methane and carbon dioxide.

In an anaerobic environment, specialized microorganisms break down complex organic matter (carbohydrates, proteins, and fats) into molecules with a smaller atomic mass that are soluble in water (sugars, amino acids, and fatty acids). Methane and carbon dioxide are the primary gaseous end products of this process, which is known as biogas. Table 18.1 lists the typical composition of biogas. More importantly, anaerobic digestion stabilizes the slurry in the digester.

The overall conversion process of complex organic matter into methane and carbon dioxide can be divided into four steps as shown in Figure 18.1, namely hydrolysis, acidogenesis, acetogenesis, and methanogenesis. The circled number in Figure 18.1 corresponds with each step described below. It should be pointed out that some researchers combine the acidogenesis and acetogenesis steps and make it a three-step conversion process.

In an anaerobic digester, the four processes occur simultaneously. When an anaerobic digester performs properly, the conversion of the products of the first three steps into biogas is virtually

Bioenergy. http://dx.doi.org/10.1016/B978-0-12-407909-0.00018-3

Table 18.1 Typical Composition of Biogas (Volumetric Percent)

Biogas Component	Composition of Biogas (%)
Methane (CH_4)	45–65%
Carbon dioxide (CO_2)	30–40%
Hydrogen sulfide (H_2S)	0.3–3%
Ammonia (NH_3)	0–1%
Moisture (H_2O)	0–10%
Nitrogen (N_2)	0–5%
Oxygen (O_2)	0–2%
Hydrogen (H_2)	0–1%

Source: Becky Larson, UW-Madison.

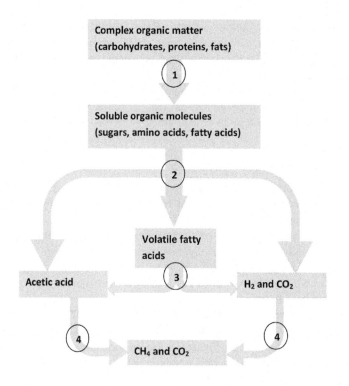

FIGURE 18.1

Four stages of the anaerobic digestion process.

(Source: Technologies for Economic Development.)

complete, so that the concentration of these products is low at any time. Biarnes (2012) describes each step as follows:

STEP 1: HYDROLYSIS

In anaerobic digestion, hydrolysis is the essential first step, as biomass is normally comprised of very large organic polymers, which are otherwise unusable. Through hydrolysis, these large polymers, namely proteins, fats, and carbohydrates, are broken down into smaller molecules such as amino acids, fatty acids, and simple sugars. Although some of the products of hydrolysis, including hydrogen and acetate, may be used by methanogens later in the anaerobic digestion process, the majority of the molecules, which are still relatively large, must be further broken down in the process of acidogenesis so that they may be used to create methane.

STEP 2: FERMENTATION OR ACIDOGENESIS

Acidogenesis is the next step of anaerobic digestion in which acidogenic microorganisms further break down the biomass products after hydrolysis. These fermentative bacteria produce an acidic environment in the digester while creating ammonia, H_2, CO_2, H_2S, shorter volatile fatty acids, carbonic acids, alcohols, as well as trace amounts of other byproducts. Although acidogenic bacteria further breaks down the organic matter, it is still too large and unusable for the ultimate goal of methane production, so the biomass must next undergo the process of acetogenesis.

STEP 3: ACETOGENESIS

In general, acetogenesis is the creation of acetate, a derivative of acetic acid, from carbon and energy sources by acetogens. Acetogens catabolize many of the products created in acidogenesis into acetic acid, CO_2 and H_2, which are used by methanogens to create methane.

STEP 4: METHANOGENESIS

Methanogenesis constitutes the final stage of anaerobic digestion in which methanogens create methane from the final products of acetogenesis, as well as from some of the intermediate products from hydrolysis and acidogenesis. There are two general pathways involving the use of acetic acid and carbon dioxide, the two main products of the first three steps of anaerobic digestion, to create methane in methanogenesis:

$$CO_2 + 4H_2 \rightarrow CH_4 + 2H_2O$$

$$CH_3COOH \rightarrow CH_4 + CO_2$$

Although CO_2 can be converted into methane and water through the reaction, the main mechanism to create methane in methanogenesis is the path involving acetic acid. This path creates methane and CO_2, the two main products of anaerobic digestion.

THE FERMENTING BACTERIA AND METHANOGEN RELATIONSHIP

In a farm-based anaerobic digester that is working properly, the digesting feedstocks (referred to as slurry) contain all of the necessary fermenting bacteria and methanogens to stabilize the manure. A feedstock is any organic material with biogas production potential. Examples of feedstocks are manure and food waste.

Stabilization occurs when the volatile solids content of a feedstock is biologically reduced via detention times, mixing, and interaction with fermenting bacteria. Volatile solids are the portion of the total solids (i.e., organic matter) that are converted to biogas. As the term "anaerobic" implies, this stabilization occurs in the absence of oxygen. Anaerobic digestion, when functioning properly, reduces odors and pathogen levels. Digestate has been described as having an earthy smell with a tinge of ammonia. Manure that is "biologically stabilized" is no longer manure but is treated effluent (Moser, 1998).

It is important to keep enough fermenting bacteria and methanogens in the digester to ensure the process continues operating efficiently. As long as the other requirements of the digestion process are met, such as pH and temperature, the only necessary control is to maintain an adequate supply of new material such that existing microbial populations within the digester can effectively treat the volume of feedstocks added to the digester each day.

When conditions change in the digester, it affects the relationship between fermenting bacteria and methanogens. A change in pH illustrates this point. Methanogens produce methane most efficiently at a pH between 6.8 and 7.2. When the pH drops below 6.8, methanogens are stressed and cannot convert the organic acids produced by fermenting bacteria into methane quickly enough. Fermenting bacteria in the first stage of biogas production are not affected by a low pH and continue to produce organic acids. The end result is the accumulation of organic acids and hydrogen. Methanogens cannot survive in this environment and die off, resulting in decreased methane production. Low methane production results in biogas that is undesirable because it is primarily comprised of carbon dioxide and is therefore not combustible.

Decreased biogas output and low slurry pH result from a condition known as a "sour" digester. The remedy for a sour digester depends on how bad the upset was and the type of digester. For example, if the digester is a plug flow digester, newly added feedstocks will eventually work their way through the digester. No reseeding is necessary. For other types of digesters, the best thing to do is to stop feeding them and mix the slurry for a period of time. Observe conditions and make adjustments based on analysis of the slurry and biogas. If biogas production does not resume after repeated efforts to correct problems, the last resort is to remove the slurry and restart the digester with new feedstocks.

SUMMARY OF SUBSECTION

Biogas is produced in an anaerobic environment by specialized microorganisms that break down complex organic matter into biogas. The production of biogas is a complex process involving four stages: hydrolysis, fermentation, acetogenesis, and methanogenesis. Biogas is typically 45–65% methane and 30–40% carbon dioxide with trace gases and moisture. Anaerobic digestion stabilizes the slurry in a digester. Methane production continues to occur as long as conditions favorable to the survival of fermenting bacteria and methanogens are maintained in the digester. If conditions change, such as a drop in pH or temperature, methane production will decline and eventually stop if not corrected. When conditions progress beyond correcting, it may be necessary to empty out the digester and start over with fresh feedstocks and digestate.

FEEDSTOCKS

A great variety of organic material can be used as feedstocks for generating biogas. Feedstocks vary in energy production potential (Figure 18.2). For example, manure has a lower biogas production

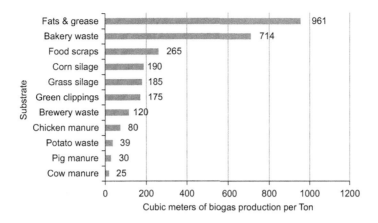

FIGURE 18.2

Gas yields of different feedstocks.

(Source: Data derived from www.biogas-energy.com, © 2007 Biogas Energy, Inc., translated from: Basisdaten Biogas. Deutschland, Marz 2005, Fachagentur Nachwachsende Rohstoffe e.V.)

potential than fats, oils, or grease. However, there are scientific, engineering, and legal limits to what and how much can be added to a digester.

FEEDSTOCK EXAMPLES

Livestock manures are low-energy feedstocks because they are predigested in the gastrointestinal tracts of livestock. Manure, however, is a prime choice for anaerobic digestion because it generally has a neutral pH and a high buffering capacity (the ability to resist changes in pH); contains a naturally occurring mix of microbes responsible for anaerobic degradation; provides an array of nutrients, micronutrients, and trace metals; is available in large quantities; and can be transferred by pumping (Kirk and Faivor, 2012). Because of these characteristics, blending energy-dense feedstocks with livestock manure is a common practice to maximize biogas production.

The use of manure as a base for farm-based anaerobic digestion is important because many of the energy-dense feedstocks, such as food processing waste and ethanol stillage, are acidic, contain little if any naturally occurring microbes, and often lack macro- and micronutrients necessary for microbial metabolism. Potentially, farms operating anaerobic digestion systems could take on additional wastes and benefit from increased gas production as well as income from tipping fees. A tipping fee is the payment a farmer receives for accepting a feedstock. Examples of feedstocks that could be added to a farm-based digester include:

- Waste feed
- Food processing wastes
- Fats, oils, and grease
- Slaughterhouse wastes
- Corn silage (energy crop)
- Syrup from ethanol production
- Glycerin from biodiesel production
- Milk-house wash-water

- Fresh produce waste
- Cafeteria waste
- Farm animal mortality

It should be understood that the volume of these feedstocks that are fed to a digester may be limited based on regulation, the characteristics of the feedstock, or both. For example, in 2011, the Michigan legislature amended the Bodies of Dead Animals Act to include anaerobic digestion as an acceptable form of mortality management. Specific rules must be followed for the inclusion of animal mortality into a digester and the use of the digestate. Another example, again from Michigan, is the volume of syrup from ethanol production and fats, oils, or grease (FOG) that can be added to a digester is limited to "a maximum of 20 percent substitution rate, by volume." This rule also defines the conditions under which digestate can be land applied or otherwise disposed of. It is widely known that the addition of highly concentrated organic materials, such as cafeteria waste, are positively correlated with high biogas production. However, at a certain level, organic overload can be reached, resulting in a sour digester. The following materials should be excluded from anaerobic digesters:

- Compounds known to be toxic to anaerobic microorganisms, such as automotive oils, greases, and paraffin. Ammonia and sulfide are toxic to these organisms at a high pH. Ammonia and sulfide toxicity are not a risk to the microbial community when the pH is controlled at the proper level.
- Poorly degradable material. These require higher retention times, meaning they must spend more time in the anaerobic digester to be broken down and converted into biogas.
- Materials such as plastics and metals remain unchanged in the digester and should be avoided altogether. Inorganic materials, such as sand, contain no carbon and cannot be converted into biogas. Sand also may cause operational problems such as pipe clogging, premature equipment wear, and digester volume reduction due to sand accumulation.

CARBON TO NITROGEN RATIO (C/N)

Fry (1973) states that anaerobic microorganisms utilize carbon for energy and nitrogen for building cell structures. These microorganisms use up carbon about 30 times faster than they use nitrogen. Anaerobic digestion proceeds best when feedstocks contain a certain amount of carbon and nitrogen together. A C/N ratio of 30 permits digestion to proceed at an optimum rate, assuming that other conditions are favorable. If there is too much carbon (high C/N ratio; 60/1 for example) in the feedstocks, nitrogen will be used up first, with carbon left over. This will make the digester slow down. On the other hand, if there is too much nitrogen (low C/N ratio; 30/15 for example), the carbon soon becomes exhausted and fermentation stops. The remaining nitrogen will be lost as ammonia gas (NH_3).

VOLATILE SOLIDS

Volatile solids are used by fermenting bacteria to produce organic acids and hydrogen—compounds methanogens convert into methane. The primary organic acid produced by fermenting bacteria is acetic acid. Feedstocks vary in their potential to produce biogas.

Fats, which are high in volatile solids, generate higher quantities of biogas, whereas manures, by comparison, generate lower. Vegetable fats and oils, such as cooking oils, are readily decomposed in anaerobic digesters; however fats, oils, and greases alone are not ideal feedstocks; they are best co-digested with other feedstocks.

SUMMARY OF SUBSECTION

Feedstocks vary in their biogas production potential. Maximum biogas production is achieved when feedstocks with the highest biogas production potential are combined with materials offering consistent operating environments (i.e., pH). Inorganic and toxic materials inhibit biogas production and should not be introduced into the digester.

THE PROCESS OF STARTING A DIGESTER

Getting any biological system to operate requires careful attention to a number of parameters. Although those parameters that assure a good start-up are not technically difficult, failure to follow the correct procedure or to properly monitor the start-up methods result in either a long delay in the anaerobic digestion start-up process or a total failure of the system to produce biogas.

START-UP

Balsam (2006) recommends the following process to prepare the digester for biogas production: Fill the digester tank with water and then heat it to the desired temperature. Add "seed" sludge from a municipal sewage treatment plant or farm digester (that is operating in the same temperature range) to 20–25% of the tank's volume. Over a 6–8 week period of time, gradually increase the amount of fresh manure that is added to the digester until the desired loading rate is reached. Assuming that the temperature within the system remains relatively constant, steady gas production should occur in the fourth week after start-up. The bacteria may require 2–3 months to multiply to an efficient population. Purging the anaerobic digester with CO_2 or another nonoxygen gas can help decrease start-up time and reduce the danger of explosion during the start-up phase.

Seeding a new digester with effluent from an operating digester is the ideal. The effluent will already contain all the fermenting bacteria and methanogens necessary to produce biogas. Inoculating the digester with effluent from an operating digester speeds up the time it takes the digester to produce burnable biogas, meaning that methane is the primary constituent of the biogas and not carbon dioxide. This assumes all other factors (temperature, pH, volatile acid concentration, etc.) are at appropriate levels. If seed effluent cannot be obtained, using preconditioned raw manure will improve the start-up time. Preconditioned manure is manure stored for at least 2 weeks in an anaerobic state before it is introduced into the digester vessel.

During start-up, it is important to monitor pH, fatty acid levels, biogas composition, and temperature. The pH of the digester should be moving toward pH 7.0. Volatile fatty acid and carbon dioxide concentrations should decrease as biogas production increases. If pH drops or the volatile acid level of the digester effluent increases, feedstocks should not be added to the digester for two days until the condition of the digestate becomes stable. Feeding should then be resumed. Temperature fluctuations, even small ones, affect anaerobic microorganisms and thereby biogas production. Mesophilic systems should be kept around 100 °F, thermophilic systems around 135 °F.

FEEDING THE DIGESTER AFTER START-UP

Once a successful digester start-up has been accomplished, the next step is daily feeding. The best feed schedule is continuous at a low rate. The next best is frequent pumping for short periods, and once a day is the worst. A digester operator can feed less volatile solids, but should not feed more. Feeding more will result in undigested material being pushed out of the digester as well as inhibit gas production.

STARTING THE DIGESTER DURING COLD WEATHER

Supplemental heating is always needed during start-up, regardless of the season. However, getting a digester from ambient temperature to mesophilic temperature during the winter requires a lot of energy, which means it is more expensive to start a digester in the winter. Typically, propane is used to heat the digester. Commissioning a digester at other times of the year other than late fall or early winter may be less expensive.

SUMMARY OF SUBSECTION

The quickest path to biogas production from a new digester is to seed it with anaerobic microorganisms found in effluent or sludge from a properly functioning digester and slowly meter in manure over a 6–8 week period of time. Monitor pH, fatty acid levels, biogas composition, and temperature and correct as necessary. Commissioning a digester in cold weather is expensive to do. It is better to commission a digester during warmer times of year.

LOADING RATE

BACKGROUND

Metering feedstocks into a digester at a regular, calculated rate over time is known as the loading rate. Fermenting bacteria are sensitive to the volume and type of feedstocks injected into a digester. Feedstock changes should be done slowly over time so as to not shock the digester, which may cause it to stop producing biogas.

FACTORS IN FEEDING THE DIGESTER

The operator must monitor and control digester feeding. The quantity and characteristics of the feedstocks affect the stability and efficiency of the digestion process. In feeding the digester, each of the following must be considered:

- The concentration of the incoming feedstocks (concentration refers to the amount of solids in a given volume of water);
- The amount of volatile solids in the incoming feedstock (volatile solids are food for fermenting bacteria and essential for the production of biogas);
- The amount of inorganic material in the feedstock, such as grit or sand, that does not convert to biogas;
- The ratio of volatile solids per unit of digester volume (this ratio is used as a loading factor);
- The hydraulic loading (hydraulic retention time) (adequate hydraulic retention time is necessary to assure microbial growth to convert volatile solids to biogas).

CALCULATING THE LOADING RATE

To determine the loading rate of a given digester, some basic information about the digester and the feedstocks coming into the digester are required. The following example illustrates how to calculate the loading rate of a given digester:

> Assume that 5000 gal of manure per day are pumped into a complete mixed digester that is 50 ft in diameter with a manure depth of 20 ft and a cone depth of 5 ft. It is operated at 100 °F. An analysis of the manure indicated it has 6.5% total solids and 69% volatile solids content. Assume the manure has a specific gravity of 1 (the density of water). What is the solids loading rate?

Calculating the manure volume

A. Manure volume in cylinder (tank) $= \pi \times$ radius squared $\times$ height
$= 3.14 \times (25\ \text{ft}^2) \times 20\ \text{ft}$
$= 39{,}250\ \text{ft}^3$
B. Manure volume in cone $= 1/3 \times \pi \times$ radius squared $\times$ height or depth
$= 1/3 \times 3.14 \times (25\ \text{ft}^2) \times 5\ \text{ft}$
$= 1/3 \times 6280\ \text{ft}^3$
$= 3271\ \text{ft}^3$
C. Total manure volume in digester $=$ volume of cylinder (tank) $+$ volume of cone
$= 39{,}250\ \text{ft}^3 + 3271\ \text{ft}^3$
$= 42{,}521\ \text{ft}^3$

Calculating the loading rate

A. Pounds of total solids per day $=$ gal/day $\times$ 8.34 lb/gal $\times$ % total solids (decimal)
$= 5000\ \text{gal/day} \times 8.34\ \text{lb/gal} \times 0.065$
$= 2710\ \text{lb/day}$
B. Pounds of volatile solids per day $=$ lb of total solids/day $\times$ % volatile solids (decimal)
$= 2710\ \text{lb/day} \times 0.69$
$= 1869\ \text{lb/day}$
C. Loading rate $=$ lb of volatile solids per day/volume of manure in digester
$= 1869\ \text{lb/day}/45{,}521\ \text{ft}^3$
$= 0.04\ \text{lb/day/ft}^3$

Based on the assumptions given above, a loading rate of 0.04 lb volatile solids/day/cubic foot falls within the acceptable loading rate range of 0.02–0.37 lb of volatile solids per cubic foot of digester (Fulhage, 1993). This loading rate range can also be expressed as 20–370 lb of volatile solids per day per 1000 cubic foot.

HYDRAULIC LOADING

Hydraulic loading (or hydraulic retention time) is the average number of days that a feedstock stays in a digester and is related to digester capacity. Hydraulic loading is calculated by dividing digester volume (gallons) by the feed volume (gal/day). Knowing the hydraulic retention time is important so as to ensure there is enough time for methanogens to convert organic acids into methane. If the time is too short, the feedstock biogas production potential is never realized and feedstocks pass through the digester not fully treated. If the time is too long, biogas production begins to tail off because methanogens run out of food. The minimum hydraulic loading varies based on the type of digester, the operating temperature, and the type and amount of solids going into the digester.

Tools that estimate biogas yield

There are several excellent tools available to estimate biogas yield. Estimating biogas yield is only one component of these tools, however. Their real purpose is to determine if a digester is appropriate for a given situation. If found appropriate, the next step would be a feasibility study. There is no cost to download and use these tools.

US EPA AgSTAR

- The AgSTAR Handbook is a comprehensive manual that provides guidance on developing biogas recovery systems at commercial farms.
- FarmWare is an expert decision support software package that can be used to conduct prefeasibility assessments.
- AD Screening Forms can be used by producers who are interested in determining whether anaerobic digestion could be feasible for their operation. Upon receipt of the form, an AgSTAR representative will follow up to discuss whether a more in-depth prefeasibility assessment should be conducted.

The University of Minnesota Extension

The Anaerobic Digester Economics spreadsheet will do initial calculations of annual costs and returns that could be expected from owning and operating an anaerobic digester on a dairy farm.

SUMMARY OF SUBSECTION

The proper loading rate ensures consistent biogas production over time.

OPERATION AND CONTROL OF A DIGESTER

Anaerobic digesters operate in a stable way when operation and control procedures are established and followed.

OPERATION AND CONTROL PROCEDURES

As has been mentioned before, anaerobic digestion is a complex process that requires strict anaerobic conditions to proceed, and depends on the coordinated activity of a complex microbial association to transform organic material into mostly carbon dioxide and methane. Problems arise when this "co-ordinated activity" is disrupted. Therefore, operating a digester to achieve a steady state basis is the goal. Zickefoose and Hayes (1976) recommend seven operational procedures to prevent possible problems and improve digestion results. These procedures are briefly summarized below. It should be noted that in their manual, Zickefoose and Hayes provide a detailed operations checklist for the first five operational procedures.

1. Set up a feeding schedule
   Keeping excess water at a minimum and feeding at regular intervals are important features of a feeding schedule.
2. Control loading
   The pounds of volatile solids fed to a digester daily and the usable volume of the digester are used to calculate the loading rate to avoid digester upset.
3. Control digester temperature
   Digester temperature must be held constant.
4. Control mixing
   The goal of mixing is to bring fermenting bacteria and methanogens in contact with food and keep scum and grit formations to a minimum.
5. Control supernatant (e.g., slurry) quality and effects
   Feedstock inputs and digester types impact slurry quality.
6. Control sludge (e.g., digestate) withdrawal
   Odors may be a concern if not digested for the proper length of time. Digestate that has been properly treated can be land applied or used for other value added purposes.
7. Use lab tests and other information for process control
   There are certain indicators that measure the progress of digestion and warn about impending upset. No one variable can be used alone to predict problems; several must be considered together. Control indicators in order of importance are:
   a. Volatile acid to alkalinity ratio.
   b. Gas production rates, both methane and carbon dioxide.
   c. pH.
   d. Volatile solids destruction.

Zickefoose and Hayes caution against looking at absolute numbers from lab analysis. The rate of change is much more significant. They point out that indicator trends are the most useful to predict the progress of digestion and as signals of process upset.

SUMMARY OF SUBSECTION

Seven operational procedures can be implemented to enable a digester to achieve a steady state basis with a minimum of problems.

ROLE OF MIXING IN BIOGAS PRODUCTION

BACKGROUND

The primary reason for mixing is to bring fermenting bacteria in contact with volatile solids and methanogens in contact with organic acids. Mixing can be done mechanically or by bubbling biogas through the slurry column. The amount and frequency of mixing is controlled by the operator.

SLURRY STABILIZATION

Mixing exposes microorganisms to the maximum amount of food, lessens temperature stratification in the digester, reduces the volume occupied by settled inorganic material (such as grit), evenly distributes metabolic waste products during the digestion process, and prevents the formation of a floating crust layer (which can slow the percolation of biogas out of the slurry). Mixing creates a homogeneous environment throughout the digester that enables the digester volume to be fully utilized (Schlicht, 1999). The benefits of mixing are:

- To speed up the process of volatile solids breakdown, and
- To increase the amount of biogas production.

Mixing can be accomplished by bubbling biogas through the slurry column in the digester or by mechanical means. An explanation of each method follows.

MIXING METHODS

Biogas

As biogas is produced, it forms in pockets and then breaks loose and rises to the surface. This action creates a boiling effect, in theory resulting in some mixing. Thermal convection currents caused by the addition of heat also cause some mixing. Mixing with biogas is controlled by the loading rate. When conditions allow a constant loading, internal mixing will occur. A loading rate of about 0.4 lb of volatile solids per cubic foot per day is needed for natural mixing (MDEQ, 2004). As long as loading can be sustained at this level, no other mixing may be required. However, if prolonged periods of low loading are experienced, mixing may be interrupted and scum blankets may form. On the other hand, increased loading may cause organic overloads resulting in slower biogas production. Conditions that cause natural mixing are somewhat unstable but do afford an inexpensive method of mixing if the operation is closely controlled.

Mechanical mixing

Many types of devices are used to mix slurry in the digester. Examples include impellers (Figure 18.3) and pumps (Figure 18.4).

Impellers

Impellers are attached to a shaft that is connected to a motor. Agitation occurs when an impeller moves at such a speed as to move the slurry. Adjusting the shaft height and/or angle allows mixing to occur throughout the profile of the slurry. A shaft can be inserted into a digester through the digester wall below the slurry level with a water-tight seal or through the gas-holder with a gas-tight seal.

Impellers can be large or small in size and respectively operate at a low speed or high speed (rpm). The size of the impeller and the speed at which it turns are functions of the purposes for agitation and mixing set forth in the "Slurry Stabilization" section.

Pumps

A strong pump should be able to put the whole contents of the digester in motion, provided that the intake and outlet of the pump are placed in a way that corresponds with the digester shape.

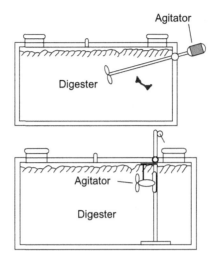

FIGURE 18.3

Impellers in various locations in a digester.

(Source: Kossmann and Pönitz.)

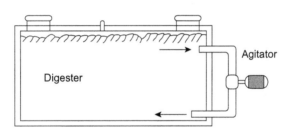

FIGURE 18.4

Hydraulic agitation with a pump.

(Source: Kossmann and Pönitz.)

EFFECT OF SPEED ON MIXING

As a general rule, agitation should be performed as much as necessary but as little as possible. Too frequent mixing with fast rotating, mechanical agitation devices can disturb the biological processes in the slurry. Additionally, a too thorough mixing of the slurry may lead to half-digested substrate leaving the digester prematurely.

Kish (2009) reports that scientists from Washington University, who were researching ways to increase digester efficiency, found that different mixing intensities did not have an effect on long-term performance, but did affect the microbial populations responsible for breaking down the waste. By controlling the microbe community, they affected the overall stability of the digestion process. The scientists determined that microbial digestion improves when mixed at low speeds during the start-up of the digester; however, higher mixing speeds during later periods may improve long-term stability. More mixing reduces the net surplus of bioenergy produced.

It should be noted that one type of digester, the plug flow digester, does not employ agitation and mixing of feedstocks. By design, a slug or "plug" of feedstock is injected into the digester and it is acted upon by microorganisms as it travels through the digester.

BENEFITS OF EXTERNAL EQUIPMENT PLACEMENT

Locating as much of the machinery needed to operate the digester on the outside of the digester is a great advantage. For example, pumps outside the digester do not require someone certified in confined space entry to work on them. Access to equipment for service and repair is easier than if that equipment was located in a digester. If something breaks down, there is no need to empty the digester.

SUMMARY OF SUBSECTION

Mixing can be done mechanically or with biogas. Its primary purpose is to bring microorganisms in contact with food necessary to create methane. Mixing speeds up the process of converting volatile solids into methane, thereby increasing the volume of gas production.

TYPES OF ANAEROBIC DIGESTERS

Lusk (2005) states that digesters can be divided into two types of systems—attached growth and suspended growth. The difference between the two systems is manifest in digester size, operating temperature, solids retention time, hydraulic retention time, total solids concentration of feedstocks fed to the digester, biogas production, ease of management, and other factors. Attached growth systems provide more surface area for anaerobes to colonize than suspended growth systems. Perforated PVC pipe and plastic balls are two examples of media that are used in attached growth systems to increase surface area for microbial colonization. Anaerobes in suspended growth systems form mats and float around in the slurry rather than attach themselves to something stationary. Examples of suspended growth system digesters include covered lagoon, complete mix digester, plug flow digester, and high solids digester. An example of an attached growth system is the fixed film digester. Table 18.2 below summarizes the main characteristics of these digesters. A brief explanation of digesters in each system follows.

SUSPENDED GROWTH SYSTEMS

Covered lagoon digester
Covered lagoon digesters (Figure 18.5) are used to treat and produce biogas from liquid manure with less than 3% solids (Roos et al., 2004). Biogas emitted from the digester is stored under an impermeable cover until it is utilized for work. These digesters are not heated and therefore are typically found in warmer areas of the United States. Hamilton (2012) lists the following advantages and disadvantages of a covered lagoon digester:

Advantages

- Inexpensive
- Easily adapted to hydraulic flushing
- Simple construction and management

Table 18.2 Characteristics of five types of digester systems

Characteristic	Covered Lagoon	Plug Flow Digester	Complete Mix Digester	Fixed Film Digester	High Solids Digester
Digestion vessel	In-ground clay or synthetically lined storage	Rectangular in-ground tank	Round/square tank in/above ground tank	In/above ground tank	Above ground chambers
Level of technology	Low	Low	Medium	Medium	High
Supplemental heat	No	Yes	Yes	Yes	Yes
Total solids	3−6%	11−13%	3−10%	2−4%	25−50%
Solids characteristics	Coarse	Coarse	Coarse	Fine	Fine to coarse
Hydraulic retention time (days)	60+	18−20	5−20	<4	About 14
Farm type	Dairy, swine	Dairy	Dairy, swine	Dairy, swine	Dairy
Optimum location	Temperate/ warm	All climates	All climates	All climates	All climates

Source: Modified by M. Charles Gould from Module 3.6 in S. Lezberg, C. Gould and M. Jungwirth (eds.). Introduction to Anaerobic Digestion Course. On-line Curriculum. Bioenergy Training Center. http://fyi.uwex.edu/biotrainingcenter/.

FIGURE 18.5

Covered lagoon digester located on Black Farms near Lillington, NC.

(Source: Mark Rice, North Carolina State University.)

Disadvantages

- Poor mixing
- Poor energy yield
- Solids settling reduces useable volume
- Bacteria can wash out if short circuiting occurs
- Limited to warmer weather or warm climates since digestion depends on temperature

Complete mix digester

A complete mix digester is a heated engineered tank with one or more mixing technologies designed to keep solids in suspension (Figure 18.6). The purpose of mixing is to increase the food-to-microbe contact to enable maximum destruction of volatile solids, methane production, and odor reduction. Complete mix digesters are designed to handle feedstocks containing 3–10% solids. Hamilton (2012) lists the following advantages and disadvantages of a complete mix digester:

Advantages

- Efficient
- Can digest different levels of dry matter content
- May take energy crops
- Good mixing
- Good solids degradation

FIGURE 18.6

Complete mix digester located at Scenic View Dairy, Fennville, MI.

(Source: M. Charles Gould, Michigan State University.)

Disadvantages

- No guarantee on how much time the material spends in the tank
- Bacteria wash out when short circuiting occurs
- Relatively expensive

Plug flow digester

Roos et al. (2004) define a plug flow digester as an engineered, heated, rectangular tank that treats feedstocks with 11–13% solids (Figure 18.7). Feedstocks are introduced to the digester as a "plug" and travel through the digester in the same way. As the plug moves through the digester, it is acted upon by anaerobes and biogas is produced. Feedstocks with less than 11% solids content do not perform well in plug flow digesters due to the lack of fiber. Hamilton (2012) lists the following advantages and disadvantages of a plug flow digester:

Advantages

- Inexpensive
- Simple to operate and fix
- Can take energy crops

Disadvantages

- Poor mixing
- Poor energy yield
- Hard top difficult to open to remove settled solids
- Membrane top subject to weather (wind and snow)

FIGURE 18.7

Plug flow digester located at Haubenschild Dairy in Princeton, MN.

(Source: M. Charles Gould, Michigan State University.)

High solids digesters

The American Biogas Council (2014) defines high solids digesters as upright, silo-style digesters made of concrete and steel with rigid covers that produce biogas from feedstocks with 20–42% total solids (Figure 18.8 and 18.9). These digesters allow operators to combine high dry matter manure, energy

FIGURE 18.8

Silo type high solids digestion system with CSTR.

(Source: Norma McDonald, OWS.)

FIGURE 18.9

Tunnel style high solids anaerobic digestion chamber.

(Source: M. Charles Gould, Michigan State University.)

crops, and crop residuals with very dilute liquid manures or cosubstrates. Hamilton (2012) lists the following advantages and disadvantages of a high solids digester:

Advantages

- Can use solid materials
- Treats under both anaerobic and aerobic conditions

Disadvantages

- Needs stackable material with at least 25% solids
- Complicated process
- Expensive

FIGURE 18.10

Fixed film digester located at the University of Florida Dairy Research Farm.

(Source: Ann Wilkie, University of Florida.)

ATTACHED GROWTH SYSTEM

Fixed film digester

Roos et al. (2004) define a fixed film digester as a tank filled with plastic media (Figure 18.10). Media facilitates microbial colonization, which enables the digester to have a smaller footprint and reduces hydraulic retention time. Fixed film digesters perform best with dilute waste streams (1–5% Total Solids). Hamilton (2012) lists the following advantages and disadvantages of a fixed film digester:

Advantages

- Efficient
- Low bacteria wash out
- High gas production per volume

Disadvantages

- Suspended solids must be removed
- Expensive
- Plugging of bacterial growth media
- Lower gas production due to removed solids

SUMMARY OF SUBSECTION

There are a wide variety of anaerobic digesters. Digesters can be divided into two systems—attached growth and suspended growth. Digesters within each system generate biogas in different ways according to the management of internal and external factors.

CONCLUSION

Specialized microorganisms in an anaerobic environment produce biogas through a series of four complex stages known as hydrolysis, fermentation, acetogenesis, and methanogenesis. Anaerobic digesters facilitate the completion of all four stages, resulting in the production of biogas. Biogas is typically 45–65% methane and 30–40% carbon dioxide with trace gases and moisture. Maximum biogas production is achieved when feedstocks with the highest biogas production potential are fed to a digester. Feeding the digester at the proper loading rate, adequate mixing of the digestate throughout the digester profile, and maintaining appropriate environmental conditions within the digester ensure consistent biogas production over time.

ACKNOWLEDGMENTS

The sections entitled "The Anaerobic Digestion Process" through "The Role of Mixing in Biogas Production" are extracted from Introduction to Anaerobic Digestion Course (2012), an on-line curriculum located at http://fyi.uwex.edu/biotrainingcenter/. Sharon Lezberg, M. Charles Gould and Maggie Jungwirth, Editors. This curriculum material was based upon the work supported by the National Institute of Food and Agriculture, U.S.

Department of Agriculture, under Agreement No. 2007-51130-03909. Any opinions, findings, conclusions, or recommendations expressed in this publication are those of the author(s) and do not necessarily reflect the view of the U.S. Department of Agriculture.

REFERENCES

American Biogas Council, 2014. What Is Anaerobic Digestion?. https://www.americanbiogascouncil.org/biogas_what.asp (verified 27.02.14.).

Balsam, J., 2006. Anaerobic Digestion of Animal Wastes: Factors to Consider. Updated by Dave Ryan in 2006. ATTRA – National Sustainable Agriculture Information Service. Purchased copy in possession of author. Available at: http://attra.ncat.org/attra-pub/anaerobic.html#six (verified 13.08.14.).

Biarnes, M., 2012. Biomass to Biogas—Anaerobic Digestion. Available at: http://www.e-inst.com/biomass-to-biogas (verified 13.08.14.).

Fry, L.J., 1973. Methane Digesters for Fuel Gas and Fertilizer. Newsletter No. 3, Spring 1973. The New Alchemy Institute, Box 432, Woods Hole, Massachusetts 02543.

Fulhage, C., Sievers, D., Fischer, J., 1993. Generating Methane Gas from Manure, fact sheet G1881. University of Missouri Extension, Columbia, MO. Available at: http://extension.missouri.edu/explore/agguides/agengin/g01881.htm (verified 13.08.14.).

Hamilton, D., 2012. Types of anaerobic digesters. Module 3. In: Lezberg, S., Gould, C., Jungwirth, M. (Eds.), Introduction to Anaerobic Digestion Course. On-line Curriculum. Bioenergy Training Center. http://fyi.uwex.edu/biotrainingcenter/ (verified 27.02.14.).

Kish, S., 2009. Making On-farm Waste Digestion Work. USDA CSREES NRI Report.

Kirk, D., Faivor, L., 2012. Feedstocks for Biogas. Available at: http://www.extension.org/pages/Feedstocks_for_Biogas (verified 27.02.14.).

Lusk, P., 2005. Anaerobic Digestion 101. WSU Anaerobic Digestion Workshop. Presentation notes in the possession of author.

Moser, M.A., 1998. Anaerobic Digesters Control Odors, Reduce Pathogens, Improve Nutrient Manageability, Can be Cost Competitive with Lagoons, and Provide Energy Too! Resource Conservation Management, Inc., Berkeley, CA 94704. Available at: http://epa.gov/agstar/documents/lib-man_man.pdf (verified 13.08.14.).

MDEQ, 2004. Digester Operator Training Handbook. Michigan Department of Environmental Quality, Lansing, Mich.

Roos, K.F., Martin Jr., J.H., Moser, M.A., 2004. A Manual for Developing Biogas Systems at Commercial Farms in the United States, second ed. AgSTAR Handbook. U.S. Environmental Protection Agency, Washington DC.

Schlicht, A.C., 1999. Digester Mixing Systems: Can You Properly Mix With Too Little Power? Available at http://www.walker-process.com/pdf/99_DIGMIX.pdf (verified 13.08.14.).

USEPA, 2013. Overview of Greenhouse Gases. http://epa.gov/climatechange/ghgemissions/gases/ch4.html (verified 27.02.14.).

Zickefoose, C., Hayes, R.B.J., 1976. Operations Manual. EPA 430/9-76-001. Anaerobic Sludge Digestion US EPA, Washington, DC.

GASEOUS FUELS AND BIOELECTRICITY SERVICE LEARNING PROJECTS AND CASE STUDIES

19

Anju Dahiya[1,2]

[1] *University of Vermont, USA;* [2] *GSR Solutions, USA*

As per the 2014 report by the World Bioenergy Association, recent years have witnessed an increased use of biogas globally—from 292 Peta Joules (PJ) in 2000 to 1103 PJ in 2011 (WBA GBS report, 2014). Biogas is increasingly being used for electricity production as the extension engineer and associate professor of Agricultural & Biological Engineering at Purdue University, Dr. Klein Ileleji et al., described in Chapter 17, "Livestock manure from concentrated livestock operations can be a source of energy production that not only provides an alternative energy source for on-farm use, but mitigates the negative consequences of odor from livestock operations. Biogas generated from manure can be used directly in a gas-fired combustion engine or a microturbine to create electricity."

The biogas field trips to the farms hosting large-scale anaerobic digesters and selling electricity to the grid provided a great learning experience to the service learning students. For instance, the Biomass to Biofuels course[1] students were able to compare different digesters they saw on the farms. In the 2012 batch, this was one student's reflection in the course discussion forum, *"It was nice to have another look at a much larger biodigester and get feedback from another source. It was good to see and hear that this farmer has a better outlook on the system than the one in Stowe did. This is probably the result of being partnered with a much larger and more successful company like Green Mountain Power. It was also nice to see how much our group has learned over the course of the semester about biomass and biofuels, as it reflects in our ability to ask more intelligent questions and have more meaningful conversations…It was clear that the technology is improving at a steady rate, and those improvements are expected to continue. With each new installation, they have learned how to make improvements."*

Another student from the same batch chimed in, *"This trip offered me a new perspective on biogas production. After working all semester with my community partner, I have gained a very narrow-minded view on how the process of producing biogas should be carried out. Monument farms had an underground system. What is different about this system is that it is harder to access the tank for cleaning. If you were to have to replace the tank, it would require ripping up the entire system as it is all housed under a thick layer of insulation in the ground. I was impressed by how centrally located the digester was. A conveyer directly dumped the bedding into a holding barn, and the manure was pushed directly into a stirring tank uphill of the digester, using gravity to get the poop into the system. It was all happening in one central area."*

---

[1] http://go.uvm.edu/7y1rr.

Bioenergy. http://dx.doi.org/10.1016/B978-0-12-407909-0.00019-5

319

The farm anaerobic digester-based biogas system producing electricity

Source: Courtesy A. Dahiya.

The cow barn at a large dairy farm visited for the biogas field trip.

Source: Courtesy A. Dahiya.

Service learning students also get an opportunity to learn about overlooked benefits of anaerobic digestion. The extension educator in Agricultural Bioenergy and Energy Conservation from Michigan State University, Charles Gould, describes these benefits in Chapter 18: "Potentially, farms operating anaerobic digestion systems could take on additional wastes and benefit from increased gas production as well as income from tipping fees. A tipping fee is the payment a farmer receives for accepting a feedstock. Examples of feedstocks that could be added to a farm-based digester include waste feed; food processing wastes; fats, oils and grease; slaughterhouse wastes; corn silage (energy crop); syrup from ethanol production; glycerin from biodiesel production; milk house wash water; fresh produce waste; cafeteria waste; farm animal mortality."

Building on these concepts, over the years many Biomass to Biofuels course students have undertaken biogas-related service learning projects in partnership with community partners engaged in biogas/bioelectricity production and anaerobic digesters implementation. In 2011, a couple of students partnered with a local company engaged in anaerobic biodigester design and operation to search for viable sources of anaerobic microbes in the local geographic area and to test their capacity to produce methane. In the same year, another student tested the feasibility of a local brewery waste byproduct to create usable methane gas. In the 2012 version of the course, eight students (Walt Auten, Craig Bishop,

Terence Boyle, Samantha Csapilla, Rose Fierman, Danika Frisbie, Anna Pirog, and Sydney Stieler) interned with a local biodigester company for their service learning project entitled "Monitoring Biogas Productivity from Source Separated Organics: A Service-Learning Project with Avatar Energy." They collected the food waste from the university cafeteria and brought it to the company's lab for monitoring biogas production from source-separated organic material. In 2013, this project was picked up by three students (Grant Troester, Adam Riggen, and Sam Grubinger) who tested the food variations for biogas production as part of their project, "Analysis of Food Waste Feedstocks for Biodigestion." In 2014, three students further investigated the potential of the past projects and demonstrated a lab-based digester that they intend to use for educating the use of food waste for biogas production at a small scale. Samantha Csapilla, with Grant Troester and Adam Riggen, combined the two reports to present as a case study (*Effect of Daily Variation in Food Waste on Biogas Production during Anaerobic Digestion*) for this chapter.

Policy plays an important role in waste to energy generation. For instance, in light of Vermont passing Act 148, the universal recycling law brought new, big changes that are estimated to take place across the waste management landscape due to the diversion of all organics from landfills. A 2014 Biomass to Biofuels student, Ariadne Brancato, decided to use this opportunity to build her service learning project (*Potential for Anaerobic Digestion in Meeting Statewide Energy Needs*) to look at the potential amount of available food waste statewide and possible methane yields from biodigestion. Her study presented in this chapter shows that anaerobic digestion can only act as an extremely limited energy source in comparison to the amount of energy Vermont uses annually.

Another high-quality gaseous fuel is produced from the process called gasification. As described by the University of Vermont engineering professor, Dr. Bob Jenkins (see Chapter 16), "The conversion of carbon-rich biomass to useful combustible gas mixtures such as carbon monoxide, hydrogen, and methane by thermal gasification processes is a proved technology for energy production. In the gasification processes, conversion is achieved by reacting biomass with air (oxygen), steam, and carbon dioxide at elevated temperatures (>600 °C)." Depending on the cost analysis, sometimes gasification could be too expensive for a facility to operate at a reasonable scale and energy return on their investment, as demonstrated by a service learning project under taken by four Biomass to Biofuels students in 2011 as described in Chapter 6, page 3.

In 2014, a Bioenergy—Biomass to Biofuels student and a graduate student in the Community Development and Applied Economics, Deandra Perruccio, for her project, *Biomass Gasification as a strategy for Rural Electrification in Developing Nations: Lessons from the Field*, studied the sustainable development through the business model and case studies of Pamoja Cleantech LLC in the Uganda environment that may be applicable elsewhere when integrated with local ethics. This report is presented in this chapter.

REFERENCE
WBA GBS report, 2014. World Bioenergy Association (WBA) Global Bioenergy Statistics. www.worldbioenergy.org.

EFFECT OF DAILY VARIATION IN FOOD WASTE ON BIOGAS PRODUCTION DURING ANAEROBIC DIGESTION

19A

Samantha Csapilla, Grant Troester, Adam Riggen

2012 & 2013 Biomass to Biofuel Course Students, University of Vermont, Burlington, VT, USA

PROJECT OBJECTIVES

The main objective of the experiment was to examine whether daily variation in food waste from breakfast, lunch, and dinner would cause variation in biogas production yields and digester stability. The purpose was to examine whether there is enough balance in nutrient composition of each meal to sustain the microorganisms responsible for biogas production during anaerobic digestion. We hypothesized that the positive control feedstock, consisting of wastes from all three meals, would perform the best, producing the largest, most consistent volume of biogas.

BACKGROUND

Anaerobic digestion (AD) is a biological process in which microorganisms break down, or digest, organic material into its various primary components in an oxygen-free (anaerobic) environment. Biogas produced as organic material is broken down, consisting of 60–70% methane, 30–40% carbon dioxide, and trace amounts of other gases. Biogas is a renewable form of natural gas energy.

The AD of food waste alone is a relatively new field of study, but research has been conducted on the subject. Chen et al. (2010) conducted an experiment examining the biogas production from various food wastes obtained from a commercial kitchen, a cafeteria, a grease trap collection service, a fish farm, and a soup processing plant. Chen et al. (2010) compared the production potential of the different food waste feedstocks and found that the nutrient content of each feedstock affected both the digestion time and gas production from each feedstock. Some feedstocks took longer to begin producing gas, and it was difficult to regulate the pH in the system. They found it necessary to use a sodium hydroxide buffer to control pH.

A previous service-learning (SL) study for this same course in 2012 (unpublished) examined food waste at Avatar Energy, LLC, a Vermont digester company as a potential feedstock for anaerobic digestion for the first time. As previous literature suggested, it was found that stable digestions of food wastes are difficult to maintain as the food waste quickly acidifies, lowers the pH of the digester, and inhibits methanogenesis. The microorganisms involved in the anaerobic digestion process can only thrive in a slim pH range of 6.8–7.6 (Rittmann and McCarty, 2001). Avatar has developed a protocol for successfully digesting food waste with the critical parameter being pH regulation in order to sustain the anaerobic microbe communities. This current study is an extension of the previously mentioned SL project, where food wastes from different meals throughout the day are investigated for biogas production performance and stability.

COMMUNITY PARTNER

Avatar Energy, LLC, our community partner, is a company that has been developing anaerobic digesters since 2005. Currently, Avatar sells digesters for the dairy industry, allowing farmers to produce their own electricity and produce a high-quality, pathogen-free bedding for their animals and a more efficient, less odorous fertilizer for their fields. Avatar is also developing equipment scaled for processing institutional organic wastes such as cafeteria food waste. During our time at Avatar, we worked under Samantha Csapilla in Avatar's R&D laboratory researching food waste digestion in South Burlington, VT on digesting food waste in South Burlington, VT.

METHODS

Approximately five gallons of food waste (FW) from breakfast, lunch, and dinner were obtained from a Champlain College cafeteria. The ingredients and proportions of FW from each meal were recorded before the FW was blended to make a uniform mixture (Figure 19A.1). Water was added during blending in order to create a "flowable" slurry. Each slurry was distributed into quart-sized bags and stored frozen until used. A feedstock analysis was conducted prior to digestion to determine the initial pH, the total solids/volatile solids (TS/VS) ratio (Table 19A.1), and the nutritional characteristics of each substrate. The final pH of each digester was also obtained at the end of the experiment (Table 19A.2).

Each feedstock mixture was tested in duplicate using Avatar Energy's lab-scale anaerobic digesters. Each digestion was started using a proprietary mixture of food waste and inoculum provided by Avatar. Control digestions were also included in this study, where all three meals mixed together with inoculum served as the positive controls, and all three meals mixed together without inoculum served as the negative controls. The digesters were sealed, placed into a 105 °F incubator, and attached to a gas collection apparatus.

Measurements of temperature, gas volume, and gas flammability were recorded twice a day for 12 days. The data was recorded and graphed to examine biogas production over time (Figure 19A.2). The digesters received one additional feeding of food waste after 8 days of digestion. Each digester was carefully mixed before and after feeding in order to achieve uniform consistency while minimizing aeration. After 12 days, the experiment was concluded.

RESULTS

The results of this study supported the initial hypothesis as the positive control digests, a mixture of food waste from all three meals, consistently produced the most biogas.. The digestions of breakfast food waste performed well initially, but the feedstock was quickly used up and biogas production quickly decreased. The digestions of dinner food waste took the longest to begin producing significant biogas, but following feeding I the digesters quickly acidified (Table 19A.2) and were unable to recover (Figure 19A.1). The digestions fed lunch food waste produced the least amount of gas but showed stable pH at the end of the experiment, indicating that the feedstock was not suitable for the microorganisms involved or there was a lack of microbes present.

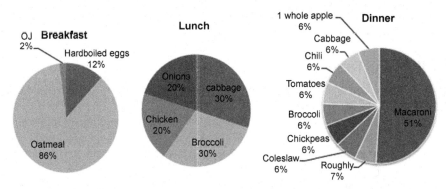

FIGURE 19A.1

Pie charts showing the breakdown of each food waste sample obtained from breakfast, lunch, and dinner preconsumer food wastes from a Champlain College cafeteria.

Table 19A.1 Initial Feedstock Characteristics of the Three Types of Food Waste and the Material Used to Inoculate Each Digester

Feedstock	Inoculum	Breakfast FW	Lunch FW	Dinner FW
pH	8.47	6.94	6.41	4.8
% Total solids	6.02	15.16	10.1	27.04
% Volatile solids	72.36	N.D.	N.D.	96.06

Table 19A.2 The final pH of each digester where A-B are the negative controls, C-D are the positive controls, E-F are the breakfast-FW digests, G-H are the lunch-FW digests, and I-J are the dinner-FW digests

Digester	A	B	C	D	E	F	G	H	I	J
Final pH	3.84	3.97	8.19	8.25	8.53	8.4	8.42	8.38	6.8	6.41

DISCUSSION

Food waste is one of the most variable feedstocks for anaerobic digestion. Many foods are lacking in many macronutrients and micronutrients required for stable digestion. This study shows that food waste as a feedstock for anaerobic digestion needs to be diverse and balanced in nutritional content in order to be a suitable substrate for the delicately balanced microbe communities. Having a well-balanced mixture of food wastes also aids in buffering the systems against dramatic changes in pH as the waste is broken down.

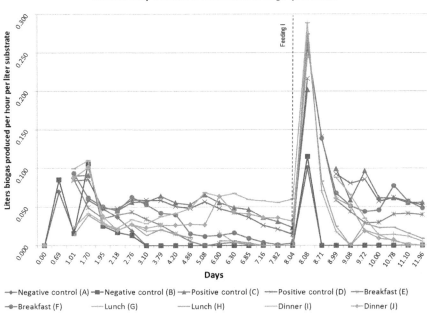

FIGURE 19A.2

Biogas yields from each laboratory-scale anaerobic digestions of cafeteria food wastes from breakfast, lunch, and dinner over 12 days of digestion.

FUTURE DIRECTIONS

Anaerobic digestion has a bright future. Methods are already being used for digestion of dairy manure, but less research has been done for food waste digestion. There will soon be a ready market for this technology in Vermont due to VT Act 148, a policy mandating the separation of organic waste from trash that has begun this year with the largest waste producers. There is already infrastructure in place to collect and consolidate this waste as is done in Burlington by waste haulers such as Casella. Currently, some of the food waste is being composted and sold as a highly valuable soil amendment. However, food waste has the potential to be a significant source of energy commercially and residentially. Biogas can be used for cooking, which is much more energy- and cost-efficient than converting it to electricity. Designs for this kind of system are currently being developed and tested by Avatar and other companies. More research and testing needs to be done to make this a viable process.

The following process flow diagram (Figure 19A.3) illustrates the possible role of an on-campus biodigester in the waste flow of a Vermont college or university.

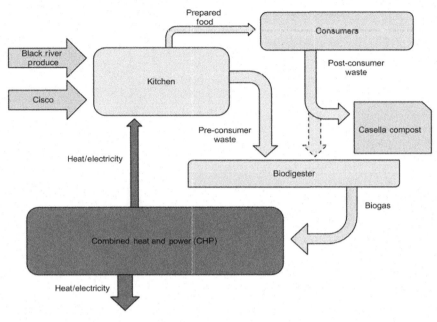

FIGURE 19A.3

Process flow diagram illustrating the possible role of an on-campus biodigester in the waste flow of a Vermont college or university.

BENEFITS TO COMMUNITY PARTNER

Avatar Energy, LLC benefitted from this experiment in that they received valuable information that they will use when designing anaerobic digestion systems for food waste.

REFERENCES

Chen, X., Romano, R.T., Zhang, R., 2010. Anaerobic digestion of food wastes for biogas production. International Journal of Agricultural and Biological Engineering 3 (4), 61–72.

Rittmann, B.E., McCarty, P.L., 2001. Anaerobic Treatment by Methanogenesis. Environmental Biotechnology: Principles and Applications. McGraw-Hill, Boston, pp. 581–584.

SUBCHAPTER

POTENTIAL FOR ANAEROBIC DIGESTION IN MEETING STATEWIDE ENERGY NEEDS

19B

Ariadne Brancato

2014 Bioenergy – Biomass to Biofuels course student, University of Vermont, Burlington, VT, USA

PURPOSE

The purpose of this report is to determine anaerobic digestion's capacity for offsetting Vermont's use of fossil fuels, in light of recent legislation, Act 148, requiring diversion of all organics from landfills.

BACKGROUND
ANAEROBIC DIGESTION: PROCESS AND SYSTEMS OVERVIEW

Anaerobic respiration is a biological process wherein microorganisms break down organic matter in an oxygen-free environment, with different organisms being active at specific stages and temperature ranges. As opposed to aerobic processes, anaerobic processes are relatively inefficient and release significantly less energy (Roberts 2014). However, anaerobic processes produce gases that can be captured, stored, and used as energy later, unlike aerobic processes.

Anaerobic Digestion (AD) technology includes a multitude of systems, all which have varying effects on methane yield as different technologies have been maximized for a variety of purposes. Historically, systems have been used for their low-tech replicability in developing countries and for minimizing solids in waste water treatment plants (American Biogas Council).

In addition to producing a gaseous mix of methane and carbon dioxide, AD also produces digested fiber that can be used as animal bedding or added to compost, depending on the feedstocks, as well as a liquid that can be recycled back into the system if the feedstock is too dry or used as a fertilizer (Bishop and Richard Shumway, 2009). The output of interest in this paper is AD's capacity for producing energy in the form of methane gas.

Sources claim varying yields of methane depending on many factors, including feedstock, total solids content, and retention time, though variation may also come from differing experimental methods. In comparison to biogas yield of manure at 25 m^3/ton of food waste, one report stated that digesting food waste could yield 376 m^3/ton, a manyfold increase (East Bay Municipal Utility District, 2008). As gas yield is measured in volume, numerous sources report yield in m^3 or ft^3 per pound or ton (Cho and Park, 1995; Zhang et al., 2007); however, for the purposes of this report, kWh will more clearly link AD yield to energy consumption rates in Vermont. Looking at sources that have measured methane produced from food waste, there is evidence of a range of kWh equivalents in methane yield: 180 kWh/ton (UK Energy Partners); 250 kWh/ton (EBMUD, 2008); and 300 kWh/ton (UK Biogas www.biogas-info.co.uk).

VERMONT ACT 148—THE UNIVERSAL RECYCLING LAW

Passed in 2012 by the Vermont Legislature, Act 148, also known as "the "Universal Recycling Law", aims to ban all organics and recyclables in landfills by 2020 in order to increase recycling, lower Vermont greenhouse gas emissions from waste processing and landfills, and create "green" jobs and conserve resources (VT DEC, 2013). Act 148 contains requirements for consumers, haulers and processors, and waste management districts in terms of how and what must get diverted and the scope of services that must be offered, but it does not mandate specific infrastructure that must be used to meet requirements. Of particular interest to this report is the information related to organics management. For organics management, Act 148 provides five priorities for waste stream diversion strategies, as follows, with highest priority strategies listed first:

Source Reduction → Food for People → Food for Animals → Composting and Anaerobic Digestion → Energy Recovery.

ACT 148 REPORT

The Act 148 Report, hereafter called "the report", conducted by DSM Environmental and commissioned by VT Agency of Natural Resources, is a systems analysis of the impact Act 148 might have on solid waste management entities through 2022, two years after the final phase-in of the ban on recyclables and organics in landfills. The analysis includes cost analyses of various iterations of an expanded Bottle Bill, costs of infrastructure build out for hauling and processing more recyclables and organics, administrative costs to Solid Waste Districts and analyses of collection rates for recyclables and organics. Of most concern to this report are the assumptions and analyses related to organics collection.

ANALYSIS
ACT 148 REPORT

The report, completed by DSM Environmental, analyzes a range of waste management issues from organics to hazardous wastes. The information most pertinent to this report are the numbers DSM Environmental pulled together for the amount of food waste Vermont generates and the amount of food waste Vermont can expect to collect based upon other states' composting programs' rates of collection.

The report modeled rates of collection from 2014 to 2022, two years after the ban is fully implemented. The model categorizes a diversity of waste, describing organics in three categories, differentiating between residential and industrial, commercial, and institutional (ICI) waste. The base food residual numbers used are taken from a previous study completed earlier in 2013, titled the "State of Vermont Waste Composition Study" (ANR, 2013). While the study had a relatively small sample size of 100 loads carrying 200–250 pounds, they reported a 90% confidence interval that the mean values lay within the stated range.

The report describes a number of assumptions the researchers made in modeling the eight years of the project's implementation. These assumptions have varying affects on the amount of potential food waste available and the quantity Vermont can expect to collect; however, this report simply uses a range of lowest projected collection rate to highest feasible collection rate to understand AD's energy potential in Vermont. Few assumptions made in the report are relevant to this study.

Table 19B.1 Range of Potential Available Food Waste in Vermont

Information Source	Available Feedstock (tons)	Possible Energy Yield (kWh)	% of Total Energy Use
Food waste, Act 148 Report	28,896–60,078	7,224,00–15,019,500	0.02–0.03%
Food waste + other organics	44,420–92,816	11,105,000–23,204,000	0.02–0.12%
Food waste, FAO	78,218–203,366	19,554,500–50,841,500	0.04–0.12%

According to numbers provided in Table 19B.1, taken from projections in the report, the range of organic wastes potentially available for anaerobic digestion is 44,420–92,816 tons. The lower number in the range is the 2022 projected tonnage of organic waste that will be collected, after higher priority management strategies are implemented and assumed 60% diversion rate is calculated. The greater number is the sum of all currently produced organic wastes, based upon DSM Environmental's calculations.

The feasibility of this range can be compared to another study. The Food and Agriculture Organization of the United Nations commissioned a world-wide study on food waste per capita in 2011, which drew upon all available research on food waste in a diversity of sectors across many cultures and countries (FAO, 2011). This study distinguished "losses" from "wastes" in that the former included edible materials that were damaged prior to reaching the retail and consumer level. This could include crops lost in the field, spoiled in storage, or damaged in transit. In the report, "waste" refers to anything thrown out by the consumer or deemed unfit for sale if, for instance, the item sat too long on the shelf and became blemished. After accounting for all sorts of variations, the report claims that in North America, the amount of food both lost and wasted, to use their terms, averages out to 650 lbs/person/year (FAO, 2011). Just looking at food wasted at the retail and consumer level, the number is 250 lbs/person/year.

According to the 2010 United States Census, the population of Vermont is 625,741. Using this number and the amount of food waste claimed in the FAO report, Vermont is responsible for 156,435,250 lbs (or 78,218 tons) of retail and consumer level food waste and 406,731,650 lbs (or 203,366 tons) of total food losses and waste.

The above described numbers estimate the amount of food lost and wasted per capita, but the actual available food waste in Vermont for anaerobic digestion might be different, as the study would have averaged out food losses across the entire population of North America, where areas with more agricultural land would see different kinds of and quantities of food waste than areas with greater population density. While this indicates that the amount of food waste available for collection in Vermont might be lower than the FAO's projections, the numbers used in the VT Act 148 Report include yard waste and compostable paper, which make up 36.5% of the total organics assumed currently available, thus increasing available feedstock for digestion.

ENERGY POTENTIAL

Using the kWh yield determined in the East Bay Municipal Utility District's study of 250 kWh/ton of food waste, Table 19B.1 shows potential energy yields and each yield's percent of total energy consumption in Vermont. Percent of total energy consumption is based off of 2011 data from the US Energy Information Association related to total nonrenewable and renewable energy consumption for each state. Vermont consumes 43,667,589,500 kWh per year.

RESULTS

According to the data, energy from AD could offset only a negligible amount of Vermont's total energy consumption, regardless of which values of food waste tonnage are used to calculate kWh yield. This finding highlights the importance of understanding the multiple values and potential sources of revenue in AD; while AD facilities may not be able to offset a significant amount of energy consumption, it may be able to pay for itself in energy returns, or at least keep itself running while it carries out its various waste management functions.

While anaerobic digestion certainly has a place in Vermont's working landscape, increased use of AD technology in light of Act 148 seems tentative. One barrier includes general lack of information about anaerobic digestion amongst decision makers. The report, whose guidance will certainly determine how Solid Waste Management entities will go about implementing Act 148, paid little attention to AD, claiming the following:

> There are simply too many unknowns about how these [AD] facilities might work and how much they might cost at this time to complete a more detailed and precise economic analysis.
>
> **(Act 148 Report)**

While there are feasibility studies available, there is no precedence in the United States for statewide use of anaerobic digestion. Furthermore, the report presented a particular focus on composting strategies, indicating the preference for composting for waste management, perhaps based as much on comfort and precedence as actual data. In light of a changing waste management landscape in Vermont, another way to look at the data in this report is that AD, as a waste management strategy, produces far more electricity than composting does.

FUTURE RESEARCH

While the initial numbers indicate that anaerobically digesting all of Vermont's food waste for a plentiful renewable energy source is a paltry selling point for AD, there are other factors worth researching in order to better understand the value of anaerobic digestion. In regard to Act 148's ban on organics in landfills, it would be particularly useful to further analyze the costs and benefits of compost facilities versus AD facilities as the legislation will force one or the other to be built in order to meet the diversion goals. This kind of analysis could include capital costs, siting, and waste transportation concerns, as well as economic opportunities for each system's finished products. While food waste alone did not produce a significant portion of Vermont's energy needs, further research into co-digestion of food wastes, dairy manure, and waste water possibilities could increase the methane output. Further information could be drawn from more in depth research on successful European AD systems, particularly focusing on the economic incentives that encourage its implementation and the extent to which methane makes up renewable energy sources.

REFERENCES

Bishop, C., Richard Shumway, C., 2009. The economics of dairy anaerobic digestion with coproduct marketing. Review of Agricultural Economics 31 (3), 394–410.

Cho, J.K., Park, S.C., 1995. Biochemical methane potential and solid state anaerobic digestion of Korean food waste. Bioresource Technology 52 (3), 245–253.

East Bay Municipal Utility District, 2008. Anaerobic Digestion of Food Waste. http://www.epa.gov/region9/organics/ad/EBMUDFinalReport.pdf.

Energy Information Administration. Vermont Data. http://www.eia.gov/state/data.cfm?sid=VT (accessed 22.04.14.).

FAO, 2011. Global Food Losses and Food Waste – Extent, Causes and Prevention. Rome. http://www.fao.org/docrep/014/mb060e/mb060e.pdf.

Roberts, G. Avatar Energy. http://www.avatarenergy.com/information.html (accessed 22.04.14.).

UK Energy Partners. http://www.ukenergypartners.co.uk/ (accessed May 2014.).

UK Biogas. www.biogas-info.co.uk/ (accessed May 2014.).

VT Department of Environmental Conservation, 2013a. State of Vermont Waste Composition Study. http://www.anr.state.vt.us/dec/wastediv/solid/documents/finalreportvermontwastecomposition13may2013.pdf.

VT Department of Environmental Conservation, 2013b. Systems Analysis of the Impact of Act 148 on Solid Waste Management in Vermont. http://www.anr.state.vt.us/dec/wastediv/solid/documents/FinalReport_Act148_DSM_10_21_2013.pdf.

Zhang, R., El-Mashad, H.M., Hartman, K., Wang, F., Liu, G., Choate, C., Gamble, P., 2007. Characterization of food waste for feedstock for anaerobic digestion. Biosource Technology 98 (4), 929–935.

BIOMASS GASIFICATION AS A STRATEGY FOR RURAL ELECTRIFICATION IN DEVELOPING NATIONS: LESSONS FROM THE FIELD

19C

Deandra Perruccio

2014 Bioenergy – Biomass to Biofuels student, University of Vermont, Burlington, VT, USA

INTRODUCTION

Gasification involves exposing biomass to high temperatures in low oxygen environments, causing pyrolysis, a process by which volatile components of a feedstock vaporize, creating a gas (producer gas) that can be used to power internal combustion engines, gas turbines, or fuel cells (Larson, 1998). These systems are able to create electricity at higher efficiencies and lower costs than boiler and steam systems of comparable size (Larson, 1998).

In developing nations, poor energy system infrastructures prevent access to electricity and hinder the productive capacity of off- grid villages, which constitute the large majority of communities. Biomass Gasification Energy (BGE) as a rural electrification strategy can build productive capacity and spur economic activity in participating communities, while also providing clean, locally produced renewable energy.

In 2012, an international group of engineers and entrepreneurs started Pamoja, Cleantech AB; a socially minded business that works with communities across Uganda to operate small scale BGE systems for productive use. Over the past two years, Pamoja has gathered funding, developed community relationships, and installed three pilot systems ranging from 10 to 32 KW in the villages of Ssekanyonyi, Tiribogo, and Opit.

METHODS

This report was compiled largely through review of relevant literature. Peer reviewed articles, Pamoja project reports, international development reports, and national level data were consulted. Interviews of Pamoja leadership were also conducted.

BIOMASS GASIFICATION AS A STRATEGY FOR SUSTAINABLE RURAL DEVELOPMENT
COMMUNITY NEED AND BUSINESS OBJECTIVES

Access to electricity for rural communities in Uganda is currently far lower than other nations. About 84% of households are located in rural areas, and less than 1% of them have access to modern energy services (Buchholz et al., 2010). The national electricity deficit in 2007 was estimated to be 165 MW. Electricity demand is increasing at around 8% (REA, 2007), and around 34% of total investment is currently put into generator backup systems (Eberhard et al., 2005). The inadequacy and lack of reliability in Uganda''s electrical supply results in lowered economic productivity and missed opportunities for development. Without access to electricity, rural communities end up paying high costs for nonrenewable, inefficient energy like kerosene and dry cell batteries for lighting and charging of cell phones (Christensen, 2013). Prices for these energy sources calculate to a high rate of $3/Kwh (SharedSolar, 2011), resulting in a scenario where many poor pay more per unit of energy than their more affluent urban counter-parts.

Barriers to large scale national grid electrification efforts include lack of infrastructure, high costs of grid connection over difficult terrain, and low demand. Costs to connect rural households to the national grid are estimated around $1000 per household (SharedSolar, 2011), an unappealing figure to national energy companies.

PAMOJA BUSINESS MODEL

Pamoja's initial business plan involved The Green Plant, a hybrid energy system that combines a 10 kW gasifier unit to turn agricultural waste into electricity, solar PV, and a diesel generator as back up. While the partnership with telecommunications companies did not materialize, Pamoja adapted their business plan and currently provides a regionally unprecedented proof of concept for a profitable business model by setting up three biomass gasification units with microgrids connecting households and agroprocessing units offering value addition to farmers. Enhancing its business model, Pamoja is developing an advanced value chain for its electricity, process heat, and charcoal byproducts through establishment of briquetting operations. Biomass briquetting increases resource efficiency by utilizing byproducts, process heat, and off-peak electricity. In East Africa, briquettes offer a high-value cooking-fuel substitute for firewood and charcoal, addressing serious forest degradation concerns.

CASE STUDIES: INVESTIGATING PAMOJA OPERATIONS

Research into the viability of Pamoja's systems concluded that in order for biomass gasification systems to financially outcompete diesel systems, they need to operate closer to capacity. Increasing output to full capacity in a pilot 150 kW system resulted in $.18/kWh (diesel was cited at $.22/kWh), and increasing output to 8 kW in the 10 kW system resulted in competitive costs at 30% minimum loads, which is typical of rural villages in Uganda (Buchholz et al., 2012). According to these studies, effective business models must balance energy demand and system capacity to be successful. Suggestions to increase economic success of these systems include creation of energy service companies, commercialization of heat energy byproducts, and feed-in tariffs to spur investment into technology.

BIOMASS GASIFICATION ENERGY SYSTEMS AND APPLICATIONS IN THE US

This venture provides many valuable lessons and insights for the biomass field. While Pamoja's ability to harness the right resources, coordinate partners, and delegate responsibilities has contributed to their success thus far, jointly building the capacities of all parties involved in their projects, there remain many challenges to developing a truly sustainable system. Success entails developing the biomass supply chain (sustainable supplies), facilitating productive use of power (creating energy demand), and building technical capacities (providing reliable power) within communities—a complicated system to develop and coordinate.

BGE systems have the potential to offer important social and environmental benefits, which need to be considered in conjunction with economic criteria. Broadly speaking, this project highlights a key element which differentiates biomass energy systems from other renewable energy options: the biomass supply chain. Developing a locally produced biomass supply chain, whether it be from current waste streams or agricultural and agroforestry systems, can create jobs and promote environmentally sustainable energy production systems contributing positively to the social, economic, and environmental well-being of involved communities.

The motivations, capacities, needs, and, therefore, applications of a BGE system in a developing country context vary greatly from the context of a developed nation like the United States. These differing contexts lead to different conclusions regarding the viability of small scale BGE system applications.

CONCLUSIONS AND CONSIDERATIONS

BGE systems have the potential to offer important social and environmental benefits, which need to be considered in conjunction with economic criteria. Currently, small scale BGE systems cannot compete economically with grid electricity. In the US, other renewable energy options have more traction and may be able to offer more context appropriate systems than the small scale BGE system; however, the environmental and social impacts of BGE systems, specifically the necessary supply chains, could work well with Vermont's working landscape ethic. Further research into the potential viability of these systems in Vermont is needed.

REFERENCES

Buchholz, T., et al., March 2010. Potential of distributed wood-based biopower systems serving basic electricity needs in rural Uganda. Energy for Sustainable Development 14 (1), 56–61.

Buchholz, T., et al., 2012. Power from wood gasifiers in Uganda: a 250 and 10 kW case study. Energy 165 (EN0).

Christensen, S., 2013. M.Sc. program "Innovative Sustainable Energy Engineering", Department of Industrial Ecology, Royal Institute of Technology (KTH), Stockholm, Sweden. Title of Research: "Development and Testing a Sustainability Assessment Framework for Biomass Supply Chains Fueling Electricity Systems in Rural Uganda".

Eberhard, A., Clark, A., Wamukonya, N., Gratwick, K., 2005. Power Sector Reform in Africa: Assessing the Impact on Poor People. World Bank Energy Sector Management Assistance Program, Washington, DC.

Larson, E.D., January 1998. Small Scale Gasification –Based Biomass Power Generation. Center for Energy and Environmental Studies. Princeton University. Prepared for Biomass Workshop Changchun, Jilin Province, China.

Rural Electrification Agency (REA), 2007. Renewable Energy Policy for Uganda, p. 27. [PDF] Available at: http://www.rea.or.ug/userfiles/RENEWABLE%20ENERGY%20POLIC9-11-07.pdf.

SharedSolar, 2011. SharedSolar Concept Note [PDF] Available at: http://sharedsolar.org/wp-content/uploads/2011/02/Concept-Note-SharedSolar-v61.pdf.

CONVERSION PATHWAYS FOR COST-EFFECTIVE BIOFUEL PRODUCTION

ANJU DAHIYA[1,2]

[1] *University of Vermont, USA,* [2] *GSR Solutions, USA*

Multiple biomass conversion pathways have been tried at commercial scales (e.g., hydroprocessing) as described in this part. A lot of research and development is in progress for converting biomass into hydrocarbon fuels and intermediates leading to "drop-in" fuels. For instance, the conversion pathways as described by the U.S. Bioenergy Technologies Office[1] are: *the Biological Conversion of Sugars to Hydrocarbons; Catalytic Upgrading of Sugars to Hydrocarbons; Algal Lipid Upgrading; Whole Algae Hydrothermal Liquefaction; Fast Pyrolysis Upgrading and Hydroprocessing; Ex-situ Catalytic Fast*

---

[1]http://www.energy.gov/eere/bioenergy/technology-pathways.

FIGURE 1

Bioconversion pathways for fuels.

Picture courtesy: A. Dahiya.

Pyrolysis; In-Situ Catalytic Fast Pyrolysis; Syngas Upgrading to Hydrocarbon Fuels. Many conversions are described here.

Part V contains seven chapters on conversion pathways including biodiesel conversion via transesterification, ethanol production, pyrolysis, multiple bioconversion technologies (hydrolysis, enzymatic, etc.), and technologies to integrate with biorefineries. The case studies include the processes used in biodiesel production, biogas and fungal breakdown of lignocellulosic biomass.

Chapter 20 (Biodiesel Production) gives an overview of the production of biodiesel and touches on a few of the highlights of biodiesel production. It covers biodiesel production processes including the selection of feedstock and catalysts used in biodiesel production, biodiesel production process such as batch processing, continuous processing, high free fatty acid systems, co-solvent systems—biox process, noncatalyzed systems—supercritical process, effects of process parameters, postreaction processing (ester/glycerol separation, ester washing, ester drying, other ester treatments, and the additization of esters). The treatment and recovery of side streams, fatty acid composition and total and free glycerol, pretreatment of high free fatty acid feedstock, and in the end it summarizes biodiesel production.

Chapter 21 (Transesterification) provides a protocol for students to make biodiesel at a lab scale essentially as it is produced industrially. The transesterification of a triglyceride (soybean oil) to yield a fatty acid methyl ester (FAME) (biodiesel) via a base-catalyzed route provides a simple and effective hands-on demonstration of the synthesis and partial characterization of an alternative fuel for students ranging from high school to college seniors. The addition of more detailed characterization methods makes the experiment suitable for college-level chemistry majors.

Chapter 22 (Whole Algal Biomass In Situ Transesterification to FAMEs as Biofuel Feedstock) discusses the yield of lipids quantified as FAMEs by using different catalysts and catalyst combinations, with the acid catalyst hydrochloric acid providing a consistently high level of conversion to FAMEs. The discussion is accompanied by a link to the large-scale application of this process as a whole biomass conversion pathway. Microalgae-focused lipid technologies for biofuel applications, renewable and biodiesel fuel properties are described. Also described are the in situ transesterification of oleaginous algal biomass, the choices of catalyst for the in situ whole biomass transesterification analytical characterization of lipid content in algal biomass using in situ transesterification.

Chapter 23 (How Fuel Ethanol Is Made from Corn) describes fuel ethanol, yeast's role in ethanol production, corn as ethanol feedstock, industrial ethanol production including wet milling, and dry-grind ethanol processing steps (milling, liquefaction, saccharification, fermentation, distillation, and recovery) are described. In the end, energy use in ethanol production is described before conclusion.

Chapter 24 (Approaches for Evaluating Biomass Bioconversion for Fuels and Chemicals) describes treatments of biomass including mechanical treatments, no pretreatment, chemical/thermal pretreatment (acidic pretreatment, neutral pretreatment, alkaline pretreatment), organic pretreatment (ionic liquid pretreatment, organosolv pretreatment), biological pretreatment. A typical protocol for biomass pretreatment and biomass analysis are included. Small-scale biomass fermentation approaches are described with the examples of small-scale pretreatment including examples of simultaneous saccharification and fermentation; examples of separate hydrolysis and fermentation; examples of consolidated bioprocessing; the identification of fermentation inhibition.

Chapter 25 (Reducing Enzyme Costs, Novel Combinations and Advantages of Enzymes Could Lead to Improved Cost-Effective Biofuels Production) presents three factsheets by the National Renewable Energy Laboratory. The first one explores the enzyme-related work with two leading enzyme companies, Genencor and Novozymes. From initial enzyme costs of $4–$5 per gallon of ethanol produced, the team achieved cost reductions. The importance of this research was recognized by *R&D Magazine* in 2004 with an R&D 100 Award, denoting it as one of the year's 100 most significant innovations. Second describes mixing disparate enzyme systems that can break down cellulose more rapidly and efficiently than either system alone. In the third one, the team isolated a highly active cellulose with a novel cellulose digestion mechanism, which represents a new and distinct paradigm for cellulose digestion.

Chapter 26 (Pyrolysis of Lignocellulosic Biomass: Oil, Char, and Gas) describes biomass pyrolysis which is a very complex system, involving simultaneous solid heat transfer, solid-phase chemical reactions, liquid evaporation and thermal ejection, liquid-phase reactions, the mass transfer of vapors through the solid matrix, and vapor-phase reactions. The best place to begin understanding these interactions is to look at the structure and chemistry of lignocellulosic material and toward that end the Lignocellulosic Structure and Chemistry for Pyrolysis is described in detail beside Biomass Pyrolysis Strategies, Slow Pyrolysis: The Path to Char; Fast Pyrolysis: The Path to Oil. Applications and Approaches for Product Use. Liquid Fuel Production, and the Upgrading of Bio-Oil; and Useful Chemicals from Bio-Oils.

Chapter 27 (Sustainable Aviation Biofuels—A Development and Deployment Success Model) begins with the aviation alternative fuel 2006 snapshot—"what if your family were an airline?"; aviation alternative fuel 2013 snapshot—acknowledged leader in sustainable transport fuels; key methodologies for sustainable progress—creation of a "new fuel dynamic"; streamlining the fuel qualification process and pathways to solutions; implementing comprehensive risk management in alternative fuel research and development; structuring and facilitating comprehensive environmental benefit assessments. It concludes that the unique and demanding requirements of sustainable aviation biofuel are indeed being met via product developments by the exciting new biofuel industry.

Chapter 28 (Cutting Edge Biofuel Conversion Technologies to Integrate into Petroleum-Based Infrastructure and Integrated Bio-Refineries) describes biofuel, biodiesel, and renewable diesel, how they are different from the petroleum diesel, the processing pathways for the conversion of biofuel into diesel fuel; the challenges from FAMEs use as an alternative to diesel fuels in the existing infrastructure, the issues due to the presence of *oxygen* in biofuel and possible solutions. The processes including hydrothermal processing, indirect liquefaction, hydroprocessing; and the fuel properties of hydroprocessed diesel products are described besides their integration with existing refineries or to form new integrated biorefineries for the commercialization of biofuel and industry standards. A case study of the largest U.S. energy company making ethanol and FAMEs shipments; and the benefits of biodiesel use are presented. The commercialization of renewable diesel; hydroprocessed renewable jet fuel; the future utilization of biocrudes and coprocessing concerns; integrated biorefineries; and colocating biorefineries are explored.

Chapter 29 (Fuel Conversion Service Learning Projects and Case Studies) includes an educational experience converting waste cooking oil to biodiesel; the use of magnetite for enhanced harvesting of wastewater biogas feedstock; and fungal breakdown of lignocellulosic biomass.

BIODIESEL PRODUCTION

20

Rudy Pruszko

Biofuels Consultant, Dubuque, Iowa, USA

INTRODUCTION

This chapter[1] provides an overview of the production of biodiesel and touches on a few of the highlights of biodiesel production based on Van Gerpen et al. (2004), where this information was extracted. It only covers a small fraction of the overall knowledge and detail needed to produce biodiesel, but it will give the reader an understanding of some of the concepts used in biodiesel production. Before attempting to produce biodiesel, a more complete understanding of biodiesel and its production will be needed and can be obtained from the book *Building a Successful Biodiesel Business* (Van Gerpen et al., 2006), which provided the insight to compose this chapter.

PRODUCTION PROCESSES

This section provides a summary of the steps in the production of biodiesel, from the preparation of the feedstock to the recovery and purification of the fatty acid esters (biodiesel) and the coproduct glycerol (also called glycerin). We will review several chemistries used for transesterification and esterification and different approaches to product preparation and purification.

The emphasis is on the choices of feedstock selection, capacity, and operating mode with the selection of the basic process chemistry, plant design, and technology to be used. While no specific process technology is favored here, an effort has been made to describe the major approaches currently in use and in development in the industry.

FEEDSTOCKS USED IN BIODIESEL PRODUCTION

The primary raw materials used in the production of biodiesel are vegetable oils, animal fats, and recycled greases. These materials contain triglycerides, free fatty acids, and other contaminants, depending on the degree of pretreatment they have received prior to conversion. Since biodiesel is a mono-alkyl fatty acid ester, the primary alcohol used to form the ester is the other major feedstock.

---

[1]The contents of this chapter represent part 4 "Types of Biodiesel Production Processes", part 7 "Pretreatment of High Free Fatty Acid Feedstocks", part 10 "Post Reaction Processing", and part 11 "Treatment and Recovery of Side Streams" from Van Gerpen et al. (2004), of the report prepared for NREL, Department of Energy, which is presented in this book with permission from NREL. See Acknowledgements below for further details.

Bioenergy. http://dx.doi.org/10.1016/B978-0-12-407909-0.00020-1

339

Table 20.1 Typical Chemical Proportions	
Reactants:	Fat or oil (e.g., 100 kg soybean oil)
	Primary alcohol (e.g., 10 kg methanol)
Catalyst:	Mineral base (e.g., 0.3 kg sodium hydroxide)
Neutralizer:	Mineral acid (e.g., 0.25 kg sulfuric acid)

Most processes for making biodiesel use a catalyst to initiate the transesterification and esterification reactions. The catalyst is required to initiate the reaction and act as a solubilizer. A solubilizer is needed because the alcohol is sparingly soluble in the oil phase and noncatalyzed reactions are extremely slow. The catalyst promotes an increase in alcohol solubility to allow the reaction to proceed at a reasonable rate. The most common transesterification catalysts are strong mineral bases such as sodium hydroxide, potassium hydroxide, and sodium or potassium methoxide. After the transesterification is complete, the base catalyst accumulates primarily in the glycerol coproduct phase. Mineral acid catalysts used in homogenous esterification reactions tend to accumulate in an acid-alcohol-water phase.

Typical proportions for the chemicals used to make biodiesel by transesterification are in the following Table 20.1.

Fats and oils: Choice of the fats or oils to be used in producing biodiesel is both a process chemistry decision and an economic decision. With respect to process chemistry, the greatest difference among the choices of fats and oils is the amount of free fatty acids that are associated with the triglycerides. Other contaminants, such as color and odor bodies, can reduce the value of the glycerin produced and reduce the public acceptance of the fuel if the color and odor persist in the fuel.

Most vegetable oils have a low percentage of associated free fatty acids. Crude vegetable oils contain some free fatty acids and phospholipids. The phospholipids are removed in a "degumming" step, and the free fatty acids can be removed in a "refining" step. Oil can be purchased as crude, degummed, or refined oil. The selection of the type and quality of the oil affects the production technology that is required.

Animal tallows and recycled (yellow) grease have much higher levels of free fatty acids. Yellow grease is limited to 15% free fatty acids and is a traded commodity that is typically processed into animal and pet food. Specifications for yellow grease are described in the chapter on feedstock preparation. Trap greases come from traps under kitchen drains. These greases can contain between 50–100% free fatty acids. There is no market for these greases at this time; most are sent to the landfill. Trap grease is not yet used for biodiesel production and has some technical challenges that have not yet been fully resolved, such as difficult to break emulsions (gels), fine silt that will cause equipment wear, high water contents and very strong color, and odor bodies that affect biodiesel and glycerin products. There are also unresolved questions about small quantities of other contaminants, such as pesticides, that might be present in the fuel.

The options for the triglyceride choice are many. Among the vegetable oils sources are soybean, canola, palm, and rape. Animal fats are products of rendering operations. They include beef tallow, lard, poultry fat, and fish oils. Yellow greases can be mixtures of vegetable and animal sources. There are other less desirable but also less expensive triglyceride sources, such as brown grease and soap stock. The free fatty acid content affects the type of biodiesel process used, and the yield of fuel from that process. The other contaminants present can affect the extent of feedstock preparation necessary to use a given reaction chemistry to produce an on-specification product.

Alcohol: The most commonly used primary alcohol in biodiesel production is methanol, although other alcohols, such as ethanol, isopropanol, and butanol, are used. A key quality factor for the alcohol is the water content (<0.08 wt%). Water interferes with transesterification reactions and can result in poor or no ester yields and high levels of soap, free fatty acids, and triglycerides in the final product. Unfortunately, all the lower alcohols are hygroscopic and, therefore, are capable of absorbing water from the air.

Many alcohols have been used to make biodiesel. As long as the product esters meet American Society for Testing and Materials (ASTM) 6751, it does not make any chemical difference which alcohol is used in the process. Other issues, such as cost of the alcohol, the amount of alcohol needed for the reaction, the ease of recovering and recycling the alcohol, fuel tax credits, and global warming, influence the choice of alcohol. Some alcohols also require slight technical modifications to the production process, such as higher operating temperatures, longer or slower mixing times, or lower mixing speeds.

Since the reaction to form the esters is on a molar basis, and we purchase alcohol on a volume basis, their properties make a significant difference in raw material price. It takes 3 mol of alcohol to react completely with 1 mol of triglyceride. Today, one gallon of methanol costs $0.61. That gallon contains 93.56 g-moles of methanol at a cost of $0.00,652 per gram-mole. By contrast, a gallon of ethanol, at the current price of $1.45 per gallon for fuel grade ethanol, costs $0.02,237 per gram-mole, or 3.4 times more.

In addition, a base-catalyzed process typically uses an operating mole ratio of 6:1 mol of alcohol rather than the 3:1 ratio required by the reaction. The reason for using extra alcohol is that it uses the law of mass action to "drive" the reaction closer to the 99.7% yield needed to meet the total glycerol standard for fuel grade biodiesel. The unused alcohol must be recovered and recycled back into the process to minimize operating costs and environmental impacts. Methanol is considerably easier to recover than ethanol because ethanol forms an azeotrope with water, making it expensive to purify the ethanol to the required dryness during recovery. If the water is not removed, it will interfere with the reactions. Methanol recycles more easily because it does not form an azeotrope.

These two factors are the reason that even though methanol is more toxic, it is the preferred alcohol for producing biodiesel. Methanol has a flash point of 10 °C, while the flash point of ethanol is 8 °C, so both are considered highly flammable. Methanol should never come into contact with skin, eyes, or through respiration, as it can be readily absorbed. Excessive exposure to methanol can cause blindness, other health effects, and death.

Methanol does have a somewhat variable pricing structure. When the production of Methyl Tertiary-Butyl Ether (MTBE) was mandated for the reduction of emissions from gasoline engines in the winter, there was a significant expansion in world capacity for the material. The excess capacity and crash in demand led to methanol prices of $0.31 per gallon in early 2002. However, in late July 2002, the production/consumption levels regained equilibrium and the methanol price doubled back to the more typical value of $0.60 ± per gallon.

The alcohol quality requirements are that it be undenatured and anhydrous. Since chemical grade ethanol is typically denatured with poisonous material to prevent its abuse, finding undenatured ethanol is difficult. Purchase ethanol that has been denatured with methanol if possible.

Catalysts and neutralizers: Catalysts may either be base, acid, or enzyme materials. The most commonly used catalyst materials for converting triglycerides to biodiesel are sodium hydroxide, potassium hydroxide, and sodium methoxide. Most base catalyst systems use vegetable oils as a feedstock. If the vegetable oil is crude, it contains small amounts (<2%) of free fatty acids that will form soaps that will end up in the crude glycerin. Refined feedstocks, such as refined soy oil, can also be used with base catalysts.

The base catalysts are highly hygroscopic, and they form chemical water when dissolved in the alcohol reactant. They also absorb water from the air during storage. If too much water has been absorbed, the catalyst will perform poorly and the biodiesel may not meet the total glycerin standard.

Although acid catalysts can be used for transesterification, they are generally considered to be too slow for industrial processing. Acid catalysts are more commonly used for the esterification of free fatty acids. Acid catalysts include sulfuric acid and phosphoric acid. Solid calcium carbonate is used as an acid catalyst in one experimental homogeneous catalyst process. The acid catalyst is mixed with methanol and then this mixture is added to the free fatty acids or a feedstock that contains high levels of free fatty acids. The free fatty acids convert into biodiesel. The acids will need neutralization when this process is complete, but this can be done as a base catalyst is added to convert any remaining triglycerides.

There is continuing interest in using lipases as enzymatic catalysts for the production of alkyl fatty acid esters. Some enzymes work on the triglycerides, converting them to methyl esters, and some work on the fatty acids. The commercial use of enzymes is currently limited to countries like Japan, where energy costs are high, or for the production of specialty chemicals from specific types of fatty acids. The commercial use of enzymes is limited because costs are high, the rate of reaction is slow, and yields to methyl esters are typically less than the 99.7% required for fuel grade biodiesel. Enzymes are being considered for fatty acid conversion to biodiesel as a pretreatment step, but this system is not commercial at this time.

Neutralizers are used to remove the base or acid catalyst from the product biodiesel and glycerol.

If you are using a base catalyst, the neutralizer is typically an acid, and vice versa. If the biodiesel is being washed, the neutralizer can be added to the wash water. While hydrochloric acid is a common choice to neutralize base catalysts, as mentioned earlier, if phosphoric acid is used, the resulting salt has value as a chemical fertilizer.

Catalyst selection: Base catalysts are used for essentially all vegetable oil processing plants. The initial free fatty acid content and the water content are generally low. Tallows and greases with free fatty acid contents greater than about 1% must be pretreated to either remove the Free Fatty Acids (FFA) or convert the FFA to esters before beginning the base-catalyzed reaction. Otherwise, the base catalyst will react with the free fatty acids to form soap and water. The soap formation reaction is very fast and goes to completion before any esterification begins.

Essentially all of the current commercial biodiesel producers use base-catalyzed reactions. Base-catalyzed reactions are relatively fast, with residence times from about 5 min to about 1 h, depending on temperature, concentration, mixing, and alcohol:triglyceride ratio. Most use NaOH or KOH as catalysts, although glycerol refiners prefer NaOH. KOH has a higher cost, but the potassium can be precipitated as K_3PO_4, a fertilizer, when the products are neutralized using phosphoric acid. This can make meeting water effluent standards a bit more difficult because of limits on phosphate effluents.

Sodium methoxide, usually as a 25% solution in methanol, is a more powerful catalyst on a weight basis than the mixture of NaOH and methanol. This appears to be, in part, the result of the negative effect of the chemical water produced in situ when NaOH and methanol react to form sodium methoxide.

Acid catalyst systems are characterized by slow reaction rates and high alcohol:TriGlycerides (TG) requirements (20:1 and more). Generally, acid-catalyzed reactions are used to convert FFAs to esters, or soaps to esters as a pretreatment step for high FFA feedstocks. Residence times from 10 min to about 2 h are reported.

Counter current acid esterification systems have been used for decades to convert pure streams of fatty acids into methyl esters at yields above 99%. These systems tend to force yields to 100% and wash water out of the system at the same time because the feedstock and the sulfuric acid/methanol mix are moving in opposite directions. Acid esterification systems produce a byproduct of water. In batch systems, the water tends to accumulate in the vessel to the point where it can shut the reaction down prematurely. The sulfuric acid tends to migrate into the water out of the methanol, rendering it unavailable for the reaction. All acid esterification systems need to have a water management strategy. Good water management can minimize the amount of methanol required for the reaction. Excess methanol (such as the 20:1 ratio) is generally necessary in batch reactors where water accumulates. Another strategy is to approach the reaction in two stages: fresh methanol and sulfuric acid is reacted, removed, and replaced with more fresh reactant. Much of the water is removed in the first round, and the fresh reactant in the second round drives the reaction closer to completion. Acid-catalyzed esterification is discussed in more detail in the chapter on pretreatment of high FFA feedstocks.

Lipase-catalyzed reactions have the advantage of reacting at room temperature without producing spent catalysts. The enzymes can be recycled for use again or immobilized onto a substrate. If immobilized, the substrate will require replacement when yields begin to decline. The enzyme reactions are highly specific. Because the alcohol can be inhibitory to some enzymes, a typical strategy is to feed the alcohol into the reactor in three steps of 1:1 mol ratio each. The reactions are very slow, with a three-step sequence requiring from 4 to 40 h, or more. The reaction conditions are modest, from 35 to 45 °C. Transesterification yields generally do not meet ASTM standards, but esterification yields can occur relatively quickly and yields are good. Excess free fatty acids can be removed as soaps in a later transesterification or caustic stripping step.

BIODIESEL PRODUCTION PROCESS OPTIONS

Batch processing

The simplest method for producing alcohol esters is to use a batch, stirred tank reactor. Alcohol to triglyceride ratios from 4:1 to 20:1 (mole:mole) have been reported, with a 6:1 ratio most common. The reactor may be sealed or equipped with a reflux condenser. The operating temperature is usually about 65 °C, although temperatures from 25 to 85 °C have been reported.

The most commonly used catalyst is sodium hydroxide, with potassium hydroxide also used. Typical catalyst loadings range from 0.3% to about 1.5% by weight of oil.

Thorough mixing is necessary at the beginning of the reaction to bring the oil, catalyst, and alcohol into intimate contact. Towards the end of the reaction, less mixing can help increase the extent of the reaction by allowing the inhibitory product, glycerol, to phase separately from the ester–oil phase. Completions of 85–94% are reported in a single step.

Some groups use a two-step reaction, with glycerol removal between steps, to increase the final reaction extent to 98+%. Higher temperatures and higher alcohol:oil ratios also can enhance the percent completion. Typical reaction times range from 20 min to more than 1 h.

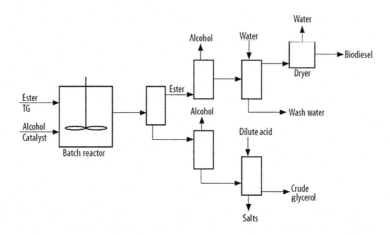

FIGURE 20.1

Reaction process.

Figure 20.1 shows a process flow diagram for a typical batch system. The oil is first charged to the system, followed by the catalyst and methanol. The system is agitated during the reaction time, then agitation is stopped. In some processes, the reaction mixture is allowed to settle in the reactor to give an initial separation of the esters and glycerol. In other processes, the reaction mixture is pumped into a settling vessel or is separated using a centrifuge.

The alcohol is removed from both the glycerol and ester stream using an evaporator or a flash unit. The esters are washed gently using warm, slightly acid water to remove residual methanol and salts and then dried. The finished biodiesel is then transferred to storage. The glycerol stream may be neutralized and then sent to the glycerol refining section.

For yellow grease and animal fats, the system is slightly modified with the addition of an acid esterification vessel and storage for the acid catalyst. The feedstock is sometimes dried (down to 0.4% water) and filtered before loading the acid esterification tank. The sulfuric acid and methanol mixture is added and the system is agitated. Similar temperatures to transesterification are used, and sometimes the system is pressurized or a cosolvent is added. Glycerol is not produced. If a two-step acid treatment is used, the stirring is suspended until the methanol phase separates and is removed. Fresh methanol and sulfuric acid is added and the stirring resumes.

Once the conversion of the fatty acids to methyl esters has reached equilibrium, the methanol/water/acid mixture is removed by settling or with a centrifuge. The remaining mixture is neutralized or sent straight into transesterification, where it will be neutralized using excess base catalysts. Any remaining free fatty acids will be converted into soaps in the transesterification stage. The trans-esterification batch stage processes as described above.

Continuous process systems

A popular variation of the batch process is the use of continuous stirred tank reactors (CSTR) in series. The CSTR can be varied in volume to allow for a longer residence time in CSTR 1 to achieve a greater extent of reaction. After the initial product glycerol is decanted, the reaction in CSTR 2 is rather rapid, with 98+% reaction completion common.

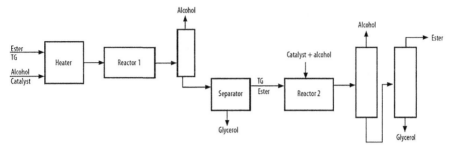

FIGURE 20.2

Plug flow reaction system.

An essential element in the design of the CSTR is sufficient mixing input to ensure that the composition throughout the reactor is essentially constant. This has the effect of increasing the dispersion of the glycerol product in the ester phase. The result is that the time required for phase separation may be extended.

There are several processes that use intense mixing, either from pumps or motionless mixers, to initiate the esterification reaction. The majority of the transesterification reaction is carried out in a pipe reactor, usually with some provision for mixing in the reactor. The reaction mixture moves through the reactor in a continuous plug, with little mixing in the axial direction. This type of reactor, called a plug-flow reactor (PFR), behaves as if it were a series of small CSTRs chained together.

The result is a continuous system that requires rather short residence times, as low as 6–10 min, for completion of the reaction. The PFRs can be staged to allow decanting of glycerol between stages. Often this type of reactor is operated at an elevated temperature and pressure to increase the reaction rate. A PFR system is shown in Figure 20.2.

High free fatty acid systems

High free fatty acid feedstocks will react with the catalyst and form soaps if they are fed to a base-catalyzed system. The maximum amount of free fatty acids acceptable in a base-catalyzed system is about 2% by weight of oil, and preferably less than 1%. Some approaches to using high free fatty acid feedstocks use this concept to "refine" the free fatty acids out of the feed for disposal or separate treatment in an acid esterification unit. The caustic is added to the feedstock and the resulting soaps are stripped out using a centrifuge or water extraction. This process is called caustic stripping.

Some triglycerides are lost with the soaps during caustic stripping. The soap mixture can be acidulated to recover the fatty acids and lost oils in a separate reaction tank. The refined oils are dried and sent to the transesterification unit for further processing. Rather than waste the free fatty acids removed in this manner, they can be transformed into methyl esters using an acid esterification process. As described earlier, acid-catalyzed processes can be used for the direct esterification of free fatty acids in a high FFA feedstock. Less expensive feedstocks, such as tallow or yellow grease, are characteristically high in FFAs. The standard for tallow and yellow grease is <15% FFA, but some lots may exceed this standard.

Direct acid esterification of a high free fatty acid feed requires water removal during the reaction or the reaction will be quenched prematurely. Also, a high alcohol to FFA ratio is required, usually

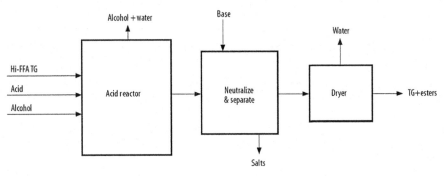

FIGURE 20.3

Acid-catalyzed direct esterification process.

between 20:1 and 40:1. Direct esterification may also require rather large amounts of the acid catalyst, depending on the process used.

The esterification reaction of FFAs with methanol produces byproduct water that must be removed, but after drying, the resulting mixture of esters and triglyceride can be used directly in a conventional base-catalyzed system. The water can be removed by vaporization, settling, or centrifugation as a methanol-water mixture. Counter-current continuous flow systems will wash out the water along with the exiting stream of acidic methanol.

One approach to the acid-catalyst system has been to use phosphoric acid as the initial catalyst, neutralize with an excess of KOH for the base step, and then neutralize with phosphoric acid upon completion. The insoluble potassium phosphate is recovered, washed, and dried for use as a fertilizer. Figure 20.3 shows a typical acid-catalyzed direct esterification process.

An alternative approach to utilization of high FFA feedstocks is to use a base catalyst to deliberately form soap from the FFAs. The soap is recovered, the oil dried, and then used in a conventional base-catalyzed system.

This strategy can lead to a false sense of economy. If the soap stock is discarded, the effective price of the feedstock is increased in inverse proportion to the percentage of remaining oil. The soap stock can, however, be converted into esters by using an acid-catalyzed reaction. The problem with this strategy is that the soap stock system contains a large amount of water that must be removed before the product esters can meet the biodiesel standard. The soap stock process is shown in Figure 20.4.

An alternative procedure for processing high FFA feeds is to hydrolyze the feedstock into pure FFA and glycerin. Typically, this is done in a counter-current reactor using sulfuric/sulfonic acids and steam. The output is pure free fatty acids and glycerin. Any contaminants in the feedstock partition mostly into the glycerin and some may leave with the steam/water effluent. Some contaminants continue with the FFA and can be removed or left in, depending on the processes and product specifications. The pure FFA are then acid esterified in another counter-current reactor to transform them into methyl esters. The methyl esters are then neutralized and dried with yields of about 99%. The process equipment must be acid-resistant, but generally, feedstock costs are extremely low.

A variation of the base-catalyzed system that avoids the problem of high FFAs is the use of a fixed bed, insoluble base. An example of this system, using calcium carbonate as the catalyst, has been demonstrated at the bench-scale. This process is depicted in Figure 20.5.

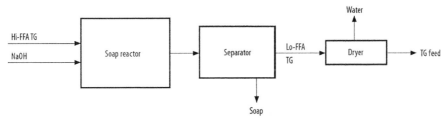

FIGURE 20.4

Preparation of soap stock from a high FFA feed.

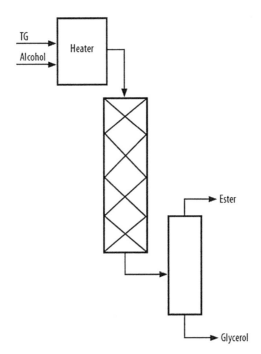

FIGURE 20.5

Fixed-bed, base-catalyzed reactor system.

NONCATALYZED SYSTEMS—BIOX PROCESS

Cosolvent options are designed to overcome slow reaction time caused by the extremely low solubility of the alcohol in the TG phase. One approach that is nearing commercialization is the Biox process. This process uses a cosolvent, tetrahydrofuran, to solubilize the methanol. The result is a fast reaction, on the order of 5–10 min, with no catalyst residues in either the ester or the glycerol phase. The TetraHydroFuran (THF) cosolvent is chosen, in part, because it has a boiling point very close to that of methanol. After the reaction is complete, the excess methanol and the tetrahydrofuran cosolvent are

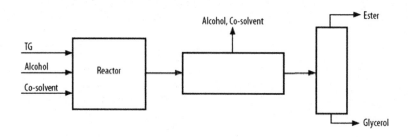

FIGURE 20.6

Biox cosolvent process.

recovered in a single step. This system requires a rather low operating temperature, 30 °C. Other cosolvents, such as MTBE, have been investigated. The ester-glycerol phase separation is clean, and the final products are catalyst- and water-free. The equipment volume has to be larger for the same quantity of final product because of the additional volume of the cosolvent. The Biox process is depicted in Figure 19. Cosolvents that are subject to the hazardous and/or air-toxic EPA list for air pollutants require special "leak proof" equipment for the entire system, including methanol/cosolvent recovery and recycling.

Fugitive emissions are tightly controlled. The cosolvent must be completely removed from the glycerin as well as the biodiesel.

The Biox process is depicted in Figure 20.6.

Noncatalyzed systems—supercritical process

When a fluid or gas is subjected to temperatures and pressures in excess of its critical point, there are a number of unusual properties exhibited. There no longer is a distinct liquid and vapor phase, but a single, fluid phase is present. Solvents containing a hydroxyl (OH) group, such as water or primary alcohols, take on the properties of superacids.

A noncatalytic approach to esters production is the use of a high (42:1) alcohol to oil ratio. Under supercritical conditions (350°–400 °C and >80 atm or 1200 psi), the reaction is complete in about 4 min. Capital and operating costs can be greater and energy consumption higher than other process options.

An intriguing example of this process has been demonstrated in Japan, where oils in a very large excess of methanol have been subjected to very high temperatures and pressures for a short period of time. The result is a very fast (3–5 min) reaction to form esters and glycerol. The reaction must be quenched very rapidly so that the products do not decompose. The reactor used in the work to date is a 5 mL cylinder that is dropped into a bath of molten metal and then quenched in water. Clearly, while the results are interesting, the scale-up to a useful process may be quite difficult. Figure 20.7 depicts one conception of a configuration for a supercritical esterification process.

SUMMARY

There are multiple operating options available for making biodiesel. Many of these technologies can be combined under various conditions and feedstocks in an infinite number of ways. The technology choice is a function of desired capacity, feedstock type and quality, alcohol recovery, and catalyst

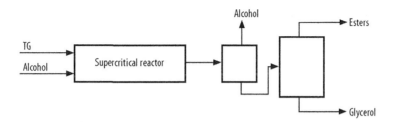

FIGURE 20.7

Supercritical esterification process.

recovery. The dominant factor in biodiesel production is the feedstock cost, with capital cost contributing only about 7% of the final product cost.

However, some reaction systems are capable of handling a variety of feedstocks and qualities, while others are not. Also, the different approaches to the esterification process result in quite different operating requirements, different water use requirements, and different operating modes.

In general, smaller capacity plants and variable feedstock quality suggest use of batch systems.

Continuous systems generally lead the operation on a 24/7 basis, requiring larger capacities to justify larger staffing requirements, and require a more uniform feedstock quality.

POST REACTION PROCESSING

The objective of this section is to describe in more detail the steps in processing the ester phase resulting from the transesterification. This section will discuss the recovery of the esters from the reaction mixture and then the refining needed to meet the requirements of ASTM D 6751. The topics include ester/glycerol separation, ester washing, ester drying, other ester treatments, and additization.

ESTER/GLYCEROL SEPARATION

The ester/glycerol separation is typically the first step of product recovery in most biodiesel processes. The separation process is based on the facts that fatty acid alcohol esters and glycerol are sparingly mutually soluble, and that there is a significant difference in density between the ester and glycerol phases. The presence of methanol in one or both phases affects the solubility of ester in glycerol and glycerol in ester.

The ester washing step is used to neutralize any residual catalyst, to remove any soaps formed during the esterification reaction, and to remove residual-free glycerol and methanol.

Ester drying is required to meet the stringent limits on the amount of water present in the final biodiesel product. In addition, there may be other treatments used to reduce color bodies in the fuel, remove sulfur and/or phosphorus from the fuel, or remove glycerides.

Additization is the addition of materials that have a specific functionality that modifies one or more fuel properties. Examples include cloud point/pour point additives, antioxidants, or other stability-enhancing agents.

Fatty acid alcohol esters have a density of about 0.88 gm/cc, while the glycerol phase has a density on the order of 1.05 gm/cc or more. The glycerol density depends on the amount of methanol, water, and catalyst in the glycerol. This density difference is sufficient for the use of simple gravity separation techniques for the two phases.

However, the rate of separation is affected by several factors. Most biodiesel processes use relatively intense mixing, at least at the beginning of the reaction, to incorporate the sparingly soluble alcohol into the oil phase. If this mixing continues for the entire reaction, the glycerol can be dispersed in very fine droplets throughout the mixture. This dispersion requires from 1 h to several hours to allow the droplets to coalesce into a distinct glycerol phase. For this reason, mixing is generally slowed as the reaction begins to progress to reduce the time required for phase separation.

The more nearly neutral the pH, the quicker the glycerol phase will coalesce. This is one reason to minimize the total catalyst use. In some batch systems, the reaction mixture is neutralized at the beginning of the glycerol/ester phase separation step.

The presence of significant quantities of mono-, di-, and triglycerides in the final mixture can lead to the formation of an emulsion layer at the ester–glycerol interface. At best, this layer represents a net loss of product, unless it is recovered and separated. At worst, the ester phase will not meet the biodiesel specification and will have to be rerun. If problems with mono-, di-, and triglycerides occur, you should reevaluate the entire reaction to see where improvements can be made to improve process yields in the preceding steps.

The esterification process is run with an excess of alcohol to ensure a complete reaction and to attain higher reaction rates. The residual alcohol is distributed between the ester and glycerol phases. The alcohol can act as a dispersant for the ester into the glycerol phase and for the glycerol into the ester phase. The result can be a need for additional processing of the products to meet specifications. Other people claim that methanol aids in phase separation, which is one reason that product is generally phase separated before methanol recovery.

PROCESS EQUIPMENT FOR THE ESTER/GLYCEROL SEPARATION

Decanter Systems: There are three categories of equipment used to separate the ester and glycerol phases. Decanter systems rely solely on the density difference and residence time to achieve the separation. For relatively small throughput, or batch processes, the 1–8 h required for complete separation of the phases may be acceptable. However, a separation that requires a residence time of 1 h requires a decanter with a volume of at least 700 gallons to affect separation for a 5,000,000 gallon per year, continuous plant. For lower extent of reaction, the separation is slower, and the decanter will have to be much larger.

The primary determinant for designing a decanter for biodiesel production is the desired residence time. This, plus the product mixture flow rate, determines the size of the unit. Decanter units should be rather tall and narrow to allow physical separation between the ester and the glycerol withdrawal points. L/D ratios of 5–10 can work well.

The temperature in the decanter affects the solubility of the alcohol in both phases and the viscosity of the two liquids. Too high a temperature in the decanter can cause residual alcohol to flash, potentially restricting the flow of the ester phase out of the tank. On the other hand, too low a temperature increases the viscosity in both phases. The increased viscosity will slow the coalescence rate in the system.

The presence of an emulsion layer is indicative of mono- and di-glycerides. The emulsion layer will form between the phases. In continuous operation, there must be a provision for removing the emulsion so it does not fill the decanter.

Centrifuge System: Many of the continuous plants use a centrifuge for the phase separation. The centrifuge creates an artificial, high gravity field by spinning at very high speeds. The separation can be completed rapidly and effectively. The disadvantage of the centrifuge is its initial cost, and the need for considerable and careful maintenance. However, centrifuges are used extensively in the food processing, rendering, and biodiesel industries. Centrifuges are high-speed, rotating devices. The artificial gravity of the centrifuge results from the rotation. Speeds of 2000–5000 rpm are not uncommon. The same high-speed rotation that creates the centrifugal effect also creates a device that requires careful and rather frequent maintenance. Although centrifuges are relatively expensive, multiple units to ensure online availability are advisable.

At smaller capacities, either a batch or a continuous centrifuge can be used. The use of a batch centrifuge in a continuous process requires a surge tank to match the batch cycle time with the continuous processing rate.

Hydrocyclone: An intriguing, relatively new device that has been considered for use in biodiesel plants is the hydrocyclone. A liquid–liquid hydrocyclone uses an inverted conical shape and the incompressibility of the liquids to accelerate the liquid entering the cyclone. The effect is similar to a centrifuge, with the heavier material being forced towards the wall and downward and the lighter material forced to the center and upward. The result is a density-based separation. Although they are now used for oil–water separations, hydrocyclones are at the experimental stage for application in biodiesel production.

A hydrocyclone operates on the Bernoulli principle of trading pressure for speed in an incompressible flowing system. The relative density of the fluids determines the separating force applied, while the relative viscosity determines the resistance to separation. The liquid mixture enters the hydrocyclone at a moderately high pressure (about 125 psig). The pressure decreases and the velocity increases as the liquid passes for the wider to the narrower part of the inverted cone. The more dense liquid is accelerated towards the outside wall, while the less dense liquid concentrates in the center. The result is a separation based upon induced g-forces.

It would appear that the presence of volatiles creates a problem in a hydrocyclone. The rapid reduction of pressure in the device will induce flashing of the volatile liquid (alcohol), disrupting or stopping the separation process. Excess methanol should be removed from the system before introducing the reaction mixture to a hydrocyclone.

ESTER WASHING

The primary purpose of the ester washing step is the removal of any soaps formed during the transesterification reaction. In addition, the water provides a medium for addition of acid to neutralize the remaining catalyst and a means to remove the product salts. The residual methanol should be removed before the wash step. This prevents the addition of methanol to the wastewater effluent. However, some processes remove the methanol with the wash water and remove it from the wash water.

The use of warm water (120–140 °F) prevents precipitation of saturated fatty acid esters and retards the formation of emulsions with the use of a gentle washing action.

Softened water (slightly acidic) eliminates calcium and magnesium contamination and neutralizes remaining base catalysts. Similarly, removal of iron and copper ions eliminates a source of catalysts that decrease fuel stability. Gentle washing prevents the formation of emulsions and results in a rapid and complete phase separation. The phase separation between esters and water is typically very clean and complete. However, the equilibrium solubility of water in esters is higher than the specified water content for B100. Therefore, after the washing step, there will be more than the equilibrium amount of water present.

Vacuum driers can either be batch or continuous devices for removing water. The system is operated at a highly reduced pressure, which allows the water to evaporate at a much lower temperature. A variation that also allows for rather high heating and evaporation rates is the falling film evaporator. This device operates at reduced pressure. As the esters pour down the inside wall of the evaporator, the direct contact with the heated wall evaporates the water rapidly. Care should be taken with high temperature evaporators to avoid darkening the fuel, which is a sign that the polyunsaturated methyl esters are polymerizing.

Because the total water burden in the esters is low, molecular sieves, silica gels, etc. can also be used to remove the water. An advantage of these systems is that they are passive. However, a disadvantage is that these units must be periodically regenerated.

OTHER ESTER TREATMENTS

There are absorbents on the market that selectively absorb hydrophilic materials such as glycerol and mono- and diglycerides (i.e, Magnesol from the Dallas Group). This treatment, followed by an appropriate filter, has been shown to be effective in lowering glycerides and total glycerol levels.

Some vegetable oils and many yellow greases and brown greases leave an objectionable color in the biodiesel. There is no color specification in ASTM D 6751, but an activated carbon bed is an effective way to remove excessive color. The fats and oil industry literature has other bleaching technologies that may also be explored for biodiesel producers.

The European specification for sulfur content is much tighter than the US requirement. As a result, a number of producers in Europe are resorting to the use of vacuum distillation for the removal of sulfur compounds from the biodiesel product. By 2006, all US biodiesel must meet new sulfur standards of 15 ppm or less. Therefore, biodiesel producers needs to be aware of the sulfur content of their fuel, and must incorporate sulfur reduction technology if needed before that date. Vacuum distillation has the added benefit of deodorization and the removal of other minor contaminants, which may provide a benefit to those firms that use highly degraded feedstocks such as trap grease.

Filtering is an essential part of all biodiesel production. While feedstocks entering the plant should be filtered to at least 100 μm, biodiesel leaving the plant should be filtered to at least 5 μm to ensure no contaminants are carried with the fuel that could damage the engine. It has been suggested that the fuel could be cooled before filtering to capture some of the saturated esters as they crystallize and thereby lower the cloud point of the fuel. The crystallized esters could be melted by heating and used within the plant as boiler fuel.

ADDITIZATION OF ESTERS

Petroleum-based diesel fuels are treated with a wide range of additives to improve lubricity, detergency, oxidative stability, corrosion resistance, conductivity, and many other properties. Additive technology for biodiesel is less advanced, so fewer additives are available to enhance performance.

One area where the biodiesel producer needs to give serious consideration is oxidative stability. Biodiesel, because it contains large numbers of molecules with double bonds, is much less oxidatively stable than petroleum-based diesel fuel. Fortunately, stability-enhancing additive technology is well-developed in the food industry, and many of these additives can be carried over to stabilizing biodiesel.

TREATMENT AND RECOVERY OF SIDE STREAMS

There are three nonester side streams that must be treated as a part of the overall biodiesel process. These streams are

1. The excess alcohol that is recycled within the process.
2. The glycerol coproduct.
3. The wastewater stream from the process.

In this module, it is assumed that methanol is the alcohol used in the process. Similar comments would apply to other alcohols.

Methanol recycle is necessary because an excess of methanol is required for an effective trans-esterification reaction system. The recovery of the unused methanol saves input costs for the process and essentially eliminates the emissions of methanol to the surroundings. The emissions reduction is needed because methanol is highly flammable and toxic. Glycerol is recovered and partially refined as a coproduct from biodiesel production. About 10% by weight of the input reactants are converted to glycerol in the transesterification reaction. On both a weight basis and a volume basis, partially refined glycerol is worth more than the biodiesel product.

Wastewater constitutes an operating cost for the plant, both because of the water consumption and because of the water treatment costs to the plant.

METHANOL MANAGEMENT

There are several physical parameters that are important to the recovery and recycle of methanol. Methanol's relatively low boiling point, 64.7 °C, means that it is fairly volatile and can largely be removed from the oil, ester, and aqueous streams by flash evaporation and recondensation. The low boiling point, along with a low flash point, 8 °C, also means the methanol is considered to be highly flammable.

Methanol is fully miscible with water and with glycerol. However, it has a low solubility in fats and oils (approx. 10% wt/wt at 65 °C in tallow). Methanol is more soluble in esters, but it is not fully miscible. The solubility in glycerol and water means that methanol will prefer these phases when there is a two-phase system present. The low solubility in fats and oils is the reason for the solubility-limited phase of the overall transesterification reaction.

When the two phases present are esters and glycerol, the methanol will distribute between the phases. At 90:10% wt/wt ester and glycerol, the methanol distributes approximately 60:40 wt% between the phases. This fact is important, since the reaction is complete at 90:10 wt%. If the methanol is allowed to remain in the system during phase separation, the methanol acts as a phase stabilizer, retarding the rate of gravity separation. It is advantageous to remove the methanol before phase separation.

Methanol can be recovered using distillation, either conventional or vacuum, or partially recovered in a single-stage flash. An alternative to distillation is a falling-film evaporator. Residual methanol in the ester phase can be removed in the water wash step in ester postprocessing. Product esters are typically washed with warm (140 °F), softened water to remove soaps and residual methanol.

GLYCEROL REFINING

The recovered glycerol from the transesterification reaction contains residual alcohol, catalyst residue, carry-over fat/oil, and some esters. The glycerol from rendered feedstocks may also contain phosphatides, sulfur compounds, proteins, aldehydes and ketones, and insolubles (dirt, minerals, bone, or fibers).

Chemical Refining: There are several factors that are important in the chemical refining of glycerol. First, the catalyst tends to concentrate in the glycerol phase where it must be neutralized. The neutralization step leads to the precipitation of salts. Also, the soaps produced in the esterification must be removed by coagulation and precipitation with aluminum sulfate or ferric chloride. The removal may be supplemented by centrifuge separation.

The control of the pH is very important because low pH leads to dehydration of the glycerol and high pH leads to polymerization of the glycerol. The glycerol may then be bleached using activated carbon or clay.

Physical Refining: The first step in physical refining is to remove fatty, insoluble, or precipitated solids by filtration and/or centrifugation. This removal may require pH adjustment. Then the water is removed by evaporation. All physical processing is typically conducted at 150–200 °F, where glycerol is less viscous, but still stable.

Glycerol Purification: The final purification of glycerol is completed using vacuum distillation with steam injection, followed by activated carbon bleaching. The advantages of this approach are that this is a well-established technology. The primary disadvantage is that the process is capital and energy intensive. Vacuum distillation of glycerol is best suited to operations >25 tons per day.

Ion exchange purification of glycerol is an attractive alternative to vacuum distillation for smaller capacity plants. The ion exchange system uses cation, anion, and mixed bed exchangers to remove the catalyst and other impurities. The glycerol is first diluted with soft water to a 15–35% glycerol-in-water solution. The ion exchange is followed by vacuum distillation or flash drying for water removal, often to an 85% partially refined glycerol. The advantage of this process is the fact that all purification takes place in the resin vessels, so the system is suited to smaller capacity operations. The disadvantages are that the system is subject to fouling by fatty acids, oils, and soaps. The system also requires regeneration of the beds, producing large quantities of wastewater. Regeneration requires parallel systems to operate and regenerate simultaneously.

WASTEWATER CONSIDERATIONS

Ester washing produces about 1 gallon of water per gallon of ester per wash. All process water must be softened to eliminate calcium and magnesium salts and treated to remove iron and copper ions. The ester wash water will have a fairly high Biochemical Oxygen Demand (BOD) from the residual fat/oil, ester, and glycerol.

The glycerol ion exchange systems can produce large quantities of low-salt waters as a result of the regeneration process. In addition, water softening, ion exchange, and cooling water blowdown will contribute a moderate dissolved salts burden.

The aggregate process wastewaters should meet local municipal waste treatment plant disposal requirements, if methanol is fully recovered in the plant and not present in the wastewater. In many areas, internal treatment and recycle of the process water may lead to cost savings and easier permitting of the process facility.

SUMMARY

Methanol affects all product recovery operations. The methanol must be fully recycled for best economy and for pollution prevention. Glycerol is an economically significant coproduct that should be as fully refined as practicable. Properly managed wastewaters can be treated in a municipal sewer system, but internal treatment and recycling should be considered.

PRETREATMENT OF HIGH FREE FATTY ACID FEEDSTOCKS

Many low-cost feedstocks are available for biodiesel production. Unfortunately, many of these feedstocks contain large amounts of free fatty acids (FFAs). As discussed elsewhere, these free fatty acids will react with alkali catalysts to produce soaps that inhibit the reaction.

The following ranges of FFA are commonly found in biodiesel feedstocks:

Refined vegetable oils	<0.05%
Crude vegetable oil	0.3–0.7%
Restaurant waste grease	2–7%
Animal fat	5–30%
Trap grease	40–100%

Generally, when the FFA level is less than 1%, and certainly if it is less than 0.5%, the FFAs can be ignored. Common catalyst amounts are shown in Table 20.2.

Table 20.2 Common Catalyst Amounts	
Sodium hydroxide	1% of triglyceride weight
Potassium hydroxide	1% of triglyceride weight
Sodium methoxide	0.5% of triglyceride weight

Soaps may allow emulsification that causes the separation of the glycerol and ester phases to be less sharp. Soap formation also produces water that can hydrolyze the triglycerides and contribute to the formation of more soap. Further, a catalyst that has been converted to soap is no longer available to accelerate the reaction.

When FFA levels are above 1%, it is possible to add extra alkali catalyst. This allows a portion of the catalyst to be devoted to neutralizing the FFAs by forming soap, while still leaving enough to act as the reaction catalyst.

Since it takes 1 mol of catalyst to neutralize 1 mol of FFA, the amounts of additional catalyst can be calculated by the following formulas.

Sodium hydroxide	[%FFA](0.144) + 1%
Potassium hydroxide	[%FFA](0.197)/0.86 + 1%
Sodium methoxide	[%FFA](0.190) + 0.5%

For example, when adding sodium methoxide to a feedstock with 1.5% FFA, the amount of catalyst would be:

$$(1.5)(0.190) + 0.5\% = 0.79\% \text{ of the triglyceride weight.}$$

Note that a factor of 0.86 has been included with the potassium hydroxide calculation to reflect that reagent grade KOH is only 86% pure. If other grades of catalyst are used, this factor should be adjusted to their actual purity.

This approach to neutralizing the FFAs will sometimes work with FFA levels as high as 5–6%. The actual limit depends on whether other types of emulsifiers are present. It is especially important to make sure that the feedstock contains no water. FFAs of 2–3% may be the limit if traces of water are present.

For feedstocks with higher amounts of FFA, the addition of an extra catalyst may create more problems than it solves. The large amount of soap created can gel. It can also prevent the separation of the glycerol from the ester. Moreover, this technique converts the FFAs to a waste product when they could be converted to biodiesel.

When working with feedstocks that contain 5–30% FFA or even higher, it is important to convert the FFAs to biodiesel or the process yield will be low. There are at least four techniques for converting the FFAs to biodiesel:

1. *Enzymatic methods*: These methods require expensive enzymes but seem to be less effected by water. At the present time, no one is using these methods on a commercial scale.
2. *Glycerolysis*: This technique involves adding glycerol to the feedstock and heating it to high temperature (200 °C), usually with a catalyst such as zinc chloride. The glycerol reacts with the FFAs to form mono- and diglycerides. Figure 20.8 shows the rate of decrease of the fatty acid level in a batch of animal fat. This technique produces a low. FFA feed that can be processed using traditional alkali-catalyzed techniques.

    The drawback of glycerolysis is the high temperature and that the reaction is relatively slow. An advantage is that no methanol is added during the pretreatment so that as water is formed by the reaction.

$$\text{FFA} + \text{glycerol} \rightarrow \text{monoglyceride} + \text{water}$$

The water immediately vaporizes and can be vented from the mixture.

3. *Acid catalysis*: This technique uses a strong acid, such as sulfuric acid, to catalyze the esterification of the FFAs and the transesterification of the triglycerides. The reaction does not produce soaps because no alkali metals are present. The esterification reaction of the FFAs to alcohol esters is relatively fast, proceeding substantially to completion in 1 h at 60 °C. However, the transesterification of the triglycerides is very slow, taking several days to complete. Heating to

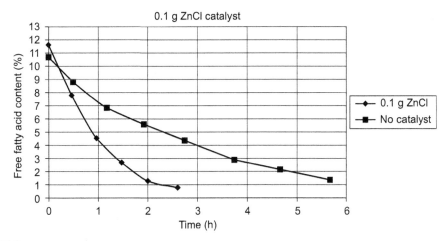

FIGURE 20.8

Reduction of fatty acids by glycerolysis.

130 °C can greatly accelerate the reaction, but reaction times will still be 30–45 min. Another problem with acid catalysis is that the water production from the following reaction

$$\text{FFA} + \text{methanol} \rightarrow \text{methyl ester} + \text{water}$$

stays in the reaction mixture and ultimately stops the reaction, usually well before reaching completion.

4. *Acid catalysis followed by alkali catalysis*: This approach solves the reaction rate problem by using each technique to accomplish the process for which it is best suited. Since acid catalysis is relatively fast for converting the FFAs to methyl esters, it is used as a pretreatment for the high FFA feedstocks. Then, when the FFA level has been reduced to 0.5% or lower, an alkali catalyst is added to convert the triglycerides to methyl esters. This process can convert high free fatty acid feedstocks quickly and effectively. Water formation is still a problem during the pretreatment phase. One approach is to simply add so much excess methanol during the pretreatment that the water produced is diluted to the level where it does not limit the reaction. Molar ratios of alcohol to FFA as high as 40:1 may be needed. The disadvantage of this approach is that more energy is required to recover the excess methanol. Another approach would be to let the acid-catalyzed esterification proceed as far as it will go until it is stopped by water formation. Then, boil off the alcohol and water. If the FFA level is still too high, then additional methanol and, if necessary, acid catalyst can be added to continue the reaction. This process can be continued for multiple steps and will potentially use less methanol than the previous approach. Again, the disadvantage is the large amount of energy required by the distillation process.

A less energy intensive approach is to let the acid-catalyzed reaction mixture settle. After a few hours, a methanol–water mixture will rise to the top and can be removed. Then, additional methanol and acid can be added to continue the reaction [Patent pending, Earl Hammond, ISU]. It is also possible to use fluids such as glycerol and ethylene glycol to wash the water from the mixture.

Procedure for high FFA feedstocks

1. Measure FFA level.
2. Add 2.25 g methanol and 0.05 g sulfuric acid for each gram *of free fatty acid* in the oil or fat. Sulfuric acid and methanol should be mixed first and then added slowly to the oil.
3. Agitate for 1 h at 60–65 °C.
4. Let the mixture settle. The methanol–water mixture will rise to the top. Decant the methanol, water, and sulfuric acid layer.
5. Take the bottom fraction and measure new FFA level.
6. If FFA is >0.5%, return to step 2 with new FFA level. If FFA is <0.5%, proceed to step 7.
7. Add an amount of methanol equal to 0.217 × (grams of unreacted triglycerides) and an amount of sodium methoxide equal to $(0.5 + [\%FFA]0.190)/100$ × (grams of unreacted triglycerides). Mix the sodium methoxide with the methanol and then add to the oil. This corresponds to a 6:1 M ratio of methanol to oil for the unreacted triglycerides. It ignores any methanol that may have carried over from the pretreatment.
8. Agitate for 1 h at 60 °C.

Example: 100 g of 12% FFA animal fat

Pretreatment: 2.25 x 12 g = 27.0 g methanol

0.05 x 12 g = 0.6 g H_2SO_4 (sulfuric acid)

Mix acid with methanol, then add mixture to fat. Agitate for one hour at 60°C. Let settle and separate bottom phase. Acid value should decrease substantially, to at least 5-6 mg KOH/g. Therefore, FFA = 2.5%.

Second step of pretreatment:

2.25 x 2.5 g = 5.6 g of methanol

0.05 x 2.5 g = 0.13 g H_2SO_4

Mix with oil, agitate at 60°C for 1 hour. FFA should be < 0.5%.Removal of upper phase is usually optional at this point.

Then add: 0.217 x (88) = 19.1 g methanol

$(0.5 + [0.5][0.190])/100$ x 88 = 0.52 g sodium methoxide

Agitate at 60°C for one hour. If glycerol and ester do not separate, add 50 g of warm distilled water to encourage separation. Wash 3-4 times.

BIODIESEL PRODUCTION SUMMARY

Biodiesel production is a deceptively complex process, rather than a simple process at first sight, especially if you want to make quality biodiesel and a product that meets ASTM D 6751 specifications and not harm a diesel engine. The type of processes and equipment used to make the biodiesel is determined by the feedstocks you plan to use in producing the biodiesel. There are a number of processes and operating parameters that affect the transesterification reaction used to produce bio-diesel and are feedstock- and other reactant-dependent. The "Biodiesel Production" chapter covers only a few of the key concepts concerning biodiesel production, and there are many other concepts that need to be considered before ever producing any biodiesel so that quality and safety are not compromised. Seek out the experts in the biodiesel industry to help you design and operate a plant and make sure quality and safety are part of the design and operation of the biodiesel plant. Research in depth the biodiesel industry and business considerations, along with the feedstock, chemical reactions, handling of materials, process parameters, materials of construction, and the type of equipment needed to produce biodiesel.

ACKNOWLEDGMENTS

The information contained in this chapter consists of excerpts from a National Renewable Energy Laboratory Subcontractor Report publication July 2004 • NREL/SR-510-36244 (Van Gerpan et al., 2004). Insights contained in this chapter, especially the introduction and summary of the chapter, are derived from the chapter "Types of Production" in the book *Building a Successful Biodiesel Business* (Van Gerpen, 2006), which contains information on all facets of biodiesel production, the industry, and how to build a biodiesel business. The 'Introduction' and the 'Biodiesel Production Summary' sections were written specifically by Rudy Pruszko for presenting the respective part from the NREL document. Rudy and the book editor are grateful to NREL, Department of Energy for granting permission to include the respective part of the report for this book.

REFERENCES

Van Gerpen, J., Shanks, B., Pruszko, R., Clements, D., Knothe, G., July 2004. Biodiesel Production Technology. Report from Iowa State University for the National Renewable Energy Laboratory, NREL/SR-510–36244.
Van Gerpen, J., Pruszko, R., Clements, D., Shanks, B., Knothe, G., 2006. Building a Successful Biodiesel Business, second ed. Biodiesel Basics, Dubuque: IA.

SYNTHESIS AND PARTIAL CHARACTERIZATION OF BIODIESEL VIA BASE CATALYZED TRANSESTERIFICATION

21

Sean M. McCarthy, Jonathan H. Melman, Omar K. Reffell, Scott W. Gordon-Wylie

Department of Chemistry, University of Vermont, Burlington, VT, USA

INTRODUCTION

Transesterification is a synthetically useful reaction that is often omitted from organic laboratories, primarily because no simple and relevant reactions exist for direct incorporation into the laboratory setting (Zanoni et al., 2001; Lindner et al., 2003; Su et al., 2003; Pedersen et al., 2005; Yadav and Lathi, 2005). We present here a laboratory aimed at teaching the concept of transesterification while simultaneously introducing students to the facile synthesis of biodiesel. Biodiesel is a renewable, alternative fuel derived in the United States, primarily from corn or soybean oil. The use of biodiesel as a fuel has been well documented (Wang et al., 2000; Sheehan et al., 1998; Tyson, 2004). Although the use of fossil fuels increases greenhouse gas levels, biodiesel is considered "CO_2 neutral" as it does not add CO_2 to the environment beyond what was recently removed via photosynthesis (Ritter, 2004). Additionally, biodiesel contains oxygen, leading to cleaner more efficient combustion, and has greater lubricity than standard diesel fuel, reducing engine wear. The large infrastructure that already exists for diesel fuel usage and the ease with which biodiesel can be substituted for diesel without the need for any engine or burner modifications makes biodiesel an attractive alternative fuel.

Biodiesel is typically synthesized via a transesterification reaction. The reaction can be catalyzed by either acid or base, and is an equilibrium reaction that must be shifted in the desired direction, Figure 21.1. In general, transesterification equilibria can be shifted in several ways. First, a lower boiling ester can be converted into a higher boiling ester by distillation of the lower boiling alcohol in the presence of the higher boiling alcohol. Second, lipases can be used to catalytically carry out the transesterification. Third, the transesterification reaction can be carried out via phase-transfer catalysis as described here.

Vegetable oils are triglycerides, composed of three fatty acid chains covalently bound via an ester linkage to a glycerol backbone, Figure 21.2. Analogous to crude oil cracking to form diesel fuel, vegetable oil is readily transesterified using basic methanolic solutions to form fatty acid methyl esters in the C_{12}–C_{22} range, i.e., biodiesel, Figure 21.3 (Wang et al., 2000; Sheehan et al., 1998; Tyson, 2004; Ebiura et al., 2005). Tables 21.1 and 21.2 list physical properties and chemical compositions of some common vegetable oils, methyl esters, and fatty acids.

Bioenergy. http://dx.doi.org/10.1016/B978-0-12-407909-0.00021-3

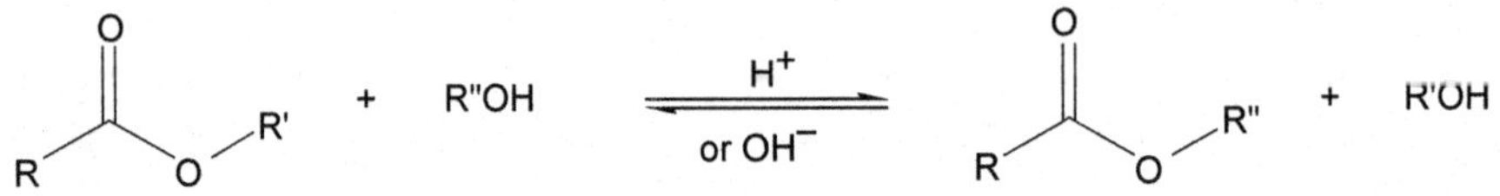

FIGURE 21.1

General reaction for transesterification. Both the forward and backward reactions can be catalyzed by acid or base. The equilibrium is generally shifted by increasing the desired esters alcohol concentration.

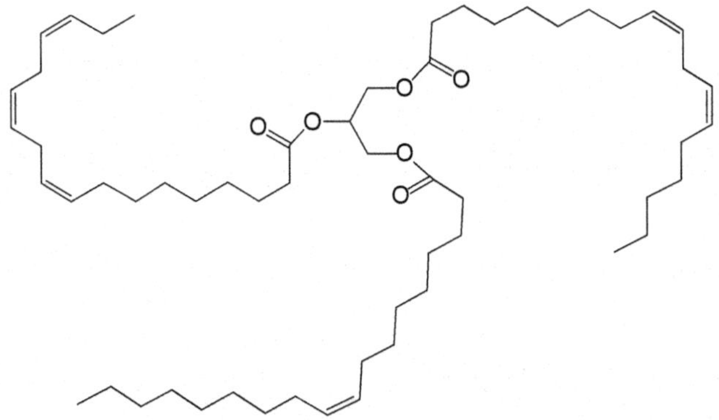

FIGURE 21.2

A hypothetical triglyceride containing oleic (mono-unsaturated), linoleic (di-unsaturated), and linolenic (tri-unsaturated) chains.

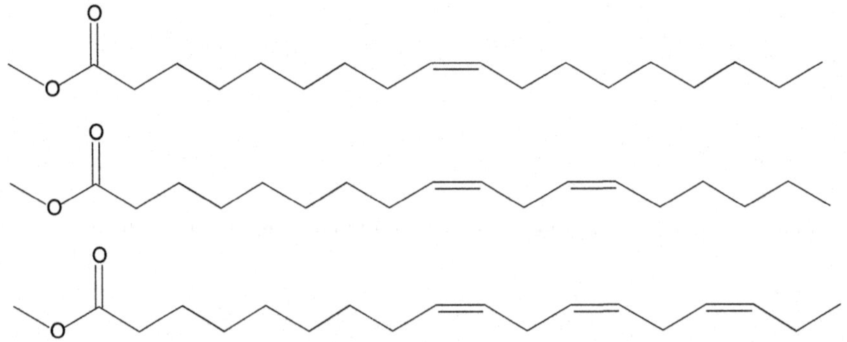

FIGURE 21.3

Sample biodiesel fatty acid esters, methyl oleate (top), methyl linoleate (middle), and methyl linolenate (bottom).

Table 21.1 Chemical Formulas and Melting Points (mp, °C) of Common Fatty Acids (CRC Handbook of Chemistry and Physics, 1992)

	Acid Linear Formula	Melting Point Free Acid	Melting Point Methyl Ester
Palmitic	$CH_3(CH_2)_{14}COOH$	63	30
Stearic	$CH_3(CH_2)_{16}COOH$	70	39
Oleic	$CH_3(CH_2)_7CH=CH(CH_2)_7COOH$	4	−20
Linoleic	$CH_3(CH_2)_4CH\ ]CHCH_2CH=CH(CH_2)_7COOH$	−5	−35
Linolenic	$CH_3(CH_2CH=CH)_3(CH_2)_7COOH$	−11	−46

Table 21.2 Percent Composition of Fatty Acids in Common Oils (CRC Handbook of Chemistry and Physics, 1992)

	Palmitic	Stearic	Oleic	Linoleic
Corn	10	3	50	34
Olive	7	2	84	5
Peanut	8	3	56	26
Sesame	9	4	45	40
Soybean[a]	10	2	29	51
Sunflower	6	2	25	66

[a]Also contains 7% linolenic acid.

MATERIALS

Reagent-grade methanol [67-56-1] was purchased from Mallinckrodt, reagent-grade sodium hydroxide [1310-73-2] was purchased from Acros, and vegetable oil (soybean oil [8001-22-7]) was purchased from the local grocery store. All materials were used without further purification.

HAZARDS

Methanol is poisonous and can cause blindness if ingested in large amounts. Methanolic sodium hydroxide solutions are caustic; gloves and safety glasses should be worn at all times. Vegetable oil is nonhazardous.

EXPERIMENTAL PROCEDURE

Biodiesel is synthesized via a base catalyzed transesterification reaction between vegetable oil and an alcohol (Figures 21.4 and 21.5). The two main variables in this reaction, other than the choice of vegetable oil, are the alcohol and the base used in the reaction. Due to cost considerations, methyl alcohol and sodium hydroxide are used.

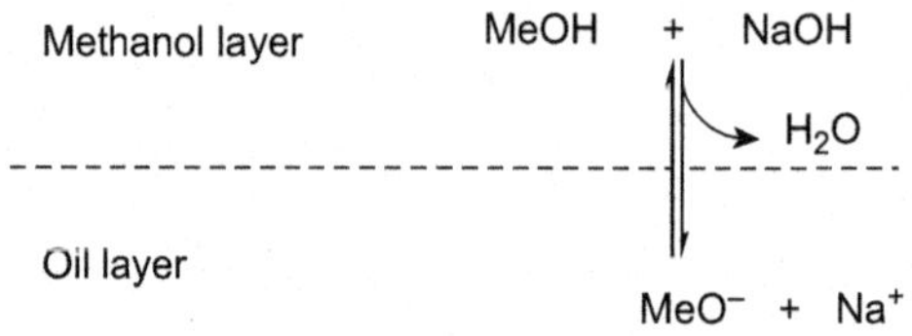

FIGURE 21.4

Formation of methoxide in the methanol layer. The methoxide is able to phase transfer into the oil and react with the triglycerides to form biodiesel.

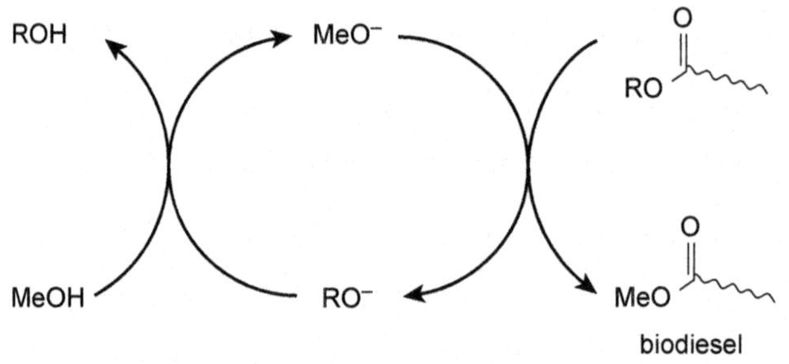

FIGURE 21.5

Mechanism for synthesis of biodiesel from the reaction of triglyceride with methoxide anions. Note that reaction is catalytic in the methoxide anion (Gottlieb et al., 1997).

Once formed, the methoxide transesterifies a fatty acid chain of the triglyceride, generating a methyl ester and an alkoxide on the glyceride backbone. This glyceroxide intermediate deprotonates methanol to form a methoxide anion and an alcohol functionality on the glyceride backbone. The process repeats until all the fatty acid chains have been converted to methyl esters and the glyceride backbone is converted to glycerol.

One part 0.3 M of methanolic sodium hydroxide, three parts vegetable oil, and a stir bar are added to a round bottom flask. The flask is fitted with a rubber septum and the mixture is vigorously stirred on a magnetic stir plate. After stirring for 15 min, the resulting mixture is allowed to settle, c. 10 min, yielding biodiesel and glycerol as two separate layers after about 10 min. The product is analyzed by Nuclear Magnetic Resonance (NMR) spectroscopy, Thin Layer Chromatography (TLC), viscosity measurements, and qualitative density observations.

RESULTS AND DISCUSSION

Qualitatively, it is simple to see that a reaction has occurred. Before vigorous stirring, the less dense (c. 0.79 g/mL) methanol solution sits above the more dense vegetable oil (0.92 g/mL). After the reaction, the resulting biodiesel layer is less dense (0.89 g/mL) than the glycerol and residual methanol (1.26 g/mL), which settle to the bottom. The synthesis of biodiesel via transesterification can also be confirmed using techniques other than density measurements. Viscosity measurements at 25 °C show that the biodiesel is significantly less viscous (7 cst) than the starting vegetable oil (57 cst). TLC on reverse phase silica provides a qualitative method for the separation and analysis of vegetable oil from methyl esters. A comparison of the ^{1}H NMR spectra of starting vegetable oil and the resultant biodiesel shows clearly that all the vegetable oil has been converted from triglyceride to biodiesel, but does not allow for an effective resolution between free fatty acids and methyl esters in the alkyl region between 0 and 2.5 ppm. Some residual methanol is observed in the NMR spectrum of the biodiesel. Industrially, residual alcohol, glycerol, and free fatty acids (as the corresponding sodium salts) are removed via water washing. Here, the main purpose of the experiment is just to form biodiesel in relatively pure form, so the water wash is omitted. If a water wash is desired, slowly dripping water through the biodiesel provides efficient removal of impurities such as methanol and deprotonated free fatty acids. The NMR comparison also shows clearly that no significant degradation of the olefinic content occurs during the base catalyzed transesterification reaction.

CONCLUSION

Students have found this experiment simple to perform, yet highly informative about alternative fuels. The hands-on aspect reinforces that biodiesel can in fact be made and, thus, real alternatives to petroleum based fuels do exist. From a pedagogical standpoint, this experiment is an excellent tool to teach the basic chemistry underlying transesterification reactions via a reaction with a real-world application.

REFERENCES

CRC Handbook of Chemistry and Physics, 73rd ed., 1992. CRC Press, Inc., Boca Raton.

Ebiura, T., Echizen, T., Ishikawa, A., Murai, K., Baba, T., 2005. Applied Catalysis, A: General 283, 111–116.

Gottlieb, H.E., Kotlyar, V., Nudelman, A.J., 1997. Journal of Organic Chemistry 62, 7512.

Lindner, E., Ghanem, A., Warad, I., Eichele, K., Mayer, H.A., Schurig, V., 2003. Tetrahedron: Asymmetry 14, 1045–1053.

Pedersen, N.R., Kristensen, J.B., Bauw, G., Ravoo, B.J., Darcy, R., Larsen, K.L., Pedersen, L.H., 2005. Tetrahedron: Asymmetry 16, 615–622.

Ritter, S.K., 2004. Chemical and Engineering News 82, 31.

Sheehan, J., Camobreco, V., Duffield, J., Graboski, M., Shapouri, H., 1998. Life Cycle Inventory of Biodiesel and Petroleum Diesel for Use in an Urban Bus. U.S. Department of Energy Office of Fuels Development and U.S. Department of Agriculture Office of Energy.

Su, Q., Beeler, A.B., Lobkovsky, E., Porco, J.A., Panek, J.S., 2003. Organic Letters 5, 2149–2152.

Tyson, K.S., 2004. Biodiesel Handling and Use Guidelines. National Renewable Energy Laboratory.

Wang, W.G., Lyons, D.W., Clark, N.N., Gautam, M., Norton, P.M., 2000. Environmental Science and Technology 34, 933–939.

Yadav, G.D., Lathi, P.S., 2005. Journal of Molecular Catalysis B: Enzymatic 32, 107–113.

Zanoni, G., Agnelli, F., Meriggi, A., Vidari, G., 2001. Tetrahedron: Asymmetry 12, 1779–1784.

WHOLE ALGAL BIOMASS *IN SITU* TRANSESTERIFICATION TO FATTY ACID METHYL ESTERS AS BIOFUEL FEEDSTOCKS

22

Lieve M.L. Laurens

National Bioenergy Center, National Renewable Energy Laboratory, Golden, CO, USA

INTRODUCTION

Technologies for food, fuels, and chemicals from microalgae are being developed at an exponential rate. The field is moving forward on a number of fronts, including upstream technologies, such as strain discovery and improvement and cultivation as well as downstream pathway development, among which converting biomass into fuel streams is an area of intense development. Some of the major challenges are associated with the characterization of the biomass and the development of novel, cost-competitive conversion technologies that emphasize the valorization of the entire biomass and thus increase the value-proposition of algal biofuels. The aliphatic chains of the fatty acids are the most prominent biofuel-precursor constituents of algal biomass and, thus, quantifying the biomass fuel yield is a prerequisite for comparing algal strains, growth conditions, and processes. There is a recent and growing emphasis on whole biomass conversion for direct access to the lipid fraction. One focus area discussed here is the application of whole biomass, direct *in situ* transesterification; in particular, the development of this technology, process and analytical application and directions of future work. Unlike lipid extraction, which can over or underestimate the lipid content and associated fuel potential, whole biomass transesterification reflects the true potential fuel yield of algal biomass. This chapter includes a discussion of the yield of lipids quantified as fatty acid methyl esters (FAME) by using different catalysts and catalyst combinations, with the acid catalyst HCl providing a consistently high level of conversion to FAME. The discussion is accompanied by a link to the large-scale application of this process as a whole biomass conversion pathway.

MICROALGAE-FOCUSED LIPID TECHNOLOGIES FOR BIOFUEL APPLICATIONS

Biofuels from microalgae can contribute to displace a significant fraction of the diesel fuel market thanks to the high lipid content of algal biomass (Wijffels and Barbosa, 2010; Williams and Laurens, 2010). In the current climate of economic challenges with algal biofuels implementation, it is important to have robust procedures in place to determine the lipid content and develop lipid- and

Bioenergy. http://dx.doi.org/10.1016/B978-0-12-407909-0.00022-5

367

whole-biomass-conversion technologies. This aspect of algal biofuels research is important and has largely been overlooked in the literature.

In the current and historical literature, wide ranges of methodologies are reported for determining lipid content and extraction yields. The definition of lipids as compounds soluble in organic solvents has been the basis for the quantification of the total lipid fraction of algae as the total quantity of compounds soluble in a chloroform:methanol solvent mixture (Christie, 2005; Iverson et al., 2001; Bligh and Dyer, 1959). A multitude of reports are present in the literature assessing the application of solvent extraction processes. For example, one recent report compares the efficiency of extracting lipids from *Synechocystis* using 15 different solvent mixes and shows that gravimetric extraction yields are highly dependent on the polarity of the solvents used and the composition of the algal lipids (Harwood, 1998; Rainuzzo et al., 1994; Guschina and Harwood, 2006).

RENEWABLE AND BIODIESEL FUEL PROPERTIES

Biodiesel is a first-generation biofuel, an oxygenated diesel fuel, defined as the mono-alkyl esters of long-chain fatty acids and is traditionally produced from triglyceride-rich, refined vegetable oils by alkali-catalyzed transesterification or from mixed feedstocks rich in free fatty acids by acid catalyst esterification (Haas et al., 2006; Knothe, 2011; Al-Zuhair et al., 2007). The reaction mechanism for transesterification is illustrated in Figure 22.1 and shows the nucleophilic attack of the alcohol group, catalyzed by either an acid or base. The alcohol used is most commonly methanol such that the product consists predominantly of fatty acid methyl esters (FAMEs). The structure of the fatty acyl chains present in the feedstock has a determining effect on many of the critical quality parameters for biodiesel. The two most important properties for biodiesel quality are (1) the degree of fatty acid unsaturation, followed by (2) the chain length. The degree of unsaturation can be quantified as the iodine value (IV), which is the number of double bonds per mass of sample. Typical terrestrial crop oils and animal fats consist almost exclusively of C16 and C18 fatty acid chains. For this narrow range of

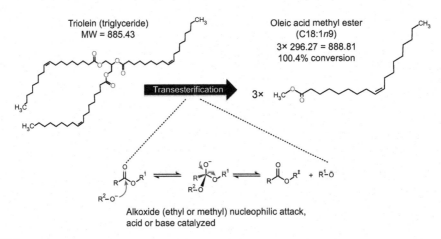

FIGURE 22.1

Illustration of acid- or base-catalyzed transesterification reaction mechanism using a triglyceride molecule as an example.

materials, IV can be correlated with many important properties such as cetane number, viscosity, density, and molar H/C ratio (McCormick et al., 2001). Algae typically have a much wider range of fatty acid constituents (Rainuzzo et al., 1994; Bigogno et al., 2002; Cohen et al., 2002; Volkman et al., 1989), and as shown in Figure 22.2, this composition and relative distribution is highly dependent on the time of harvest throughout a cultivation process. For example in a *Scenedesmus* sp. strain of algae, oleic acid (C18:1*n*9) contributes ∼15% of the total fatty acids in the early and mid-stage harvests, while it becomes the major contributor (∼57%) of the lipid fatty acid makeup at the late-stage harvest. Therefore, the study of the biodiesel quality parameters relative to the fatty acyl composition of the lipids is a hot topic of study, in particular in relation to typical terrestrial sources of lipids (Knothe, 2011).

Another critical quality parameter for diesel fuel is cetane number, which is a measure of the ignitability of the fuel in a diesel engine. Minimum cetane numbers of 40 and 50 are required in the United States and European Union, respectively. Fully saturated FAME have a high cetane number, and all saturated FAME with 10 or more carbons in the fatty acid chain will easily exceed the minimum US value of 40 (Graboski and McCormick, 1998). Fully saturated FAMEs have high melting points and lower solubility in a biodiesel or hydrocarbon diesel fuel matrix at low temperatures. Thus, a biodiesel too high in saturated FAME content will not be useful, even as a blend with petroleum diesel, in cold winter climates. A blend with more highly unsaturated FAME may improve the cold-flow properties, however, the fuel oxidative stability may decrease (Knothe, 2011). Therefore, a significant fraction of mono- and polyunsaturated FAME (PUFA) is desirable in biodiesel. The PUFA has much lower cetane number but also much lower melting point (and much greater solubility at cold

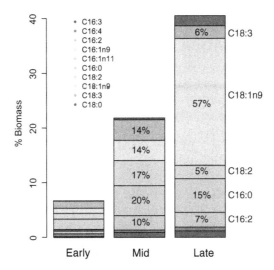

FIGURE 22.2

Total lipid content as FAME on a biomass basis and FAME constituent composition in an early, mid-, or late-harvested culture stage for *Scenedesmus* sp. biomass sample. Naming convention: C18:1n9 = fatty acid methyl ester of which the fatty acid chain contains 18 carbons with an unsaturated bond between the ninth and tenth carbon from the terminal carbon end of the fatty acyl chain.

temperatures). There have been concerns that unsaturated acyl chains are not adequately stable to oxidation, however this problem could be mitigated by the use of antioxidant additives. A more challenging area is the impact of impurities. For biodiesel made from conventional terrestrial crop oils and animal fats, these impurities are mono- and diglycerides, plant sterols and steryl glucosides, free fatty acids, and residual metals from the transesterification process (Chupka et al., 2011).

As an alternative to biodiesel, hydrocarbon renewable diesel or green diesel (or jet) can be produced from lipid feedstocks by processes involving hydrogenation, decarboxylation, and isomerization over heterogeneous catalysts. Hydrogenation is used to saturate the double bonds, and in some processes to remove the oxygen. Because of the relatively high cost of hydrogen, other process configurations remove a large fraction of the oxygen as CO_2 by decarboxylation. Using a C16/C18 feedstock, the products from these reactions are C15 to C18 normal alkanes, which are likely to be solid at room temperature. Therefore, an isomerization catalyst is also needed to introduce branching, which can produce a dramatic lowering of cloud point and a moderate reduction in cetane number (Kalnes et al., 2007; Smith et al., 2009; Smagala et al., 2013). These materials typically consist of more than 80% isoalkanes, and fuels with a cloud point of −25 °C will have a cetane number above 70 (Smagala et al., 2013). These fuels have the advantage of being very similar to petroleum-derived fuels and containing few impurities that might cause operational issues.

Since the acyl chains will determine the fuel potential of algal biomass in both a biodiesel and green diesel conversion pathway, and in the context of this discussion, a definition of lipids as the sum of their fatty acid constituents is appropriate. Similarly, in the context of biofuels, food, or feed applications, defining lipids as fatty acids and their derivatives is appropriate. In particular in light of this suggested definition, and in published work, *in situ* transesterification is a novel concept and applicable specifically for algal biomass, with varying levels and mixtures of algal lipids.

IN SITU TRANSESTERIFICATION OF OLEAGINOUS ALGAL BIOMASS

In situ transesterification refers to the conversion of biomass lipids without an initial lipid extraction. Using this method as the basis of quantification, all fatty acids can be measured as fatty acid methyl esters (FAMEs), and most importantly this conversion is not dependent on lipid extraction efficiency (Carrapiso and Garcia, 2000). This process is gaining traction as a lipid measurement and biomass conversion procedure for algae (Haas et al., 2006; Bigelow et al., 2011; Griffiths et al., 2010; McNichol et al., 2012; Laurens et al., 2012; Haas and Wagner, 2011; Ehimen et al., 2010), and the economics and process strategy of this technology as a strategy for simplification of biodiesel production has been explored (Haas et al., 2006). Because of the highly complex nature of algal lipids, catalysis of the transesterification reaction or any other biofuel-relevant reaction is not straightforward (Smith et al., 2009; Huber et al., 2006) A comprehensive study of the reaction yields with different catalysts and a comparison with standard methods on different algal strains was recently reported (Laurens et al., 2012). One of the major technological advantages of *in situ* transesterification is the elimination of reliance on an extraction procedure, which can have limitations due to the co-extraction of impurities, which can inhibit downstream catalytic lipid upgrading. Alternatively, extraction can suffer from incomplete solvent transfer of the biomass due to cell wall recalcitrance. The application of *in situ* transesterification combines an effective "biomass pretreatment" step with the simultaneous catalytic conversion of the lipid fraction to FAMEs, which due to their nonpolar nature lend themselves to extraction with organic solvents, such as hexane.

CHOICE OF CATALYST FOR *IN SITU* WHOLE BIOMASS TRANSESTERIFICATION

Between both acid and alkaline catalysis for transesterification, it has been shown that an alkaline catalyzed reaction is much faster, but it is also more limited with regard to types of lipids. For example, free fatty acids are notoriously hard to convert to fatty acid methyl esters with a base catalyst (Nagle and Lemke, 1990). If the algal biomass sample contains high levels of free fatty acids, the overall FAME yield obtained by alkali-catalyzed *in situ* transesterification may underestimate the actual FAME yield of the biomass due to partial saponification (Al-Zuhair, 2007). When the effect of acid and base catalysts were compared on the basis of conversion of algal oils, it was found that acid catalysis resulted in consistently higher yields, albeit with longer reaction times (Carrapiso and Garcia, 2000; Nagle and Lemke, 1990).

Different catalysts for *in situ* transesterification are reported in the literature. When using the overall FAME yield obtained with different catalysts, a like-for-like performance comparison between acid and base catalysis can be made (or a combination of both). The goal of the work described in Laurens et al. (Laurens et al., 2012) was to find an easy one-step procedure that is robust across different strains and conditions. A simplification of reagent preparation based on dissolution of concentrated HCl in methanol and, therefore, eliminating the use of anhydrous acetyl chloride greatly aided with the preparation steps without affecting the conversion yields. The procedure for quantification and conversion has been used by Ichihara (Volkman et al., 1989) and was shown to not affect the conversion efficiency of fatty acids, as might be predicted because of the lack of a completely anhydrous environment. An additional prior solubilization step of the biomass lipids with chloroform: methanol can aid with access of the reagents to the lipids embedded in the algal cell matrix. This method compared well with previously established AOAC method for oil transesterification (AOAC 991.39, also referred to as NaOMe:BF_3). Even though the original AOAC 991.39 method was designed to use ~ 20 mg of oil, the method was adapted to reduce the overall sample requirement to 4–10 mg of biomass for analytical scale. When only BF_3 was used as the catalyst in the reaction, the conversion yield was shown to drop significantly. Furthermore, when using only NaOMe, no FAME conversion was measured, indicating that the two stages (NaOMe and BF_3) are necessary to obtain yields comparable to the one-step HCl procedure. Interestingly, neither of the two base-catalysis methods investigated in Laurens et al. (2012) showed FAME yields comparable to the yields obtained by acid catalysis. Even after optimizing the reaction conditions over a range of different time and temperature intervals (data not shown), there is still a considerable lack of conversion efficiency. Possible explanations are the lack of penetration of the catalyst through the algal cell walls and the presence of high levels of free fatty acids leading to a lower efficiency by this base catalyst. The modified AOAC 922.06 and 989.05 procedures reduced the overall FAME yield compared with the HCl:MeOH or combined base/acid (NaOMe:BF_3) procedures. The observations support the fact that the published standard procedures, although they incorporate a significant amount of pretreatment of the protein and carbohydrate components in the biomass, do not allow for complete recovery of the fatty acids in algal biomass. Future work would include a comprehensive study of process-relevant conditions with different catalyst loadings on the effectiveness of the conversion of whole biomass lipids to FAMEs. This will allow for the extrapolation of some of the small-scale data obtained to a biofuel-scale process.

The ability to catalyze the transesterification of a wide range of lipid types with acid catalysis, made for an ubiquitous implementation of acid catalysis for transesterification of lipids and free fatty acids. Of all catalysts, hydrogen chloride (HCl:MeOH) is, according to some authors, the best general purpose esterifying agent, and it is the most widely mentioned catalyst being used for *in situ* procedures (Haas et al., 2006; Carrapiso and Garcia, 2000; Griffiths et al., 2010; Laurens et al., 2012; Nagle and Lemke, 1990; Ichihara et al., 2010; Lepage and Roy, 1986). A modified HCl-catalyzed procedure was developed to suit algal biomass hydrolysis and transesterification in a simple, one-step reaction on an analytical scale.

As an example of larger scale implementation of *in situ* transesterification, Ehimen et al. (Ehimen et al., 2010) have studied the variables affecting the effectiveness of the conversion of algal lipids and analyzed the effect of alcohol volume, temperature, and reaction time and moisture on transesterification with sulfuric acid as the acid catalyst and found a requirement for high catalyst loading and methanol consumption and a high sensitivity to water in the biomass. For Ehimen et al., the process catalyst choice, catalyst loading, and methanol requirement parameters were considered important economical parameters. It is important in this process to retain the exact compositional makeup of the fatty acids and minimize degradation through oxidation of the fatty acids and be able to account and report on the total fatty acid content and profile because of the importance of the fatty acid profile for downstream fuel quality (Knothe, 2011). A comprehensive study by Haas and Wagner (Haas and Wagner, 2011) investigated the application of *in situ* transesterification of whole biomass as a way to reduce the complexity and associated costs of feedstock and processing in a biodiesel production process. Similar to Ehimen et al. the authors also identified the cost implications of catalyst loading and methanol requirement as major economic factors. Improvements in biomass pretreatment have reduced the methanol requirements for *in situ* transesterification as a process option, and it is most likely a more economical manner to produce biodiesel directly from oleaginous biomass, such as algae (Haas and Wagner, 2011).

ANALYTICAL CHARACTERIZATION OF LIPID CONTENT IN ALGAL BIOMASS USING *IN SITU* TRANSESTERIFICATION

In situ transesterification can also be applied to the quantification of lipid content, or biofuel potential, in small quantities of biomass (typically 20 mg or less) to rapidly screen and evaluate the potential of different sources of lipids or biomass feedstock quality. Biomass concentrations of typical algal cultures are low (<1 g/L) and, thus, only small quantities of biomass samples can be generated in laboratory-scale shake flask cultures. Setting a minimum sample size at the milligram level allows for replicates and multiple time points using small volumes of algal cultures. Therefore, adaptation of a procedure to small amounts of biomass is a priority in recently published work. There have been reports in the literature recently that list small scale adaptations of the *in situ* transesterification as the preferred method for lipid quantification in microalgae (Laurens et al., 2012; Bigelow et al., 2011; Lohman et al., 2013), in particular, a 4–10 mg application of an *in situ* transesterification procedure was demonstrated and can greatly increase the throughput of fuel-potential evaluation experiments (Laurens et al., 2012).

In the context of accurate and precise quantification of lipid content, and by definition, the fuel potential in algae, there are several methods published in the literature for quantification of fatty acids. The methods listed by AOAC (Association of Official Analytical Chemists) are routinely used in the food and agricultural industries (e.g., AOAC 922.06, 989.05, 991.39) for fatty acid determination, however, no methodology specifically tailored or optimized for algal biomass has been reported by AOAC, though some of the listed methods, or modifications of those, have been applied to algal biomass (Griffiths et al., 2010; Bigelow et al., 2011). Modifications of these methods for analysis, and a protein hydrolysis step could liberate additional FAMES from a complex protein matrix. Similarly, the 922.06 method was designed around applications to flour fatty acid determinations and includes a concentrated HCl-hydrolysis step prior to transesterification to hydrolyze the carbohydrates and release FAMEs. Because of the reported rigid, recalcitrant cell walls in algae, this method was included as an option for yield comparisons in reference (Laurens et al., 2012). The AOAC 991.39 method was published as an effective transesterification method for quantification of FAMEs and typically used for fish oil transesterification. This method consists of a two-step catalyzed reaction first by NaOMe followed by BF_3. Modifications of this method are also referred to as NaOMe:BF_3 methods and are often compared against a single-step acid catalysis method (Laurens et al., 2012; Lohman et al., 2013).

Lipid extraction is often reported as a quantitative measure of biofuel precursor yields. However, total yield of lipid extraction is highly dependent on the solvent system and lipid extraction parameters used. The pressurized fluid extraction systems, e.g., Thermo Fisher's Accelerated Solvent Extractor (ASE), have potential to provide higher throughput extractions of algal biomass. However, the gravimetric yields are dependent on the temperature and pressure. For example, raising the temperature and the pressure during pressurized fluid extraction increases the gravimetric recovery for both solvent systems, and using hexane:isopropanol as a solvent reduces the overall extraction yield by ~70% or ~50% (Laurens et al., 2012). The gravimetric yields and extraction efficiencies vary with the severity of the extraction (Laurens et al., 2012; Folch et al., 1957). These observations demonstrate the uncertainties associated with quantification of lipid content based on a solvent-extraction process. An example of variable lipid extractability quantification data is shown in Figure 22.3 (adapted from Laurens et al. (2014)), where for three different microalgal strains, the quantification of extractable lipids was based on pressurized fluid extraction with a hexane:isopropanol solvent system, relative to the fuel potential determined through *in situ* transesterification. The data illustrate that, for all three organisms, the initial extraction-based quantification overestimates the fuel potential, whereas at the later time points, the extractable lipids either match the fuel potential (*Nannochloropsis* sp. and *Scenedesmus* sp.) or underestimate (*Chlorella* sp.). The closer matching at the later time points can be attributed to the increase in triglyceride accumulation of the lipid fraction, which have a higher conversion efficiency to FAME, whereas the underestimation for *Chlorella* sp. indicates an increased recalcitrance of the cells to solvent extraction, as reported before (Laurens et al., 2012).

Another consideration with fuel production based on lipids resulting from solvent extraction is that those extracts also often contain nonfuel components (e.g., chlorophyll, pigments, proteins, and carbohydrates that make up part of the glycolipids, e.g., galactose from galactolipids). Fuel potential in our context is defined as the fraction of the lipids composed of fatty acids that are amenable to upgrading. To investigate whether gravimetric lipid extraction yield reflects the true fuel potential, the fuel fraction of extracts should be measured by conversion of fatty acids to FAMEs. Different lipid types can be converted to fuel to varying degrees. For example, triglyceride

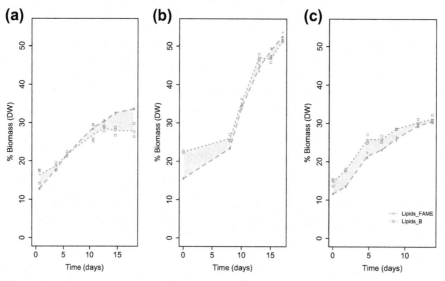

FIGURE 22.3

Illustration of extractable lipid quantification (Lipids_B) relative to total fuel potential, reflected in the total FAME content by *in situ* transesterification (Lipids_FAME) for three microalgal strains; (a) *Chlorella* sp., (b) *Nannochloropsis* sp., and (c) *Scenedesmus* sp.

lipids convert to 100% FAME on a gravimetric basis since the addition of the methyl group balances the loss of a glycerol in the hydrolysis step. Some of the most common lipids found in algae are shown in Figure 22.4 with their respective theoretical conversion efficiency calculations. Due to a larger proportion of the mass associated with the carbohydrate functionality, a glycolipid such as digalactosyl diglyceride (DGDG) only converts to 63% FAME. Thus, the relative composition of the lipids affects the fuel potential of the extracted lipids (Nagle and Lemke, 1990). Including both a low- and high-lipid-containing biomass sample (assumed to have a low and high concentration of triglycerides, respectively) from one strain allows for the investigation of lipid extraction and conversion efficiency for different samples. Similar to the data shown in Figure 22.3, for nutrient-deprived *Chlorella vulgaris* biomass, the fuel yield of the extract almost matches the weight of the lipid extract (93.3% of the lipids can be converted into FAME), but the overall extracted fraction contains fewer fatty acids as compared with the whole biomass *in situ* transesterification yields, as demonstrated in the fatty acid extraction mass balance shown in Laurens et al. (2012). For both nutrient-replete-grown *Chlorella vulgaris* and *Nannochloropsis* sp., the gravimetric extraction yield far exceeds the actual fuel yield; only 30.9% and 51.4% of the lipids can be converted to FAME. This supports the notion of limitation of lipid quantification through extraction and gravimetric recovery and the adoption of an *in situ* acid-catalyzed transesterification procedure for lipid quantification in algal biomass.

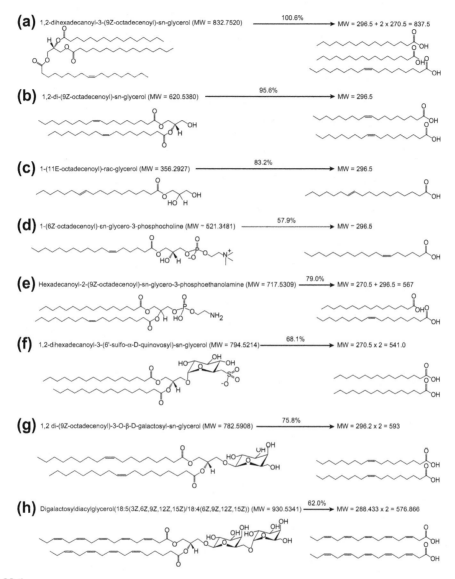

FIGURE 22.4

Overview of the chemical structures of the most common representatives from seven lipid classes, tri-acylglycerides (a), diacylglycerides (b), monoglycerides (c), phospholipids (d and e), sulfo-lipids (f), and glycolipids, monogalactosyldiglyceride (MGDG) and digalactosyl diglyceride (DGDG) (g and h). Structures from www.LipidMAPS.org. Illustration of quantitative theoretical calculation of conversion efficiency for each of the lipid classes to fatty acid methyl esters (note: free fatty acid forms are shown, but fatty acid methyl esters (FAME) molecular weights were used in the calculations).

CONCLUSIONS

To aid with the more accurate and complete recovery of whole biomass fuel potential in algae, alternatives to extraction-based lipid quantification and conversion have been established. One example of this is the one-step, acid-catalyzed procedure for quantitative determination and complete conversion of algal lipids in the biomass as FAMEs through *in situ* whole biomass transesterification. This process has the potential to greatly simplify the whole biomass utilization in a conversion process rather than depend on a two-step extraction and lipid-upgrading pathway. A one-step procedure, based on *in situ* HCl-catalysis of algal biomass applicable to small quantities of material with yields comparable to the AOAC 991.39 two-stage NaOMe:BF$_3$ procedure, was validated and implemented for high-throughput screening of different algal strains or tracking lipid productivity throughout a cultivation process scenario. The small-scale application on 4–7 mg of biomass allows for rapid screening of multiple strains for lipid content or selection of the optimal harvesting scenario for a particular strain and cultivation condition. The high analytical precision of this technology has been shown and allows for rapid tracking of strain and fuel productivity increments.

For industrial processing applications, additional work would be required to demonstrate large-scale implementation, with perhaps a cheaper acid source (such as sulfuric acid). This demonstration work might also include an improvement (reduction) of the methanol requirements and water tolerance of the reaction yields, as well as including a demonstration of different sources of algal biomass. The efficiency of transesterification of different catalysts has been compared using the total fatty acid yield as the main metric, as well as the distribution or profile of the individual fatty acids, and was shown to be potentially process relevant.

REFERENCES

Al-Zuhair, S., 2007. Production of biodiesel: possibilities and challenges. Biofuels, Bioproducts and Biorefining 1, 57–66.

Al-Zuhair, S., Ling, F.W., Jun, L.S., 2007. Proposed kinetic mechanism of the production of biodiesel from palm oil using lipase. Process Biochemistry 42, 951–960.

Bigelow, N.W., Hardin, W.R., Barker, J.P., Ryken, S.A., MacRae, A.C., Cattolico, R.A., 2011. A comprehensive GC-MS sub-microscale assay for fatty acids and its applications. Journal of the American Oil Chemists' Society 88, 1329–1338.

Bigogno, C., Khozin-Goldberg, I., Boussiba, S., Vonshak, A., Cohen, Z., 2002. Lipid and fatty acid composition of the green oleaginous alga Parietochloris incisa, the richest plant source of arachidonic acid. Phytochemistry 60, 497–503.

Bligh, E.G., Dyer, W.J., 1959. A rapid method of total lipid extraction and purification. Canadian Journal of Biochemistry and Physiology 37, 911–917.

Carrapiso, A.I., Garcia, C., 2000. Development in lipid analysis: some new extraction techniques and in situ transesterification. Lipids 35, 1167–1177.

Christie, W.W., 2005. Lipid Analysis: Isolation, Separation, Identification and Structural Analysis of Lipids, third ed. American Oil Chemists Society.

Chupka, G.M., Yanowitz, J., Chiu, G., Alleman, T.L., McCormick, R.L., 2011. Effect of saturated monoglyceride polymorphism on low-temperature performance of biodiesel. Energy and Fuels 25, 398–405.

Cohen, Z., Bigogno, C., Khozin-Goldberg, I., 2002. Accumulation of arachidonic acid-rich triacylglycerols in the microalga *Parietochloris incisa* (Trebuxiophyceae, Chlorophyta). Phytochemistry 60, 135–143.

Ehimen, E.A., Sun, Z.F., Carrington, C.G., 2010. Variables affecting the in situ transesterification of microalgae lipids. Fuel 89, 677–684.

Folch, J., Lees, M., Sloane-Stanley, G.H., 1957. A simple method for the isolation and purification of total lipids from animal tissues. Journal of Biological Chemistry 226, 497–509.

Graboski, M.S., McCormick, R.L., 1998. Combustion of fat and vegetable oil derived fuels in diesel engines. Progress in Energy and Combustion Science 24, 125–164.

Griffiths, M.J., van Hille, R.P., Harrison, S.T.L., 2010. Selection of direct transesterification as the preferred method for assay of fatty acid content of microalgae. Lipids 45, 1053–1060.

Guschina, I. a., Harwood, J.L., 2006. Lipids and lipid metabolism in eukaryotic algae. Progress in Lipid Research 45, 160–186.

Haas, M.J., Wagner, K., 2011a. Simplifying biodiesel production: the direct or in situ transesterification of algal biomass. European Journal of Lipid Science and Technology 113, 1219–1229.

Haas, M.J., Wagner, K.M., 2011b. Substrate pretreatment can reduce the alcohol requirement during biodiesel production via in situ transesterification. Journal of the American Oil Chemists' Society 88, 1203–1209.

Haas, M.J., McAloon, A.J., Yee, W.C., Foglia, T.A., 2006. A process model to estimate biodiesel production costs. Bioresource Technology 97, 671–678.

Harwood, J.L., 1998. Membrane lipids in algae. In: Siegenthaler, P.-A., Murata, N. (Eds.), Lipids in Photosynthesis. Structure Function and Genetics. Springer, Netherlands, pp. 53–64.

Huber, G.W., Iborra, S., Corma, A., 2006. Synthesis of transportation fuels from biomass: chemistry, catalysts, and engineering. Chemical Review 106, 4044–4098.

Ichihara, K., Yamaguchi, C., Araya, Y., Sakamoto, A., Yoneda, K., 2010. Preparation of fatty acid methyl esters by selective methanolysis of polar glycerolipids. Lipids 45, 367–374.

Iverson, S.J., Lang, S.L.C., Cooper, M.H., 2001. Comparison of the Bligh and Dyer and Folch methods for total lipid determination in a broad range of marine tissue. Lipids 36, 1283–1287.

Kalnes, T., Marker, T., Shonnard, D.R., 2007. Green diesel: a second generation biofuel. International Journal of Chemical Reactor Engineering 5.

Knothe, G., 2011. A technical evaluation of biodiesel from vegetable oils vs. algae. Will algae-derived biodiesel perform? Green Chemistry 13, 3048.

Laurens, L., Quinn, M., Van Wychen, S., Templeton, D., Wolfrum, E.J., 2012. Accurate and reliable quantification of total microalgal fuel potential as fatty acid methyl esters by in situ transesterification. Analytical and Bioanalytical Chemistry 403, 167–178.

Laurens, L.M.L., Van Wychen, S., McAllister, J.P., Arrowsmith, S., Dempster, T.A., McGowen, J., et al., 2014. Strain, biochemistry, and cultivation-dependent measurement variability of algal biomass composition. Analytical Biochemistry 452, 86–95.

Lepage, G., Roy, C.C., 1986. Direct transesterification of all classes of lipids in a one-step reaction. Journal of Lipid Research 27, 114–120.

Lohman, E.J., Gardner, R.D., Halverson, L., Macur, R.E., Peyton, B.M., Gerlach, R., 2013. An efficient and scalable extraction and quantification method for algal derived biofuel. Journal of Microbiological Methods 94, 235–244.

McCormick, R.L., Graboski, M.S., Alleman, T.L., Herring, a M., Tyson, K.S., 2001. Impact of biodiesel source material and chemical structure on emissions of criteria pollutants from a heavy-duty engine. Environmental Science and Technology 35, 1742–1747.

McNichol, J., MacDougall, K.M., Melanson, J.E., McGinn, P.J., 2012. Suitability of soxhlet extraction to quantify microalgal fatty acids as determined by comparison with in situ transesterification. Lipids 47, 195–207.

Nagle, N., Lemke, P.R., 1990. Production of methyl-ester fuel from microalgae. Applied Biochemistry and Biotechnology 24, 355 361.

Rainuzzo, J.R., Reitan, K.I., Olsen, Y., 1994. Effect of short and long term lipid enrichment on total lipids, lipid class and fatty acid composition in rotifiers. Aquaculture International 2, 19–32.

Smagala, T.G., Christensen, E., Christison, K.M., Mohler, R.E., Gjersing, E., McCormick, R.L., 2013. Hydro-carbon renewable and synthetic diesel fuel blendstocks: composition and properties. Energy and Fuels 27, 237–246.

Smith, B., Greenwell, H.C., Whiting, A., 2009. Catalytic upgrading of tri-glycerides and fatty acids to transport biofuels. Energy and Environmental Science 2, 262–271.

Volkman, J.K., Jeffrey, S.W., Nichols, P.D., Rogers, G.I., Garland, C.D., 1989. Fatty acid and lipid composition of 10 species of microalgae used in mariculture. Journal of Experimental Marine Biology and Ecology 128, 219–240.

Wijffels, R.H., Barbosa, M.J., 2010. An outlook on microalgal biofuels. Science 329 (80), 796–799.

Williams, P.J.L.B., Laurens, L.M.L., 2010. Microalgae as biodiesel & biomass feedstocks: Review & analysis of the biochemistry, energetics & economics. Energy and Environmental Science 3, 554–590.

HOW FUEL ETHANOL IS MADE FROM CORN

23

Nathan S. Mosier, Klein E. Ileleji

Department of Agricultural and Biological Engineering, Purdue University, West Lafayette, IN, USA

INTRODUCTION

Fuel ethanol has become a very important agricultural product over the past two decades. In 2005, more than 13% of US corn production went toward making this fuel additive/fuel extender, which lessens U.S. dependence on foreign oil imports, is cleaner for the environment, and has substantial impact on the rural economy and agriculture production.

FUEL ETHANOL

Ethanol is an alcohol produced by yeast from sugars. It is the same alcohol produced by yeast in beer, wine, and spirits. Fuel ethanol is ethanol that has been highly concentrated to remove water and blended with other compounds to render the alcohol undrinkable. Fuel ethanol can be used alone as a fuel, such as in Indy Racing League cars, or can be blended with gasoline and used as fuel. All cars and trucks on the road today can use gasoline/ethanol blends of up to 10% ethanol (90% gasoline), also called "E10." Blends of up to 85% ethanol, also known as "E85," can be used as transportation fuel by cars and trucks with slight modifications (approximately $100 per vehicle). These flexible fuel vehicles can use either gasoline or ethanol blends, including E85.

YEAST'S ROLE IN ETHANOL PRODUCTION

All ethanol production is based upon the activity of yeast (*Saccharomyces cerevisiae*), an important microorganism to humans. Through a process called "fermentation," yeast eat simple sugars and produce carbon dioxide (CO_2) and ethanol as waste products. For each pound of simple sugars, yeast can produce approximately ½ pound (0.15 gallons) of ethanol and an equivalent amount of carbon dioxide.

CORN AS ETHANOL FEEDSTOCK

In 2005, approximately 11 billion bushels of corn were produced in the United States. Indiana corn production in 2005 was approximately 889 million bushels (USDA, 2006). Ethanol production in the United States topped 4 billion gallons in 2005 and consumed 1.4 billion bushels of corn, valued at $2.9 billion (NCGA, 2005). This represents the third largest demand for US corn after animal feed and export markets. With additional construction of ethanol plants and increasing ethanol demand, fuel ethanol production is expected to exceed 7.5 billion gallons before the year 2012 target set forth in the Energy Policy Act of 2005 (EPACT05).

The value of corn as a feedstock for ethanol production is due to the large amount of carbohydrates, specifically starch, present in corn (Table 23.1). Starch can be rather easily processed to break it down into simple sugars, which can then be fed to yeast to produce ethanol. Modern ethanol production can produce approximately 2.7 gallons of fuel ethanol per bushel of corn.

Table 23.1 Composition of Corn

Component	Percent (Average) Dry Matter
Carbohydrates (total)	84.1%
Starch	72.0%
Fiber (NDF)	9.5%
Simple sugars	2.6%
Protein	9.5%
Oil	4.3%
Minerals	1.4%
Other	0.7%

Source: From Corn: Chemistry and Technology, *1987.*

INDUSTRIAL ETHANOL PRODUCTION

Commercial production of fuel ethanol in the United States involves breaking down the starch present in corn into simple sugars (glucose), feeding these sugars to yeast (fermentation), and then recovering the main product (ethanol) and byproducts (e.g., animal feed). Two major industrial methods for producing fuel ethanol are used in the United States: wet milling and dry grind. Dry-grind ethanol production represents the majority of ethanol processing in the U.S. (>70% of production), and all newly constructed ethanol plants employ some variation on the basic dry-grind process because such plants can be built at a smaller scale for a smaller investment.

WET MILLING

Wet milling is used to produce many products besides fuel ethanol. Large-scale, capital-intensive, corn-processing wet mills produce such varied products as high fructose corn syrup (HFCS), biodegradable plastics, food additives such as citric acid and xanthan gum, corn oil (cooking oil), and livestock feed.

Wet milling is called "wet" because the first step in the process involves soaking the grain in water (steeping) to soften the grain and make it easier to separate (fractionate) the various components of the corn kernel. Fractionation, which separates the starch, fiber, and germ, allows these various components to be processed separately to make a variety of products. The major byproducts of wet mill ethanol production are two animal feed products, corn gluten meal (high protein, 40%) and corn gluten feed (low protein, 28%), and corn germ, which may be further processed into corn oil.

DRY GRIND

In the dry-grind ethanol process, the whole grain is processed, and the residual components are separated at the end of the process. There are five major steps in the dry-grind method of ethanol production.

DRY-GRIND ETHANOL PROCESSING STEPS

1. Milling
2. Liquefaction
3. Saccharification
4. Fermentation
5. Distillation and recovery

MILLING

Milling involves processing corn through a hammer mill (with screens between 3.2 and 4.0 mm) to produce a corn flour (Rausch et al., 2005). This whole corn flour is slurried with water, and heat-stable enzyme (α-amylase) is added.

LIQUEFACTION

This slurry is cooked, also known as "liquefaction." Liquefaction is accomplished using jet-cookers that inject steam into the corn flour slurry to cook it at temperatures above 100 °C (212 °F). The heat and mechanical shear of the cooking process break apart the starch granules present in the kernel endosperm, and the enzymes break down the starch polymer into small fragments. The cooked corn mash is then allowed to cool to 80–90 °C (175–195 °F), additional enzyme (α-amylase) is added, and the slurry is allowed to continue liquefying for at least 30 min.

SACCHARIFICATION

After liquefaction, the slurry, now called "corn mash," is cooled to approximately 30 °C (86 °F), and a second enzyme (glucoamylase) is added. Glucoamylase completes the breakdown of the starch into simple sugar (glucose). This step, called "saccharification," often occurs while the mash is filling the fermentor in preparation for the next step (fermentation) and continues throughout the next step.

FERMENTATION

In the fermentation step, yeast grown in seed tanks are added to the corn mash to begin the process of converting the simple sugars to ethanol. The other components of the corn kernel (protein, oil, etc.) remain largely unchanged during the fermentation process. In most dry-grind ethanol plants, the fermentation process occurs in batches. A fermentation tank is filled, and the batch ferments completely before the tank is drained and refilled with a new batch.

The upstream processes (grinding, liquefaction, and saccharification) and downstream processes (distillation and recovery) occur continuously (grain is continuously processed through the equipment). Thus, dry-grind facilities of this design usually have three fermentors (tanks for fermentation) where, at any given time, one is filling, one is fermenting (usually for 48 h), and one is emptying and resetting for the next batch.

Carbon dioxide is also produced during fermentation. Usually, the carbon dioxide is not recovered and is released from the fermenters to the atmosphere. If recovered, this carbon dioxide can be compressed and sold for carbonation of soft drinks or frozen into dry ice for cold product storage and transportation.

After the fermentation is complete, the fermented corn mash (now called "beer") is emptied from the fermentor into a beer well. The beer well stores the fermented beer between batches and supplies a continuous stream of material to the ethanol recovery steps, including distillation.

DISTILLATION AND RECOVERY

After fermentation, the liquid portion of the slurry has 8–12% ethanol by weight. Because ethanol boils at a lower temperature than water does, the ethanol can be separated by a process called "distillation." Conventional distillation/rectification systems can produce ethanol at 92–95% purity. The residual water is then removed using molecular sieves that selectively adsorb the water from an ethanol/water vapor mixture, resulting in nearly pure ethanol (>99%).

The residual water and corn solids that remain after the distillation process are called "stillage." This whole stillage is then centrifuged to separate the liquid (thin stillage) from the solid fragments of the kernel (wet cake or distillers' grains). Some of the thin stillage (backset) is recycled to the beginning of the dry-grind process to conserve the water used by the facility.

The remaining thin stillage passes through evaporators to remove a significant portion of the water to produce thickened syrup. Usually, the syrup is blended with the distillers' grains and dried to produce an animal feed called "distillers' dried grains with solubles" (DDGS). When markets for the feed product are close to the plant, the byproduct may be sold without drying as distillers' grains or wet distillers' grains.

ENERGY USE IN ETHANOL PRODUCTION

It is true that the laws of physics dictate that energy will be lost in converting one form of energy to another. Thus, ethanol does have less energy than the corn used to produce it. However, this is also true for converting crude oil to gasoline and coal to electricity. The important questions about ethanol production are "is ethanol truly a renewable fuel?" and "how much fossil fuel is used?" Yes; ethanol is a renewable fuel. The energy used to produce ethanol includes fuel for tractors, combines, and transportation of the grain to the ethanol plant, as well as the energy in processing the corn to ethanol. However, the largest portion of the total energy present in corn is solar energy captured by the corn plant and stored in the grain as starch. When these amounts are totaled, the energy in the ethanol exceeds the fossil fuel energy used to grow and process the corn by 20–40% (Farrell et al., 2006).

Most of the energy for processing corn to ethanol is spent on the distillation and DDGS drying steps of the process. When wet distillers' grain can be fed to livestock close to the ethanol plant, the savings in natural gas for drying can be as high as 20% of the total energy cost for processing corn to ethanol.

CONCLUSIONS

Modern dry-grind ethanol plants can convert corn grain into ethanol (2.7–2.8 gallons per bushel) and DDGS (17 pounds per bushel). This rather energy-efficient process produces a renewable liquid fuel that has significant impacts on the agricultural economy and energy use in the U.S.

Increasing ethanol production presents many opportunities and challenges for U.S. agriculture as demands on corn production for feed, fuel, and export markets increase. Additionally, advances in biotechnology and engineering are opening possibilities for new raw materials, such as switch grass and corn stover, to be used for even greater fuel ethanol production into the future.

ACKNOWLEDGMENT

This paper was prepared for Purdue University Extension BioEnergy Series (technical ID# ID-328).

REFERENCES

Farrell, A.E., Plevin, R.J., Turner, B.T., Jones, A.D., O'Hare, M., Kammen, D.M., 2006. Ethanol can contribute to energy and environmental goals. Science 311 (5760), 506–508.

National Corn Growers Association (NCGA) Annual Report, 2005.

Rausch, K.D., Belyea, R.L., Ellersieck, M.R., Singh, V., Johnston, D.B., Tumbleson, M.E., 2005. Particle size distributions of ground corn and DDGS from dry grind processing. Transactions of the ASAE 48 (1), 273–277.

U.S. Department of Agriculture. National Agriculture Statistics Service. http://www.nass.usda.gov. Accessed in 2014.

Watson, S.A., 1987. Structure and composition. In: Watson, S.A., Ramstad, P.E. (Eds.), Corn: Chemistry and Technology. American Association of Cereal Chemists, Inc, pp. 53–82.

SMALL-SCALE APPROACHES FOR EVALUATING BIOMASS BIOCONVERSION FOR FUELS AND CHEMICALS

24

Jonathan R. Mielenz

White Cliff Biosystems, Rockwood, Oak Ridge National Laboratory, Department of Energy (retired) Oak Ridge, USA

INTRODUCTION

Plant biomass is a valuable resource for renewable materials whether they are transportation fuels, chemical intermediates/bulk chemicals, or sources of heat and generated power. Transportation fuels include gasoline additives or replacements such as oxygenates like ethanol, or so-called *drop-in fuels*. Similarly, petrochemical diesel can be replaced by biodiesel from plant oil conversion or with newly developing *drop-in molecules*. On the chemical side, the opportunities are wide open; the various substituents in biomass can be converted to many types of chemical molecules by both fermentation and thermochemical catalysis. Heat and power generation involve the use of biomass as a partial replacement for coal in power plants, including boiler combustion as well as advanced gasification technologies. Use of biomass for heat and power generation is largely a mature technology, yet accessing biomass for chemical conversion requires complex, still-developing technologies categorized broadly as biochemical or thermochemical conversion. Although both types of conversion technology have their own benefits and challenges (Lynd et al., 2009), this chapter describes in some detail approaches to evaluate biomass sources efficiently and to determine preferred biochemical conversion schemes. Also, many published methods described in peer-reviewed publications have a minimum level of detail because of space considerations, so this chapter provides a more detailed description of laboratory procedures to evaluate the bioconversion of biomass.

TYPES OF BIOMASS

Biomass refers to multiple types of plant matter, each with their own specific composition and characteristics. Biomass is classified broadly as woody species, herbaceous species, and special plant matter such as algae. This chapter reviews the technology of utilization of terrestrial biomass, so algae technology is not included. Structurally, biomass contains two broad types of complex carbohydrates: cellulose and hemicellulose. Cellulose is composed primarily of polymeric glucose, both in crystalline form and unorganized amorphous cellulose. Hemicellulose is composed of a highly variable

Bioenergy. http://dx.doi.org/10.1016/B978-0-12-407909-0.00024-9

385

polymer of the five-carbon-sugar xylose, which is highly substituted with other sugars such as arabinose, galactose, mannose, as well as glucose (Dodd and Cann, 2009). In addition, hemicellulose has multiple acetate esters on its backbone. These esters of acetate, called *O-acetylated polymers*, can constitute up to 4% of the weight of dry hardwoods (Samara, 1992). Hemicellulose also has linkages to the third major chemical substituent: lignin. Lignin is a complex polyphenolic structure with numerous *o*-methyl side groups (Boerjan et al., 2003; Ralph et al., 2004), and it is often referred to as an important structural component of plant biomass, acting as a linkage to the hemicellulose complex (Zhao et al., 2012). Lignin is produced by plants from the aromatic amino pathway, specifically from precursors for the phenylalanine pathway, and is a polymer of three monomer chemicals derived from phenylalanine: coniferyl, coumaryl, and sinapyl alcohol (Vanholme et al., 2010). Each plant species has its own process to polymerize these monomers, so lignin composition and content varies by species.

Woody sources include both softwood and hardwood trees, plus more bushy species such as willow. Softwood species include pine and fir trees, which are grown in abundance for the paper and structural wood industries. The composition of these species vary, but in general contain elevated levels of pitch and resin materials that must be removed or controlled by any utilization process (except combustion) to eliminate them from the final products. These materials can possibly constitute a source of by-products such as adhesives and purified resins. Otherwise softwood (pine) trees have about 42% cellulose, 21% hemicellulose, and 26% lignin. In addition to these proportions, the individual sugars found in pine are found to contain elevated levels (about 11%) of the simple six-carbon-sugar mannose (NREL biomass database).

Hardwood trees including oak, maple, eucalyptus, and poplar species, as well as exotic species such as mahogany and teak, are common to many parts of the world. In the United States, the black cottonwood tree (*Populus trichocarpa*) has been chosen by the U.S. Department of Energy as a model hardwood energy crop because of its fast growth and relative ease of management in a forest, including harvesting (Sannigrahi et al., 2010). Hardwood trees commonly have about 44–50% cellulose, 15–19% hemicellulose, and 18–26% lignin; *P. trichocarpa* fits this composition average with ~49% cellulose, ~15% hemicellulose, and ~22% lignin (Zhao et al., 2012; NREL biomass database).

Herbaceous species include all food and feed crops such as corn, soybeans, wheat/oats, alfalfa, and other hays, as well as potential dedicated energy crops such as switchgrass and other high-yielding prairie grasses. Although the food crop itself (grains and seeds) is not considered for biomass applications, each of these crops has a portion of the plant that is nonfood and yields agricultural residues that are important sources of biomass for conversion to additional products. In the United States, the prairie grass switchgrass (*Panicum virgatum*) has been chosen by the U.S. Department of Energy as a model hardwood energy crop because of its fast growth (Sokhansanj et al., 2009; Mclaughlin and Kszos, 2005). Herbaceous agricultural residues (corn stover, sugarcane bagasse, wheat straw) have about 38–41% cellulose, 21% hemicellulose, and 17–25% lignin (Zhao et al., 2012; NREL biomass database). Additional types of biomass include postprocessing wastes from paper mills, such as sludges and mill wastes, construction and demolition debris from the building industry, forestry residue (slash), and postconsumer waste such as municipal solid waste, plus segregated recycled waste. Each of these classes of waste is highly heterogeneous and varies depending on the source, but the technologies described in this chapter apply to these materials after special treatments have been completed. Special treatments include removal of nonbiomass matter such as metal and nonwood construction debris, as well as removal of plastic matter from municipal solid waste, for example.

TREATMENT OF BIOMASS

Biomass is naturally recalcitrant to breakdown because plant species evolved to be able to resist the impacts of weather, insects, and microbial attack as they grew in the open environment. In addition, the need for structural strength, especially with woody species, contributed to their resistances to rapid breakdown to individual chemical substituents. As a result, technologies have been developed during the past couple decades to produce simple components that are amenable to bioconversion or possibly a combined biochemical/chemical conversion to value-added fuels and chemicals. This overall technology is called *pretreatment* and it comprises more than a half dozen well-developed technologies, each with specific benefits and shortcomings, and here the leading methods are reviewed.

MECHANICAL TREATMENTS

The diverse types of biomass described earlier arrive from the source in a variety of forms and sizes. To be able to process these materials in a commercial facility or in the laboratory requires size reduction, or milling. Commercial size requirements are less stringent than in the laboratory because of the large scale. Additional size reduction contributes to added costs, estimated at $1.60/metric ton (Sokhansanj et al., 2009). Laboratory evaluation of any source of biomass requires milling to an acceptable size. A common process is milling the dried biomass through a mill that includes a screen. Specifically, we have used drying at 45 °C to completion before milling through a Wiley mill with a 20-mesh (0.84-mm) screen. The plant matter can be screened further to remove very small particles if desired, because the smallest particles are more susceptible to biochemical degradation and may bias the biomass process conversion results. Specifically, this is recommended if the plant material is particularly dusty, indicating the presence of considerable levels of fine particles. Other milling approaches can be used that yield different size distributions, but when an acceptable protocol is determined, the milling process should not be varied because it is an important variable in any conversion process as a result of the need for consistent biomass particle size and surface area.

NO PRETREATMENT

The elimination of the pretreatment step during initial processing of biomass is highly beneficial because of the high cost of this step (Lynd et al., 2008; Foust et al., 2009), but this can only be possible from an economic standpoint if certain conversion goals are met. Of course, there are numerous microorganisms in nature that readily consume unpretreated biomass as a source of food and energy, but this is accomplished without regard to rate or degree of carbohydrate conversion. A common goal for initial industrial processes might be utilization of 80% of the biomass' carbohydrates, with a future target being 90% (USDOE MYPP, 2011). Despite this requirement, a few publications have claimed selected microorganisms are capable of conversion of biomass without pretreatment (Yang et al., 2009; Kataeva et al., 2013). Reported results include 26% conversion of switchgrass-insoluble carbohydrates and 15% of *P. trichocarpa*-insoluble carbohydrates by *Caldicellulosiruptor bescii* (Yang et al., 2009). This degree of biomass conversion is completely unacceptable with regard to commercialization. Although *Caldicellulosiruptor* sp. microorganisms have an interesting portfolio of biomass-degrading enzymes, until biomass is developed that lacks typical recalcitrance, pretreatment is an important key requirement for a commercially viable process for production of fuels and chemicals from biomass.

CHEMICAL/THERMAL PRETREATMENT

Although some pretreatments at ambient temperatures have been developed, such as lime pretreatment, these are the exception. Pretreatment typically comprises the exposure of biomass to elevated temperatures under a variety chemical conditions, including acidic, neutral, and basic. Pretreatment methods include steam explosion, dilute acid, autohydrolysis, ammonia fiber expansion (AFEX), aqueous ammonia, liquid hot water, lime, sulfite, phosphoric acid, and ionic liquid pretreatment (Grous et al., 1986; Grethlein and Converse, 1991; Balan et al., 2009; Kim et al., 2009a, 2009b; Yang and Wyman, 2009; Yang et al., 2009; Sierra et al., 2009; Ewanick et al., 2007; Zhang et al., 2007; Kilpeläinen et al., 2007). These pretreatment methods were developed during past decades and most of them have been evaluated carefully as a collaborative effort to discern the best approach for particular biomass conversion challenges (Mosier et al., 2005; Wyman et al., 2005, 2009). The challenge for all these methods is whether they can be scaled up efficiently and economically, and the determination of the impact of each of these pretreatment processes on the overall economics of the process.

ACIDIC PRETREATMENT

One of the most common pretreatments is dilute acid pretreatment, which involves exposure of biomass to 0.1–1% sulfuric acid at elevated temperatures of 140–200 °C for 5–30 min or longer. This method yields highly digestible biomass solids plus a liquid fraction that contains most of the hemicellulose sugars as monomeric or oligomeric (small chains of sugars) molecules. Conditions must be determined carefully for each biomass source, because excessive pretreatment (i.e., 1% acid, 200 °C, 30 min) yields acid degradation products of both glucose and xylose, and also produces degradation products of lignin (Klinke et al., 2004; Larsson et al., 1999). The generation of these degradation products not only decreases the yield of product such as ethanol or butanol, but also the sugar degradation products are not fermentable. Furthermore, these acid degradation products also inhibit the microorganism—whether it is a yeast, bacterium, or fungi—performing the bioconversion. The concern of overexposure of biomass to harsh conditions is true for all pretreatments methods, so each must be fine-tuned for the biomass selected for bioconversion.

NEUTRAL PRETREATMENT

Neutral or near-neutral pH pretreatments such as hot-water pretreatment have been developed to minimize the inhibitor impact characteristic of highly acidic pretreatments, but often they also are acidic (pH 4–6) because of the presence of the aforementioned acetyl groups that, on release from the hemicellulose, generate acetic acid. These hot-water pretreatments are milder than acidic pretreatments but still require elevated temperatures and additional incubation time to yield the same degree of digestibility of the solids (Allen et al., 2001). The steam explosion pretreatment is commonly accomplished by exposure to steam followed by a rapid release of pressure. The rapid decompression of the biomass permits expansion of the plant material structure not unlike popcorn created from solid starch. This approach is readily conducted in the pilot plant and at full scale, but it is less attractive at small scales because of equipment and safety concerns. Therefore, a common practice is to cool the biomass after exposure to high temperatures, thus reducing the high pressures generated by the high temperature. Special apparatuses and test reactors are used, which are described later.

ALKALINE PRETREATMENT

Alkaline pretreatment methods include lime pretreatment, which is an ambient-temperature process during which biomass is exposed to lime for extended periods of time, permitting a slow, controlled breakdown. Laboratory methods to evaluate this process have been developed, as mentioned earlier (Sierra et al., 2009). Other pretreatment methods using elevated pH conditions include aqueous ammonia pretreatment, which is similar to the dilute acid and neutral hot-water pretreatment methods, except that ammonium hydroxide is the chemical catalyst (Kim et al., 2009a). More complicated but elegant is AFEX, which uses gaseous ammonia as a catalyst combined with the rapid decompression process mentioned previously (Balan et al., 2009). This combination facilitates removal of most of the gaseous ammonia, permitting its recovery, while generating biomass with improved susceptibility to bioconversion (Lau et al., 2008). Tests have shown AFEX generates fewer inhibitory compounds than other pretreatments, with the additional benefit that any residual ammonia can act as a nitrogen source for the subsequent fermentation. AFEX test equipment has the same limitations as the steam explosion apparatus, but can be used in a dedicated laboratory with properly trained personnel.

ORGANIC PRETREATMENT
IONIC LIQUID PRETREATMENT

Ionic liquids (ILs) are organic salts that have an ionic organic portion plus as inorganic counterion and, as the name suggests, they are liquid at moderate temperatures. These new chemicals are currently being evaluated carefully as a potential method of disassembling biomass, because it has been shown that these unique chemicals can dissolve biomass (Swatloski et al., 2002; Sun et al., 2009). For example 1-butyl-3-methylimidazolium chloride and 1-allyl-3-methylimidazolium chloride have been shown to dissolve wood biomass up to 8% when the material is milled (Kilpeläinen et al., 2007). A technoeconomic analysis of IL pretreatment (Klein-Marcuschamer et al., 2011) provided an important evaluation of the benefits and significant challenges facing the use of ILs on a large scale. For example, it takes considerable amounts of IL to dissolve biomass (Kilpeläinen et al., 2007), they are very expensive (Mora-Pale et al., 2011), and their presence inhibits biomass enzymes and fermentation microorganisms (Turner et al., 2003; Docherty and Kulpa, 2005). Last, their removal is difficult, yielding significant levels of ILs in wastewater, which are difficult to remove by waste treatment systems (Gericke et al., 2012; Zhu et al., 2012, p. 2013a,b). These issues limit the economic utility of these otherwise interesting chemicals until significant progress is made in reducing the costs and increasing their recovery, plus improving significantly the processes using these chemicals.

ORGANOSOLV PRETREATMENT

Organosolv processes involve removal of lignin from the carbohydrates in plant matter, and have been developed in support of the pulping industry (Kleinert, 1974). The National Renewable Energy Laboratory investigated organosolv processing (Chum et al., 1985, p. 1988, 1990) for both biofuel development and for developing uses of lignin. Regarding biomass processing, organosolv processing does improve cellulose and hemicellulose processing, but there are significant costs associated with the solvents purchased, as well as very costly capital and processing costs associated with extensive removal of the solvents from the residual biomass while controlling volatile organic emissions.

Therefore, as reviewed by Zhao et al. (Zhao et al., 2009), organosolv pretreatment is too expensive to be commercialized on a large scale for biofuel production.

BIOLOGICAL PRETREATMENT

Biological pretreatment involves minimal energy input and depends on incubation of the biomass with selected microorganism that produce extracellular enzymes that modify the biomass, improving its ability to be used for either biological or thermochemical processing (Chen et al., 2010). The largest use of biological pretreatment involves biopulping with indigenous microorganism (Akhtar et al., 1998). However, early work showed the advantage of adding fungal organisms to accelerate and control the process (Eriksson and Vallander, 1982). Preferred inoculums are various fungal microorganisms, with white rot fungi having significant deligninification activities (Zeng et al., 2011). However, the time for processing varies from 28 to 60 days, as reviewed by Chen et al. (2010); therefore, biological pretreatments are viewed as too long to be economically viable for biofuel production.

EXAMPLES OF SMALL-SCALE PRETREATMENT

Acidic, neutral, or aqueous alkaline pretreatment of biomass can be conducted readily on small amounts of biomass using a combination of methods, permitting evaluation of new biomass sources (Kim et al., 2009b; Yang et al., 2009; Balan et al., 2009). A consideration for pretreatment planning and eventual fermentation is how much material is needed for all steps in the process. Specifically, carbohydrate and possibly the lignin content of the biomass needs to be determined before and after pretreatment, along with the composition of the residues after fermentation. The biomass carbohydrate and lignin composition can be determined routinely with 0.1 g biomass on a dry basis (db) in triplicate (requiring 0.3 g). Pretreatment often results in a 40–50% loss of solid material, depending on the method, with acidic conditions typically losing more solids. Last, if the fermentation converts 50% of the solids to product, additional starting material is needed to complete a full evaluation. Methods developed at Oak Ridge National Laboratory (ORNL) Biosciences Division for pretreatment based on published protocols have determined the minimum amount of dry biomass for pretreatment, fermentation, and residue analysis is 10 g (db) (Mielenz et al., 2009; Fu et al., 2011; Yee et al., 2012).

TYPICAL PROTOCOL FOR BIOMASS PRETREATMENT

1. Specifically, pretreatment is started by soaking 10 g (db) of the subject biomass overnight in 100 mL of the appropriate liquid, such as 0.5% sulfuric acid, water, or other chemical solution to ensure penetration of the liquid into the dried biomass.
2. The biomass is separated from the liquid by filtration or centrifugation. The liquid is placed in a weighed (tared) container, as are the solids, to be able to track the materials.
3. The wet biomass is divided into three equal portions for pretreatment. The reactors for small-scale pretreatment are 0.5-inch-diameter Hastelloy C276 steel tubes sealed by a stainless steel Swagelok threaded coupling (Yang et al., 2009). The caps are not available in Hastelloy metal, so to protect the couplings from acid the ends are sealed with a Teflon plug (0.5 in × 3 mm) cut from a 0.5-inch-diameter Teflon rod. The tube length varies, depending on the size of the sand bath

used to heat the reactors, but this example uses a 4-inch tube. The tube is sealed at one end in a shop vise with an adjustable crescent wrench, then the other end cap is attached loosely and the reactor is weighed. The biomass is added with a funnel and is pressed in with a wood rod to form a solid plug with minimal liquid. The end is sealed with the end coupling in the vise. The seal must be tight because of the high pressure obtained at elevated temperatures. The reactor is reweighed to determine the biomass wet weight.

4. The reactors are placed securely in a wire holder, with up to four reactors containing the 10 g prepared as stated. Fluidized sand baths are required to reach temperatures of 140–200 °C safely, and ORNL uses an Omega FSB1 (Techne Co.) sand bath with a temperature regulator able to hold 4-inch tubes. Larger sand baths are available. The reactors are preheated in boiling water for 2 min to accelerate the heating in the sand bath. If available, a second sand bath is set 20 °C above the target temperature and the reactors are transferred into the bath for 1 min before placement in the primary reactor for the time required (Yang et al., 2009). This procedure minimizes the expected drop in the temperature of the sand bath as the reactors heat up to the target temperature. If desired, the temperature of the sand bath is recorded manually or automatically with a recorder.

5. After the desired incubation time in the primary sand bath, the reactors are plunged into ice water for 2 min with agitation initially to quickly quench the thermal treatment. The contents are removed from the dried reactors with a wood rod and are place into 50-mL disposable centrifuge tubes. Interestingly, some biomass (such as switchgrass) is removed easily whereas woody biomass may require effort to eject the solids because of swelling.

6. The pretreated biomass is either used directly for fermentation after any required pH adjustment or it is used for washing. Evaluation of the solids in the absence of free sugars can be beneficial to evaluate differences in the solid substrate, so washing removes solubilized sugars and other materials that might impact fermentation. The initial wash involves addition of 10 mL water/g starting material, so typically 25 mL is added to the contents of one reactor in the tube. After centrifugation (10,000 g for 10 min), the liquid is removed for analysis. The biomass is washed further with a total of 100 mL/g biomass, conveniently in a larger centrifuge bottle or other container if filtration is used. Filtration of biomass uses three layers of milk filters cut to the size needed (KenAg filter disks).

7. After pretreatment, a portion can be removed for compositional analysis. Samples can be taken before and after washing. It is convenient to combine the biomass for the four-reactor set, mix, and then remove three approximately 0.5-g wet-weight samples into individual drying pans for compositional analysis after drying at 45 °C. This yields approx. 0.1 g db each, which generates db data as well as samples for carbohydrate and/or lignin analysis (discussed next).

There are numerous other approaches and scales for pretreatment, so the previous example is intended to provide a guide for a small-scale pretreatment that is flexible yet effective for the evaluation of different types and sources of biomass.

BIOMASS ANALYSIS

A critical aspect to working with solid substrates such as biomass is the analysis of the composition of the substrate both at the beginning and as the processing proceeds, especially regarding the impact of pretreatment. One vital analysis is the determination of the db of the biomass, which is simple to perform either manually or via automatic moisture detections apparatuses. The reason this is important is that, without moisture determination during processing, the material flow cannot be followed. Packed, wet biomass, after either centrifugation or filtration to the point where no free liquid is visible, typically has, at best, 15% db and, often, only 20% dry matter. Although the methods are simple enough not to be described, it is important to remember as one proceeds in processing and analyzing the biomass that the 80–85% liquid portion contains soluble substituents (sugars and so forth) of the bulk free liquid, and these materials should be included in compositional evaluation. Also, if the content of the solids is to be determined only, extensive washing with water or buffer is necessary to remove the soluble substituents from the bulk fluid.

As part of the analysis, it is critical to weigh both the liquid and solids as the various fractions are generated, so the data are available if analysis of material balance is needed. The general rule is track weights even if you do not think the information is required, so retracing the process and material flow is possible should results require further analysis. The value of evaluating the material balance was demonstrated when multiple pretreatment methods were analyzed in a collaborative effort (Garlock et al., 2011), and such analysis should be instructive to any research effort. Determination of the composition of biomass substrates as they are processed is important because the composition is used to evaluate the effectiveness of processing steps as well as to determine the dose of enzymes. The key analytical methods were developed by the National Renewable Energy Laboratory (NREL) and were adopted by the American Society for Testing and Materials (ASTM) as method ASTM E 1758–01 (ASTM 2003) and high-performance liquid chromatography (HPLC) method NREL/TP 51–42623, which is the method to determine the composition of biomass titled "Determination of Structural Carbohydrates and Lignin in Biomass." Although the referenced material has significant detail, some specific approaches are provided here in simplified form.

The process is, essentially, the acid hydrolysis of carbohydrates and lignin followed by the determination of each of the carbohydrate sugars produced by this quantitative saccharification analyzed by HPLC. Lignin is analyzed by determining acid-insoluble lignin by muffle furnace degradation, and determining the acid-soluble lignin by absorbance with an ultraviolet (UV)-visible spectrophotometer. For much of the research aimed at fermentation, the critical data are the carbohydrate composition determined by this quantitative saccharification method. Work at ORNL and elsewhere has confirmed the carbohydrate method can be scaled down from 300 mg of dry biomass to only 100 mg of dry biomass, especially if the analysis is done in triplicate, which is routine. This is particularly helpful when the fermentation is scaled to 10 g starting material (db), as is common when biomass materials are analyzed throughout the process. The issue becomes even more important when new plant variants are evaluated by fermentation, and the total biomass supply from greenhouse-grown lines is limited to tens of grams only.

It has been found the quantitative saccharification method can be completed easily for up to five samples in triplicate with standards, although more processing manually is possible. As mentioned before, weights of the samples at the various steps are recorded using premade data templates, again to assist in material balance tracking. Safety considerations are of utmost importance and require the notification of other researchers that the autoclave, which is used for 1 h for the final hydrolysis, is reserved and to be opened only by the person performing the tests to avoid others handling glass pressure tubes containing sulfuric acid. Last, it is important to neutralize the samples carefully with calcium carbonate, which does not modify the sample volume but releases carbon dioxide (CO_2). So, the calcium carbonate must be added in small batches (50 mg), because of bubbling, until neutral pH is obtained. The sugar content was determined by HPLC analysis with the Bio-Rad Aminex HPX-87P column equipped with the appropriate guard column for sugar determination. This column is run with a mobile phase of degassed distilled water with neutralized samples as acid content in the mobile phase or sample can be deleterious. We have determined that both the quality and longevity of the guard columns and the main analysis column can contribute to sugar peak shape deterioration, so both parts of the column train need to be well maintained. When operating well, we are able to track and analyze quantitatively five to six sugars in biomass hydrolysates: glucose, xylose, arabinose, galactose, mannose, and fructose, if present. In addition, the sugar breakdown compounds hydroxymethylfurfural (HMF) from glucose and furfural from xylose acid degradation can be quantified with the same Aminex HPX-87P column separation, so evaluation of their levels in the hydrolysate is important. It should be mentioned that the standards for these compounds are light sensitive, so they are prepared in a darkened room (shades drawn and the lights off). Notably, it is important to note that fructose does not survive the quantitative saccharification analysis, as shown by work from Penner's laboratory (Nguyen, 2009), so its quantification must be determined after milder hydrolysis methods. Last, one should run multiple concentrations routinely of the expected sugars and furfurals both before and after the samples are analyzed to obtain proper quantification of the sugars of interest as well as to confirm the HPLC sugar separation is running properly because sequences can run for multiple days.

Quantitative saccharification analysis is required after pretreatment to determine the degree of hemicellulose release and the resulting increase in the less susceptible cellulosic materials. Figure 24.1 shows the result for biomass before and after dilute acid pretreatment and washing of the solids. The results show a significant decrease in xylose in the solids on a weight basis, and an apparent compensating increase the level of cellulose. This occurs as a result of the high degree of susceptibility of the hemicellulose xylose linkages compared with cellulose glucose linkages. The increase in cellulose content is particularly important if use of cellulases is anticipated, because enzyme dosing is typically done on a cellulose-content basis. In addition, analysis with the P column detects furfural and HMF levels, and a severe pretreatment generates these compounds. If elevated levels of furfural and HMF are detected in the quantitative saccharification analysis of the solids, the pretreatment liquid can be analyzed by HPLC with the P column (after neutralization) to confirm the severity of pretreatment. The ideal pretreatment generates low levels of various acid degradation products (Palmqvist E, and Hahn-Hägerdal, 2000).

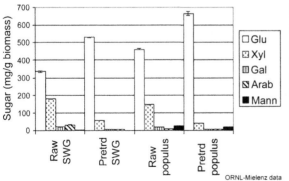

ORNL-Mielenz data

FIGURE 24.1

Sugar composition of raw and pretreated (Pretrd) switchgrass (SWG) and *Populus*. One standard deviation symbol is shown. arab, arabinose; gal, galactose; glu, glucose; mann, mannose; xyl, xylose.

SMALL-SCALE BIOMASS FERMENTATION APPROACHES

Small-scale fermentation of biomass is most useful when the supply of the feedstock is limited since larger scale, ie: multiple liter fermentations, for example, may require more substrate than available. The best examples of this are newly developed transgenic or natural-variant lines grown in the greenhouse on a small scale. Research at ORNL has been evaluating biomass sources for the ORNL-led BioEnergy Science Center funded by the U.S. Department of Energy Office of Science. These sources are primarily of two types—switchgrass (*P. virgatum*) and black cottonwood (*P. trichocarpa*), including both transgenic and natural variants of these species. As part of the research, BioEnergy Science Center has access to high-throughput pretreatment and enzymatic screening, developed at NREL and the University of California Riverside (Selig et al., 2010; Demartini et al., 2011), using a robotic high-throughput protocol needed for analysis of large numbers of samples (Sykes et al., 2009). These methods have been used to evaluate more than 1100 natural-variant poplar samples taken from the northwestern United States and Canada, and these high-throughput methods have determined lignin content has an impact on in vitro enzymatic sugar release (Studer et al., 2011). Similarly, generation of transgenic plants such as alfalfa and switchgrass has produced numerous variant plants grown in the greenhouse that have been subjected to sugar release assays (Chen and Dixon, 2007; Dien et al., 2009). The rapid analysis results with high-throughput screening are important to begin to select promising plant variants to develop for eventual tests in the field. A required next step is the testing of the plant material by fermentation to confirm the selection based on physical characteristics such as carbohydrate and lignin content, and in vitro enzymatic sugar release. Fermentation provides important information on both biomass conversion productivity and identification of potential inhibitors in the biomass extractives and possibly produced during fermentation.

The starting biomass is from either transgenic or natural-variant switchgrass or *Populus* lines, and these materials are available initially in small quantities, with 50–100 g being the levels provided for any given transgenic line. With the natural-variant *Populus*, more material is available, but only after a longer cultivation time as a result of slower growth rates for the woody material. The limited amounts of biomass necessitated development of small-scale fermentation processes that were reliable and reproducible, and that would work with the aforementioned pretreatment protocols. The overall goal was to be able to ferment the biomass with existing methods in triplicate to obtain solid data regarding the performance of the specific biomass samples relative to standard materials, and this has been accomplished using three different fermentation processes.

Fermentation of biomass carbohydrates requires the hydrolysis of the polymeric cellulose and hemicellulose to fermentable sugars. This hydrolysis can occur essentially three ways: (1) use of industrial enzymes such as cellulases, hemicellulases, β-glucosidases, and pectinases either during the fermentation with the fermentative microorganism (termed *simultaneous saccharification and fermentation* [SSF]), at a temperature comfortable for the microorganism, (2) use of these enzymes in the absence of the fermentative microorganism at a higher temperature closer to the temperature optimum of the enzymes. After hydrolysis, the temperature of the mixture is lowered to match the optimum of the fermentation microorganism. This process is called *separate hydrolysis and fermentation* (SHF). For ethanol production from biomass, currently readily available microorganisms include various *Saccharomyces* yeast strains, or genetically improved strains of *Escherchia coli* or *Zymomonas mobilis*, all are capable of producing ethanol from (most) biomass sugars. The third (3) approach is fermentation of the solids by a microorganism capable of producing the aforementioned enzymes during the fermentation process, resulting in the formation of carbohydrate hydrolysis products. This approach is termed *consolidated bioprocessing* (CBP) or possibly *direct microbial conversion*. The advantage of this approach is that the fermentation temperature is acceptable for both the viability and growth of the microorganism, and the production and activity of the enzyme portfolio. Also, no separate source of industrial enzymes is needed because they are produced by the microorganism in vivo. All three of these approaches provide useful information regarding the evaluation of biomass feedstocks for commercial production of fuels and chemicals. As with the description of the scaled-down pretreatment, details are provided here regarding these scaled-down fermentation approaches. It should be noted that the biomass used can be either unpretreated or pretreated direct microbial conversion (DMC).

EXAMPLE OF SIMULTANEOUS SSF

1. Biomass should be milled, as discussed previously, to a set size (see Mechanical Treatment). This biomass can be provided either dry or wet with a known db so the amount of dry biomass is known in the beginning. The latter is common for pretreated biomass produced previously. In addition, the carbohydrate—in particular, the cellulose content—must be known for the samples if the cellulase is to be dosed based on a cellulose level (i.e., units or enzyme weight per gram cellulose). Dose of the enzymes based on total biomass is possible, but substrates with different cellulose levels based on inherent structural differences or susceptibility to pretreatment may cause misleading results unless the cellulose level is considered. Fermentation occurs in small, sealable containers such as 100-mL serum vials or sealable bottles. Fermentation can be tracked routinely by weight loss by venting liberated CO_2, so containers should have a serum seal capable of being punctured by a needle. Biomass equivalent to 1 g (db) is added to previously weighed vials or bottle, and the weight is recorded again to obtain exact biomass weight in each bottle. Bottles are prepared in triplicate for each biomass sample. Always include a no-biomass control in triplicate that contains all the components except the biomass to permit detection of fermentable substrates in the enzymes or other issues with the fermentation, such as residual sugars or product in the inoculum.

2. For yeast SSF, the stock buffer is 1 M sodium citrate (pH 4.8), and buffer and water are added to the container based on the final volume chosen to reach a 50-mM concentration of the buffer. For the SSF approach with other microorganisms requiring a different pH, such as *E. coli* or *Zymomonas*, use a concentrated buffer stock at the desired pH. In addition, the supporting medium of other microorganisms must be determined individually. Use of a concentrated source of nutrients is recommended to minimize the final volume. A convenient final volume is 20 mL, which yields a 5% biomass loading using 1 g. The bottles are sealed loosely and autoclaved for 30 min under standard conditions. Before autoclaving, care must be taken to minimize biomass clinging to the sides of the container because the process of autoclaving causes the biomass to be stuck to the sides and is not available for fermentation. Use a spatula, if needed, to remove the clinging biomass before autoclaving or, better yet, do not tip over the containers.

3. The cooled containers are dosed with the following ingredients (for yeast SSF) to initiate the fermentation: 10% yeast extract, which has been sterilized, added to 0.5%; 0.5 mL of an overnight culture of the fermentative microorganism; and any additional water added if water was

lost during autoclaving. The industrial enzymes are added as a bulk mixture prepared just before use. Typical enzyme mixtures contain 15 filter paper units/g cellulose. Other protocols add enzymes based on protein weight, but the key for comparative studies is to add the cellulase at a consistent dose because its level impacts fermentation results significantly. Hemicellulose, β-glucosidases, and pectinases (if used) can be added at one-quarter the volume of the cellulase, or as the manufacturer recommends. The enzyme mix can be altered by the addition of other components. For yeast fermentations, streptomycin to 62.5 μg/mL (50 μL of a stock 25 mg/mL into 20 mL) can be included as a precautionary measure to minimize mesophilic anaerobic growth, but this can be skipped if desired. The inoculum size for the fermentative microorganism can be varied as desired. Determine routinely the optical density at 600 nm (OD) of the culture before use. Yeast cultures grown overnight in YPD broth can reach more than the 10 OD600-nm units. Make sure to prepare and freeze a portion of the inoculum broth after cell removal to determine the residual sugar and product concentration in the inoculum.

4. Seal the fermentation containers and record the time zero (T_0) weight before starting the incubation at the desired temperature. To assist in providing full access of all the solids to the enzymes and fermentative microorganism, the containers are kept upright to minimize coating of the sides with biomass, because as it is difficult for it to drop down into the bulk liquid for fermentation. Shaking at 100–125 rpm is the normal range.

5. Track the weight loss routinely by piercing the serum top with a 25-gauge needle and permitting the gas (CO_2 for yeast) to escape. Allow the venting to go 20 s or longer, initially, if the venting noise is still present. Do not allow the containers to vent longer than this because, as the container cools, the internal gases contract, drawing in air. After venting, weigh and record the results. To minimize differential cooling, vent and weigh containers in groups of six or less. The fermentation is most active during the first 24–48 h, but continue venting and weighing until the fermentation weight loss profile has essentially flattened out. Venting at approximately18 h and 24 h, and then daily after that should provide excellent data on the progress of the fermentation. Figure 24.2 shows a fermentation profile for switchgrass and *Populus* during SSF bioconversion

FIGURE 24.2

Fermentation time course for switchgrass and *Populus* using saccharification and fermentation processing. End point ethanol concentration in fermentation is shown.

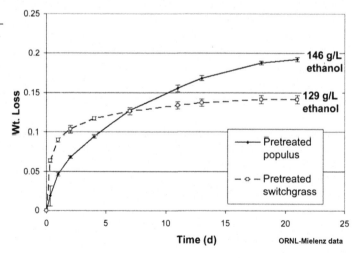

ORNL-Mielenz data

and, as can be seen readily, these two substrates respond differently in the bioconversion process. Without tracking the weight and relying just on end point data, we would not have identified when the end point of the bioconversion occurred nor would we be able to provide an understanding of how the fermentation progressed.

6. Upon completion of the fermentation, mix and pour the contents into an appropriate centrifuge container. Use of a tared, 50-mL disposable centrifuge tube is recommended. Centrifuge at 10,000 g for 20 min to separate the solids from the broth, and pour off the liquid into a separate tared centrifuge tube. Record the weights of all tubes and either process immediately or freeze for later analysis. As a side note, ORNL determined that, after pretreatment, biomass can be frozen and thawed at least five times without impacting the SSF fermentation results.

EXAMPLE OF SHF

1. The preparation of the fermentation containers is essentially the same when hydrolysis occurs before the fermentation, except the requirement for venting is eliminated during the first step. Prepare the containers with biomass, buffer, and water before autoclaving, including taking care not to suspend biomass on the sides of the hydrolysis container. After autoclaving and cooling, add the desired enzyme cocktail, with dosing again based on cellulose or possibly total biomass if preferred. Incubate the containers at the desired temperature at 80 rpm to avoid foaming. The concern regarding biomass sticking on the sides of the vessel is less important, so incubate the containers on their side to provide for better mixing.

2. Incubate for up to 5 days. With Novozymes enzymes, company literature suggests that hydrolysis is expected to be completed in 4–5 days. On a daily basis, examine the bottles and swirl gently off the sides any clinging biomass. Within 24 h, the viscosity of the hydrolysate drops significantly. After the incubation time has elapsed, remove and cool the containers. A well-mixed sample (1 mL) should be removed and frozen for analysis of the free sugar content. For specific biomass sources, optimal hydrolysis time can be determined by setting up 10 identical hydrolysis bottles and removing a pair of them sequentially, daily, and determining the concentration of free sugar in the liquid portion after removal of the solids.

3. Fermentation of the biomass can occur using either total hydrolysate or hydrolysate of the liberated sugars with exclusion of the solids. If the hydrolysis is complete, the former approach is very rapid and provides results quickly. If the substrate is more difficult to hydrolyze, initiating fermentation with the solid provides additional time for the enzymes to continue to act while the fermentative microorganism converts the free sugars to product, which may provide better results. As shown in Figure 24.2, it is possible switchgrass is more amenable to hydrolysis and fermentation of the liquid only, because hydrolysis occurs rapidly, whereas a woody substrate—here, *Populus*—may be best fermented with the solids present, permitting additional enzyme hydrolysis. Regardless, preparation for fermentation is the same; a portion or all of the hydrolysate (with or without solids removed by centrifugation) is transferred to a sterile, weighed fermentation container that has a ventable top, such as serum vials. For yeast conversion, yeast extract is added to 0.5% volume/volume, and 0.5 mL of an overnight yeast culture is added along with the recommended streptomycin and water as desired to reach the preferred volume. As mentioned earlier, other microorganisms need their own concentrated nutrients to be added. Determine the T_0 weight as done during the SSF process to be able to track progress.

4. Incubation should be at the optimum temperature for the fermentative microorganism, with shaking to keep any solids and the microorganism suspended. Vent at 6–12 h because the fermentation is very rapid, and continue to monitor until the fermentation is complete. On completion of the fermentation, mix and pour the contents into an appropriate centrifuge container. Prepare the fermentation broth and any solids as described in step 6 for SSF.

EXAMPLE OF CBP

CBP requires the fermentative microorganism to be able to produce cellulases, hemicellulases, and accessory enzymes needed to access polymeric biomass carbohydrates as well as use the simple sugars it generates to produce fermentation products. There are a number of natural microorganisms that have this capability, although modern genetic modification techniques are developing additional CBP candidates. The natural CBP microorganisms used thus far for biofuels production are anaerobic bacteria, and although these are capable CBP microorganisms, they use the biomass breakdown products for their own growth and metabolism. The tests described in the following example use the natural CBP microorganism *Clostridium thermocellum*, which is capable of rapid cellulose and biomass degradation while producing acetic acid, ethanol, and small amounts of lactic acid. Other candidate microorganisms include *Caldicellulosiruptor* and *Thermoanaerobium/Thermoanaerobacterium* species, among others. In addition, progress has been made developing yeast stains as CBP-capable microorganisms (Ilmén et al., 2011; Khramtsov et al., 2011; Olson et al., 2012). The primary difference between the SSF and SHF approaches is the CBP microorganisms are strict anaerobic bacteria that require addition preparation for successful fermentation in the absence of oxygen. Among the important requirements is oxygen-free nitrogen for sweeping out oxygen as the fermentation bottles are prepared, and additional details are provided by Strobel (2009).

1. Prepare serum bottles as described in the SSF description, step 1. Be prepared to use substantial rubber stoppers, such as black rubber, to improve the airtight seal for strict anaerobes. An inoculum should be prepared and available that has reached the near-stationary phase but is still growing. You may need to modify this step based on specific requirements of the chosen microorganism.

2. Add water to the biomass based on the planned final volume, and the anticipated medium and cell inoculum volume. The medium for the specific microorganism should be prepared at two times strength at least to facilitate preparation of the bottles on a small scale. After adding the water, seal the vials and exchange or blow out the air using either a vacuum or fill station with oxygen-free nitrogen, or by passing oxygen-free nitrogen through the bottle with a venting needle for 10 min. This begins the process of oxygen removal. After gas exchange, autoclave the bottles for 20 min. While the bottles are still hot from the autoclaving, using thermal gloves repeat the oxygen removal step just described by passing oxygen-free nitrogen through the bottle with a venting needle for 20 min. Autoclaving forces gases out of the liquid, making oxygen more easy to remove.

3. When cool, inject the preplanned volume of medium and an inoculum. If an anaerobic chamber is available, you can mix the inoculum carefully with the concentrated medium and inject them together using a vent needle to avoid back-pressure. Be sure to protect the work surface with an adsorbent pad in case of a spill of the selected inoculum. It is beneficial if the medium contains a

reducing agent such as cysteine to reduce any residual oxygen present. Use of resazurin is helpful to note oxygen presence (Yee et al., 2012).

4. Place the completed vials in an incubator shaker at the preferred temperature after measuring the weight because, as with the aforementioned fermentations, the weight loss can be used to follow the fermentation. Initiate venting as discussed in SSF step 5. The primary difference is the venting should be done in an anaerobic chamber to eliminate the addition of oxygen, possibly from the needle itself. This is especially important when working with thermophiles because considerable cooling cannot be avoided. It is convenient to heat the vials to running temperature at the start and vent in the anaerobic chamber, thus equalizing the pressure that occurs as a result of heating at the beginning, avoiding an apparent weight loss resulting from heating from room temperature. When the fermentation is complete, sample the fermentation for analysis, and separate the solids and liquid as described in SSF step 6.

ANALYSIS OF FERMENTATION RESULTS

Analyses of the fermentation results are essentially identical regardless of the hydrolysis and fermentation approach used to characterize the plant material. Critical data include the carbohydrate and possibly the lignin content of the starting substrates, the same data for the substrate if it is subjected to pretreatment, and then, possibly, depending on the experimental questions being asked, the same data for the residues after fermentation. These data come from the quantitative saccharification analysis (with possible lignin analysis) described earlier using primarily HPLC equipped with a Bio-Rad Aminex HPX-87P column. Characterization of the fermentation results uses similar approaches—namely, analysis of the fermentation broth with HPLC equipped with a Bio-Rad Aminex HPX-87H column with an acidic mobile phase (usually 5 mM H_2SO_4). Fortunately, this analysis can generate quantitative data based on standards for ethanol, butanol, butyric acid, lactic acid, acetic acid, and other organic acids such as formic acid, as well as data on glucose and xylose. The other biomass sugars do not separate well enough with this column, so a separate analysis using the HPX-87P column (and distilled-water mobile phase!) is required. Therefore, it is recommended you complete analysis of the fermentation broth with both separation methods (Aminex HPX-87H and Aminex HPX-87P or equivalent columns) to extract as much data as possible regarding product yield and substrate use during the fermentation process.

IDENTIFICATION OF FERMENTATION INHIBITION

One of the benefits of conducting a fermentation analysis of the potentially interesting transgenic or natural-variant feedstock lines selected by the high-throughput screening is that an in vivo test such as fermentation can detect possible inhibitory effects of the selected biomass or its pretreatment preparation on the fermentation process or the fermentative microorganism. Such inhibitory effects cannot be detected by examining in vitro structural characteristics or the degree of cell-free enzyme hydrolysis. Such inhibition can be specific to certain fermentative microbes, as has been found during switchgrass fermentation tests at ORNL.

Research by ORNL and the Samuel Roberts Noble Foundation, funded initially by the U.S. Department of Agriculture and then by the U.S. Department of Energy, examined the fermentation of a set of lines of switchgrass modified in a single gene within the lignin pathway: the caffeic-O-methyl transferase (COMT) gene (Fu et al., 2011). Modification of this gene, when compared with wild-type switchgrass, decreased lignin about 16% in the best lines, and in vitro tests with enzyme digestion detected improved hydrolysis. These materials are good examples why small-scale pretreatment, fermentation, and analysis methods are needed, because initial samples were available in about 100-g lots, requiring small-scale methods. So, analysis at ORNL using the pretreatment and SSF conversion approaches described earlier showed that the transgenic COMT lines produced more ethanol on a biomass weight basis than the wild-type lines. In addition, the amount of added cellulase enzymes could be decreased up to fourfold with the COMT feedstock without reducing the ethanol yield produced from wild-type switchgrass with no enzyme reduction (Fu et al., 2011). The fermentation process was straightforward, with no indication of inhibition of the yeast used for the fermentations. The analysis was extended to determine whether CBP conversion yielded the same results. Fermentation of the wild-type and COMT switchgrass was undertaken with the robust, cellulose-degrading CBP microorganism *C. thermocellum*. Using the methods described earlier, CBP fermentation was started with no added cellulase enzymes. An example of a fermentation profile for *C. thermocellum* is shown in Figure 24.3. The feedstock substrate was COMT transgenic and wild-type switchgrass after dilute acid pretreatment and extensive washing to remove the acid. An initial weight loss value at T_0 should be determined by venting the bottles after a brief incubation to warm the bottles to the incubation temperature ($58\,°C$) and to remove pressure resulting from the heating. In this case (Figure 24.3), this was not done, so the impact of not pre-venting is shown at 18 h (open triangles in Figure 24.3), followed by weight loss resulting from fermentation. The improvement in yield at 36% is shown for the transgenic switchgrass, which is the same magnitude obtained with the yeast-based fermentations.

However, when the same biomass was used for fermentation after no pretreatment or only milder hot-water pretreatment, the fermentation results, based on the weight loss data, were a surprise; the COMT feedstock yielded poor results when compared with the wild-type feedstock. As shown in Figure 24.4, the transgenic COMT biomass (triangles dashed line) performed significantly poorer than the wild type (squares solid line), regardless of whether the biomass was unpretreated (open symbols) or pretreated with hot water only (180 °C, 25 min; solid symbols). The analysis of the fermentation broth provided supportive data, as shown in Figure 24.5. *C. thermocellum* produces ethanol, acetic

FIGURE 24.3

Time course comparison of fermentation of switchgrass biomass by *Clostridium thermocellum*. Weight loss includes no biomass and all were not pre-vented so a weight loss is shown at 18 h as a result of temperature differences. End point ethanol concentration in fermentation is shown. COMT, caffeic acid *O*-methyltransferase transgenic switchgrass; W-T, wild-type switchgrass.

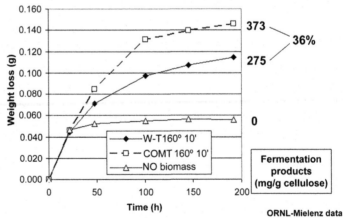

ORNL-Mielenz data

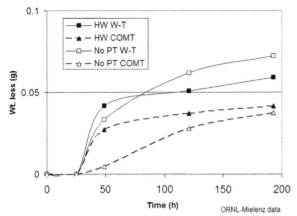

FIGURE 24.4

Time course fermentation of unpretreated (No PT) and hot-water pretreated (HW) switchgrass for wild-type (W-T) and caffeic acid *O*-methyltransferase (COMT; transgenic) switchgrass.

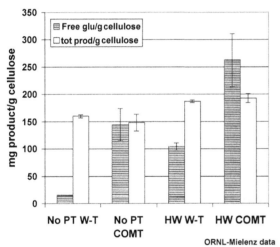

FIGURE 24.5

Productivity of *Clostridium thermocellum* fermentation of unpretreated (No PT) and hot-water pretreated (HW) switchgrass for wild-type (W-T) and caffeic acid *O*-methyltransferase (COMT; transgenic) switchgrass. Residual glucose (glu) and fermentation product are shown in milligrams per gram cellulose. tot prod, total production.

acid, and lactic acid during fermentation of glucose, galactose, and mannose, but does not ferment xylose or arabinose. Examination of the total products (ethanol, acetic acid, lactic acid) for these four fermentations showed all conditions produced nearly identical total products per gram of input cellulose for either the no pretreatment or the hot-water pretreatment, with the pretreatment producing approximately 23% more total products. The expected improvement for the COMT switchgrass did not occur. However, the explanation is apparent when the free unfermented glucose level is determined. As shown in Figure 24.5, there are significant levels of free glucose left unfermented in the unpretreated switchgrass sample and, similarly, there is unfermented glucose in both hot-water pretreated fermentation samples, but significantly more in the COMT transgenic substrate. The presence of free fermentable sugars from solid complex carbohydrate substrates indicates the enzymes were active, but the fermentation of these available sugars by *C. thermocellum* was inhibited. This is important because the enzymes were produced by *C. thermocellum*; so, apparently, production of the hydrolytic enzyme was not impaired as significantly because free sugars were liberated by their action. Indeed, although the products of fermentation were very similar for each type of substrate, the COMT transgenic substrate liberated more free sugar, which supports early data regarding superior in vitro hydrolysis of the COMT switchgrass by industrial enzymes when compared with the wild type (Fu et al., 2011).

Subsequent research has shown *C. thermocellum* is sensitive to the COMT switchgrass unless the soluble substituents are washed out of the solids for either unpretreated or pretreated samples. When the soluble substances are removed, the COMT switchgrass shows superior yield based on substrate (Yee et al., 2012). Interestingly, this work identified a second CBP thermophilic anaerobe— *Caldicellulosiruptor*—as being extremely sensitive to the COMT switchgrass regardless of the degree of washing to remove soluble substituents. It is hypothesized the inhibitor substances are generated by the action of the enzyme's portfolio found in *Caldicellulosiruptor*, which is different than *C. thermocellum*. Additional analysis with gas chromatography–mass spectrometry (GC-MS) mass spectroscopy of the composition of the fermentation broth from *C. thermocellum* detected a previously unknown intermediate in the lignin pathway for switchgrass—an iso-sinapyl alcohol (Tschaplinski et al., 2012)—but tests showed this molecule was not present in sufficient levels to explain the inhibition. It is likely that blocking an intermediate step in lignin production caused a backup of pathway substrates and possibly resulted in new side reactions. A takeaway lesson is, as mentioned earlier, that the complete process of bioconversion of solid substrates is complex, and examination of all aspects of the processes, such as full knowledge of the composition of the biomass and the kinetics of conversion shown by weight loss, and examination of the residual substrate and products can yield valuable insight into the conversion process.

CONCLUDING THOUGHTS

After more than three decades of research into bioconversion of biomass into fuels and chemicals, the various bioprocesses that have emerged are efficient and well developed to the point they can now be used as assays for evaluating the input biomass substrates. Important bioprocess procedures needed to evaluate new lines and varieties of biomass include biomass size reduction, pretreatment, and fermentation. Critical for using such unit operations is their reproducibility, robustness, and ability to differentiate between closely related biomass sources being developed either by modern plant genetic engineering or from natural variants. This chapter provides additional unpublished details that are being practiced with transgenic and natural-variant switchgrass and woody biomass (*Populus*) to facilitate other laboratories to evaluate their biomass sources. These evaluation approaches are flexible and can be modified, yet should be standardized based on individual laboratory capabilities, specific biomass requirements, and specific research goals. The overall intent is, hopefully, to accelerate development and identification of superior feedstocks needed for the emerging biomass-based economy.

ACKNOWLEDGMENTS

The author thanks ORNL for allowing his data to be used in figures after his retirement. The author thanks Choo Hamilton and Miguel Rodriquez Jr. for technical assistance obtaining data used in the figures. The COMT transgenic and wild-type switchgrass was provided by the Samuel Roberts Noble Foundation and, in particular, by

Dr. Z.- Y. Wang and Dr. C. Fu. The data in the figures are possible as a result of research funded by the BioEnergy Science Center, which is a U.S. Department of Energy Bioenergy Research Center supported by the Office of Biological and Environmental Research in the Department of Energy's Office of Science.

REFERENCES

Akhtar, M., Blanchette, R.A., Myers, G., Kirk, K.T., 1998. An overview of biomechanical pulping research. In: Young, R.A., Akhtar, M. (Eds.), Environmentally Friendly Technologies for the Pulp and Paper Industry. John Wiley & Sons, New York, pp. 309–340.

Allen, S.G., Schulman, D., Lichwa, J., Antal Jr, M.J., Jennings, E., Elander, R., 2001. A comparison of aqueous and dilute-acid single-temperature pretreatment of yellow poplar sawdust. Industrial and Engineering Chemistry Research 40 (10), 2352–2361.

ASTM E 1758-01, 2003. Determination of carbohydrates in biomass by high performance liquid chromatography. Annual Book of ASTM Standards, 2003, vol. 11.05. ASTM International, West Conshocken, PA.

Balan, V., Bals, B., Chundawat, S.P., Marshall, D., Dale, B.E., 2009. Lignocellulosic biomass pretreatment using AFEX. Methods Molecular Biology 581, 60–78.

Boerjan, W., Ralph, J., Baucher, M., 2003. Lignin biosynthesis. Annual Review of Plant Biology 54 (1), 519–546.

Chen, F., Dixon, R.A., 2007. Lignin modification improves fermentable sugar yields for biofuel production. Nature Biotechnology 25, 759–761.

Chen, S., Zhang, X., Singh, D., Yu, H., Yang, X., 2010. Biological pretreatment of lignocellulosics: potential, progress and challenges. Biofuels 1 (1), 177–199.

Chum, H.L., Douglas, L.J., Feinberg, D.A., Schroeder, H.A., 1985. Evaluation of Pretreatments for Enzymatic Hydrolysis of Cellulose. http://www.nrel.gov/docs/legosti/old/2183.pdf.

Chum, H.L., Johnson, D.K., Black, S., Baker, J., Grohmann, K., Sarkanen, K.V., Wallace, K., Schroeder, H.A., 1988. Organosolv pretreatment for enzymic hydrolysis of poplars: I. Enzyme hydrolysis of cellulosic residues. Biotechnology Bioengineering. 31, 643–649.

Chum, H.L., Johnson, D.K., Black, S., 1990. Organosolv pretreatment of poplars, 2: catalyst effect and the combined severity parameter. Industrial and Engineering Chemistry Research 29, 156–162.

DeMartini, J.D., Studer, M.H., Wyman, C.E., 2011. Small-scale and automatable high-throughput compositional analysis of biomass. Biotechnology and Bioengineering 108 (2), 306–312.

Dien, B.S., Sarath, G., Pedersen, J.F., Sattler, S.E., Chen, H., Funnell-Harris, D.L., Nichols, N.N., Cotta, M.A., 2009. Improved sugar conversion and ethanol yield for Forage Sorghum (*Sorghum bicolor* L. Moench) lines with Reduced lignin contents. BioEnergy Research 2 (3), 153–164.

Docherty, K.M., Kulpa, C.F., 2005. Toxicity and antimicrobial activity of imidazolium and pyridinium ionic liquids. Green Chemistry 7, 185–189.

Dodd, D., Cann, I.K.O., 2009. Enzymatic deconstruction of xylan for biofuel production. GCB Bioenergy 1 (1), 2–17.

Eriksson, K.-E., Vallander, L., 1982. Properties of pulps from thermomechanical pulping of chips pretreated with fungi. Svensk Papperstiding 85 (6), R33.

Ewanick, S.M., Bura, R., Saddler, J.N., 2007. Acid-catalyzed steam pretreatment of lodgepole pine and subsequent enzymatic hydrolysis and fermentation to ethanol. Biotechnology and Bioengineering 98 (4), 737–746.

Foust, T.D., Aden, A., Dutta, A., Phillips, S., 2009. An economic and environmental comparison of a biochemical and a thermochemical lignocellulosic ethanol conversion processes. Cellulose 16, 547–565.

Fu, C., Mielenz, J.R., Xiao, X., Ge, X., Hamilton, C., Rodriguez Jr, M., Chen, F., Foston, M., Ragauskas, A., Bouton, J., Dixon, R.A., Wang, Z.-Y., 2011. Genetic manipulation of lignin reduces recalcitrance and improves ethanol production from switchgrass. Proceedings of the National Academy of Sciences USA 108 (9), 3803–3808.

Garlock, R.J., Balan, V., Dale, B.E., Pallapolu, V.R., Lee, Y.Y., Kim, Y., Mosier, N.S., Ladisch, M.R., Holtzapple, M.T., Falls, M., Sierra-Ramirez, R., Shi, J., Ebrik, M.A., Redmond, T., Yang, B., Wyman, C.E., Donohoe, B.E., Vinzant, T.B., Elander, R.E., Hames, B., Thomas, S., Warner, R.E., 2011. Comparative material balances around pretreatment technologies for the conversion of switchgrass to soluble sugars. Bioresource Technology 102, 11063–11071.

Gericke, M., Fardim, P., Heinze, T., 2012. Ionic liquids-promising but challenging solvents for homogeneous derivatization of cellulose. Molecules 17, 7458–7502.

Grethlein, H.E., Converse, A.O., 1991. Common aspects of acid prehydrolysis and steam explosion for pretreating wood. Bioresource Technology 36 (1), 77–82.

Grous, W.R., Converse, A.O., Grethlein, H.E., 1986. Effect of steam explosion pretreatment on pore-size and enzymatic-hydrolysis of Poplar. Enzyme and Microbial Technology 8 (5), 274–280.

Ilmén, M., den Haan, R., Brevnova, E., McBride, J., Wiswall, E., Froehlich, A., Koivula, A., Voutilainen, A.P., Siika-aho, M., la Grange, D.C., Thorngren, N., Ahlgren, S., Mellon, M., Deleault, K., Rajgarhia, V., van Zyl, W.H., Penttilä, M., 2011. High level secretion of cellobiohydrolases by *Saccharomyces cerevisiae*. Biotechnology for Biofuels 4 (30). http://dx.doi.org/10.1186/1754-6834-4-30.

Kataeva, I., Foston, M.B., Yang, S.-J., Pattathil, S., Biswal, A.K., Poole II, F.L., Basen, M., Rhaesa, A.M., Thomas, T.P., Azadi, P., Olman, O., Saffold, T.D., Mohler, K.E., Lewis, D.L., Doeppke, C., Zeng, Y., Tschaplinski, T.J., York, W.S., Davis, M., Mohnen, D., Xu, Y., Ragauskas, A.J., Ding, S.-Y., Kelly, R.M., Hahn, M.G., Adams, M.W., 2013. Carbohydrate and lignin are simultaneously solubilized from unpretreated switchgrass by microbial action at high temperature. Energy and Environmental Science 6, 2186–2195.

Khramtsov, N., McDade, L., Amerik, A., Yu, E., Divatia, K., Tikhonov, A., Minto, M.A., Kabongo-Mubalamate, G., Markovic, Z., Ruiz-Martinez, M., Henck, A., 2011. Industrial yeast strain engineered to ferment ethanol from lignocellulosic biomass. Bioresource. Technology 102 (17), 8310–8313.

Kilpeläinen, I., Xie, H., King, A., Granström, M., Heikkinen, S., Argyropoulus, D.S., 2007. Dissolution of wood in ionic liquids. Journal of Agriculture Food Chemistry 55 (22), 9142–9148.

Kim, T.H., Gupta, R., Lee, Y.Y., 2009a. Pretreatment of biomass by aqueous ammonia for bioethanol production. Methods in Molecular Biology 581, 79–91.

Kim, Y., Hendrickson, R., Mosier, N.S., Ladisch, M.R., 2009b. Liquid hot water pretreatment of cellulosic biomass. Methods in Molecular Biology 581, 93–102.

Klein-Marcuschamer, D., Simmons, B.A., Blanch, H.W., 2011. Techno-economic analysis of a lignocellulosic ethanol biorefinery with ionic liquid pre-treatment. Biofuels, Bioproducts. and Biorefining 5 (5), 562–569.

Kleinert, T.N., 1974. Organosolvent pulping with aqueous alcohol. TAPPI 57 (8), 99–102.

Klinke, H.B., Thomsen, A.B., Ahring, B.K., 2004. Inhibition of ethanol-producing yeast and bacteria by degradation products produced during pre-treatment of biomass. Applied Microbiology and Biotechnology 66, 10–26.

Larsson, S., Reimann, A., Nilvebrant, N.-O., Jönsson, L.J., 1999. Comparison of different methods for the detoxification of lignocellulose hydrolysates of spruce. Applied Biochemistry and Biotechnology 77, 91–103.

Lau, M.W., Dale, B.E., Balan, V., 2008. Ethanolic fermentation of hydrolysates from ammonia fiber expansion (AFEX) treated corn stover and distillers grain without detoxification and external nutrient supplementation. Biotechnology and Bioengineering 99 (3), 529–539.

Lynd, L.R., Laser, M.S., Bransby, D., Dale, B.E., Davison, B., Hamilton, R., Himmel, M., Keller, M., McMillan, J.D., Sheehan, J., Wyman, C.E., 2008. How biotech can transform biofuels. Nature Biotechnology 26, 169–172.

Lynd, L.R., Larson, E.D., Greene, N., Laser, M., Sheehan, J., Dale, B.E., McLaughlin, S., Wang, M., 2009. The role of biomass in America's energy future: Framing the analysis. Biofuels, Bioproducts, and Biorefining 3, 113–123.

Mclaughlin, S.B., Kszos, L.A., 2005. Development of switchgrass as a bioenergy feedstock in the United States. Biomass and Bioenergy 28, 515–535.

Mielenz, J.R., Bardsley, J.S., Wyman, C.E., 2009. Process for fermentation of soybean hulls to ethanol while preserving protein value. Bioresource Technology 100, 3532–3539.

Mora-Pale, M., Meli, L., Doherty, T.V., Linhardt, R.J., Dordick, J.S., 2011. Room temperature ionic liquids as emerging solvents for the pretreatment of lignocellulosic biomass. Biotechnology and Bioengineering 108 (6), 1229–1245.

Mosier, N., Wyman, C., Dale, B., Elander, R., Lee, Y.Y., Holtzapple, M., Ladisch, M., 2005. Features of promising technologies for pretreatment of lignocellulosic biomass. Bioresource Technology 96 (6), 673–686.

Nguyen, S.K., Sophonputtanaphoca, S., Kim, E., Penner, M.H., 2009. Hydrolytic methods for the quantification of fructose equivalents in herbaceous biomass. Applied Biochemistry and Biotechnology 158 (2), 352–361.

NREL biomass database, http://www.nrel.gov/biomass/data_resources.html.

Olson, D.G., McBride, J.E., Shaw, A.J., Lynd, L.R., 2012. Recent progress in consolidated bioprocessing. Current Opinion Biotechnology 23 (3), 396–405.

Palmqvist, E., Hahn-Hägerdal, B., 2000. Fermentation of lignocellulosic hydrolysates. II: inhibitors and mechanisms of inhibition. Bioresource Technology 74, 25–33.

Ralph, J., Lundquist, K.T., Brunow, G., Lu, F., Kim, H., Schatz, P.F., Marita, J.M., Hatfield, R.D., Ralph, S.A., Christensen, J.H., Boerjan, W., 2004. Lignins: natural polymers from oxidative coupling of 4-hydroxyphenylpropanoids. Phytochemistry Reviews 3 (1–2), 29–60.

Samara, M. [M.S. thesis], Colorado State University; 1992.

Sannigrahi, P., Ragauskas, A.J., Tuskan, G.A., 2010. Poplar as a feedstock for biofuels: a review of compositional characteristics. Biofuels, Bioprod, and Biorefining 4, 209–226.

Selig, M.J., Tucker, M.P., Sykes, R.W., Reichel, K.L., Brunecky, R., Himmel, M.E., Davis, M.F., Decker, S.R., 2010. Lignocellulose recalcitrance screening by Integrated high throughput Hydrothermal pretreatment and enzymatic saccharification. Industrial Biotechnology 6, 104–111.

Sierra, R., Granda, C.B., Holtzapple, M.T., 2009. Lime pretreatment. Methods in Molecular Biology 581, 115–124.

Sokhansanj, S., Mani, S., Turhollow, A., Kumar, A., Bransby, D., Lynd, L., Laser, M., 2009. Large-scale production, harvest and logistics of switchgrass (*Panicum virgatum L.*) - current technology and envisioning a mature technology. Biofuels, Bioproducts. and Biorefining 3, 124–141.

Strobel, H.J., 2009. Basic laboratory culture methods for anaerobic bacteria. Methods Mol. Biol. 581, 247–261.

Studer, M.H., DeMartini, J.D., Davis, M.F., Sykes, R.W., Davison, B., Keller, M., Tuskan, G.A., Wyman, C.E., 2011. Lignin content in natural *Populus* variants affects sugar release. Proceedingsof the National Academy of Sciences USA 108 (15), 6300–6305.

Sun, N., Rahman, M., Qin, Y., Maxim, M.L., Rodriguez, H., Rogers, R.D., 2009. Complete dissolution and partial delignification of wood in the ionic liquid 1-ethyl-3-methylimidazolium acetate. Green Chemistry 11 (5), 646–655.

Swatloski, R.P., Spear, S.K., Holbrey, J.D., Rogers, R.D., 2002. Dissolution of cellulose with ionic liquids. Journal of the American Chemical Soceiety 124, 4974–4975.

Sykes, R., Yung, M., Novaes, E., Kirst, M., Davis, M., 2009. High throughput screening of plant cell Wall composition using Pyrolysis Molecular Beam mass spectroscopy. Methods in Molecular Biology 581, 169–183.

Tschaplinski, T.J., Standaert, R.F., Engle, N.L., Martin, M.Z., Sangha, A.S., Parks, J.M., Smith, J.C., Samuel, R., Pu, Y., Ragauskas, A.J., Hamilton, C.Y., Fu, C., Wang, Z.-Y., Davison, B.D., Dixon, R.F., Mielenz, J.R., 2012. Down-regulation of the caffeic acid *O*-methyltransferase gene in switchgrass reveals a novel monolignol analog. Biotechnology forBiofuels 5 (71).

Turner, M.B., Spear, S.K., Huddleston, J.G., Holbrey, J.D., Rogers, R.D., 2003. Ionic liquid salt-induced inactivation and unfolding of cellulase from *Trichoderma reesei.* Green Chemistry 5, 443–447.

US DOE EERE Biomass Multi-Year Program Plan (MYPP), April 2011, www1.eere.energy.gov/biomass/pdfs/mypp_april_2011.pdf

Vanholme, R., Demedts, B., Morreel, K., Ralph, J., Boerjan, W., 2010. Lignin biosynthesis and structure. Plant Physiology 153 (3), 895–905.

Wyman, C.E., Dale, B.E., Elander, R.T., Holtzapple, M., Ladisch, M.R., Lee, Y.Y., 2005. Coordinated development of leading biomass pretreatment technologies. Bioresource Technology 96 (18), 1959–1966.

Wyman, C.E., Dale, B.E., Elander, R.T., Holtzapple, M., Ladisch, M.R., Lee, Y.Y., Mitchinson, C., Saddler, J.N., 2009. Comparative sugar recovery and fermentation data following pretreatment of poplar wood by leading technologies. Biotechnology Progress 25 (2), 333.

Yang, B., Wyman, C.E., 2009. Dilute acid and autohydrolysis pretreatment. Methods in Molecular Biology 581, 103–114.

Yang, S.J., Kataeva, I., Hamilton-Brehm, S.D., Engle, N.L., Tschaplinski, T.J., Doeppke, C., Davis, M., Westpheling, J., Adams, M.W., 2009. Efficient degradation of lignocellulosic plant biomass, without pretreatment, by the thermophilic anaerobe "Anaerocellum thermophilum" DSM 6725. Applied Environmental Microbiology 75, 4762–4769.

Yee, K.L., Rodriguez Jr, M., Tschaplinski, T.J., Engle, N.L., Martin, M.Z., Fu, C., Wang, Z.-Y., Hamilton-Brehm, S.D., Mielenz, J.R., 2012. Evaluation of the bioconversion of genetically modified switchgrass using simultaneous saccharification and fermentation and a consolidated bioprocessing approach. Biotechnology for Biofuels 5, 81.

Zeng, Y., Yang, X., Yu, H., Zhang, X., Ma, F., 2011. The delignification effects of white-rot fungal pretreatment on thermal characteristics of moso bamboo. Bioresource Technology 114, 437–442.

Zhang, Y.H., Ding, S.Y., Mielenz, J.R., Cui, J.B., Elander, R.T., Laser, M., Himmel, M.E., McMillan, J.R., Lynd, L.R., 2007. Fractionating recalcitrant lignocellulose at modest reaction conditions. Biotechnology and Bioengineering 97, 214–223.

Zhao, X., Cheng, L., Liu, D., 2009. Organosolv pretreatment of lignocellulosic biomass for enzymatic hydrolysis. Appl Microbiol Biotechnol. 82 (5), 815–827. http://dx.doi.org/10.1007/s00253-009-1883-1.

Zhao, X., Zhang, L., Liu, D., 2012. Biomass recalcitrance. Part I: the chemical compositions and physical structures affecting the enzymatic hydrolysis of lignocellulose. Biofuels, Bioproducts, and Biorefining 6, 465–482.

Zhu, S., Yu, P., Tong, Y., Chen, R., Lv, Y., Zhang, R., Lei, M., Ji, J., Chen, Q., Wu, Y., 2012. Effects of the ionic liquid 1-butyl-3-methylimidazolium chloride on the growth and ethanol fermentation of Saccharomyces cerevisiae AY92022. Chemical and Biochemical Engineering Quarterly 26, 105–109.

Zhu, S., Yu, P., Lei, M., Tong, Y., Zhang, R., Ji, J., Chen, Q., Wu, Y., 2013a. Influence of the ionic liquid 1-butyl-3-methylimidaxolium chloride on the ethanol fermentation of Saccharomyces cerevisiae AY93161 and its kinetics analysis. Energy Education Science and Technology Part A: Energy Science and Research 30 (2), 817–828.

Zhu, S., Yu, P., Wang, Q., Cheng, B., Chen, J., Wu, Y., 2013b. Breaking the Barriers of lignocellulosic ethanol production using ionic liquid technology. BioResources 8 (2), 1510–1512.

REDUCING ENZYME COSTS, NOVEL COMBINATIONS AND ADVANTAGES OF ENZYMES COULD LEAD TO IMPROVED COST-EFFECTIVE BIOFUELS PRODUCTION

25

National Laboratory of the U.S. Department of Energy, Office of Energy Efficiency and Renewable Energy, USA

REDUCING ENZYME COSTS INCREASES THE MARKET POTENTIAL OF BIOFUELS

(NREL/FS-6A42-59013 | August 2013 bioenergy)

QUICK FACTS

- NREL provided two leading enzyme companies, Genencor and Novozymes, with access to its innovative biomass characterization, pretreatment, and process integration research, which led to lower enzyme costs. Genencor is now part of DuPont Industrial Biosciences, headquartered in Wilmington, Delaware, while Novozymes North America is based in Franklinton, North Carolina.
- Novozymes and Genencor worked with NREL scientists to identify approaches for reducing cellulase costs; those strategies included both decreasing the cost of enzyme production and increasing enzyme efficiency. The research also led to improvements in sugar yields from cellulosic biomass.
- The project made tremendous progress in lowering the projected cost of cellulase enzymes. From initial enzyme costs of $4–5 per gallon of ethanol produced, the team achieved cost reductions that exceeded the subcontract goal of decreasing the enzyme cost by a factor of 10.
- This reduction in enzyme cost dramatically reduced the projected cost of ethanol production and represents a major step toward commercializing large-scale biomass-to-ethanol production.
- The importance of this research was recognized by *R&D Magazine* in 2004 with an R&D 100 Award, denoting it as one of the year's 100 most significant innovations.

Cellulosic ethanol prices depend heavily on the cost of the cellulase enzymes used to break down the biomass into fermentable sugars. To reduce these costs, the National Renewable Energy Laboratory (NREL) partnered with two leading enzyme companies, Novozymes and Genencor (now part of

Bioenergy. http://dx.doi.org/10.1016/B978-0-12-407909-0.00025-0

407

DuPont Industrial Biosciences), to engineer new cellulase enzymes that are exceptionally good at breaking down cellulose. The work was funded in part by the Office of Energy Efficiency and Renewable Energy at the U.S. Department of Energy.

Ethanol is produced by releasing the sugars from biomass and then fermenting those sugars into alcohol. Starch-based biomass, such as corn grain, is easily converted to glucose. But breaking down cellulose-based biomass, such as crop residues or forestry residues, is much more difficult. This requires pretreatment with dilute acid or other techniques to make the cellulose vulnerable to enzymatic hydrolysis, which involves using cellulases and other enzymes to convert the cellulose into glucose and other five- and six-carbon sugars. Before the development of advanced cellulases, the process for hydrolyzing cellulose to sugars was very expensive—too expensive to compete with the technology commonly used to break down the starch in corn kernels to sugars, so cellulosic ethanol could not compete with corn-based ethanol.

To hydrolyze the cellulose, NREL and its partners developed a technology that employs a cocktail of predominantly three types of cellulase enzymes: endoglucanases, exoglucanases, and beta-glucosidases. The endoglucanases are thought to break the cellulose chains, creating two new chain ends at each break. Next, the exoglucanases attach to the exposed chain ends and move the cellulose chains away from the crystal structure by a complex process still under study today. The exoglucanase enzymes then proceed to work their way down the chains, liberating cellobiose (a sugar composed of two glucose molecules) as they proceed. Finally, the beta-glucosidases split each cellobiose molecule into two separate glucose molecules, making them available for processing into chemicals or fuels.

This research to engineer cheaper and more efficient cellulases, combined with advances in other aspects of biomass conversion technology, has been critical in progressing cellulosic ethanol technology towards its ultimate goal: becoming cost-competitive with gasoline.

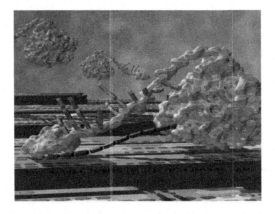

FIGURE 25.1

An exoglucanase cellulase enzyme attaches to a cellulose molecule in this computer-generated conceptual image. Successfully reducing the cost of enzymes that break down cellulose to fermentable sugars is key to cutting the cost of producing ethanol and other products from nonedible, cellulosic biomass, such as trees, grasses, and agricultural and forestry residues.

NOVEL COMBINATION OF ENZYME SYSTEMS COULD LOWER BIOFUEL COSTS

(NREL/FS-2700-60026 | August 2013) *Highlights in Science*

KEY RESEARCH RESULTS

Achievement

Researchers have demonstrated that mixing disparate enzyme systems can break down cellulose more rapidly and efficiently than either system alone.

Key result

Although free cellulases and cellulosomes employ very different physical mechanisms to break down recalcitrant polysaccharides, when combined these systems display dramatic synergistic enzyme activity on cellulose.

Potential impact

This study indicates new opportunities for mixing free enzymes and cellulosomes in an industrial setting, with the potential for an optimal synergy between two natural mechanisms for biomass deconstruction that further enables cost-effective biofuels production.

Two biomass-degrading enzyme systems that work in very different ways are shown to be more effective at releasing plant sugars when used together.

Two natural enzyme systems—one produced by fungi and the other by bacteria—break down cellulose faster if used in combination. The resulting process shows promise for less expensive biofuels. Researchers from the National Renewable Energy Laboratory (NREL) and their partners studied a cocktail of individual fungal enzymes that depolymerize biomass, and an alternative bacterial system in which multiple biomass-degrading enzymes, termed the cellulosome, are linked together by a protein scaffold. This study suggests that two of the most thoroughly studied and distinct paradigms of biomass degradation, namely free fungal enzymes and multi-enzyme bacterial cellulosomes, function together in an unexpected way to efficiently break down polysaccharides.

A large barrier to reaching the goal of producing low-cost biofuels is the high cost of enzyme treatment, a crucial step in turning biomass—switchgrass, energy trees, corn stover, and the like—into liquid fuels. A number of enzymatic strategies are used to degrade polysaccharides in a plant cell wall into sugars for conversion to biofuels. Free enzymes are more active on pretreated biomass; in contrast, cellulosomes are much more active on purified cellulose. In this research, free enzymes and cellulosomes were compared. When the two enzyme systems were combined, cellulose was broken down to sugar faster and more efficiently than with either system alone. Physical changes to the substrate suggest synergistic deconstruction mechanisms.

Transmission electron microscopy revealed evidence that free enzymes and cellulosomes employ different physical mechanisms to degrade cellulose microfibrils. The individual fungal enzyme system demonstrates an "outside-in" degradation pattern where the biomass is broken down sequentially from the outer surface. The bacterial cellulosomal system shows evidence of "splitting" the cellulose into smaller pieces by hydrolyzing down the middle of the cellulose macrofibril. When combined, these systems display dramatic synergistic enzyme activity on cellulose, hinting at a means for faster

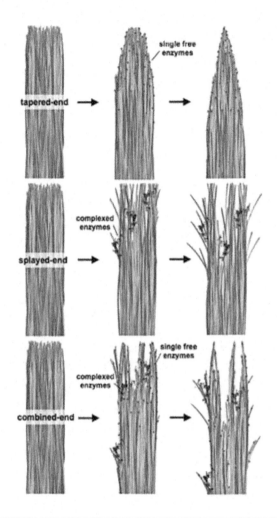

FIGURE 25.2

Illustration of the mechanisms by which free enzymes (top) and cellulosomes (middle) differ in their action on cellulose microfibril bundles and act synergistically to degrade cellulose (bottom).

Image by Bryon Donohoe, NREL.

and more efficient conversion of biomass, which would lead to lower costs for biomass-derived renewable fuels.

Technical contact: Michael Resch, michael.resch@nrel.gov

References: Resch, M.G., Donohoe, B.S., Baker, J.O., Decker, S.R., Bayer, E.A., Beckham, G.T., Himmel, M.E., 2013. Fungal cellulases and complexed cellulosomal enzymes exhibit synergistic mechanisms in cellulose deconstruction. Energy and Environmental Science (6), pp. 1858–1867.

ADVANTAGES OF ENZYME COULD LEAD TO IMPROVED BIOFUELS PRODUCTION HIGHLIGHTS IN SCIENCE

(NREL/FS-2700-61022 | January 2014)

KEY RESEARCH RESULTS

Achievement

The research team isolated CelA, a highly active cellulase with a novel cellulose digestion mechanism. The X-ray structures of the primary protein components of CelA were also determined, advancing the understanding of the mode of action of this cellulase.

Key result

CelA was shown to retain high activity at all temperatures tested, converting 60% of glucan compared to 28% glucan conversion for the more common exo/endo cellulase standard mixture, Cel7A/Cel5A, at its optimal temperature of 50 °C.

Potential impact

CelA and similar multifunctional cellulases represent a new and distinct paradigm for cellulose digestion. This mechanism is fundamentally different from conventional cellulases and could help increase intercellulase synergy in consolidated bioprocessing microorganisms, as well as in commercial cellulase formulations used for biofuels production.

Cellulase *Caldicellulosiruptor bescii* CelA, a highly active and stable enzyme, exhibits a new cellulose digestion paradigm promoting intercellulase synergy. *C. bescii* CelA, a hydrolytic enzyme with multiple functional domains, may have several advantages over other fungal and bacterial cellulases for use in biofuels production: very high specific activity, stability at elevated temperatures, and a novel digestion mechanism. A research team from the U.S. Department of Energy's Bio-Energy Science Center, which comprised scientists from the National Renewable Energy Laboratory (NREL) and the University of Georgia, isolated the thermophilic cellulase CelA from *C. bescii*. A comparison was conducted of its cellulolytic activity with that of a binary mixture containing both *Trichoderma reesei* Cel7A exoglucanase and *A. cellulolyticus* Cel5A endoglucanase on several substrates. The researchers also compared the cellulose digestion mechanisms of these two enzyme systems using electron microscopy and modeling. CelA was shown to retain high activity at all temperatures tested, converting 60% of glucan at 85 °C compared to 28% glucan conversion for the common exo/endo cellulase standard mixture, Cel7A/Cel5A, at its optimal temperature of 50 °C. This difference in activity translates to a seven-fold increase in activity for CelA at the molecular level.

Transmission electron microscopy studies of cellulose following incubation with CelA suggest that CelA is capable of not only the common surface ablative mechanism driven by general cellulase processivity, but also of excavating extensive cavities into the surface of the substrate. Additionally, during the digestion experiments, CelA achieved 60% conversion of xylan in native switchgrass, showing its potential for industrial processes using mild or no pretreatment.

Technical Contact: Yannick Bomble, yannick.bomble@nrel.gov

References: Brunecky, R., Alahuhta, P., Xu, Q., Donohoe, B., Crowley, M., Kataeva, I, Yang, SJ, Adams, M, Lunin, V., Himmel, M., Bomble, Y., 2013. Revealing Nature's cellulase diversity: the

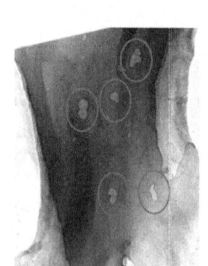

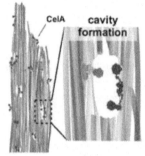

FIGURE 25.3

Transmission electron micrographs and schematic of partially digested small Avicel particles. Particles digested to approximately 65% conversion with CelA display surface cavities of various sizes. All scale bars are 500 nm.

Image by Bryon Donohoe, NREL.

digestion mechanism of *Caldicellulosiruptor bescii* CelA. Science 342:6165, pp. 1513–1516. http://dx.doi.org/10.1126/science.1244273.

NREL is a national laboratory of the U.S. Department of Energy, Office of Energy Efficiency and Renewable Energy, operated by the Alliance for Sustainable Energy, LLC. 15013 Denver West Parkway Golden, CO 80,401 303-275-3000 | www.nrel.gov

ACKNOWLEDGMENTS

The editor is grateful to National Renewable Energy Laboratory (NREL) (Kristi Theis Communications, Fuels & Efficiency Section Manager) for permission to include the three factsheets presented here.

PYROLYSIS OF LIGNOCELLULOSIC BIOMASS: OIL, CHAR, AND GAS

26

Brennan Pecha, Manuel Garcia-Perez

Biological Systems Engineering, Washington State University, Pullman, WA, USA

INTRODUCTION

Most people have never heard the word "pyrolysis" unless they are a renewable energy researcher or a venture capitalist. However, this technology for "distilling wood" has been around since before the Egyptians and is leaping back to popularity as a method to produce renewable fuels, chemicals, and carbonaceous products.

The word "pyrolysis" is a combination of two Greek words: $\pi \upsilon \rho$ (*pýr*), meaning "fire," and $\lambda \upsilon \sigma \iota \varsigma$ (*lýsis*), meaning "splitting" or "cracking." A pyre is an old term for a pile of wood or other substance that has been prepared for burning. Based on the etymology, pyrolysis means "fire splitting," or better yet, "heat splitting." Practically speaking, pyrolysis is currently used to convert a solid material, such as wood or rubber tires, into carbonaceous char, condensable oil, and gases. In some concepts, the final outcome is char and heat production.

Pyrolysis char is still used today for indoor heat or cooking fuel in developing countries because it releases less smoke than when burning wood directly. The pyrolysis oil (also known as tar or pyroligneous water) is a mixture of chemicals that are released in vapor phase and are typically condensed for further use. Some of the chemicals are soluble in water (aqueous), whereas others are not (organic), so depending on the processing conditions (fast or slow pyrolysis), the oil can be formed by a single phase or by several phases.

Historians and art lovers can trace the use of pyrolytic char back to the origins of civilization, seeing it used to create the paintings on the walls of France's Lascaux Caves (28,000 BC). Figure 26.1 illustrates a timeline of some of the important developments in the history of pyrolysis. A more elaborate account can be seen in detail in a publication listed in the References section by Antal and Grønli (2003), but we will look at some notable events up to the present day.

In the 1600s, Johann Rudolf Glauber discovered that the acid found in the aqueous phase of pyrolysis oil is acetic acid (vinegar). In the late 1700s, England commercialized the use of the noncondensable gas to illuminate cities. In the 1800s, new reactors were created that did not require the input of oxygen. During this time, methyl alcohol was discovered to be contained in the aqueous phase of the oil. The demand for methyl alcohol for a dye called aniline purple spring-boarded the pyrolysis industry in the late nineteenth century. Similarly, acetone was also produced from pyrolysis for the production of smokeless gun powder.

Bioenergy. http://dx.doi.org/10.1016/B978-0-12-407909-0.00026-2

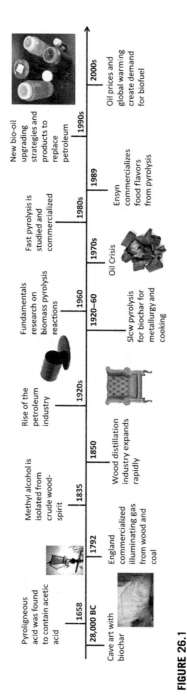

FIGURE 26.1

Timeline with some important milestones in the development and use of pyrolysis.

From the 1920 to the 1950s, pyrolysis was replaced by the petroleum industry, which proved more economical at the time on the production of acetic acid and acetone. In the 1960s, some researchers pushed on with the technology of pyrolysis, exploring fast pyrolysis in fluidized beds. This technology found acceptance in the western world as a result of the oil crisis in the 1970s. Researchers in the 1980 and 1990s developed new reactors, uses for pyrolysis oil, and chemical product conversion methods that greatly advanced the viability of pyrolysis technologies. Finally, international turmoil in the 2000s reminded the world of the instability of the petroleum industry, and more funds flowed to research renewable fuels and the development of bio-oil refineries.

When pyrolysis is used to make char and the biomass particle is heated very slowly, it is called "slow pyrolysis." The slow heating rates maximize the production of the solid carbonaceous material and production of water from dehydration reactions. Slow pyrolysis was the only technologically available method until the 1960s, at which point it was found that if you heat up your feed material very fast using very small particles (less than 2 mm in fluidized or circulating beds), you form less char but more oil. This method was called "fast pyrolysis." Although much of the literature suggests that the high yields oil achieved by this technology are the result fast heating, some researchers are calling the attention of the importance of reduced mass transfer of the oligomeric fractions when using small particles (Shen et al., 2009).

The pyrolysis of lignocellulosic biomass, such as wood and straw, is now considered a promising way to produce renewable fuels and chemical products to replace petroleum and stimulate agricultural economies. The next sections will illustrate slow pyrolysis, fast pyrolysis, and how biomass properties affect the outcome of these processes.

TYPES OF THERMOCHEMICAL CONVERSION

To truly understand pyrolysis, it is crucial to see that it is just one class of thermochemical conversion technologies. Thermochemical conversion is a term that means exactly what it sounds like: converting something by using the chemical reactions induced by heating.

Let us start from a point that everyone understands: burning wood. Think about the last campfire you were at: the flames coming off the top of the wood, the smells, the heat, the crackling, the char, and (if you are a good Boy Scout) the ashes after the fire died down. The combustion of wood involves all five important types of thermochemical conversion: (1) evaporation of water and other volatile small molecules (up to 200 °C), (2) torrefaction (between 225 and 300 °C), (3) pyrolysis (between 300 and 650 °C), (4) gasification (between 700 and 850 °C), and (5) combustion (450–2000 °C) (Note that the range of temperature given should only be used as a first approximation). These phenomena occur in the order that they were numbered if the heating rate is very slow and an oxidizing agent is present, as illustrated in Figure 26.2. If the heating rate is very fast (a few 100 °C/s) all these events are thought to happen at the same time.

Evaporation of biomass moisture is typically performed at temperatures below 200 °C. This step is highly important because the latent heat of vaporization of water is high (2230 J/g) and will hold the biomass near 100 °C until most of the water is released. The biomass density and porosity is important to consider because mass transfer limitations can slow the drying of dense materials like hardwood. Heat and mass transfer limitations can be minimized through chopping and grinding. The vapors produced in this step are mainly water and tend to be white.

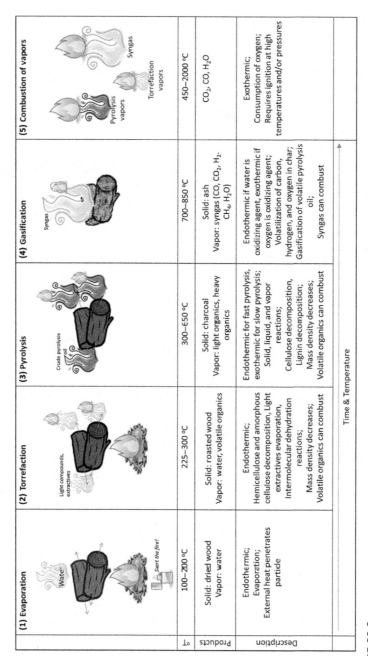

	(1) Evaporation	(2) Torrefaction	(3) Pyrolysis	(4) Gasification	(5) Combustion of vapors
T	100–200 °C	225–300 °C	300–650 °C	700–850 °C	450–2000 °C
Products	Solid: dried wood Vapor: water	Solid: roasted wood Vapor: water, volatile organics	Solid: charcoal Vapor: light organics, heavy organics	Solid: ash Vapor: syngas (CO, CO₂, H₂, CH₄, H₂O)	CO₂, CO, H₂O
Description	Endothermic; Evaporation; External heat penetrates particle	Endothermic; Hemicellulose and amorphous cellulose decomposition, Light extractives evaporation, Intermolecular dehydration reactions; Mass density decreases; Volatile organics can combust	Endothermic for fast pyrolysis, exothermic for slow pyrolysis; Solid, liquid, and vapor reactions; Cellulose decomposition, Lignin decomposition; Mass density decreases; Volatile organics can combust	Endothermic if water is oxidizing agent, exothermic if oxygen is oxidizing agent; Volatilization of carbon, hydrogen, and oxygen in char; Gasification of volatile pyrolysis oil; Syngas can combust	Exothermic; Consumption of oxygen; Requires ignition at high temperatures and/or pressures

Time & Temperature

FIGURE 26.2

Thermochemical reactions in the combustion of wood.

The next thermochemical reaction to occur is torrefaction, also known as roasting. This is the exact same process used to make coffee beans. (Coffee beans are green when they leave the plants!) Torrefaction can be thought of as low-temperature pyrolysis. At the temperature range between 225 and 300 °C, extractives and nonstructural light compounds degrade and evaporate, and hemicellulose decomposes and evaporates. In this temperature range, lignin and amorphous cellulose starts to depolymerize and form liquid intermediates on the surface of biomass cell walls. The decomposition (think splitting/lysis) of a major macromolecule like hemicellulose is what characterizes torrefaction.

Although torrefaction can be one of the steps of pyrolysis, gasification, or combustion reactors, it can be also the main step of a stand-alone technology. Torrefaction is important for the transport and combustion of biomass because it typically increases the combustion heating value of the biomass (in mass bases) by about 25% and reduces the grinding energy by up to 10 times, mainly due to the lysis of the amorphous cellulose zones that contributed to the formation of brittle fibers (Phanphanich and Mani, 2011). In other words, burning torrefied biomass releases 25% more heat than nontorrefied biomass, relative to weight, and it takes less energy to grind it after torrefaction. Furthermore, the organic vapors released from torrefaction can be combusted. Torrefaction vapors tent to be acidic due to the release of acetic acid attached to the hemicellulose structure.

Pyrolysis follows torrefaction at temperatures between 300 and 650 °C. The rest of this book chapter will explore pyrolysis in depth. In a short spiel, pyrolysis first breaks down the hemicelluloses, cellulose, lignin, and other residual organic macromolecules into smaller compounds. The smaller molecules (monomers or oligomers) could form a liquid intermediate, then vaporize (or are thermally ejected) to escape the porous biomass particles. Pyrolysis products that do not escape the biomass either fragment into even smaller molecules, or polymerize into char. It is important to note that the porosity of the biomass increases as pyrolysis progresses to form char; in this way, the solid's density decreases. Pyrolysis vapors can combust at these temperatures when in contact with oxygen. The rich smell that is released from burning wood comes from pyrolysis vapors that evaded the combustion zone!

Gasification follows pyrolysis at temperatures between 700 and 850 °C. During gasification, reactions convert the carbonaceous solid remaining from the pyrolysis step (also call char) and the pyrolysis vapors into CO, H_2O, Chapters $_4$, and H_2, a gas mixture called "syngas" (short for "synthesis gas"). A syngas is typically produced in oxygen-starved environments with between 15 and 28% of stoichiometric ratio for complete combustion. In the chemical and energy industries, gasification is used as a stand-alone technology to create high-value hydrogen gas, which can be used for various purposes. The important reactions in gasification are the carbon–oxygen reaction, Boudouard reaction, carbon–water reaction, hydrogenation reaction, water–gas shift reaction, and the methanation reaction. When oxygen is present, the heat required to perform these reactions is provided by combustion—this is called authothermal gasification. For higher hydrogen content, the biomass can be heated externally (water is added)—this is called indirectly-heated gasification. Gasification is important to consider for pyrolysis, especially if there is inhomogeneous heating where some of the biomass reaches gasification temperatures while other parts of the biomass are still warming up. Steam gasification is used to enhance the surface area of the biochar produced in the physical activation processes for the production of activated carbon. (Kumar et al., 2009).

Finally, there is combustion. As you well know, combustion needs three things: carbon, oxygen, and a high temperature ignition source. Flames from biomass combustion can exceed 2000 °C, depending on moisture content, heating value, the amount of air added, and the gas flow patterns in the reactor. During slow-pyrolysis or directly heated gasification, oxygen is provided such that some of the pyrolysis vapors or syngas are combusted in vapor phase to provide heat back to the solid biomass or char. Alternatively, the vapors can be channeled to a combustion unit that burns the pyrolysis vapors and sends the exhaust into the reactor to heat the biomass. There are various methods for heating pyrolysis reactors, but almost all of them use combustion due to its simplicity. Combustion is an essential component of pyrolysis, even when it does not occur within the pyrolysis reactor. Many slow pyrolysis reactors (or other autothermal systems) combust some of the products of pyrolysis (vapor or gases) to produce the heat needed to sustain the process. Combustion also happens when the oxygen attack the carbonaceous residue formed in the pyrolysis step. The white powder left after the complete combustion of the biomass is called ash and is typically formed by the inorganic constituents.

LIGNOCELLULOSIC STRUCTURE AND CHEMISTRY FOR PYROLYSIS

Biomass pyrolysis is a very complex system, involving simultaneous solid heat transfer, solid phase chemical reactions, liquid evaporation and thermal ejection, liquid phase reactions, mass transfer of vapors through the solid matrix, and vapor phase reactions. The best place to begin understanding these interactions is to look at the structure and chemistry of lignocellulosic material. Lignocellulosic material is plant dry matter and encompasses trees, bushes, grasses, corn stover/cobs (what is left of the plant after you remove the corn kernels), sugarcane bagasse (what is left after you wash out the sugar), and even paper mill discards. Further reading about wood structure and chemistry can be found in a book by Sjöström (1993).

STRUCTURE OF LIGNOCELLULOSE AND IMPACT ON PYROLYSIS

To explore the structure of lignocellulose, let's start with an upright fiber plant like flax, which is grown for food (seed), oil (for varnishes), and fiber (for textiles). Wood stems and trunks are very similar to flax. In plants, long plant cells travel up through the body of the plant to provide structure, nutrients, and protection. To do this, there are different types of cells, as illustrated in Figure 26.3.

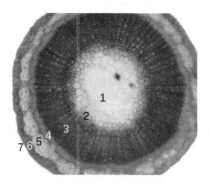

FIGURE 26.3

Flax plant stem cross-section. 1: pith, 2: protoxylem, 3: xylem, 4: phloem, 5: sclerenchyma (bast fiber), 6: cortex, 7: epidermis.

(Source: wikipedia.org/wiki/File:Stem-histology-cross-section-tag.svg)

In the middle of the stem is the pith (1). Pith is soft and spongy, made up of parenchyma cells that store and transport nutrients through the plant. Encircling the pith is the protoxylem (2), which is a mixture of pith and xylem cells. The xylem (3) functions to transport water and minerals throughout the plant. Wood is primarily made of xylem inside. Xylem cells are tubular like vessels, and provide some structural support; xylem also contains parenchyma and fiber cells. Around the xylem is the phloem (4), which carries organic nutrients throughout the plant and is made up of parenchyma cells, conducting cells, and supportive fibers. Most importantly, it distributes sugars like sucrose, and organics made during photosynthesis. Outside the phloem is the sclerenchyma (5), or bast fiber, which provides support to plants. The sclerenchyma, sometimes also called the "inner skin," is made up of fiber cells and sclereid cells, and has very thick cell walls. Next out is the cortex (6), which is just under the epidermis, and is mostly undifferentiated cells. Finally, the outer layer is the epidermis (7), which protects the plant from water loss, absorbs water and minerals, controls gas exchange, and secretes organic compounds. The epidermis has various layers and differs plant by plant, but is mostly made up of specialized parenchyma cells. In wood, the epidermis is surrounded by the periderm (bark) for protection.

The structure and arrangement of the plant cells greatly dictates the heat and mass transfer properties during pyrolysis. Figure 26.4 illustrates how heat enters into a lignocellulose particle through the sides and the ends, and the vapors produced during pyrolysis first leave through the axial ends of the particle through the cell centers, and the through the sides as natural pits between cells expand while the wood pyrolyzes.

If the aerosols produced during pyrolysis hit the solid cell walls, they are likely to be retained, and secondary reactions are likely to convert those molecules into char or smaller, less desirable molecules. During slow pyrolysis, this is fine because we want to maximize char production. However, during fast pyrolysis, this is undesirable because larger pyrolysis vapor and aerosol products are more valuable than char or lighter compounds. We will see what those molecules are in the next sections.

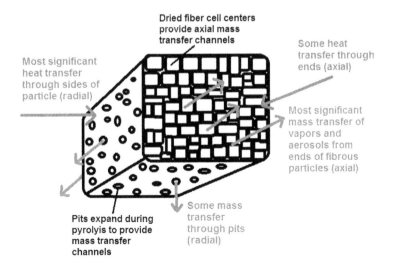

FIGURE 26.4

Wood particle heat and mass transfer during pyrolysis.

So in general, pyrolysis reactions can be classified into: (1) primary and (2) secondary reactions. Primary reactions happen on cell walls and are responsible for the formation of liquid intermediates or volatile products. Secondary reactions happen when the products of primary reaction are exposed to the action of natural catalysts and heat on their way from the surface of biomass cell walls to the condenser where the pyrolysis vapors are cooled down and the secondary reactions are dramatically slowed down. In general, secondary reactions can be classified in homogeneous and heterogeneous reactions, depending whether they happen on the bulk of the vapor phase or on the surface of the converting biomass particle.

Depending on the size of the particle used, the fast pyrolysis can be divided in two main regimes: (1) single cell walls and (2) particles that conserve cell wall structures as illustrated in Figure 26.5. The main mechanism for the removal of the monomeric products of biomass pyrolysis reactions is evaporation. The oligomeric products cannot evaporate, but can leave through thermal ejection. Thermal ejection happens due to the rupture of bubbles (spitting) formed by volatile species in the liquid intermediates (Teixeira et al., 2011). In the case of relatively large biomass particles with cell wall structures, the thermally ejected oligomeric products hit the internal cell wall and are retained (Zhou et al., 2013; Kersten and Garcia-Perez, 2013).

CELLULOSE: 40–45 DRY WT%

Cellulose is the main structural component in lignocellulose, and makes up the cell walls, as illustrated in Figure 26.6. The cell walls are made up of a mesh of fibers that are composed of microfibrils. The

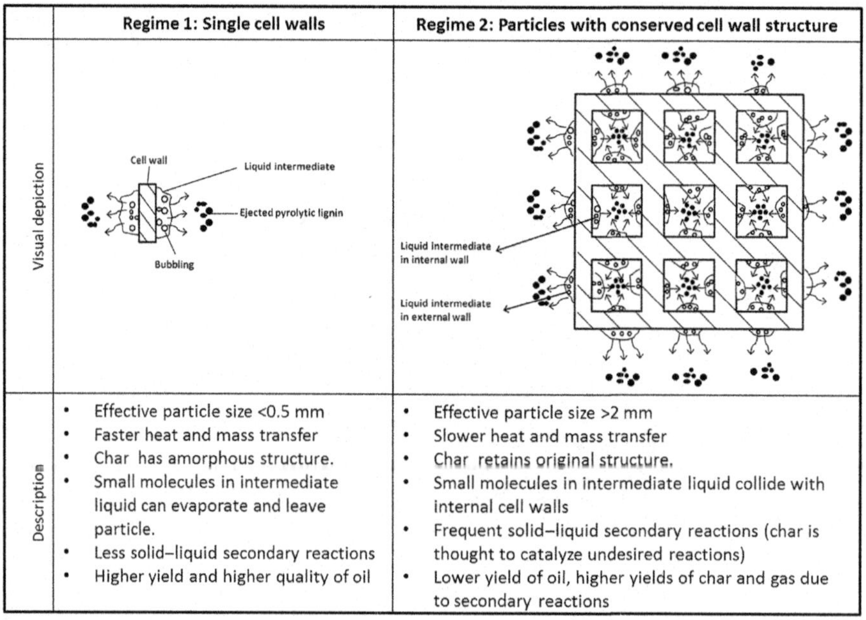

	Regime 1: Single cell walls	**Regime 2: Particles with conserved cell wall structure**
Visual depiction		
Description	• Effective particle size <0.5 mm • Faster heat and mass transfer • Char has amorphous structure. • Small molecules in intermediate liquid can evaporate and leave particle. • Less solid–liquid secondary reactions • Higher yield and higher quality of oil	• Effective particle size >2 mm • Slower heat and mass transfer • Char retains original structure. • Small molecules in intermediate liquid collide with internal cell walls • Frequent solid–liquid secondary reactions (char is thought to catalyze undesired reactions) • Lower yield of oil, higher yields of char and gas due to secondary reactions

FIGURE 26.5

Two significant pyrolysis regimes based on the effective biomass particle size (Zhou, 2013).

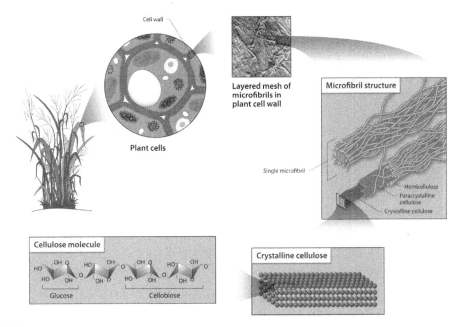

FIGURE 26.6

Cellulose: the structural backbone of lignocellulosic material.

(Image source: Biological and Environmental Research Information System, Oak Ridge National Laboratory)

microfibrils are made up of crystalline cellulose at the center, then paracrystalline cellulose, and surrounded by hemicellulose, which acts to connect the microfibrils together like glue.

Cellulose is made up of $(1 \rightarrow 4)$ linked β-D-glucopyranose; this is just glucose linked together by ether bonds into straight chains of between 10,000 and 15,000 units long. Cellulose is the most abundant chemical species in wood, making up 40–45 dry weight%.

The most desirable product from the pyrolysis of cellulose is the sugar levoglucosan, shown below. In current fast pyrolysis reactors, approximately 10–20 wt.% conversion of cellulose into valuable sugar products is typical. However, the theoretical maximum conversion of cellulose into hydrolysable sugars (levoglucosan, cellobiosan, and oligo-anhydrosugars) can be nearly 100 wt.%. During pyrolysis, cellulose breaks down into its smaller sugar units through degradation reactions. The exact chemistry behind this reaction mechanism is difficult to pinpoint, but the general scheme is fairly well understood: primary reactions from cellulose produce sugars (levoglucosan, cellobiosan, etc.), and secondary reactions (1) break the primary products into smaller molecules and (2) convert primary products into larger molecules and char, as explained well by Mamleev et al. (2009).

There is a continuous debate among cellulose pyrolysis researchers as to what the lump-reaction scheme is for cellulose–pyrolysis. It is difficult to verify each pyrolysis scheme experimentally, but they are important for fitting data and approximately predicting product yields. Further complicating conclusive results is the fact that different kinetic data can be collected with each reactor setup due to fluid mechanics and heat transfer effects. Nevertheless, it is important to be aware of what is in the literature, and Figure 26.7 provides common examples with some kinetic data where available.

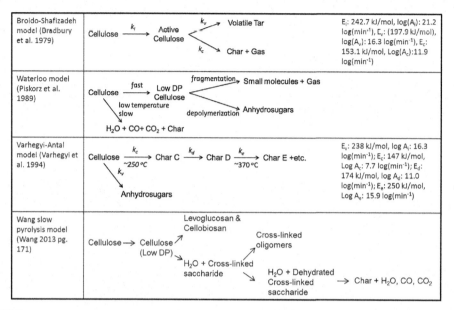

FIGURE 26.7

Common cellulose pyrolysis lump-reaction schemes and a recent one for slow pyrolysis (Bradbury et al., 1979; Piskorz et al., 1989; Varhegyi et al., 1994).

Kinetic data follows the Arrhenius rate law equation, which is itself a debatable equation to use for pyrolysis kinetics.

Primary cellulose pyrolysis reactions

During cellulose primary reactions, the $1 \rightarrow 4$ ether bond between glucopyranose monomer units is broken, and intramolecular chemical rearrangement internalizes the $1 \rightarrow 4$ to produce levoglucosan, shown in Figure 26.8.

During fast pyrolysis, both the amorphous and crystalline cellulose first melt into a liquid phase formed by levoglucosan and oligo-anhydrosugars, then the products are able to evaporate or are thermally ejected. It is in this liquid phase that many of the char-forming reactions take place. Figure 26.9 shows some of the proposed reaction mechanisms for the important reactions in cellulose pyrolysis that we will discuss.

The depolymerization of cellulose into levoglucosan under fast heating rate conditions takes place above 300 °C. Kinetic parameters vary depending on the reaction scheme used.

Under slow heating rate conditions, while the amorphous cellulose melts at temperatures over 300 °C, the crystalline cellulose remains as a solid mostly due to presence of cross-linking reactions (Wang et al., 2013). Remember, plant cells vary throughout the organism, and there are multiple types of cellulose. In the cell wall, cellulose oscillates between crystalline and noncrystalline cellulose. It is generally understood that the noncrystalline (amorphous cellulose) is the first to pyrolyze, followed by crystalline cellulose, which requires some heat to break the interchain bonds.

The primary products of cellulose include not only levoglucosan, but also cellobiosan (dimer), cello-triosan (trimer), and some larger sugars. However, the larger sugars are less likely to evaporate at pyrolysis

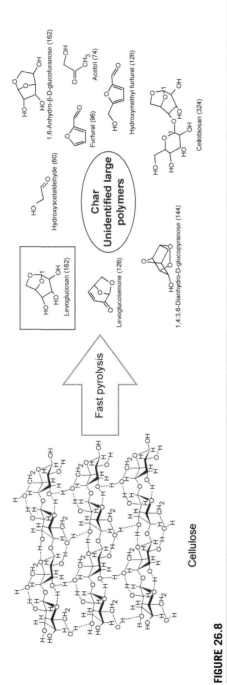

FIGURE 26.8

Important primary and secondary products of cellulose pyrolysis.

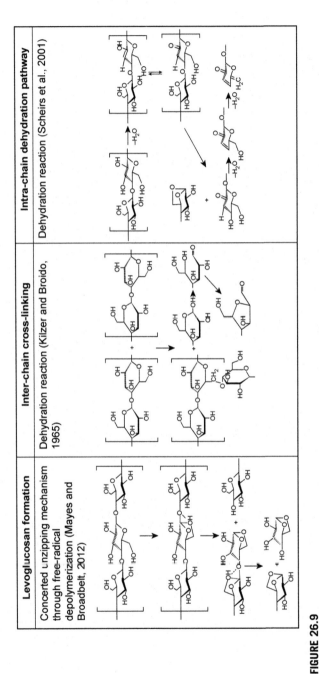

FIGURE 26.9

Cellulose pyrolysis reaction examples (Kilzer and Broido, 1965; Mayes and Broadbelt, 2012; Scheirs et al., 2001).

conditions; their removal has to happen through a mechanism called "thermal ejection." Levoglucosan has a boiling point of 290–300 °C, cellobiosan $\sim$581 °C, and cellotriosan $\sim$792 °C. Therefore, the larger sugars above levoglucosan are typically not observed under atmospheric pressure pyrolysis conditions because they simply remain in the liquid phase and turn into char or smaller undesired products.

It is important to realize that levoglucosan is typically observed in fast pyrolysis, or at slow pyrolysis under vacuum. During slow pyrolysis, dehydration reactions predominate, as was illustrated in Figure 26.9. Furthermore, high yields of levoglucosan are only seen with small particle sizes and with rapid cooling of the vapors or in tests under vacuum. One reason for this is that the slow heating rate achieved in the large particle sizes used in slow pyrolysis enhances dehydration reactions.

In the presence of alkalines cellulose, fragmentation reactions responsible for the formation of C1–C4 are catalyzed (Patwardhan et al., 2010). The main products of these reactions are: acetol and hydroxyacetaldehyde. Figure 26.10, shows the mechanism proposed by Richards 1987 to explain the formation of glycoaldehyde. Further reaction mechanisms for cellulose fragmentation reactions can be found in pages 20–31 of Wang (2013).

Secondary cellulose pyrolysis reactions

Secondary reactions can occur in the solid phase, the liquid intermediate phase, or the vapor phase; they produce char small undesired products including like acetol, hydroxyacetaldehyde, furfural, hydroxymethylfurfural, and some rearranged or dehydrated sugars. Char formation is always seen simultaneously with the production of CO_2, H_2O, and H_2 (Antal and Varhegyi, 1995).

The secondary reactions in the liquid phase have already been discussed as being a source for char formation by a series of reactions starting with (1) intermolecular dehydration and (2) polycondensation to form polyaromatic solid carbonaceous material.

HEMICELLULOSE: 20–30 DRY WT% IN WOOD

Hemicellulose is similar to cellulose in that it is made up of sugars. However, hemicellulose is amorphous and much more diverse, made up of glucose, mannose, galactose, xylose, arabinose, O-methyl-glucuronic acid, and galacturonic acid. This makes it a heteropolysaccharide. Typically, hemicellulose has a degree of polymerization of around 200, meaning that each chain of hemicellulose has around 200 monomers linked together. Furthermore, the chains of hemicellulose can be branched out, rather than straight chains.

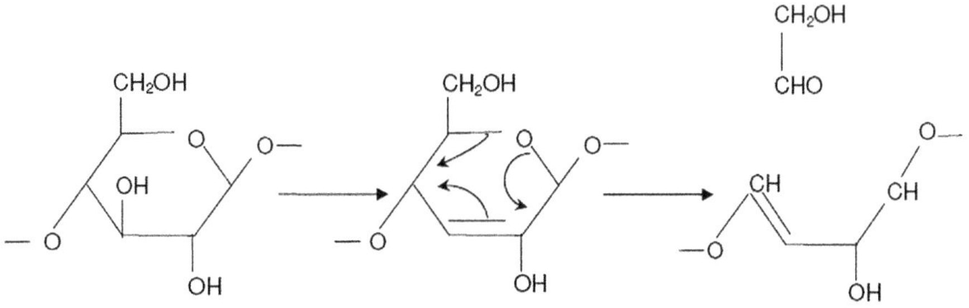

FIGURE 26.10

Mechanism for the fragmentation of levoglucosan to form glycoaldehyde (Richards, 1987).

Pectin is another component in lignocellulose, consisting of galacturonic acid. Pectin resides throughout the cell, working like hemicellulose as an adhesive. It is more abundant in fruit than in lignocellulose, so pectin is often roped into the hemicellulose content during biomass chemical characterization.

Hemicellulose undergoes weight loss (primary degradation reactions) at close to the same temperatures as amorphous cellulose, 250–300 °C. Its chemical products are difficult to study due to the fact that it is hard to extract native hemicellulose, so the sugar xylan is often used as a surrogate for experiments. It is found that hemicellulose is a source for acetic acid, formic acid, acetol, xylose furfural, and other small products during pyrolysis (Patwardhan et al., 2011).

LIGNIN: 15–36% DRY WT% IN WOOD

Lignin is a complex phenolic polymer that serves the purposes of mechanical support, water transport, and microbial defense. Lignin is the primary constituent of the middle lamella (the space between cell walls) and of the secondary cell wall, shown in Figure 26.11. In the cell wall, lignin binds with the cellulose–hemicellulose fibers to protect them during water transport up through the plant (Campbell and Sederoff, 1996).

Specific lignin chemistry varies from plant to plant, but the base is common: lignin is a network of phenylpropanoid units. The main alcohol base units are *p*-hydroxyphenyl, guaiacyl, and syringyl lignins, making the monolignols of paracoumaryl alcohol, coniferyl alcohol, and sinapyl alcohol, respectively. These monolignols are linked together by ether bonds (C–O–C) or carbon–carbon bonds (C–C). An example of lignin and the three primary monolignols is shown in Figure 26.12. Pyrolysis of lignin occurs over a broad temperature range, and is often classified into low-temperature reactions (180–300 °C) and high-temperature reactions (300–600 °C).

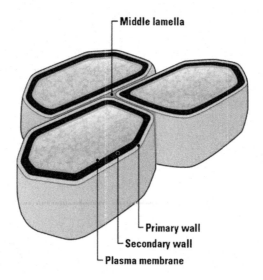

FIGURE 26.11

High lignin content in the middle lamella and the secondary wall.

(Image source: Biological and Environmental Research Information System, Oak Ridge National Laboratory)

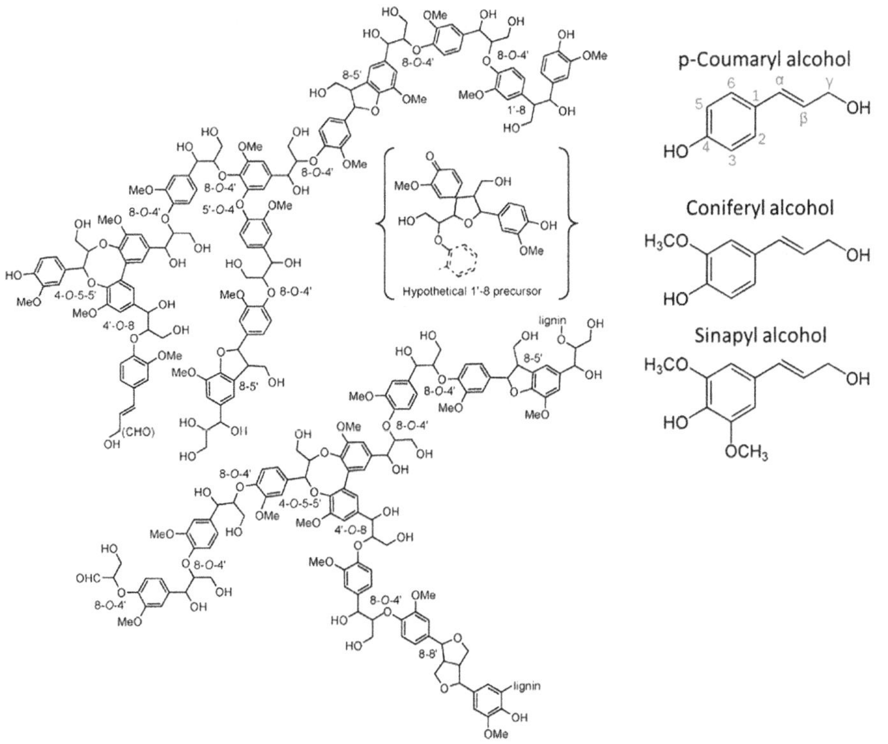

FIGURE 26.12

Lignin structure example and three primary mono-lignols.

(Image source: Biological and Environmental Research Information System, Oak Ridge National Laboratory)

Low-temperature lignin pyrolysis reactions

The low-temperature reactions between 180 and 300 °C are thought to produce the lignin oligomers that further crack to form lignin monomers.

The ether bonds are the first to break. Like in cellulose and hemicellulose pyrolysis, a liquid intermediate forms and persists when large molecules are not able to evaporate under the temperature and pressure conditions. As the liquid lignin remains heated at these temperatures, they will eventually cross-link and form char. The lignin oligomers thermally ejected by the liquid intermediate can further crack to form lignin monomers during vapor phase secondary reactions, as illustrated in Figure 26.13 (Zhou et al., 2013).

High-temperature lignin pyrolysis reactions

At higher temperatures, the lignin in the liquid intermediate phase will cross-link and subsequently polycondense to create polyaromatic carbonaceous solid. Important side products from the formation of char are the release of methanol and formaldehyde. In the early twentieth century, an entire industry was built around pyrolysis to produce these chemicals. At these high temperatures, vapor phase lignin oligomers will crack to form monomers as the vapors travel out of the reactor.

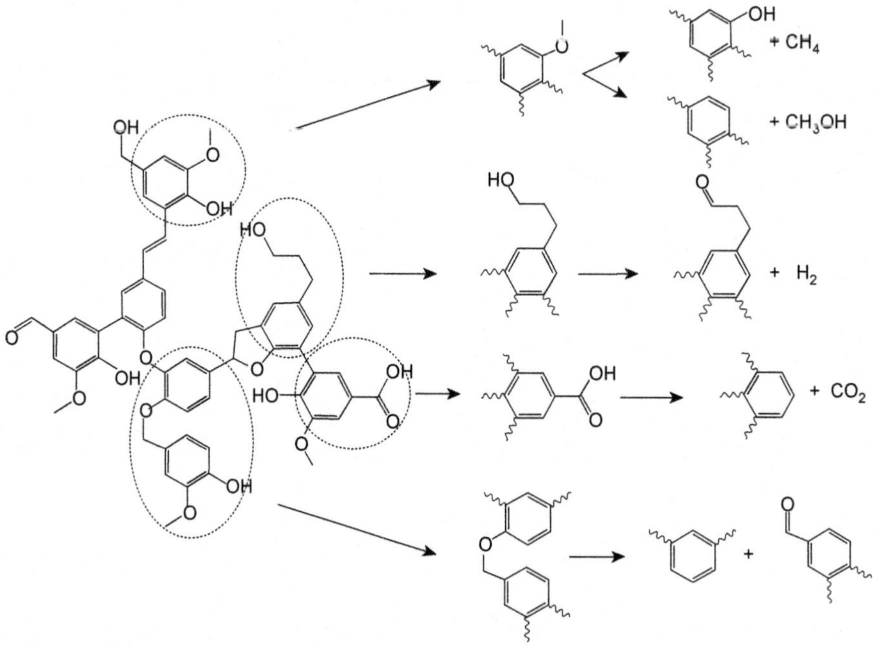

FIGURE 26.13

Monomer formation from lignin oligomers during secondary reactions (Zhou et al., 2013).

EXTRACTIVES: 2–10 DRY WT% IN WOOD

Extractives are the nonstructural components of lignocellulose, including fats, phenolics, resin acids, waxes, and inorganics. In wood, phenolic extractives are found in the bark and heartwood. Waxes are found in the parenchyma cells. The phenolic extractives serve the purpose of protecting lignocellulose from fungal and microbial attack, and can be washed out as "resin" from live trees. Turpentine was an important chemical industry as a source for paints, solvents, and varnishes. Phenolic extractives have a chemical base similar to lignin, and contain a large portion of guiacol and syringol. The fats provide energy storage in the biomass and are stored inside paranchyma cells. Note that the content and character of the extractives varies from biomass to biomass, and even varies between different parts of each plant.

Organic extractives, which you may know are quite sticky, cause clogging and produce low-quality oil during pyrolysis. Because a large source for resin is at the bark level of live trees, trees are typically de-barked before being ground and sent into a pyrolysis reactor. Furthermore, storing the wood for a period of time reduces some of the issues with extractives as they are naturally oxidized into smaller compounds.

Lignocellulosic biomass also contains inorganic material such as calcium, potassium, sodium, magnesium, iron, and manganese. In pyrolysis, the inorganics are known to decrease the yield of sugars from cellulose and change the properties of lignin products. Grasses and smaller biomass also contain significant amounts of soil and dust by nature. Since these inorganics tend to increase the char yield, washing the biomass with mild acids is often a good way to improve the product yield.

BIOMASS PYROLYSIS STRATEGIES

There are several strategies for producing and utilizing the products of pyrolysis: biochar, bio-oil, and gas. Operating parameters can be altered to increase or decrease the creation of certain desired products. For example, to maximize the production of char, it is better to use chopped wood than sawdust. Under these conditions, relatively slow heating rates are achieved. The pyrolysis reactors can be coupled with other reactors to take advantage of heat recovery, combustion, etc. Four general examples of pyrolysis strategies can be seen in Figure 26.14.

To engineer pyrolysis reactors, we need to incorporate some way to physically model the process. Numerical values and dimensionless analysis can tell us the controlling mechanism for pyrolysis, based on various parameters. Single particle models are often used to describe the competing effect of internal and external heat transfer and reaction kinetics on the outcome of the pyrolysis process. Phenomena that can be accounted for include external heat transfer, internal heat transfer, molecular diffusion, chemical reaction rates, particle shrinking, pore expansion, etc. (Pyle and Zaror, 1984; Paulsen et al., 2013; Levenspiel, 1999).

Modeling multiple phenomena simultaneously can become a very laborious project. Therefore, quick calculations are typically done by simplifying the model to a one-dimensional model, based on a controlling regime (the slowest phenomenon). The typical controlling regimes are (1) external heat transfer, (2) kinetic, and (3) internal heat transfer. Dimensionless numbers can guide us to which regime to select for the model, or whether we cannot select a regime at all. Table 26.1 describes the important dimensionless values for one-dimensional pyrolysis models (Di Blasi, 2008).

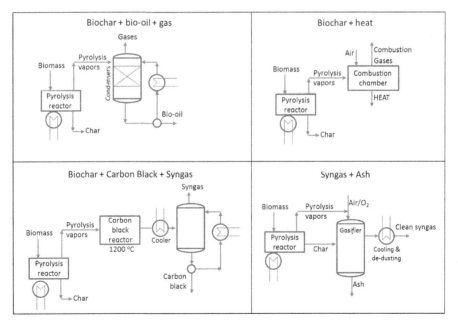

FIGURE 26.14

Four strategies for pyrolysis reactors (Paláez-Samaniego et al., 2008).

Table 26.1 Dimensionless Values for One-Dimensional Pyrolysis Models (Pyle and Zaror, 1984)

Dimensionless Number	Value	What It Compares
Biot number (Bi)	hR/K	External and internal heat transfer
Pyrolysis number (py)	$K/(k\rho c_p r^2)$	Internal heat transfer and reaction rate
Pyrolysis number' (py')	$h/(k\rho c_p R)$	External heat transfer and reaction rate

Note: h = heat transfer coefficient, k = apparent rate constant, R = particle radius, ρ = bulk particle density, K = thermal conductivity, c = bulk heat capacity.

The dimensionless values in Table 26.1 are calculated using empirical values for the constants within the numbers, or averages for constants that vary throughout the pyrolysis reaction like density, and the radius. Once these values are calculated, the range of validity can be determined according observed values, shown in Table 26.2 with descriptions of the controlling regimes. Common values for the thermal conductivity of wood are between 0.1 and 0.05 W/mK.

SLOW PYROLYSIS: THE PATH TO CHAR

Slow pyrolysis maximizes char production. Large particles (>2 mm) and slow heating rates (<10 °C/min) to temperatures between 400 and 600 °C are common operational parameters of slow pyrolysis reactors. Typically, liquid yields of 30–50 wt% and char yields of 25–35 wt% are observed. The vapors do not escape rapidly, and tend to remain in the reactor for 5–30 min.

There are two important types of slow pyrolysis: carbonization (heating for days) and conventional pyrolysis (heat for 5–30 min). In carbonization, the vapors are typically vented to the atmosphere or flared to produce more char, and ignore the vapor products. Carbonization is an environmental issue, and is often only allowed in developing countries for cooking charcoal, as evident in Figure 26.15 (Garcia-Perez et al., 2010). For the design of carbonization units, the reader is recommended to read the Handbook of Charcoal Making of Emrich (1985).

Conventional pyrolysis is what the early pyrolysis industry was built around, and allows for the collection of char, oil, and noncondensable vapors. The oil separates into an organic phase and an aqueous phase when cooled; the aqueous phase is where the methanol and acetone are found. In most reactor setups, the vapor (oil and noncondensable vapors) are combusted to recover process heat, electricity, or both. The heating value of the oil from slow pyrolysis is around 25 MJ/kg (Garcia-Perez et al., 2010).

Table 26.2 Regions of Validity for One-Dimensional Pyrolysis Models (Pyle and Zaror, 1984.)

Model	Range of Validity			Description
	Bi	Py	Py'	
Noncontrolling conditions		*All values*		Needs mathematical modeling
External heat transfer	<1	>1	>1	Low heat transfer coefficient, small particle
Kinetics	<1	>10	>10	High heat transfer coefficient, small particle (i.e., Fast pyrolysis)
Internal heat transfer	>50	$<10^{-3}$	<<1	Large particle size (i.e., Slow pyrolysis)

FIGURE 26.15

Carbonization of biomass in Poland, illustrating the environmental impact of releasing vapors into the air.

(Photo by ecksunderscore @ flickr)

Biochars are a desirable product for applications in (1) soil amendment-fertilizer (Lehmann and Joseph, 2009), (2) nutrient adsorption, (3) filtration, (4) gasification or combustion, and (5) atmospheric carbon sequestration. Figure 26.16 shows how biochar can enhance soil quality for agricultural purposes. Heating values for the char vary from 15 to 30 MJ/kg, depending on the feedstock. The problem with biochar is that it has a low density, which often limits the economic feasibility of transport.

Based on the fact that slow pyrolysis typically has lower yields of oil (30–50 wt%), the 2-phase oil produced by slow pyrolysis is typically combusted or gasified. However, it is possible, and often desirable, to extract some valuable chemicals from the aqueous phase of slow-pyrolysis oil: acetone/ketones ($\sim$5% yield dry wood basis), methanol (1–2%), formic/acetic acid (5–8%).

In a bioprocess using slow pyrolysis, there are many setup variations, depending on the economics. Some gasify the char, others gasify the pyrolysis vapors, and others burn all the vapors for process heat or electricity.

There are many types of reactors used for slow pyrolysis (see Table 26.3) (Garcia-Perez et al., 2010). Kilns were the earliest in use, and can be made by simply covering chopped wood with dirt, igniting, and partially combusting some of the wood for heat. Then, retorts were developed. Retorts heat the biomass through the walls of the reactor vessel, so the resulting oxygen content of the oil and the char will be much lower than in kilns. A shelf reactor is a continuous version of the kiln reactor, with mobile modules for the biomass. Beyond kilns and retorts, there are more innovative reactor setups. The Reichert converter uses separate reactors for drying, carbonization, and cooling. A Lambiotte reactor is a downdraft style reactor where biomass is added at the top and there are different regions for drying, carbonization, and cooling. A Herreshoff reactor is a multilevel reactor with blades to push biomass down to the subsequent levels. Rotary drum, auger, agitated bed, and paddle kiln reactors physically push or shake biomass through the reactors to increase heat transfer.

FAST PYROLYSIS: THE PATH TO OIL

Fast pyrolysis maximizes oil production. Small particles (desirably <2 mm) and fast heating rates (>100 °C/s) to temperatures between 400 and 650 °C are common characteristics of fast pyrolysis. Typically, liquid yields of 60–75 wt%, char yields of 15–25 wt%, and noncondensable gas

FIGURE 26.16

Soil amended by biochar (terra preta) on the right improves the crop yield compared to unamended soil on the left.

(Image source: http://www.biochar-international.org/biochar/soils)

yields of 10–15 wt.% are observed. Reactors are designed to remove and condense vapors in less than 2 s. Excellent reviews on fast pyrolysis can be found elsewhere (Meier and Faix, 1999; Czernik and Bridgwater, 2004, Bridgwater, 2012; Kersten and Garcia-Perez, 2013).

The oil typically contains between 10 and 30% water, depending on the moisture of the biomass. It is typically dark brown (or closer to red or dark green depending on the feedstock), very viscous, and smells like a campfire crossed with barbecue sauce (Mohan et al., 2006; Czernik and Bridgwater, 2004). Table 26.4 shows some of the most common and notable properties of bio-oil from fast pyrolysis. Just-condensed bio-oil from the fast pyrolysis of clean wood is typically one main phase; however, oils derived from bark, leaves, or materials with high alkaline contents exhibit a multiphase structure with char particles, nonpolar extractive rich materials, waxy material, aqueous droplets, organic droplets, and micelles forming separate phases (Garcìa-Pérez et al., 2006; Frantini et al., 2006). Furthermore, if the oil is reheated, polymerization occurs. All of these properties combined make the oil difficult to transport. Because of that, strategies to stabilize the oil are typically used, including catalytic hydrogen treatment, esterification, or simply solvent addition.

Table 26.3 Slow Pyrolysis Reactors and Details (Garcia-Perez et al., 2007; 2010; Beaumont, 1985)

Reactor Type	Size	Mode of Operation	Heating Method	Construction Materials	Reactor Details	Comments
Earth kiln	Chopped wood	Batch, 7–30 days	Partial combustion, foliage combustion	Earth, often in a pit	Stationary; size varies; piles of wood covered in dirt	Slow cooling, high emissions. Short lifetime
Brick kiln	Chopped wood	Batch, ~30 days	Partial combustion; hot gas contact	Cinder block, brick, iron hardware	Stationary; 3–300 m^3; longer lifespan than earth kilns	Slow cooling, no vapor recovery
Large "American" kiln	Chopped wood	Semibatch, ~30 days	Partial combustion; hot gas contact	Firebrick/iron	Stationary; 300–1000 m^3; vapor collection common	Slow cooling
Small retorts without liquid recovery	Chips, chopped wood	Semibatch, 8–12 h	Indirect through walls via burning of pyrolysis gases and vapors	Oil drums, cinder block	Portable; 1/5 m^3; vapors are led to bottom of drum for combustion	
Small retorts with liquid recovery	Wood chips, chopped wood	Semibatch	Indirect through walls via burning of pyrolysis noncondensable gases	Steel, brick	Stationary; 1–2 m^3; vapor condenser system	Collection of organic acids, methanol, and acetone is possible
Wagon retort	1-m Long logs, chopped wood	Continuous	Indirect through walls via burning of pyrolysis noncondensable gases	Iron, train cars	Stationary; (1) carbonization chamber with vapor condensers, (2) char cooling chambers	20–200 cords per day
Shelf reactor	Chips	Semicontinuous	Indirect heating by vapor combustion	Metal/cinder block	Trucks carry around shelves with biomass, move them from drying to carbonization to cooling units	Fairly fast operation, good for medium-to-small processes
Reichert converter	Chopped wood	Semicontinuous	Direct contact with recirculating hot gases	Iron, steel	Stationary or mobile; multichambers for drying, carbonization, and cooling; combustion of vapors for heat	Higher equipment costs, lower labor costs; no bio-oil collection

Continued

Table 26.3 Slow Pyrolysis Reactors and Details (Garcia-Perez et al., 2007; 2010; Beaumont, 1985)—cont'd

Reactor Type	Size	Mode of Operation	Heating Method	Construction Materials	Reactor Details	Comments
Lambiotte reactor	Chopped wood, wood chips	Continuous	Direct contact with recirculating hot gases	Steel, metal	Stationary or mobile; vertical "down draft" setup: (1) drying region, (2) carbonization, (3) cooling in series	One of the best slow-pyrolysis technologies; vapor condensation is possible
Herreshoff multiple-hearth furnace	Chips, fine particles	Continuous	Direct contact with recirculating hot gases	Metal	Stationary; vertical, multiple levels with rotating blades to move biomass around	Vapors can be condensed; potential use for fast pyrolysis as well
Rotary drum converter	Chips	Continuous	Indirect heating through walls/pipes	Metal	Stationary or mobile; horizontal drum rotates inside a heated vessel	Vapors can be condensed; potential use for fast pyrolysis as well
Auger reactor	Chips, fine particles	Continuous	Indirect heating through walls, hot sand, and/or hot gases	Metal	Stationary or mobile; screw pushes biomass through kiln	One of the best slow pyrolysis technologies; vapors can be condensed; potential use for fast pyrolysis as well
Moving agitated bed	Chips, fine particles	Continuous	Indirect heating via hot gas or molten salt	Metal	Stationary; angled bed shakes to move product through reactor	Also suitable for fast pyrolysis and can have vacuum capabilities
Paddle pyrolysis kiln	Chips	Continuous	Indirect heating through walls	Metal	Paddles inside reactor mix biomass to increase heat transfer; vapor collection is possible	Fairly robust setup

Table 26.4 Some Notable Properties of Bio-Oil (Garcìa-Pérez et al. 2006, 2007; Mohan et al., 2006)

Property	Description
Appearance	Brown to black, depending on the feedstock
Structure	Multiphase structure at room temperature due to the presence of char particles, waxy material, aqueous droplets, organic droplets, micelles, and water. Greater homogeneity is observed above 60 °C. In poor-quality oil, the oil separates into heavy (organic) and light (aqueous) layers.
Density	~1.2 kg/L at 20 °C
Kinematic viscosity	Varies greatly: 50–672 cst (20 °C), 35–300 cst (40 °C), 5–200 cst (50 °C)
Water content	15–30 wt% From dry biomass; up to 50 for moist biomass
High heating value (HHV)	20–24.3 MJ/kg (anhydrous), 15–18 (as produced)

Discounting the water, the elemental mass composition of the oil is typically: 44–47% carbon, 6–7% hydrogen, 46–48% oxygen, and 0–0.2% nitrogen.

Table 26.5 lists some of the most commonly observed small chemicals in bio-oil from fast pyrolysis of lignocellulosic material. However, because of its large content of organic molecules, fast pyrolysis bio-oil can be upgraded in a variety of ways to produce fuels and chemicals. There are over 3000 unique compounds found in bio-oil.

Reactors

As discussed in the beginning of this section, fast pyrolysis ideally occurs in the kinetically-controlled region with small particle sizes and high heating rates. This section describes reactors commonly used in fast pyrolysis that maximize heat transfer. Furthermore, in fast pyrolysis, it is important to condense the vapors rapidly or risk excessive secondary reactions while the vapors travel through the hot reactor. In order to limit secondary pyrolysis reactions, vapors should be inside the reactor less than 2 s. This is

Table 26.5 Abundant Chemical Species Observed in Bio-Oil (of Around 3000 Compounds) (Garcia-Perez et al., 2010)

Cellulose/Hemicellulose Derived Compounds	Lignin-Derived Compounds	Common Derivatives
Levoglucosan	Isoeugenol	Char
Hydroxyacetaldehyde	2,6-dimethoxyphenol	Water
Acetic acid	Phenol	CO, CO_2, CH_4, C_2H_6
Acetol	4-ethyl phenol	
Furfural	2-ethyl phenol	
Furfuryl alcohol	p-cresol	
Cellobiosan	o-cresol	
2-methyl-2-cyclopenten-1-one	m-cresol	
3-methyl-2-cyclopenten-1-one		

accomplished by applying a carrier gas to drag the vapors out of the reactor. In the laboratory, the carrier gas is often nitrogen, but large-scale systems typically use combustion off-gases, and typically those gases are preheated. In fluidized-bed reactors, the carrier gas is extremely important because it also acts to lift the solids and fluidize the biomass/sand slurry.

Most of the heat transfer in fast pyrolysis reactors is through solid–solid conduction (contact between the biomass and the sand), but some is also from gas–solid and radiation. Due to the low thermal conductivity of biomass, a temperature increase of 10,000 °C/s can be achieved at the surface. If large particles are used, the particles might be cool in the center. This highlights the importance of small particle sizes for fast pyrolysis, or ablative methods to simultaneously break down the biomass.

Some notable issues with fast pyrolysis reactors are the following: (1) high alkaline content causes problems, (2) conversion of cellulose into sugars is typically not effective, (3) the oils are unstable due to acetic acid content and multiphase nature, (4) carrier gases dilute pyrolysis gases, (5) small molecules with <5 carbons have limited value, (6) hydrotreatment of crude bio-oils also causes catalyst deactivation through coking, and (7) high amounts of hydrogen are required for hydrotreatment (Table 26.6).

APPLICATIONS AND APPROACHES FOR PRODUCT USE

CHAR COMBUSTION

Biochar is commonly combusted for process heat, steam, or electricity. In fluidized bed reactors, this is the main use for the char, which is often mixed with sand and cannot be easily sold for other purposes. Biochar has heating values from 15 to 30 MJ/kg but a low bulk density, ranging from 200–1000 kg/m^3 depending on the feedstock.

The residual ash from char combustion can be used for cement production or other purposes.

CHAR AS SOIL AMENDMENT AND NUTRIENT ADSORBER

It is widely accepted that biochar can greatly improve the properties of soil, through water and nutrient retention. In South America on the Amazon basin, a type of soil called "Terra Preta" has very high carbon content and some of the best growing properties in the world. Therefore, markets are being developed for biochar as a soil amendment for farmers.

Char has also been recently explored as a cheap method for recovering nutrients from liquid runoff from field fertilization or effluents from anaerobic digestion. Once the nutrients are absorbed, they can potentially be reapplied to fields for slow release of those nutrients (Antal and Grønli, 2003).

BIO-OIL COMBUSTION

Table 26.4 touched upon some of the fuel properties of bio-oil: density of around 1.2 kg/L, relatively high viscosity of 5–200 (50 °C), solid content, corrosiveness, alkaline metal content, high ignition temperature, heating value 15–18 MJ/kg including water, and chemical instability. The properties of bio-oils do not exactly match those of petroleum heating oils, and the acids/aldehydes and water devalue the bio-oil (Chiaramonti et al., 2007). Oil classifications and storage strategies can be seen elsewhere, notably in reports by Oasmaa and Czernik (1999) and Czernik and Bridgwater (2004).

Table 26.6 Fast Pyrolysis Reactors

Reactor Type	Size	Mode of Operation	Heating Method	Construction Materials	Reactor Details	Comments
Bubbling fluidized bed	<0.2 mm diameter, predryed	Continuous	Hot circulating sands, hot reactor walls, and hot gas; external combustion for heat	Metal (with sand fluidizing medium)	The carrier gas fluidizes sand and biomass enters through the side of the reactor. As char is produced, it is blown off the top of the reactor and the sand falls back down	Very common in fast pyrolysis; easy control of temperature; mobile units available; low quality char due to sand; plugging is common
Circulating fluidized bed	~6 mm	Continuous	Hot circulating sands, hot reactor walls, and hot gas; external combustion for heat	Metal (with sand fluidizing medium)	Like fluidized bed, but more heat transfer is done by the walls with fast circulation of the solids within the reactor	Allows for larger particles than fluidized bed; rapid condensation of pyrolysis vapors; complex design and operation
Ablative reactors (e.g., cone reactor)	Chips	Continuous	Hot, moving metal surface	Metal	Biomass is pressed against a moving, heated surface to melt biomass surface and evaporate oil. Carrier gas not necessarily required. Examples: Spinning cone reactor, pyrolysis mill, ablative plate, or tube reactor	Complex to design, wear and tear on hot, moving metal. High particulate matter in oil
Auger reactor	<0.2 mm diameter	Continuous	Indirect heating through walls, hot sand, and/or hot gases	Metal	Stationary or mobile; screw pushes biomass through kiln. Some carrier gas is useful	Small particle sizes can achieve fast-pyrolysis effectively. Simple design and operation

Combustion of bio-oils is advantageous due to the efficiencies of the Rankine cycle for producing electricity. However, bio-oil can cause problems in gas turbines and diesel engines due to property variation, ash content, and low cetane numbers (0–27). The combustion of bio-oil droplets have been extensively studied (Chiaramonti et al., 2007; Wornat et al., 1994; D'Alessio et al. 1998, Garcia-Perez et al., 2006).

LIQUID FUEL PRODUCTION AND UPGRADING OF BIO-OIL

One of the most popular research fields in pyrolysis today is in the production of liquid fuels. Because of the wide variety of products in bio-oil, various fractions can be used for different purposes. Furthermore, there are a wide variety of pathways to reach liquid fuels. The main advantage of producing renewable fuels from pyrolysis is that the wide range of molecule sizes allows for production of gasoline equivalents, jet fuel equivalents, asphalt equivalents, and more. Jet fuel, for example, it very difficult to produce economically through biological methods.

Upgrading pyrolysis oil is crucial for developing valuable products. The following upgrading strategies are currently considered important (Proceedings from tcbiomass2013): (1) Dilution/ solubilization with alcohol for stabilization, (2) Esterification or acetilization to eliminate acids/ carbonyls, (3) Hot gas filtration, (4) Catalytic hydro-deoxygenation (HDO) of oil after pyrolysis, (5) Catalytic pyrolysis, and (6) Aqueous phase fermentation.

1. Dilution and solubilization of bio-oil can greatly improve its properties. Through this dilution, viscosity decreases and aging properties are improved. Examples of solvents include biodiesel, methanol, ethers, and other alcohols. Because this adds a large volume of externally produced chemicals (up to 60%), it can become quite expensive to improve oil through this path.
2. Another method for stabilizing oils is by chemically reacting the acid and carbonyl groups that normally produce most of the negative properties in bio-oil. These chemical reactions are called esterification and acetalization reactions. This is similar to dilution/solubilization, but involves chemical reactions (Li et al., 2011; Hu et al., 2011).
3. Hot gas filtration can also greatly improve the quality of bio-oil. Recall that crude bio-oil contains char particles and alkalines that blow out with the oil. The pyrolysis oil can be filtered as it comes out of the reactor through candle filters with ceramic elements. Successful tests have shown that this method improves the bio-oil physicochemical properties, stability, and ash content.
4. Crude bio-oil is 46–48 wt.% oxygen, or about 22 mol% oxygen, in the form of alcohols, ether, carbonyl groups, acids, etc. It is desirable to remove the oxygen and simultaneously stabilize and upgrade the oil to fuel-quality, with a final oxygen content of less than 0.5 wt%. Yields of around 40 wt.% total usable liquids are achieved, of which around 40 wt.% is gasoline, 40 wt.% diesel, and 20 wt% fuel oil (with $\sim$29% JetA overlapping between fuel-oil and diesel). Researchers have discovered that catalytic reactions with hydrogen gas and bio-oil are an effective way to achieve this goal. Most researchers perform a 2-step hydrotreatment with two different catalysts and different temperatures (Elliott, 2007; Elliott et al., 2009). The first hydrotreatment is low temperature ($\sim$170 °C), and the second is more severe ($\sim$400 °C). Oxygen leaves as H_2O and carbonyl groups are converted to alcohols. Gasoline, diesel, and jet fuel equivalents have been produced at the lab scale and small pilot scale with this method. Issues include catalyst

deactivation, high costs for hydrogen gas (consumed at 0.04–0.07 g/g oil), and heat inefficiencies from condensing, reheating, compressing, and cooling the oil.

5. Catalytic pyrolysis could avoid some of the heat loss inefficiencies of cooling the bio-oil before catalytic upgrading. It does this by either adding catalyst pellets into a fluidized bed reactor or setting up a fixed catalyst bed in-line with the pyrolysis reactor. At the catalyst, the pyrolysis vapors react with the goal of reducing oxygen, conserving the carbon content, and producing a bio-oil with lower oxygen content. Typically, the products of catalytic pyrolysis are aromatics, which can be useful fuel additives or intermediates for the production of chemicals (Bridgwater, 1996; Williams and Nugranad, 2000).

6. The first five methods focus mainly on producing fuels from the organic, nonsugar portion of the bio-oil. In fact, the sugars in pyrolysis oil are thought to cause problems during catalytic reactions and are wasted by converting them into char. Alternatively, the sugars that are produced from the cellulose in biomass (predominantly levoglucosan) can be collected by fractional condensation and fermented into ethanol or lipids. Direct fermentation of levoglucosan has been reported, as well has hydrolysis of the sugars to produce glucose followed by fermentation (Prosen et al., 1993; Bennett et al., 2009; Lian et al., 2010; Jarboe et al., 2011). As discussed previously, sugar yields can be improved by washing the biomass with mild acids before pyrolysis. Issues with this strategy include the negative effects of bio-oil aldehydes and ketones on microbes. Therefore, strategies need to be developed to detoxify the aqueous liquid before fermentation.

USEFUL CHEMICALS FROM BIO-OILS

There are hundreds of compounds in bio-oil, and many more can be produced. Fuel alone cannot economically support a pyrolysis refinery, and higher valuable chemicals are crucial for keeping pyrolysis profitable. Some chemicals can be extracted from the oil through separation techniques, and others can be produced by combining all or parts of the oil with other compounds. More research needs to be explored in this field, but the following are some examples, similarly iterated before by Garcia-Perez et al. (2010) and Czernik and Bridgwater (2004):

1. Calcium salts and phenolates are useful for SO_x capturing in coal combustion, and can be produced by reacting the carboxylic acids ($\sim$1.2–2.1 mol/kg organics) and phenols ($\sim$1.8–2.1 mol/kg organics) with lime.
2. Terpenoids and phenols can replace creosotes as wood preservatives.
3. The heaviest fractions of pyrolysis oil can be used as tar for roofing or roads, as well as glues and sealants.
4. Fertilizers can be produced by reacting carbonyls (1.8–6.2 mol/kg organic) with ammonia, or by spraying biochar with pyrolysis oil.
5. Aldehydes and phenolics naturally present in the aqueous phase are useful for meat browning.
6. Road deicers can be produced by reacting the aqueous phase of bio-oil with calcium salts.
7. Resins and plastics can be produced from oligomeric lignin and sugars.
8. Methanol, acetic acid, and acetone can be recovered from slow-pyrolysis oil, as discussed before.

CONCLUSION

Biomass pyrolysis is an ancient technology that has had its ups and downs throughout history. Since the 1920s, it has generated interest for researchers who seek a simple way to gain wealth by converting biomass into liquid, char, and gases. Although there are many, many technologies available, hurdles need to be overcome to develop a sustainable method for producing fuels and chemicals from the oil, as well as higher-value products from the char and gases. Some companies have begun scaling up fast pyrolysis by using catalytic pyrolysis (Kior) or hydrotreatment (UOP-Honeywell) for the production of fuel. The future of pyrolysis will likely lie in a combination of simple technological innovations and heavy science to be able to predictably process the naturally heterogeneous feedstock of lignocellulose.

The following issues require particular attention in future research: (1) clogging issues at the entrance and exit of pyrolysis reactors and in condenser systems, (2) bio-oil stabilization, (3) catalyst deactivation, (4) development of strategies for selective chemical production, (5) identifying markets for high-value products from pyrolysis, and (6) developing pyrolysis technologies that can utilize the lowest quality feedstocks available.

REFERENCES

Antal, M.J., Grønli, M., 2003. The art, science, and technology of charcoal production. Industrial and Engineering Chemistry Research 42, 1619–1640.

Antal Jr., M.J., Varhegyi, G., 1995. Cellulose pyrolysis kinetics: the current state of knowledge. Industrial and Engineering Chemistry Research 34, 703–717.

Beaumont, E., 1985. Industrial Charcoal Making. FAO Forestry Paper, 63.

Bennett, N.M., Helle, S.S., Duff, S.J.B., 2009. Extraction and hydrolysis of levoglucosan from pyrolysis oil. Bioresource Technology 100 (23), 6059–6063.

Bradbury, A.G.W., Sakai, Y., Shafizadeh, F., 1979. A kinetic model for pyrolysis of cellulose. Journal of Applied Polymer Science 23, 3271–3280.

Bridgwater, A.V., 1996. production of high grade fuels and chemicals from catalytic pyrolysis of biomass. Catalysis Today 29, 285–295.

Bridgwater, A.V., 2012. Review of fast pyrolysis of biomass and product upgrading. Biomass and Bioenergy 38, 68–94.

Campbell, M.M., Sederoff, R.R., 1996. Variation in lignin content and composition (mechanisms of control and implications for the genetic improvement of plants). Plant Physiology 110, 3.

Chiaramonti, D., Oasmaa, A., Salantausta, Y., 2007. Power generation using fast pyrolysis from biomass. Renewable and Sustainable Energy Reviews 11 (6), 1056–1086.

Czernik, S., Bridgwater, A.V., 2004. Overview of applications of biomass fast pyrolysis oil. Energy and Fuels 18, 590–598.

Di Blasi, C., 2008. Modeling chemical and physical processes of wood and biomass pyrolysis. Progress in Energy and Combustion Science 34, 47–90.

D'Alessio, J., Lazzaro, M., Massoli, P., Moccia, V., 1998. In: Thermo-optical Investigation of Burning Biomass Pyrolysis Oil Droplets. Twenty-seventh Symposium on Combustion. The Combustion Institute, pp. 1915–1922.

Elliott, D.C., 2007. Historical developments in hydroprocessing bio-oils. Energy Fuels 21 (3), 1792–1815.

Elliott, D.C., Hart, T.R., Neuenschwander, G.G., Rotness, L.J., Zacher, A., October 2009. Catalytic hydroprocessing of biomass fast pyrolysis bio-oil to produce hydrocarbon products. Environmental Progress and Sustainable Energy 28 (3), 441–449.

Emrich, W., 1985. Handbook of Charcoal Making. The Traditional and Industrial Methods. Kluwer Academic Publisher.

Frantini, E., Bonini, M., Oasmaa, A., Solantausta, Y., Teixeira, J., Baglioni, P., 2006. SANS analysis of the Microstructural evolution during the aging of pyrolysis oils from biomass. Langmuir 22 (1), 306–312.

Garcìa-Pérez, M., Chaala, A., Pakdel, H., Kretschmer, D., Rodrigue, D., Roy, C., 2006. Multiphase structure of bio-oils. Energy and Fuels 20, 364–375.

Garcia-Perez, M., Lappas, P., Hughes, P., Dell, L., Chaala, A., Kretschmer, D., Roy, C., May 2006. Evaporation and combustion characteristics of biomass vacuum pyrolysis oils. Article number 200601 IFRF Combustion Journal. ISSN 1562–479X.

Garcia-Perez, M., Chaala, A., Pakdel, H., Kretschmer, D., Roy, C., 2007. Characterization of bio-oils in chemical families. Biomass and Bioenergy 31, 222–242.

Garcia-Perez, M., Lewis, T., Kruger, C.E., 2010. Methods for Producing Biochar and Advanced Biofuels in Washington State. Part 1: Literature Review of Pyrolysis Reactors. First project report. Department of Biological Systems Engineering and the Center for Sustaining Agriculture and Natural Resources, 87–8. available at: https://fortress.wa.gov/ecy/publications/publications/1107017.pdf.

Proceedings from tcbiomass2013, 2013, Gas Technology Institute, Chicago IL, http://www.gastechnology.org/tcbiomass2013/Pages/2013-Presentations.aspx (Last checked December 31, 2013).

Hu, X., Lievens, C., Larcher, A., Li, C.-Z., 2011. Reaction pathways of glucose during esterification: effects of reaction parameters on the formation of humin type polymers. Bioresources Technology 102 (21), 10104–10113.

Jarboe, L.R., Wen, Z., Choi, D.W., Brown, R.C., 2011. Hybrid thermochemical processing: fermentation of pyrolysis-derived bio-oil. Applied Microbiology and Biotechnology 91, 1519–1523.

Kersten, S.R.A., Garcia-Perez, M., 2013. Recent developments in fast pyrolysis of lignocellulosic materials. Current Opinion in Biotechnology 24 (3), 414–420.

Kilzer, F., Broido, A., 1965. Speculations on the nature of cellulose pyrolysis. Pyrodynamics 2, 151–159.

Kumar, A., Jones, D.D., Hanna, M.A., 2009. Thermochemical biomass gasification: a review of the current status of the technology. Energies 2, 556–581.

Lehmann, J., Joseph, S., 2009. Biochar for Environmental Management: Science and Technology. Earthscan Publishers Ltd.

Levenspiel, O., 1999. Chemical Reaction Engineering. Chapter 18: Solid Catalyzed Reactions, third ed. John Wiley and Sons, Hoboken, NJ.

Li, X., Gunawan, R., Lievens, C., Wang, Y., Mourant, D., Wang, S., Wu, H., Garcia-Perez, M., Li, C.-Z., 2011. Simultaneous catalytic esterification of carboxylic acids and acetilisation of aldehydes in fast pyrolysis from malee biomass. Fuel 90 (7), 2530–2537.

Lian, J., Chen, S., Zhou, S., Wang, Z., O'Fellon, J., Li, C.-Z., Garcia-Perez, M., 2010. Separation, hydrolysis and fermentation of pyrolytic sugars to produce ethanol and lipids. Bioresources Technology 101 (24), 9688–9699.

Mamleev, V., Bourbigot, S., Le Bras, M., Yvon, J., 2009. The facts and hypotheses relating to the phenomenological model of cellulose pyrolysis interdependence of the steps. Journal of Analytical and Applied Pyrolysis 84, 1–17.

Mayes, H.B., Broadbelt, L.J., 2012. Unraveling the reactions that unravel cellulose. Journal of Physical Chemistry a 116, 7098–7106.

Meier, D., Faix, O., 1999. State of the art of applied fast pyrolysis of lignocellulosic materials – a review. Bioresource Technology 68 (1), 71–77.

Mohan, D., Pittman, C.U., Steele, P.H., 2006. Pyrolysis of wood/biomass for bio-oil: a critical review. Energy & Fuels 20, 848–889.

Oasmaa, A., Czernik, S., 1999. Fuel oil quality of biomass pyrolysis oils-state of the art for the end users. Energy & Fuels 13, 914–921.

Patwardhan, P.R., Satrio, J.A., Brown, R.C., Shanks, B.H., June 2010. Influence of inorganic salts on the primary pyrolysis products of cellulose. Bioresource Technology 101 (12), 4646–4655.

Patwardhan, P.R., Brown, R.C., Shanks, B.H., 2011. Product distribution from the fast pyrolysis of hemicellulose. ChemSusChem 4, 636–643.

Paulsen, A.D., Mettler, W.S., Dauenhauer, P.J., 2013. The role of sample dimension and temperature in cellulose pyrolysis. Energy Fuels 27 (4), 2126–2134.

Peláez-Samaniego, M., Garcia-Perez, M., Cortez, L., Rosillo-Calle, F., Mesa, J., 2008. Improvements of Brazilian carbonization industry as part of the creation of a global biomass economy. Renewable and Sustainable Energy Reviews 12, 1063–1086.

Phanphanich, M., Mani, S., 2011. Impact of torrefaction on the grindability and fuel characteristics of forest biomass. Bioresource Technology 102, 1246–1253.

Piskorz, J., Radlein, D.S., Scott, D.S., Czernik, S., 1989. Pretreatment of wood and cellulose for production of sugars by fast pyrolysis. Journal of Analytical and Applied Pyrolysis 16, 127–142.

Prosen, E., Radlein, D., Piskorz, J., Scott, D.S., Legge, R.L., 1993. Microbial utilization of levoglucosan in wood pyrolysate as a carbon energy source. Biotechnology and Bioenegineering 42 (4), 538–541.

Pyle, D.L., Zaror, C.A., 1984. Heat transfer and kinetics in the low temperature pyrolysis of solids. Chemical Engineering Science 39, 147–158.

Richards, G.N., 1987. Glycolaldehyde from pyrolysis of cellulose. Journal of Analytical and Applied Pyrolysis 10, 251–255.

Scheirs, J., Camino, G., Tumiatti, W., 2001. Overview of water evolution during the thermal degradation of cellulose. European Polymer Journal 37, 933–942.

Shen, J., Wang, X.S., Garcia-Perez, M., Mourant, D., Rhodes, M.J., Li, C.Z., 2009. Effects of particle size on the fast pyrolysis of oil mallee woody biomass. Fuel 88, 1810–1817.

Sjöström, E., 1993. Wood Chemistry: Fundamentals and Applications, second ed. Elsevier, San Diego.

Teixeira, A., Mooney, K.G., Kruger, J.S., Williams, C.L., Suszynski, W.J., Schmidt, L.D., Schmidt, D.P., Dauenhauer, P.J., 2011. Aerosol generation by reactive boiling ejection of molten cellulose. Energy Environ. Sci. 4, 4306–4321.

Varhegyi, G., Jakab, E., Antal Jr., M.J., 1994. Is the Broido-Shafizadeh model for cellulose pyrolysis true? Energy & Fuels 8, 1345–1352.

Wang, Z., 2013. Understanding Cellulose Primary and Secondary Pyrolysis Reactions to Enhance the Production of Anhydrosaccharides and to Better Predict the Composition of Carbonaceous Residues [Ph.D. dissertation], Washington State University, Pullman.

Wang, Z., McDonald, A., Cuba-Torres, C., Ha, S., Westerhof, R., Kersten, S., Pecha, B., Garcia-Perez, M., March 2013. Effect of cellulose crystallinity on the formation of a liquid intermediate and on product distribution during pyrolysis. Journal of Analytical and Applied Pyrolysis 100, 56–66.

Williams, P.T., Nugranad, N., June 2000. Comparison of products from the pyrolysis and catalytic pyrolysis of rice husks. Energy 25 (6), 493–513.

Wornat, M.J., Porter, B.G., Yang, N.Y.C., 1994. Single droplet combustion of biomass pyrolysis oil. Energy and Fuels 8, 1131–1142.

Zhou, S., 2013. Understanding Lignin Pyrolysis Reactions on the Formation of Mono-phenols and Pyrolytic Lignin from Lignocellulosic Materials [Ph.D. dissertation], Washington State University, Pullman.

Zhou, S., Garcia-Perez, M., Pecha, B., McDonald, A.G., Kersten, S.R.A., Westerhof, R.J.M., 2013. Secondary vapor phase reactions of lignin-derived oligomers obtained by fast pyrolysis of Pine Wood. Energy and Fuels 27, 1428–1438.

SUSTAINABLE AVIATION BIOFUELS: A DEVELOPMENT AND DEPLOYMENT SUCCESS MODEL

Richard Altman

Commercial Aviation Alternative Fuels Initiative (CAAFI)

From 2006 to 2013, the area of sustainable aviation alternative fuels transitioned from a stagnant, research-focused project to a multidimensional technology development and deployment thrust commanding the attention of airline and defense buyers, advanced biofuel producers, and governments worldwide. These successes have moved aviation to the forefront of transport sectors with regard to interest in advanced alternative fuels by seeking to achieve sustainable growth in a scant 7 years.

This chapter focuses on two aspects of aviation biofuels. It seeks to both inform of the unique fuel characteristics and to explain the process by which success came to pass.

The first aspect focuses on how technology and process developments have been successful in qualifying biofuels for safe and environmentally favorable operation in jet aircraft, which is a necessary prerequisite to enable acceptance and successful deployment.

The second and equally important focus for Sustainable Aviation biofuels is the implementation of processes to enable deployment. Sustainable biofuels for aviation should, of course, be viewed as a "work in progress." New developments are occurring on a regular basis as we move forward toward large-scale usage. That said, sustainable renewable sources to replace nonrenewable liquid fuels will ultimately be critical to the long-term viability of aviation. Aviation remains as the only viable means of long-distance travel and is dependent on the availability of high power density liquid fuels for the foreseeable future.

Preceding the description of the "what and how" (the first and second part above) of development and deployment are snapshots of the industries position in the sustainable renewable market at the initiation of the focused industry efforts organized by the Commercial Aviation Alternative Fuels Initiative (CAAFI), along with subsequently formed sister-development activities around the world that have propelled developments since their CAAFI's formation at the end of 2006 and the dawn of 2013.

AVIATION ALTERNATIVE FUEL 2006 SNAPSHOT: "WHAT IF YOUR FAMILY WERE AN AIRLINE?"

The Genesis of what has become the CAAFI coalition started from a simple recommendation from the Research and Development Advisory Committee (REDAC) at the FAA Office of Environment and Energy during the second Quarter of 2005. The committee asked a seemingly simple question: What was the Office doing to expedite developments with the second (Energy) in its Charter?

Bioenergy. http://dx.doi.org/10.1016/B978-0-12-407909-0.00027-4

443

The problem statement was brought to the newly formed National Academy of Sciences, Transportation Research Board Committee on Aviation Effect on the Environment, colorfully named "AV030." The subject was discussed at the January 2006 at the annual Transportation Research Board annual meeting in Washington.

Entering the TRB meeting, the generally accepted premise was that aviation was a poor candidate for Alternatives to Petroleum-based fuels. Only a handful of specialists who had toiled for decades since the first Oil shock thought that alternative fuels for aviation could work at all.

What the committee found was an industry that:

- Was completely dependent on high power density hydrocarbon fuels and had no sustainable options to replace oil.
- Required replacements limited to "drop-in" fuels compatible with existing vehicles and fueling infrastructure, as equipment modification to assets in excess of $100 million each, as well as billions in infrastructure investment, were clearly not affordable. What this meant was that replacements needed to target a normal distribution of molecules in the C12 to C18 range and meet as well a broad series of form, fit, and function requirements ranging from lubricity to electrical conductivity in a narrow range.
- Was portrayed by environmentalists as having a high growth rate and increasing significance as a polluter for both emerging greenhouse gas issues as well as criteria air quality pollutants such as small particles.
- Had a qualification process that required 10 years to complete and, even then, was limited to specific production facilities. Long term investment commitments short of those from embargoed countries such as South Africa were not viable.
- Represented only 10% of transport demand and had view relationships with biofuel suppliers.
- Featured airline buyer balance sheets that left few buyers with investment grade ratings.
- Was headed toward scenario in which fuel costs would become the single largest airline expense—a scenario that came to pass a scant 2 years later.

Putting it in terms that most of us can relate to…what would you do if your family viewed itself in these desperate straits? The collective wisdom was that Aviation being dependent on liquids would get the last ounce of oil produced from what ultimately would be a dying infrastructure of a fixed resource. Surely the 10% of transport demand represented by aviation could not lead the way to the development of new fuels sources. If Aviation did not lead the way, fuel delivery infrastructure would be tailored to address the needs of other modes; it would be at high risk of having investment directed elsewhere.

Such dire circumstances among a close-knit family unit can also result in the family coming together, pooling their resources and unique skills, and building upon their unique strengths to alter their fate. That is exactly what happened over the course of the 7 years since January 2006 in the aviation family.

AVIATION ALTERNATIVE FUEL 2013 SNAPSHOT: ACKNOWLEDGED LEADER IN SUSTAINABLE TRANSPORT FUELS

By the first quarter of 2013, Commercial & Military Aviation had moved into a leadership position for advanced alternative Transport fuels in the United States. Nothing more fundamentally marked this change than remarks made by U.S. President Obama in a March 30, 2011 speech on Energy security policy. In the speech, the president specifically called our commercial aviation as a user of Advance Biofuels to be developed by the military, Energy Department (DOE), and Agriculture Department (USDA), specifically stating that

> "I'm directing the Navy and the Departments of Energy and Agriculture to work with the **Private sector** to create advanced biofuels that can power...not just fighters...but trucks and **commercial airliners**."

In reality, this statement in part was a recognition of what had been accomplished by the CAAFI public/private coalition. Specific, major accomplishments in the short span of 7 years included:

- The passage of the first all new aviation fuel specification (ASTM D7566) in nearly 20 years in September 2009 by using the Fischer–Tropsch (FT) process for all facilities and feedstocks, including biofuel blends of up to 50%. The last prior specification change (from JP4 to JP8) addressed aviation safety concerns via CAAFI's certification team.
- July 2011: Qualification of Hydrotreated Esters and Fatty Acid (HEFA) fuel under ASTM D7566 by using, among others, nonfood crops such as camelina, jatropha, and algae.
- The creation of a gated risk management approach to govern the development and deployment of alternative fuels called "Fuel Readiness Level" or FRL.[1] In November 2009, the process developed by CAAFI's R&D and Certification teams was approved as International best practice by an International Civil Aviation Organization, the United Nations' aviation governing body. This systems engineering methodology, long used by the Air Force and NASA to conduct evaluation of complex system development, included all elements used to reduce qualification time to 3 years from 10 years.
- The development of the Feedstock Readiness Level (FSRL)[2,3] in 2011 through collaboration between FAA/DOT and the Agriculture Department (USDA) that uses Aviation Systems Risk management processes to identify the steps required to address novel feedstock development and commercialization.
- The development of an aviation specific "ground to wake" carbon Life Cycle analysis process and specific fuel life-cycle GHG evaluations[4] as an outcome of CAAFI Environmental Team. FAA team leadership, through its MIT-led Aviation/Environment Center of Excellence (PARTNER), led this effort that has contributed to and been aligned with the efforts of the Department of Energy and United States Air Force.[5]
- An approach for "environmental progression" proposed by CAAFI's Environmental team[6] to parallel FRL to ensure that environmental certainty follows technical and feedstock readiness.
- The execution of some 30 different operators of demonstration and operation programs including commercially scheduled flight for multimonth periods by Lufthansa, KLM, and others.

---

[1]http://www.caafi.org/information/fuelreadinesstools.html#FRL.
[2]http://www.caafi.org/information/fuelreadinesstools.html#FeedstockReadinessLevel.
[3]Steiner et al. (http://link.springer.com/article/10.1007%2Fs12155-012-9187-1.
[4]Stratton et al., 2011: http://web.mit.edu/aeroastro/partner/reports/proj28/partner-proj28-2010-001.pdf.
[5]http://web.mit.edu/aeroastro/partner/reports/proj28/greenhs-gas-ftprnts.pdf.
[6]www.caffi.org.

- Alliances with the Defense Departments fuel purchasing arm, DLA Energy through a unique partnership of private sector (airline) and public (Defense Department) buyers in March of 2010.
- The creation of the Public/Private "Farm to fly" initiative in July 2010 between airlines, Boeing, the Agriculture and Energy Departments established a methodology for local developments in the Pacific Northwest under the SAFN (Sustainable Alternative Fuel Northwest) that could be applied and adapted to other States and regions.
- Some 25 U.S. State and regional initiatives locally led in consultation with CAAFI, its sponsors, and stakeholders.
- Global public/private partnerships supported by detailed work plans with Australia (signed September 2011) and Germany (signed September 2012), as well as U.S. government agreements signed with Brazil (February 2011) and Spain (February 2013).
- Aviation biofuels have been featured at the Paris, Farnborough, Berlin, and Australian Air shows over the last 4 years with over 30 sponsors and stakeholders exhibiting their approaches to Aviation biofuels.
- A poll taken at a major conference in the spring of 2012 showed that 68% of the largely fuel producer attendees believed that aviation would be the first to adopt advanced biofuels solutions among all potential customers.

These achievements in themselves form an impressive transformation from no more than an afterthought in Biofuels space to the Cutting Edge of innovation.

The technical and operational details describing the fundamentals about how these achievements have been accomplishments have been executed form the remainder of this treatise.

Specifically,

- What are the key methodologies that are enabling sustainable progress in aviation?
- Which processes and feedstocks are achieving sustainable outcomes?
- What approaches are being used to facilitate development and deployment in varying geographic locations?
- How is the difficult subject of biofuel economics being approached?

An understanding of all of these can lead to the adoption of these approaches to achieve similar results for other end customers.

KEY METHODOLOGIES FOR SUSTAINABLE PROGRESS: CREATION OF A "NEW FUEL DYNAMIC"

Most new endeavors either consciously or purposefully begin with an SWOT (strength, weakness, opportunity, threat) assessment.

Although the description of 2006 status clearly notes the weaknesses of aviation as a lead transport market, the more difficult task for those engaged at the time was to recognize sector strengths, and more importantly, how those strengths could be turned into opportunities.

In the case of commercial aviation strengths, these are realized by examining what is unique to the industry, including constraints, and establishing whether and how these strengths can be translated into distinctive competencies that can strengthen the pursuit of sustainable biofuel development and deployment. Specifically:

- Aviation is limited to high power density liquid fuel use. Electrification for main propulsion is not an option. Hence, investors can be assured that the industry will not shift to another alternative to liquid fuels—a positive for fuel suppliers.
- The demanding safety requirements of aviation, set by agencies such as the U.S. Federal Aviation Administration (FAA) and its European equivalent EASA and implemented by the standard setting organizations, most commonly ASTM International, create a "barrier to entry" for all but the most serious fuels producers.
- The small market size, when consolidated into a limited and informed group of buyers, can facilitate group decision-making that serious producers can be assured will be data-based.
- Distribution for aviation is concentrated at relatively few airports. In the United States, 80% of all traffic flows through 35 destinations. Suppliers need not establish a complex distributions network with high development and operational costs.
- Jet fuel producers will execute multiyear off-take (purchase) agreements. This is not the case with diesel buyers operating in the similar fuel space.
- Systems integration and gated risk management of product and process development are ingrained in aviation, and in fact are requirements of technology and product development by the military, NASA, and defense contractors.
- Research in Aviation is well supported by both Commercial and Military sources. Aviation contractors have great expertise in meeting the requirements of government research contracts at the outset of technology development.
- Aviation environmental regulations and rules for safety are governed globally by the United Nations, International Civil Aviation Organization, and not by individual States and Nations.[7]

Collectively, the set of solutions developed for Qualification, Environmental, R&D, and Business have sought to utilize these strengths to create what we in Commercial Aviation and the CAAFI refer to as a "new fuel dynamic." The solutions have been approached via the use of four disciplinary process, owner-led teams tackling the challenges of fuel qualification, environmental acceptance, aggressive R&D, and customer-led supply chain economics development. Each functional area is explained by illustrating the challenge that the team faced, the solution pathway, and the results to date.

STREAMLINING THE FUEL QUALIFICATION PROCESS
THE CHALLENGE

At the dawn of jet propulsion in the late 1930s, engines were designed to operate on kerosene. Kerosene was readily available (military aircraft relied on more volatile gasoline), and its properties best suited the Brayton cycle combustion process used by jets. Over time, engine and aircraft designers realized that they needed to more tightly control the properties of kerosene for both commercial and military operations to ensure both safe operation and consistent performance. With the fuel properties known and understood, engineers could then incorporate technological advances into turbine engine designs to achieve significant gains in fuel efficiency and durability. Thus were born the aviation fuel

---

[7]Note that regulations and policies regarding alternative fuels are generally governed by individual countries or states (e.g., the US EPA's Renewable Fuel Standard, the EU's Renewable Energy Directive, etc.).

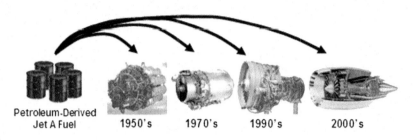

FIGURE 27.1

Legacy jet fuel adoption process.

specifications used to control the formulation, manufacture, and distribution of jet fuel. These fuel specifications specify "performance-based" properties that are designed to control the known variation of crude oil-derived or petroleum-derived jet fuel.

Early in the twenty-first century, the aviation fuel industry leaders who were strategizing how to deploy alternatives to petroleum-derived jet fuel quickly realized that the designs and performance of the many thousands of existing jet engines and aircraft produced and certificated over the ensuing 70 years had all been optimized for this existing, petroleum-derived jet fuel (see Figure 27.1).

They faced a seemingly insurmountable challenge to figure out how to design and certify a new alternative fuel for this existing fleet of aircraft (see Figure 27.2). The only significant change in fuel specifications that had occurred was a move by military aviation to JP8 fuel from volatile JP4 in the late 1980s. In the commercial world, JP8 equivalent, Jet A, required 10 years and millions of dollars to accommodate a Coal to Liquid alternative fuel from the Fischer–Tropsch from a single production facility in South Africa. Clearly, a robust process to allow multiple processes across all facilities to use that process in a much speedier time frame was required.

Many different design approaches had been used to accommodate the specified jet fuel properties, and this resulted in a myriad of different designs existing on these products. How could each of the many thousands of engines and aircraft with these many different designs be evaluated and tested to ensure that the new fuel was safe and performed in a similar manner to the existing fuel?

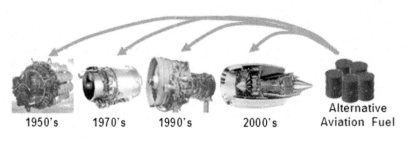

FIGURE 27.2

Adapting existing products to alternative fuels.

PATHWAYS TO SOLUTIONS

The CAAFI Qualification Panel quickly realized that the key to solving this challenge would depend on the ability to prove that the new fuel was essentially identical to the existing, petroleum-based jet fuel. If this could be proven with a thorough technical investigation, then the FAA regulations would not require any certification at all. This was based on the existing, FAA-approved operating limitations for all aircraft and engines that specify the aviation fuel permitted for use. If an aviation fuel qualification process could be established that could prove that the alternative fuel was not a "new fuel," but rather the "same fuel" produced from different raw materials and/or processes, then the alternative fuel would fit under these existing operating limitations. These fuels would be called "drop-in" fuels to reflect their seamless entry into the distribution infrastructure once approved.

The Qualification Panel worked with the key aviation fuel-specification writing organization, ASTM International, to expedite the development and approval of a qualification process for new jet fuels. In parallel with that effort, an ASTM Task Force was formed to apply this qualification process to the approval of Fischer–Tropsch (FT) fuel, the initial alternative-aviation fuel.

THE RESULTS

On September 1, 2009, ASTM International approved the world's first semisynthetic aviation fuel specification. This specification, number D7566, entitled "Standard Specification for Aviation Turbine Fuel Containing Synthesized Hydrocarbons," was a significant milestone toward the CAAFI's goal of promoting the deployment of alternative aviation fuels in the commercial aviation world because it allowed the use of D7566 fuels in all existing engines and aircraft. Specification D7566 is considered the "drop-in" fuel specification, because any new fuel added to that specification will have been proven to be essentially identical to petroleum-derived jet fuel. The specification is structured to define each new fuel in an annex, with FT fuel included as the first annex at publication. The process from initial tests begun on and Air Force B-52 with Gas-to Liquid fuel had taken 3 years and qualified the entire family of fuels from Coal, Gas, and Biomass by using the Fischer–Tropsch process.

Also, in October of 2009, ASTM International standard D4054, "Guideline for the Qualification and Approval of New Aviation Turbine Fuels and Fuel Additives," was issued to provide candidate alternative-aviation fuel producers with a guide to evaluate their new fuel. Figure 27.3 shows how ASTM D7566 and D4054 work together to solve the "certification challenge" that at one time seemed insurmountable.

ASTM 4054 documented the experience developed from learning from both the Sasol qualification experience, as well as the steps needed to achieve FT fuel qualification under the new ASTM 7566.

Building on both these standards, the CAAFI and the ASTM fuel committee in which its members participate took on the challenge of fuels from other processes. In July 2011, fuels types from lipids or fats from oil seed plants and tallow from animal fats labeled Hydrotreated Esters and Fatty Acids (HEFA) were qualified and published as a second annex to ASTM 7566. From the time of initial research via production of small samples were published under the Defense Advanced Research Projects Agency (DARPA) until the mid-2011s, approximately 3 years had passed.

With the achievement of HEFA qualification, the focus of ASTM moved to an added group of processes and feedstocks that promises to greatly increase sustainable supplies. As of this writing, research reports (the precursor to fuel approval documenting results versus ASTM 7566 and 4054 requirements) are being developed and circulated for both Alcohol to Jet (ATJ) and Thermochemical/Pyrolysis (HDCJ) pathways. Several other pathways (see Figure 27.4 below), along with the

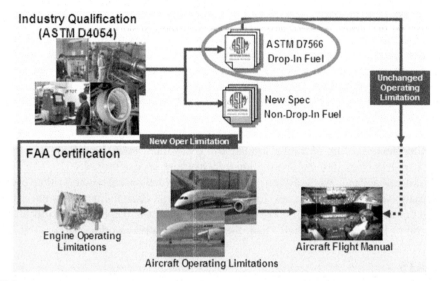

FIGURE 27.3

ASTM advanced fuels process integration.

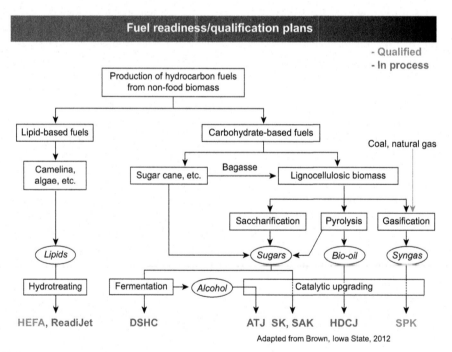

FIGURE 27.4

Fuel/readiness qualification pathway candidates.

(Source: Adapted from Brown, Iowa State, 2012.)

coprocessing of biocrude, are now under consideration and are candidates for fuel qualification during the current decade.

Perhaps the greatest success of qualification process implementation of this process in the case of both FT and HEFA processes is that they removed the fuel safety qualification fuel from the critical time-path to development and deployment. In so doing, a major barrier to investment in Aviation alternative fuels had been maintained while retaining the rigor of the process.

By opening new pathways and opportunities to grow, supply has the potential to emerge.

IMPLEMENTING COMPREHENSIVE RISK MANAGEMENT IN ALTERNATIVE FUELS RESEARCH AND DEVELOPMENT
THE CHALLENGE

Aviation needs as many different sources of alternative fuels as possible to reduce environmental impacts and stabilize both price and energy security. With the success of FT fuels and the template for success that they have provided, there has been a proliferation of potential fuel production pathways. While efforts were being exerted by multiple sources within government and biofuels producers, and with the experience that research agencies (Air Force, NASA, FAA, among others) had in systems-level R&D, the need for risk management tools specific to fuel development was conceived. In this Manner, the status of pathway development and the technical suitability of fuels could be communicated. Such tools would also provide a mechanism for tracking research and development (R&D) efforts and identifying gaps.

PATHWAYS TO SOLUTIONS

As a result of this challenge, the CAAFI R&D team in partnership with the Air Force Fuels lab sought to develop a "Fuel Readiness Level" (FRL) scale[8] that adapts the previously existing Technology Readiness Level (of NASA and DOD) to more explicitly cover alternative aviation fuel development. It did so by merging traditional technology measures (TRL) and manufacturing Readiness levels (MRL) for fuels that were significantly different than those used in hardware production.

Because the FRL process was sufficiently mature and proven in the FT process developments and the early stages of HRJ development, in 2009, it was proposed for global approval to a United Nations, International Civil Aviation Organization Committee (ICAO) Fuels Conference after consultation with European interests who use similar tools.

Like its predecessors, the FRL is a gated risk management process. As such, it has an associated check-list of pass/fail criteria to be applied first by producers for subsequent ratification by customer sources such the Air Force or Commercial research authorities to move to the next step.

With FRL risk management framework defined, the CAAFI R&D team, at its September 2009 meeting held at the United States Department of Agriculture, recommended to CAAFI leadership that its efforts be expanded to encompass feedstock readiness in collaboration with USDA. The goal was to identify if the pacing of fuel suitability for aviation use and the ability to produce that fuel could be better matched in schedule. In addition, the use of Aviation-level system risk-management techniques by agricultural researchers for energy crops could lead to rapid development in those areas.

---

[8]www.caafi.org/information/fuelreadinesslevel.html.

USAF/CAAFI Fuel Readiness (FRL) D7566

FRL	Description	CAAFI Toll Gate	Fuel Qty
1	Basic principles observed and reported	Feedstock/process *principles* identified.	
2	Technology concept formulated	Feedstock/*complete* process identified.	
3	Proof of concept	Lab scale fuel sample produced Basic fuel properties validated	500 ml
4.1 4.2	Preliminary technical evaluation	System perf. & integration studies Entry criteria/specification properties Evaluated (MSDS/D1655/MIL 83133)	10 gal
5	Process validation	Sequential scaling from laboratory to pilot plant	80 gal to 225 K gal
6	Full-scale technical evaluation	Fitness, fuel properties, rig testing, and engine testing	80 gal to 225 K gal
7	Fuel approval	Fuel class/type listed in int'l fuel standards	
8	Commercialization validated	Business model validated for production airline/military purchase agreements	
9	Production capability established	Full scale plant operational	

	Legend:	R&D	Certification/qualification	Business & economics	CAAFI

FIGURE 27.5

The fuel readiness process as approved by ICAO.[9]

THE RESULTS

- In November 2009, the ICAO Conference on Alternative Fuels in Rio De Janeiro accepted the CAAFI Fuel Readiness Level process (Figure 27.5).
- The Air Force Research Lab has received hundreds of small fuel samples to execute the initial FRL-scale laboratory testing.
- The FRL risk management tool is of particular use in the legitimization of new fuel pathways for which work was initiated following the inception of FRL. This scale is being used to help increase the readiness of various other processes, such "alcohol to jet." This involves sugar fermented to alcohols, which is then dehydrated and oligomerized into hydrocarbon jet fuel (Figure 27.5).

With the speeding of fuel-processing readiness, the critical path in fuel development and deployment had become the time and quality of readiness for feedstocks to be introduced to use in the qualified processes. For this reason, it was agreed with the USDA that a gated risk management process using Aerospace gated risk management principles was required. In November 2011, the execution of Feedstock Readiness (FSRL) between FAA and USDA's newly formed research centers implementing a process of feedstock development to parallel process development for several added feedstock candidates (Figure 27.6 (a,b,c)).

---

[9]Declaration and Recommendations, in Conference on Aviation and Alternative Fuels 2009, ICAO Secretariat: Rio de Janeiro, Brazil. p. B-3, Appendix A: Item 25 (adoption of FRL).

(a)

Feedstock Readiness Level (FSRL) Gates 1- 4

Technology Readiness	Feedstock Readiness	Production (PRO) Component Gateway
1	Basic Principles	• Identify potential feedstock for a specific conversion technology
2.1	Concept Formulated	• Estimate likely range of production environments and competing land uses
2.2		• Identify production system components
2.3		• Develop enterprise budget for candidate feedstock
2.4		• Identify possible consequences of expanded production, articulate response to trade-off's
3.1	Proof of Concept	• Screen candidate genetic resources for feedstock yield
3.2		• Screen candidate genetic resources for biofuel conversion potential
4.1	Preliminary Technical Evaluation	• Perform coordinated regional feedstock trials to determine potential for yield improvement and dependability of feedstock supply
4.2		• Compare performance of candidate feedstock with alternative feedstock choices
4.3		• Implement agricultural extension and education programs to promote feedstock production

(b)

Feedstock Readiness Level (FSRL) Gates 5 - 9

Technology Readiness	Feedstock Readiness	Production (PRO) Component Gateway
5.1	Production System Validation	• Define range of adaptation for feedstock and identify production uncertainties
5.2		• Conduct on-farm, field-scale production and cost trials and assess production impacts on resources concerns
5.3		• Establish partial budget costs and returns
5.4		• Establish price points for feedstock market competitiveness with competing land uses
6.1	Full-Scale Production Initiation	• Establish source material nurseries and begin feedstock production scale-up process
6.2		• Produce feedstock planting materials to meet demand
7	Feedstock Availability	• Commercial-scale production and feedstock delivery to conversion facility – payments made for feedstock
8	Commercial- ization	• On-going monitoring and research to improve production system performance while managing multiple resource concerns
9	Sustainable Feedstock Production Capability Established	• Full array of private services support feedstock production sector – understanding of feedstock sector evolves – made adjustments as commercial-scale biofuel production expands

FIGURE 27.6

Feedstock readiness (a, b) and feedstock/process candidates for jet fuel (c).

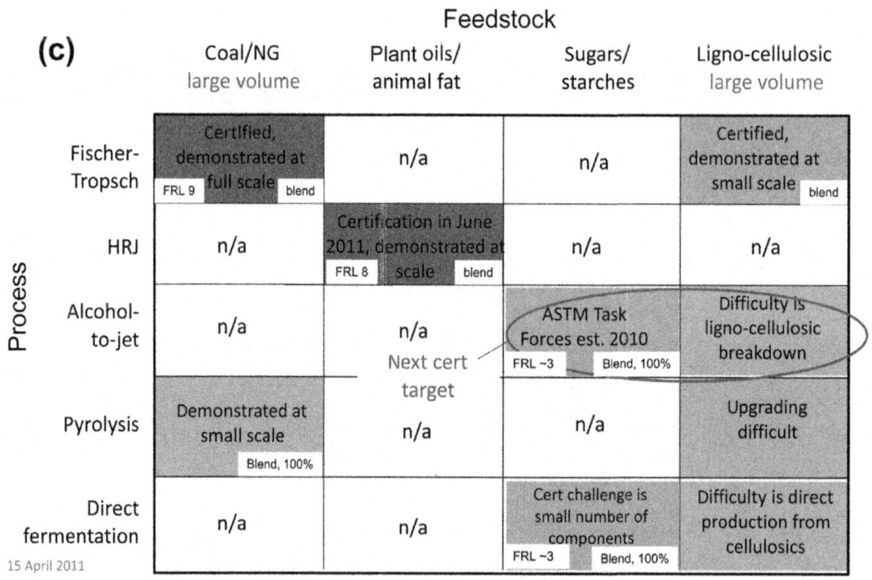

Feedstock

(c)	Coal/NG large volume	Plant oils/ animal fat	Sugars/ starches	Ligno-cellulosic large volume
Fischer-Tropsch	Certified, demonstrated at full scale FRL 9 · blend	n/a	n/a	Certified, demonstrated at small scale blend
HRJ	n/a	Certification in June 2011, demonstrated at scale FRL 8 · blend	n/a	n/a
Alcohol-to-jet	n/a	n/a Next cert target	ASTM Task Forces est. 2010 FRL ~3 · Blend, 100%	Difficulty is ligno-cellulosic breakdown
Pyrolysis	Demonstrated at small scale Blend, 100%	n/a	n/a	Upgrading difficult
Direct fermentation	n/a	n/a	Cert challenge is small number of components FRL ~3 · Blend, 100%	Difficulty is direct production from cellulosics

15 April 2011

(left axis label: Process)

FIGURE 27.6 Cont'd

STRUCTURING AND FACILITATING COMPREHENSIVE ENVIRONMENT BENEFITS ASSESSMENTS

THE CHALLENGE

With the advent of climate change/global warming concerns, it became clear that the industry needed to find additional ways to reduce its GHG footprint. Sustainable alternative aviation fuels are one of the most promising opportunities. In fact, both studies of the multigovernment and industry team working on implementing growth in the "Next Gen" Aviation system and the International Air Transport Association identified that equipment improvements, offering as much as $1\frac{1}{2}$% fleet efficiency gains annually and further gains from air traffic efficiency of new management systems, could not achieve the goal of achieving GHG reductions at a rate that would prevent growth from its 2005 base, a metric of success for many.

Policies arbitrarily tied to specific calendar achievements in the case of aviation fail to credit achievements of the industry. Even before greenhouse gas (GHG) emissions' potential contribution to climate change became an environmental concern, the aviation industry was achieving tremendous GHG emissions savings. In fact, the U.S. airlines improved their fuel efficiency by 110% between 1978 and 2009, saving over 2.9 billion metric tons of carbon dioxide (CO_2), an amount roughly equivalent to taking 19 million cars off the road each of those years. In addition, despite its growth, U.S. commercial aviation accounts for only 2% of the nation's man-made CO_2 (and the global commercial aviation sector likewise accounts for 2% of global CO_2).

Further gains in greenhouse gas emissions were to occur over the entire "ground to wake" life cycle for fuels production and use—not only end use itself (Figure 27.7).

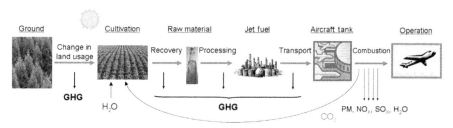

FIGURE 27.7

"Ground to wake" analysis required for greenhouse gas emissions.

Quantification of gains in real, discrete, and auditable terms was brought into sharp focus in the Energy Independence and Security Act of 2007. This act requires government purchasers to demonstrate that alternative fuel purchases comply with the Act's section 526, demonstrating that on a life-cycle basis, the fuel purchased would be better in GHG Life-cycle than fuel purchased from an oil refinery—not a given for any fuel.

Beyond Greenhouse gases, the industry also was experiencing added challenges to in the area of small particle control. (PM2.5, or particle material of less than 2.5 μm in diameter). The PM2.5 concern cited in National Ambient Air Quality Standards is viewed as a precursor to control measures and regulations of that pollutant under the Clean Air Act as amended in 1990. Sixty percent of all U.S. Airports are found to be in "nonattainment" areas for this pollutant. However, a globally accepted means of measuring PM2.5 for aviation-specific challenges has yet to be finalized.

Lastly, the overall subject of sustainability (combining environmental, economic, and social factors to gain acceptance), and the "who and how" approvals to assure that sustainability criteria are being met by new projects, are critical components of alternative fuels success.

Whether it be GHG life-cycle, PM2.5, or sustainability certification, the uncertainty of the outcomes for alternatives in these areas is a true barrier to assuring investors that they can proceed with projects knowing that environmental factors will be positive assets for projects rather than barriers to timely decision making.

PATHWAYS TO SOLUTIONS

The challenges, although daunting, do allow for clarity in needed steps. These pathways can be categorized as follows.

- Establish universally accepted goals for Greenhouse Gas control by airlines.
- Set means of quantifying carbon and GHG accounting for projects that are specific to aviation.
- Communicate options and certification techniques for sustainability certification.
- Obtain means of quantifying PM2.5 benefits that are unique to aviation and obtain adequate data to assess benefits.
- Incorporate all algorithms in a comprehensive set of tools that can be utilized to assess project benefits by all stakeholders.

THE RESULTS
Universally accepted goals
While section 526 of the Energy Independence and Security Act of 2007 provided a legislated goal and constraint for government purchasers, the airlines were quick to follow with voluntary measures.

On Earth Day 2008, the Air Transport Association in the United States put in place a policy for U.S. airlines that paralleled section 526, declaring that: *"we believe it is incumbent upon all segments of the transport sector to take voluntary measures to limit their impact on the environment. As combustion of traditional, petroleum-based jet fuel is a source of such emissions, we seek alternative fuel sources having a reduced emissions profile relative to traditional fuels."*[10]

Following these declarations of goals, IATA (the International Air Transport Association) and the Air Transport Action Group (ATAG), which includes both manufacturers and airlines, decided to go further to establish calendar-based goals to achieve Carbon neutral growth. Specifically, IATA adopted the goal of achieving Carbon neutral growth for the airline industry starting in 2020, with a reduction of 50% in GHG emissions by 2050 from the 2005 benchmark level. The approach outlined for achieving these goals included assumptions on benefits incorporated from equipment efficiency gains and improved operational efficiency from Air Traffic control gains. The remainder can be at least partially fulfilled by sustainable alternative fuels insertion (Figure 27.8).

Quantifying real terms carbon and GHG calculation for projects
Initiated under research executed by the FAA-funded, MIT-led PARTNER center of Excellence (Partnership for Air Transport Noise and Emissions Reduction[11]), GHG reductions for a variety of processes and feedstocks in an effort to quantify progress against carbon neutral growth goals. Furthermore, the MIT process bounds projects identifying uncertainties in land-use questions for various projects and the processes themselves where uncertain. For example, there is considerable issue with processes such as Algae-based hydrotreated renewable jet. Energy use in water extraction and temperature retention in open ponds could add to energy requirements and limit or even eliminate GHG benefits for that process/feedstock combination (Figure 27.9). MIT researchers, other CAAFI leaders, and others from a variety of agencies and organizations (academic and governmental) came together under an interagency working group headed by the USAF and NETL to synthesize an agreed upon approach and put together the report "Framework and Guidance for Estimating Greenhouse Gas Footprints of Aviation Fuels" (Final Report) (2009, AFRL-WP-TR-2009-2206).[12] This document builds on ISO Standard 14040 and augments and applies the ISO Standard approach to life-cycle analysis to aviation fuels. It identifies the steps associated with life-cycle greenhouse gas analysis for aviation fuels and makes recommendations for dealing with open issues.

Work continues on the actual case studies of three different processes, using the rules and tools. In addition, PARTNER continues its work expanding feedstocks and processes for review and enabling its team members to expand efforts for regional feedstocks and process.

---

[10]https://www.iata.org/pressroom/pr/Pages/2009-06-08-03.aspx.
[11]web.mit.edu/aeroastro/partner/index.html.
[12]The full title of this report is: "Propulsion and Power Rapid Response research and Development (R&D) Support – Delivery Order 0011: Advanced Propulsion Fuels Research and Development Subtask: Framework and Guidance for Estimating Greenhouse Gas Footprints of Aviation Fuels (Final Report)."

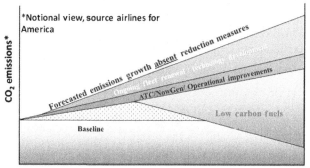

FIGURE 27.8

International aviation industry carbon reduction pathway.

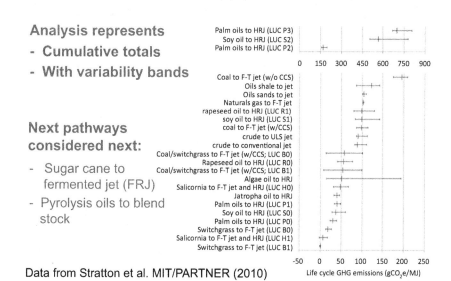

FIGURE 27.9

Summary of PARTNER Greenhouse Gas life-cycle analysis results (as of 03/10).

(Source: Data from Stratton et al. MIT/PARTNER (2010).)

Communicate options and certification techniques for sustainability certification

Greenhouse Gas life-cycle quantification from "well to wake" is a subset of the larger issue of full environmental sustainability development and certification. CAAFI efforts to define an aviation specific path for sustainability certification are not as advanced as GHG quantification by are proceeding. At its August 2010 meeting, the CAAFI Environment team put in place a panel of sustainability practitioners to at a minimum communicate options for certification of sustainability and to identify processes that might be used for aviation unique activities.

Several candidate organizations/processes are engaged in the definition of approaches to sustainability up to the point of certification. Of particular interest are:

- Development emerging from the G-20 nations Global BioEnergy Partnership (GBEP)[13]
- The use of elements of processes developed by the Roundtable on Sustainable Biofuels (RSB), a Swiss-based consortium[14]
- Sustainability definition options under consideration by The International
- Organization for Standardization (ISO)[15]

Such sustainability quantification will include such factors as water usage and quality, biodiversity, and the use of chemicals in crop propagation. Whether energy crops are harmful to animals or humans and a multitude of other factors must be considered.

Quantifying PM2.5 benefits

As they are dramatically reduced in sulfur content, generally a precursor of small particle formation, alternatives qualified under the Fischer–Tropsch process produce significant benefits in air quality for small particles (PM2.5). Early indications of this progress were obtained in measurements particle reduction obtained by the Air Force Research Laboratory (Figure 27.10).

The execution of further particle measurements and benefit studies is organized under PARTNER Project 20, entitled Emissions characteristics of alternative fuels. The project seeks to expand the data base for such fuels for applications and benefit assessments of both policy and project benefits.

Comprehensive set of tools to assess project benefits by all stakeholders

CAAFI's role has been to ensure that analysis capability for use in evaluating potential projects. The approach to developing project evaluation tools has concentrated on mechanisms provided by the Airport Cooperative Research Program of the Transportation Research Board of the U.S. National Academy of Science. ACRP is in the process of executing three projects that target alternative fuel benefits assessments for aviation. Specifically, these projects are:

- ACRP 02-18: Guidelines for Integrating Alternative Jet Fuel into the Airport Setting, led by Metron Aviation
- ACRP 02-23: Alternative Fuels as a Means to Reduce PM2.5 Emissions at Airports, led by AEA of the UK via its U.S. affiliate PPC
- ACRP 02-36: Assessing Opportunities for Multi-Modal Alternative Fuel Deployment, led by Metron Aviation

[13]http://www.globalbioenergy.org.
[14]www.rsb.org.
[15]www.iso.org.

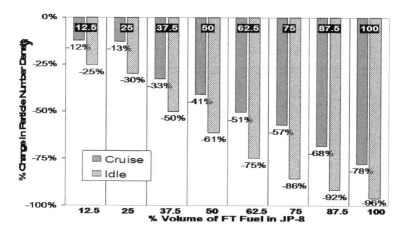

FIGURE 27.10

Measurements of PM2.5 benefits for Fischer–Tropsch jet fuels.[16]

Together, these projects provide handbooks that are intended to structure the use of specific analysis (defined above) into the overall project assessments.

DEPLOYING A "A NEW FUEL DYNAMIC" THROUGH PUBLIC/PRIVATE PARTNERSHIP, AND MULTIPLE-SUCCESS MODELS
THE CHALLENGE

Corresponding to a period of consolidation and stress for the U.S. Commercial airlines, the period from 2005 to 2013 displayed an unprecedented roller coaster of unstable cost drivers. Although extraordinary measures were taken to control labor, the cost of fuel, in 2006, for the first time exceeded labor as a percent of airline operating costs to become the highest cost element of airline operations—as much as 40%.

The cost run-up from 2005 to 2008 was the most severe the industry had seen. Even with a 13% drop in consumption from in 2007 (20 billion gal/year) to 2011s (17B plus projected performance), the outcome has proven similar in recent years to that shown in 2008.

Investigation of the underlying cause of this performance reveals an inherent flaw in the current factors leading to jet fuel production. Owing to typical refinery process that produces jet fuel in the middle distillate range, the total output of jet fuel is generally not more than 10% of a barrel of oil to jet (Figure 27.11). The result in high-demand times has been added increases to jet fuel price, leading to crack spreads (the difference between crude and jet fuel prices) in the $25 to $30 per barrel range, an amount higher than the total cost of fuel in the pre-2005 time periods of fuel price stability. Outcomes seen in 2008 were repeated during the economic recovery that began in 2009 with crack spreads rising

[16]Corporan, Edwin. Particulate emissions data supplied for an Allison engine using Fischer-Tropsch Fuel. Air Force Research Lab. 2007.

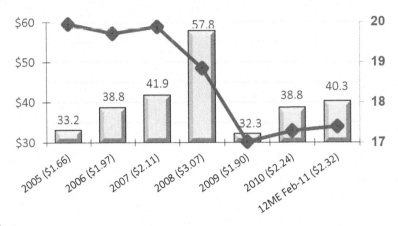

FIGURE 27.11

Airline industry fuel cost (left scale)/consumption (Bgals).

to the same level as 2008 in the 2012 time period. As a result, it can assumed that this pricing pattern will continue in the future and require a response if the industry is to remain economically viable under growing economic conditions.

Exogenous factors such as weather shocks (Hurricane Katrina) and dependence on imported oil from politically volatile regions were also factors in the price volatility.

Notable in the period was also the fact that the beginnings of work on biofuels in the United States were exclusively dedicated to ethanol for gasoline, a fuel that is completely unsuitable to use in jet aircraft. Aviation was not even thought to be capable of using alternatives to petroleum. Government interest in investing in aviation alternative was limited to a few military sources (notably the Air Force) and the remnants of efforts initiated during earlier oil shocks.

PATHWAY TO SOLUTIONS

Given the multiple challenges, the construct of a solution required multiple parallel pursuits of solutions building upon and in many cases interacting with the qualification and interacting with the rationale and progress and understanding generated from qualification success, R&D progress and environmental need described in prior paragraphs. These pathways required simultaneous efforts to:

- Convince airline buyers and executives that the current pathway for fuel purchases was structurally flawed, exclusive of the short term price, and that a "new fuel dynamic" involving alternative options is the only way to achieve sustainable industry success from both an economic and environmental perspective.
- Work with government to prove that not only did aviation deserve status as a candidate for alternatives, but in fact could be a "first mover" and lead the nation. That outcome would in fact lead to supportive relationships with the energy supply and agricultural industry in government. In fact, the term "new fuel dynamic" first appeared in a letter from CAAFI Airline and manufacturer sponsors and Advance biofuel producers to then President elect Obama on January 16, 2009. The letter declared *the aviation industry is eager for an entirely new fuel dynamic and*

will be an enthusiastic purchaser. Publication of the letter itself suggested that the job of convincing airline buyers had been largely accomplished.

- Through the combined efforts of multiple airlines, encourage the development of many success models across the country and around the world. Through the development of these success models and through government support for alternative aviation fuel, private sector investment, including that from major oil companies, could be obtained.
- Team with both government buyers and with global efforts to ensure that fuel suppliers viewed aviation supply as a single, rational global buyer eager for long-term commitments (not the nature of other fuel types) with an assured global market for liquid fuels and high standards suited to serious suppliers with strong technical and business models.

The pursuit of these multiple pathways also involves significant interdependencies. Government declarations embolden private sector researchers and producers to build new pathways, knowing that there will be a ready buyer. The qualification authorities respond to "demand pull" by expediting the testing and analysis that leads to qualification. Knowledge that policy makers have been emboldened indicates to buyers that their efforts will be rewarded. Airline commitments encourage more states and nations to be involved in airline supply. The visualization of the aviation alternative-fuel snowball rolling down a hill gathering mass and, in the ultimate case, producing an avalanche of activity, is then possible. The seeds of such growth are now evident in the United States and elsewhere.

THE RESULTS

In October 2010, the prestigious and conservative U.K. periodical "The Economist" stated the best summary of Aviation's budding success. In a larger story about the recovering prospects of biofuels, it rightly noted that, *There is no realistic prospect for widespread electric air travel: …if you want low-carbon flying, drop-in biofuels are the only game in town.*[17] It concluded with the apt double-entendre, *Over the long run, the future for biofuels may be looking up.*

The trend continued through 2012 in a poll of attendees at the World Biofuels Market conference in Rotterdam in March. Sixty-eight percent of the attendees believed that Aviation would be the first to deploy sustainable biofuels.

The progress in the acceptance of Aviation as a lead transportation customer extended to recognition of key customers as critical to biofuels success. In a December 2012 poll, nine individuals from the U.S. Navy, CAAFI, and even CAAFI's overseas partner in Australia, AISAF, had been recognized among the top 100 people in Biofuels and Bioenergy. No other transportation mode had even a single participant recognized on the list.

Another measure of success is the representation of Biofuels in Aviation global forums. In 2009, a single fuel producer (Rentech) had exhibited at the prestigious Paris Air show. By 2013, Exhibitors had expanded to Air shows in Berlin, Germany, Avalon, Australia, and Farnborough in the United States. The number of exhibitors had expanded to over 30 at the CAAFI annual meeting in November 2011. In addition, many high government officials had appeared at the Air shows to tour the exhibitor pavilion. At Paris in 2011, three U.S. Government cabinet-level officials (Agriculture, Transportation, and Commerce) participated and toured the Pavilion.

---

[17]http://www.economist.com/node/17358802, "The post-alcohol world: Biofuels are back. This time they might even work" (October 28, 2010).

These reports in the financial and biofuel, not aviation, publications and actions of the fuel companies themselves in becoming paid subscribers to global aviation customer events give apt indication of the results of the measures described above.

Although polls are a good overall indicator, discrete progress is best shown in four areas:

- the number of entities actually flying Biofuels,
- the concurrent rise of Government focus in the United States and with programs at both the State and Federal level,
- the emergence of Global partnerships,
- the focus on both cost reduction and financial option development.

Flying biofuels

As of 2013, over 30 different commercial and military customers had flown aircraft on biofuels blends. Several operations have entered commercial service. The most significant of these are:

- Regular commercial service initiated in 2011–2012 by Lufthansa between Frankfurt and Hamburg. That operation demonstrated that fuel quality and reliability could be met in Airline Operations. Fuel was HEFA product from Finland's Neste producer.
- Regular commercial service was initiated in January 2013 by KLM from the JFK airport in New York. Fuel was HEFA product sourced initially from Dynamic Fuel, cooking oil–based product based in Louisiana by SkyNRG of the Netherlands.

U.S. federal focus

- At the end of March 2010, in an Energy address at Georgetown University, President Obama focused a major push in the area of transport biofuels on Aviation, both Commercial and Military. This focus took several forms and has grown ever since. Specifically:
- In July 2010, USDA, the Airlines for America (then the Air Transport Association) and Boeing initiated a program called "Farm to Fly." Farm to Fly focused on specific deployment initiatives in the Pacific Northwest in a project referred to as SAFN (Sustainable Aviation Fuels Northwest).
- In 2011, the Navy, USDA, and Energy Department launched a program to evaluate and build up to four facilities in the United States under Title III of the Defense Production Act.
- In April of 2013, "Farm to Fly 2.0" was signed by USDA, the DOT U.S. Department of Transportation, and CAAFI Airport (Airports Council International, North America), manufacturer (Aerospace Industries Association), A4A (Airlines for America). Both the as well as by the General Aviation (General Aviation Manufacturers Association, GAMA) and Business Aircraft Industry (National Business Aircraft Association, NBAA) joined for the first time as signatories. 2.0 is designed as a national program to leverage USDA's existing programs in U.S. States and Regions. Farm to Fly 2.0 targets production of 1 B gal of alternative jet fuel by 2018 to potentially meet carbon goals.

U.S. state focus

"Farm to Fly 2.0" was as much a recognition of the scope of Aviation State and Regional efforts as a new Federal activity. At the time of writing, in excess of 20 U.S. states are engaged as leaders of CAAFI-motivated efforts in their jurisdictions (Figure 27.12).

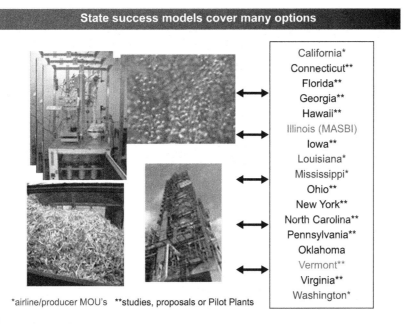

*airline/producer MOU's   **studies, proposals or Pilot Plants

FIGURE 27.12

Alternative fuel efforts in U.S.

State focuses range from the large regional focuses such as the United, Boeing, and UOP-sponsored Midwest Aviation Sustainable Biofuels Initiative (MASBI), to the development of Research initiatives to scale-up new pathways and feedstocks, such as the GSR algae project[18] using Waste streams proposed in Vermont.

What the projects have in common is that they are run by the "opinion leaders" with strong public/private connections.

In many cases, the projects do leverage existing programs at USDA such as Value Added Producer Grants (VAPG) and Rural Business Enterprise Grants (RBEG), as well as companion programs at the State level.

Global public/private partnerships

An attraction to fuel producers of working with Aviation is that it is truly a global market.

Seventy-five percent of up to 80 billion gal of jet fuel consumed annually are burned outside of the United States.

In addition to the markets, the research and development burden of as many as six new pathways to qualify requires a global commitment from all entities that can benefit from the industries development.

[18]www.gensysresearch.com.

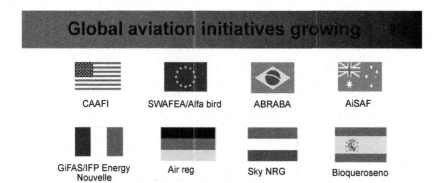

- **Brazil/U.S. Bilateral (03/11)**
- **Australia/U.S. Bilateral (09/11)**
- **Germany/U.S. Aireg/CAAFI Bilateral (9/12)**
- **Spain/U.S. Bilateral (2/13)**
- **Discussions with EC R&D Directorate**

FIGURE 27.13

Public/private global partnerships growing.

A key approach to address the goals of development and deployment is through the use of bilateral public/private partnerships between coalitions of private entities and their governments across the globe (Figure 27.13).

By forming success "templates" currently between four continents, resources to develop have grown. Often these public/private successes have seeds sown in relationships forged initially by aviation manufacturers such as Boeing (via its SAFUG initiative) and Airbus, or the global trading skills manifested by Netherlands-based SkyNRG.

Global success is also fed by efforts led by the public/private World Economic Forum (WEF), who have mapped public/private partnerships across the globe for aviation biofuels.

Production cost reduction, debt financing, and project analysis tools

As complements to the aviation industries, development of the biofuels options as well as the supply chain to airport supply hubs are developments to support the most often cited challenges to fuels implementation: cost, financing, and reliable project evaluation mechanisms.

- Cost Reduction: start-up costs and need for production learning are common to virtually all new technologies. Although feedstock cost is being tackled through the feedstock-readiness protocols, issues such as capital cost and operating costs must be tackled if the long-term competitiveness is to be assured. One key to cost reduction is in the United States is the Department of Energy Programs, which bring down these cost elements. DOE supported initiatives were successful in bring "learned out" ethanol production cost from $7 to $2 when normalized over a decade of targeted spending by DOE labs.

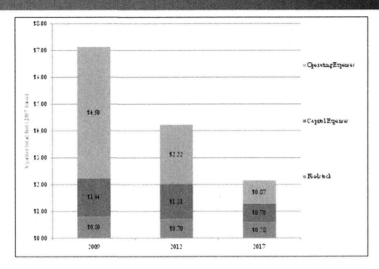

*DOE - Current funded pathway to $2.00 /Gallon Pyrolysis Oil

FIGURE 27.14

Production cost reduction by using DOE methodology.

The application of this methodology to HDCJ thermochemical pathways by improving catalyst life is projected to allow cost parity for this pathway with petroleum sources by the second half of the current decade (Figure 27.14).

Similar plans are in the works for other pathways. Yet another public/private partnership, DOE with the aerospace producers allows this methodology to progress to make rapid process by using industry-developed processes that have tackled similar problems in aircraft production.

- Debt Financing Mechanisms: although equity financing has been raised by several of the most successful biofuels companies, the challenge in any capital-intensive venture, such as biofuels production, is to establish debt financing, which must complement the support of venture capitalists. Institutions such as Westar, Broadway Capital, and Stern Brothers have developed new bond financing mechanisms that can be attractive to new producers and potentially draw resources from capital sources such as Pension funds.
- Project Analysis Tools: developments from the ACRP Projects discussed in the prior section have now been applied to real cases at multiple airports in the United States. Using these tools and other more subjective means, a matrix of 12 factors that explains why a project can make sense to airports and how and when to undertake such evaluations has been developed and posted on the website of CAAFI's airport sponsor, the Airports Council International of North America. With this tool, airports, their communities, and airline partners can evaluate projects as stakeholders.

CLOSING SUMMARY

The unique and demanding requirements of Sustainable Aviation Biofuel are indeed being met via product developments by the exciting new biofuel industry. The challenges of managing qualification for aviation safety, environmental safety, and adequate supplies at acceptable costs are progressing at a far faster pace than could have been expected at the dawn of the Aviation biofuels age less than a decade ago.

Although the formula is set for success and goals for Carbon neutral growth (by 2020) with acceptable cost and delivery models in place, there is much work to be done. Global developments and deployment models are in place that is growing to address the challenge.

The process has produced extraordinary progress and has exceeded the most optimistic of projections for progress after CAAFI's first 5 years.

Beyond the field of aviation, the efforts to date also represent an important success model for both other transportation fuels and new energy sources that seek similar development and deployment goals.

CUTTING-EDGE BIOFUEL CONVERSION TECHNOLOGIES TO INTEGRATE INTO PETROLEUM-BASED INFRASTRUCTURE AND INTEGRATED BIOREFINERIES

28

Anju Dahiya[1,2]

[1] *University of Vermont, USA;* [2] *GSR Solutions, USA*

BIODIESEL AS RENEWABLE DIESEL

"Where are we now?" is a question that frequently comes up during my presentations and lectures on biofuels, and to get onto that right away, I guess *"Biodiesel: Advanced Biofuel—Here, Now"* by National Biodiesel Board (NBB), (2013) is a good starting point. *"The US biodiesel industry produced a record of nearly 1.1 billion gallons in 2011, supporting almost 41,000 jobs. The industry is projected to grow sustainably in the coming years, producing a target of nearly 2 billion gallons in 2015 and supporting more than 74,000 jobs and some $7.3 billion in GDP."* In fact, the biodiesel and renewable diesel industries reached a record production of 1.8 billion gallons in 2013, supporting over 62,200 jobs (NBB, 2014). And, these figures are growing with innovations.

Looking back, in the year 1900, Rudolf Diesel demonstrated his engine at the Paris World Exposition by using peanut oil. Eventually, cheap petroleum took over too-viscous vegetable oils, making the modern engines fossil fuel dependent. In the United States, the 1970s oil crisis scenario brought attention back to biofuels, and even with recent heightened interest of oil companies to produce biofuel at commercial scales, we still have a long way to go.

The term *"biofuels"* is widely used to address different types of fuels derived from living matter e.g., ethanol, biodiesel, and biogas, whereas *"biodiesel"* is generally used for *mono-alkyl esters* of long chain fatty acids as an alternative to petroleum diesel that can be used as blends in diesel engines. Biodiesel derived from plant or animal sources contains fats called triglycerides (triacylglycerol or TAG) made up of oxygen, hydrogen, and carbon atoms—with three fatty acid chains attached to the glycerol backbone (Figure 28.1). The biodiesel component added to conventional diesel fuels as blends (e.g., B5 to meet ASTM D975) are fatty acid methyl esters (FAME). *"Renewable diesel"* is the hydrocarbon-like fuel produced in refineries by hydrotreating the biomass-derived oil.

Bioenergy. http://dx.doi.org/10.1016/B978-0-12-407909-0.00028-6

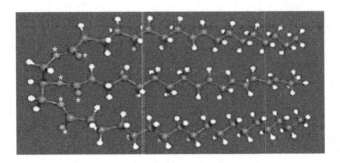

FIGURE 28.1

Chemical structures of triglycerides—gray = carbon atoms, white = hydrogen atoms, 6 atoms marked by a star on the left = oxygen.

FAME could be derived from feedstocks such as *vegetable oil* (sunflower, canola, soy, corn, jatropha, etc.), *animal fats* (e.g., tallow), nonfood sources (algae), and recycled cooking oil. These fatty acids are different based on the number of carbon atoms in the fatty acid chains (8–22).

BIODIESEL VERSUS PETROLEUM DIESEL

The US specification for biodiesel is ASTM D6751, and the European specification is EN 14,214. The key difference between the two is that the former applies to fatty acid extraction via any type of alcohol (e.g., methanol and ethanol), whereas the latter applies to the extraction of fatty acids only via methanol. Irrespective of extraction mode, both are different from fossil fuel-based petroleum diesel. Table 28.1 below summarizes the main differences between biodiesel and petroleum diesel. "The quality of petroleum diesel fuel tends to be more uniform and reliable, especially when compared to small-scale production of biodiesel where quality control may or may not have been good. Petroleum diesel can vary in quality from plant to plant or from region to region, but the variations are typically much smaller. Poor-quality biodiesel fuel can lead to many problems in engine performance, and care should be taken to ensure that your fuel is of good quality. Biodiesel that conforms to ASTM standard D6751 should be of a consistent, high quality" (Ciolkosz, 2009).

PROCESSING PATHWAYS FOR CONVERSION OF BIOFUEL INTO DIESEL FUEL
TRANSESTERIFICATION

The chemical process, transesterification, is a well-known process for the conversion of vegetable oil to biodiesel in which triglycerides are reacted with an alcohol (methanol or ethanol) in the presence of an alkaline catalyst (potassium hydroxide), and the reaction forms a

Table 28.1 Properties of Biodiesel versus. Petroleum Diesel, Based on CONCAWE (2009) and Ciolkosz (2009)

Property	Petroleum Diesel	Biodiesel	Pros/Cons of Biodiesel
Similarities:			
Molecule size		Approx. the same as petroleum.	None
Differences:			
Chemical structure	About 95% saturated hydrocarbons and 5% aromatic compounds	Consists of chemicals called fatty acid methyl esters (FAME) & unsaturated "olefin" components.	Different fuel properties.
Lubricity	Lower	Higher than petroleum diesel.	*Pro*: High lubricity is reduces engine wear.
Sulfur content	High sulfur	No sulfur.	*Pro*: Expected to result in reduced pollution from engines using biodiesel.
Oxygen content	Low	Higher oxygen content (usually 10–12%) than petroleum diesel.	*Con*: Relative to petroleum diesel, higher oxygen slightly reduces peak in engine power ($\sim 4\%$).
Gel up	Does not "gel up"	Biodiesel tends to "gel up" at low temperatures compared to petroleum diesel.	*Con*: A concern, especially for the cold winters.
Oxidize	Does not	Biodiesel more likely to oxidize (react with oxygen) to form a semisolid gel-like mass.	*Con*: A concern, for extended fuel storage and while using engines occasionally.
Chemically active	Does not	Chemically active as a solvent.	*Con*: More aggressive to some materials normally considered safe for diesel fuel.
Toxicity	Unsafe for environment	Biodiesel less toxic than petroleum diesel.	*Pro*: A real benefit for spill cleanups.

mixture of fatty acids, including fatty acids and glycerol as byproducts (first processing pathway in Figure 28.2(a)). From the chemistry standpoint, it is an exchange of an organic group of an ester with the organic group of an alcohol that takes place preferably with a catalyst. Biodiesel primarily contains fatty acid methyl esters (FAME) molecules. (Note: FAME and related extraction processes are described in detail in two respective chapters of this book—see chapters 22 and 20.)

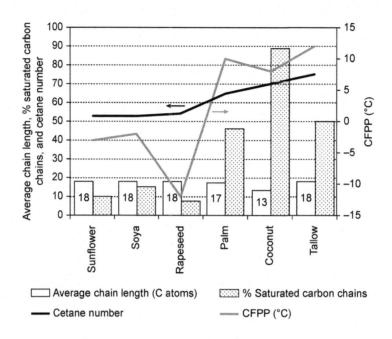

FIGURE 28.2

FAME from different feedstocks—average chain length, percentage saturated carbon chains, cetane number, and cold filter plugging point (CFPP) (CONCAWE, 2009).

CHALLENGES FROM FAME USE AS AN ALTERNATIVE TO DIESEL FUELS IN EXISTING INFRASTRUCTURE

FAME is different from hydrocarbon-only fuels since the chemistry and fatty acid composition determine and impact the properties of a FAME product (its cetane number, cold flow, filterability, and oxidation stability) (Table 28.2, Figure 28.2). As a result, blending FAME into hydrocarbon fuels introduces some specific challenges that must be carefully addressed in the production, blending, distribution, and supply of diesel fuels as follows (CONCAWE, 2009):

- Oxidation stability, under both thermal and longer-term storage conditions.
- Cold flow properties and filterability behavior. Cold filter plugging point (CFPP) is a measure of cold flow properties useful for judging the low temperature performance of conventional diesel fuels.
- Propensity for supporting microbiological growth.
- Tendency to increase the dissolved water content and degrade the water-shedding ability of diesel fuels.
- Compatibility with materials used in refinery, distribution, and fuel supply systems.
- Removal of dirt, rust, and other solid contaminants in the supply and distribution system.

Table 28.2 Impact of Fatty Acid Composition on the Properties of the FAME Product (CONCAWE, 2009)

	Impact on Oxidation Stability	Impact on Cold Flow Properties	Impact on Cetane Number
Increasing the number of carbon atoms in the fatty acid chain	Not significant	Poorer	Better
Increasing the number of unsaturated double bonds in the fatty acid chain	Poorer	Better	Poorer
Increasing the number of double bonds in the fatty acid chain: • None (saturated) • One (mono-unsaturated) • More than one (poly-unsaturated)	Relative oxidation rate: Low Medium High		

- Transport of FAME/diesel blends in multiproduct pipelines and other distribution systems.
- Safety, firefighting, and waste handling measures.
- Performance and compatibility of additives commonly used in distillate fuels.

ISSUES DUE TO THE PRESENCE OF OXYGEN IN BIOFUEL AND POSSIBLE SOLUTIONS

As shown in Table 28.1, one of the major challenges with the biological feedstock-derived oil use as diesel fuel replacement is the presence of "oxygen content." The biofuels (such as biodiesel and ethanol) contain oxygen that increases the affinity of these fatty acid fuel molecules for water. Water being a polar molecule (one end positively charged and the other negatively charged) is the reason it does not mix with the nonpolar oil molecules. The polarity of ethanol and FAME can also make the separation of dirt and water more difficult or slower than for petroleum fuels (ORNL, 2010). Some of the fatty acid chains are fully saturated while others may be mono-unsaturated or poly-unsaturated (i.e., contain one or more than one carbon–carbon double bond, respectively), and the fatty acid chains having more than one double bond, especially those having conjugated (adjacent) double bonds, are usually more chemically reactive and can be susceptible to oxidative degradation (CONCAWE, 2009). Petroleum diesel, on the other hand, consists of about 95% saturated hydrocarbons and 5% aromatic compounds, whereas biodiesel consists of FAME, which contains unsaturated "olefin" components (Ciolkosz, 2009).

The majority of the petroleum distribution infrastructure, mainly pipelines, tanks, and related equipment, is composed of low-carbon and low-alloy steels, and controlling rust and corrosion is of primary importance. However, certain fixes can deal with distribution problems due to the water in the FAME product (ORNL, 2010):

1. *Fuel fix*: use of the less polar form of biofuel, such as butanol vs. ethanol or converting oils or fats to hydrocarbons rather than esters. The less polar form of biofuel will also solve the problem of dirt in the product and cross-contamination of the products. To fix the corrosion and rust of mild- or low-alloy steel inhibitors makes the fuel more petroleum-like, reduces dissolved water, reduces fuel polarity and selection and use of effective corrosion inhibitors.
2. *Equipment fix*: design tankage to minimize contact between water and product, use separators to remove water, and use corrosion resistant steels.
3. *Operating fix*: drain water bottoms more frequently, combined with careful maintenance and monitoring, especially. The current practices may need to be executed more frequently with biofuels.

The most efficient fix would be the "*fuel fix.*" In order for the biomass liquids to be useful as transportation fuels, they require a chemical transformation to increase *volatility* and *thermal stability* and *reduce viscosity* through oxygen removal and molecular weight reduction (Elliott, 2007). Although, biodiesel's performance can be improved by adding additives such as cold-flow improvers, fuel stabilizers that can work as antioxidants, antimicrobial additives, detergent additives, and corrosion inhibitors (Ciolkosz, 2009), also demonstrated at the commercial scale by Kinder Morgan (ORNL, 2010). Alternatively, the biologically derived oil can be transformed into "*renewable diesel*" via *hydrogenation* (treatment with addition of hydrogen) or *hydrocracking* (catalytic breakdown of large molecules under pressurized hydrogen).

Hydroprocessing is one of the conversion pathways explored by national laboratories and the oil industry. The Pacific Northwest National Laboratory (PNNL) operated for the Department of Energy (DoE) has focused on heterogeneous catalytic hydroprocessing (Elliott, 2007) and has proven that the hydrothermal liquefaction (HTL) of biomass provides a direct pathway for liquid biocrude production (Elliott et al., 2014). As per PNNL, two types of methods are possible for conversion of fatty acids to renewable diesel: "high-pressure liquefaction" and "atmospheric pressure fast pyrolysis." These and other processes are described below under renewable diesel processing.

Integrating biofuel production into the existing infrastructure of petroleum refineries has been long recognized as the key step toward the reduction of fossil fuels, especially in the transportation industry. "A new technology can use renewable feedstocks derived from vegetable oils to produce a high-quality diesel fuel. And this new diesel process has possible advantages over other renewable-fuel technologies such as FAME also known as biodiesel—while FAME has many desirable qualities such as high cetane, there are other issues associated with its usage, such as poor stability and high solvency, leading to filter plugging problems" (*UOP LLC* (Honeywell's Limited Liability Company) and *Eni S.p.A*: Holmgren et al., 2007).

RENEWABLE DIESEL PROCESSING

According to the EPA (Act 2005, Section 45K c3), renewable diesel is diesel fuel derived from biomass, using the process of thermal depolymerization that meets the following (NREL, 2006):

1. Registration requirements for fuels and chemicals established by the Environmental Protection Agency under Section 211 of the Clean Air Act (42 U.S.C. 7545).
2. Requirements of the American Society for Testing and Materials (ASTM) D975 or D396.

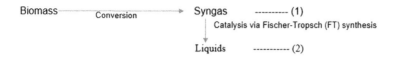

FIGURE 28.3

Indirect liquefactiom.

(Source: Courtesy: A. Dahiya.)

"Renewable diesel" can be produced via the following three processes (NREL, 2007).

A. Hydrothermal processing: In this process, the biomass is reacted on the order of 15–30 min in water at a very high temperature (typically 570° to 660 °F) and pressure (100–170 atm standard atmosphere), enough to keep the water in a liquid state to form oils and residual solids. The organics and distillate suitable for diesel use are separated (NREL, 2006).

B. Indirect liquefaction: This is a two-step process to produce ultra-low-sulfur diesel (Figure 28.3). The Fischer-Tropsch (FT) diesel has the advantage of being nonoxygenated with a comparable cloud point to diesel, and as per current producers, it is compatible with the existing oil pipeline (ORNL, 2010). First, biomass is converted to a syngas, a gaseous mixture rich in hydrogen and carbon monoxide, then the syngas is catalytically converted to liquids. The production of liquids is accomplished using FT synthesis (applied to coal, natural gas, and heavy oils (NREL, 2007).

(Syngas/Pyrolysis and FT synthesis processes are covered in detail in two respective chapters in this book—see chapters 16 and 26.)

C. Hydroprocessing: The renewable diesel referred to as "green diesel" can be produced from fatty acids by traditional hydroprocessing technology used to remove impurities by treating feeds with hydrogen (conversion temperatures typically 600° to 700 °F, pressures 40–100 atm, the reaction times are on the order of 10–60 min, and catalysts are used), for which the starting biomass-derived oils can be the same as for biodiesel or renewable diesel. The triglyceride-containing oils can be hydroprocessed either as a co-feed with petroleum or as a dedicated feed resulting into a premium diesel fuel in which the triglyceride molecule is reduced to four hydrocarbon molecules under hydroprocessing conditions (a propane molecule and three hydrocarbon molecules in the C12 to C18 range). It contains no sulfur and has a cetane number of 90–100 (NREL, 2006).

Honeywell UOP (with Eni from Italy) developed processes for the production of biodiesel fully compatible with petroleum-derived diesel fuel by developing the conventional hydroprocessing technology already used in the oil refineries that use hydrogen for the removal of oxygen from triglyceride molecules to produce "green diesel" (the processing pathway in Figure 28.4(a)).

In order to implement hydroprocessing in its process design, UOP (with Eni) considered two options (Holmgren et al., 2007): one) co-processing in an existing distillate hydroprocessing unit, or two) building a standalone unit (as shown in Figure 28.4(b)). The first option was found to be problematic due to the presence of trace elements as contaminants (e.g., phosphorus, sodium, potassium, and calcium), and their removal required a pretreating reactor, which led to the second option, which was more cost-effective and facilitated combining the vegetable oil with hydrogen. After bringing it to the required reaction temperature, it is sent to the reactor where vegetable oil is converted to green diesel via fractionation (Figure 28.4(c)) (for a detailed account see Holmgren et al., 2007, also published online by UOP).

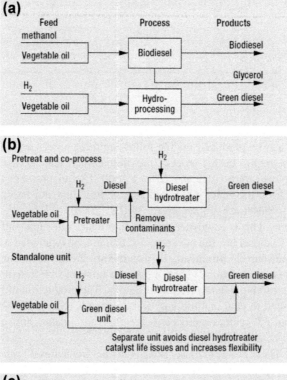

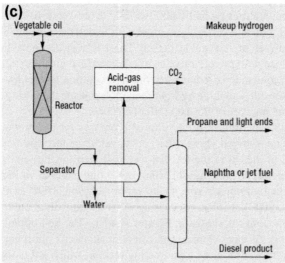

FIGURE 28.4

(a) Vegetable-oil processing routes for transportation fuels. (b) Alternative vegetable-oil hydroprocessing routes to transportation fuels. (c) The new green diesel process converts vegetable oil into fuels.

FUEL PROPERTIES OF HYDROPROCESSED DIESEL PRODUCT

First, the product properties of green diesel obtained from most of the vegetable oils are found to be similar (ensuring continuous feedstock availability). Second, the properties are comparable to petroleum diesel produced via Fisher-Tropsch processes (Table 28.3(a)), with the benefits of higher cetane and lower density (Table 28.3(b)) allowing the blending of low-value hydrotreated (LCO) into the typical refinery diesel pool and meeting the required standards at affordable and variable operating costs (Holmgren et al., 2007), paving the way for integration with existing refineries.

Table 28.3 Green Diesel Fuel Properties[1], Blending Costs and Economics Study Costs

A. Green Diesel Fuel Properties (Based on Alternative Fuels Comparison Chart, NREL[1])

	Mineral ULSD	Biodiesel, FAME	Green Diesel
O, %	0	11	0
Specific gravity	0.84	0.88	0.78
Sulfur content, ppm	<10	<1	<1
Heating value, MJ/kg	43	38	44
Cloud point, °C	−5	−5 to +15	−10 to +20
Distillation, °C	200 to 350	340 to 355	265 to 320
Cetane number	40	50 to 65	70 to 90
Stability	Good	Marginal	Good

B. Green Diesel Blending Benefits

Diesel-Pool Components	Barrels in Pool	Cetane Index
Kerosine	500	41
Hydrotreated straight-run diesel	7500	52
Hydrotreated LCO	2000	20
Green diesel	2346	74
Blended product cetane		50

C. Economic Study Costs

	Palm Oil	Soybean Oil
ISBL cost, $MM	33.9	33.9
OSBL cost, $MM	6.8	6.8
Variable cost, $/bbl	5.40	6.94

O: Oxygen percentage.
Source: Reprinted with permission from Hydrocarbon Processing, by Gulf Publishing Company, copyright 2012, all rights reserved.

[1]"Alternative Fuels Comparison Chart," NREL: http://www.eere.energy.gov/afdc/altfuel/fuel_comp.html.

INTEGRATION WITH EXISTING REFINERIES OR TO FORM NEW INTEGRATED BIOREFINERIES
COMMERCIALIZATION OF BIOFUEL

ASTM International, formerly known as the American Society for Testing and Materials (ASTM), is a globally recognized organization in the development and delivery of international voluntary consensus standards to improve product quality, enhance safety, facilitate market access and trade, and build consumer confidence[2].

Industry standards

Biodiesel commonly used as a blend with petroleum diesel is a legally registered fuel and fuel additive with the U.S. Environmental Protection Agency (EPA) that meets the ASTM biodiesel specification, ASTM D6751. The standard for biodiesel, further specified by a number following the "B" indicates the percentage of biodiesel in a gallon of fuel (the remainder of the gallon can be No. 1 or No. 2 diesel, kerosene, jet A, JP8, heating oil, or any other distillate fuel) as follows (NREL, 2009):

- At concentrations of up to 5 vol% (B5) in conventional diesel fuel, the mixture will meet the ASTM D975 diesel fuel specification and it is compatible to pure petroleum diesel.
- For home heating oil, B5 will meet the D396 home heating oil specification.
- At concentrations of 6–20%, biodiesel blends can be used in many applications that use diesel fuel with minor or no modifications to the equipment.
- B20 is the most commonly used biodiesel blend in the United States because it provides a good balance between material compatibility, cold weather operability, performance, emission benefits, and costs. It is also the minimum blend level allowed for compliance with the Energy Policy Act of 1992, which requires the use of renewable fuels and/or alternative fuel vehicles by certain covered fleets. Equipment that can use B20 includes compression-ignition (CI) engines, fuel oil and heating oil boilers, and turbines.
- The pure biodiesel is known as B100. B100 or other higher blend levels such as B50 require special handling and may require equipment modifications. These issues can potentially be managed with heaters and/or changing engine seal and gasket materials. However, because the level of special care needed is high, it is not recommend to use the high-level biodiesel blends, except where human exposure to diesel particulate matter (PM) is elevated and health concerns merit the additional attention to equipment and fuel handling.

Case study of the largest US energy company making ethanol and FAME shipment

Kinder Morgan (KM), one of the leading pipeline transportation and energy storage companies in North America, operating more than 28,000 miles of pipelines and 170 terminals, delivered approximately 100,000 barrels of blended biodiesel through its 115-mile Oregon pipeline (EcoSeed, 2009) in 2009. After demonstrating the biofuel transport through the petroleum pipeline infrastructure, in 2011 the KM Partners (KMP) invested approximately $550 million in the renewable fuels handling business (Pipeline and Gas Journal, 2011), including the building of a new ethanol unit-train facility capable of handling 14,000 barrels per day with space for multiple unit trains, an offloading rail rack for

---

[2]http://www.astm.org/ABOUT/aboutASTM.html.

unit-trains of approximately 100 railcars, and an 80,000-barrel storage tank at its terminal. In 2013, KMP's Products Pipelines segment handled more than 10.3 million barrels of biofuels, and according to their quarterly report, KMP continues to handle approximately 30% of the ethanol used in the United States[3] using their proprietary mix of corrosion inhibitors.

This case study demonstrates that while the water solubility, cleaning effect, contamination, and stress corrosion issues associated with ethanol and FAME pipeline shipping are significant, they are not insurmountable, and the proper cleaning, chemical additives, segregation/batch sequencing, parallel pipelines, and monitoring can either eliminate or greatly mitigate these problems and make ethanol and FAME pipeline shipment possible (ORNL, 2010).

Benefits of biodiesel use in compliance with industry standards

Biodiesel is renewable, energy efficient, displaces petroleum-derived fuel, can be used as a 20% blend in most diesel equipment, can reduce tailpipe emissions, and is nontoxic, biodegradable, and suitable for sensitive environments. These benefits with the respective industry standards are as follows (NREL, 2009):

Easy to use: Blends of B20 or lower are literally a "drop in" technology, and no new equipment or modifications to existing ones are necessary because B20 can be stored in diesel fuel tanks and pumped with diesel equipment.

Improves engine operation: Biodiesel improves fuel lubricity (required to save moving parts like fuel pumps from wearing prematurely) and raises the cetane number of the fuel even in very low concentrations. Reducing the allowable fuel sulfur to only 15 ppm by federal regulations also reduced the lubricity of petroleum diesel, as the hydrotreating processes used to reduce fuel sulfur and aromatic contents also reduces polar impurities such as nitrogen compounds, which provide lubricity.

Positive effects of biodiesel on air toxics and human health improvement: Some of the particulate matter and the hydrocarbon emissions from diesel fuel combustion are toxic or even carcinogenic. Using B100 can eliminate as much as 90% of these air toxics, and using B20 would reduce air toxics by 20–40%.

Biodiesel provides a high energy return and displaces imported petroleum: Life-cycle analyses show that biodiesel contains 2.5–3.5 units of energy for every unit of fossil energy input in its production, and because very little petroleum is used in its production, its use displaces petroleum at nearly a one-to-one ratio on a life-cycle basis (Sheehan et al., 1998; Hill et al., 2006; Huo et al., 2008). This value includes energy used in diesel farm equipment and transportation equipment (trucks, locomotives); fossil fuels used to produce fertilizers, pesticides, steam, and electricity; and methanol used in the manufacturing process. Because biodiesel is an energy-efficient fuel, it can extend petroleum supplies (NREL, 2009).

Biodiesel reduces harmful emissions including tailpipe emissions: Biodiesel contains 11% oxygen by weight that helps in reduction of harmful emissions, such as the tailpipe particulate matter (PM), hydrocarbon (HC), and carbon monoxide (CO) emissions from most modern four-stroke CI or diesel engines (NREL, 2009), and Nitrogen oxide (NOx), methane and carbon dioxide emissions are estimated to be reduced by 41% (Sheehan et al., 1998). EPA (2002) reviewed 80 biodiesel emission tests on CI engines and concluded that the benefits are real and predictable over a wide range of biodiesel blends (Figure 28.5).

---

[3]http://ethanolproducer.com/articles/10081/kinder-morgan-highlights-ethanol-handling-in-q2-results.

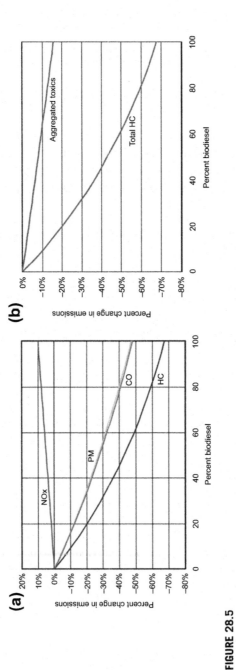

FIGURE 28.5

Biodiesel (percent) effect on HC (a) Emissions (NOx, PM, CO) (b) Aggregated toxics.

(Source: EPA, 2002.)

COMMERCIALIZATION OF RENEWABLE DIESEL

Hydrothermal processing technology is being commercialized in the United States by Changing World Technologies (CWT). CWT states that its product meets the requirements of ASTM D975 and uses the term "thermal depolymerization" to describe the process (NREL, 2007).

Indirect liquefaction for the production of FT liquids is a commercial technology, but most FT diesel today is produced using natural gas. FT diesel fuel can meet the requirements of ASTM D975 (NREL, 2006).

Hydroprocessing has been explored widely, and as such, renewable diesel is available commercially, for instance:

- Honeywell's UOP has commercialized the UOP/Eni Ecofining™ process to convert nonedible, second-generation natural oils to Honeywell Green Diesel™—a drop-in diesel fuel for use in any percentage in the existing fuel tanks. UOP (with Eni) has done extensive process performance testing of green diesel to determine optimum process conditions, catalyst stability, and product properties (Table 28.2(c)), and their pilot-plant tests have shown that there is no measurable deactivation after over 2000 h on stream (Holmgren et al., 2007).
- The product called H-Bio is produced by Petrobras in Brazil.
- Neste in Finland has commercialized NExBTL, the renewable diesel by Nestle.

"Green diesel's major advantage over FAME is the deoxygenation that occurs during processing, so that green diesel does not display a cleaning effect and should exhibit improved storage stability. It also has a comparable cloud point to diesel, making it compatible with the existing oil pipeline in the opinions of several of its current producers. This potential, coupled with the current lack of extensive research on its compatibility and fungibility, makes green diesel a good candidate for future study" ORNL (2010).

HYDROPROCESSED RENEWABLE JET FUEL (HRJ)

The low-level FAME contamination in jet fuel is of current concern (ORNL, 2010). The most common biofuel types, currently in use in road transport, are not suitable for use as aviation fuels because they do not meet jet fuel specification requirements (e.g., freezing point, thermal stability, etc.). Due to FAME (from bio-blended-diesels) contamination in jet fuel, it can have the following issues which are of concern for aircraft operations (Froment, 2010):

- Corrosion: formic and acetic acids, glycerin, water and methanol can be present.
- Cracking or softening of Elastomer seals.
- Presence of alkaline earth metals with an effect on engine components.
- High freezing point (freezing at -5C).
- Thermal stability: polymerization can occur, leading to a filter blockage.

HRJ identified as Hydrotreated Esters and Fatty Acid (HEFA) fuel was qualified under ASTM D7566, the specification for nonpetroleum jet fuels developed in 2009 as a 50% blendstock with conventional ASTM 1655 jet fuel A1 fuel. The 2011 qualification of HEFA followed 2009 approval of Fisher Tropsch fuels as a 50% blend. Hydroprocessed Esters and Fatty Acids (HEFA) were successfully shown to meet the requirements of a "drop-in" blend with ASTM 1655 jet fuel through a series of tests specified in ASTM 4054. Successful qualification of HEFA was in large part expedited due to hydroprocessing and distillation producing a fuel blending product similar to the range of aliphatic compounds found in jet fuels (see details about Aviation Renewable Biofuels in the chapter in this book by Richard Altman, executive director emeritus of CAAFI).

According to the Commercial Aviation Alternative Fuels Association (CAAFI) (Lakeman et al., 2013): "the testing and approval of Fischer-Tropsch fuels (synthetic paraffinic kerosene – SPK) has given the manufacturers and the industry as a whole the substantiation to allow the acceptance of fuel blends that meet all traditional jet fuel specifications and which contain a range of aliphatic compounds similar to what is found in FT fuels. The approvals of SPK and HEFA as blending fuels have been accepted because the resultant blends have been deemed indistinguishable from petroleum distillate jet fuels, albeit low aromatic jet fuels." According to Altman, "more than 1500 commercial flights using renewable jet fuels since HRJ/HEFA were approved for use in July of 2014. Operational flight trials by Lufthansa (from Frankfurt) and KLM from John F. Kennedy International Airport in New York have proved ongoing scheduled operations. By the end of 2014 the first facility dedicated to Aviation HEFA production operated by Altair fuels is scheduled to open operating under multiyear purchase agreements with United Airlines. While issues of feedstock approvals, carbon regulatory uncertainty and cost and financing issues remain the aviation maintains its target of approximately 5% fuel supply from renewables viable by 2020 to achieve the industries goal of carbon neutral growth. With additional facilities planned to come on line in multiple US states and globally, and with as many as six new process pathways qualification possible over the next two to three years achievement of the goal, while difficult, is possible."

FUTURE UTILIZATION OF BIOCRUDES AND CO-PROCESSING CONCERNS

"One potential hurdle to the future utilization of biocrudes is the RFS2 legislation concerning co-processing. If a biocrude is co-processed with petroleum crude and the resulting mixture is refined to produce a diesel–biodiesel mixture, this fuel will not qualify as a biomass-based diesel fuel. Biomass-based diesel fuels can be blended with petroleum diesel, but they cannot, by RFS2 definition, be a product of co-processing operations. If the appropriate greenhouse gas emission requirements are met, the resulting fuel would qualify as an advanced biofuel or cellulosic biofuel. This could potentially hinder the flexibility of the biocrude medium, specifically as it pertains to biodiesel production and distribution" (ORNL, 2010).

For a bio-derived hydrocarbon market, ORNL further recommends that bio-derived blending streams can be utilized via three paths to market:

1. They can be manufactured and used within a specific refinery fuel-blending operation.
2. The biofuel blend streams could be transported to a specific refinery or blending site by a proprietary or common carrier pipeline or other means such as truck or barge for private sale to an energy company.
3. Certain biofuel blending streams can be sold or traded on the open market.

In conjunction with integrating the new bio-derived hydrocarbon sources with existing petroleum-based refineries, developing the new integrated biorefineries would move the commercialization of biofuels forward.

INTEGRATED BIOREFINERY

The "guiding truth" is that if biofuel production is considered to be the primary goal, the generation of other co-products must be correspondingly low since their generation will inevitably compete for carbon, reductant, and energy from photosynthesis. Hence, the concept of a biorefinery for utilization of every component of the biomass raw material must be considered as a means to enhance the economics of the process (Figure 28.6) (DoE, 2009). A biorefinery is a facility that integrates biomass conversion processes and equipment to produce fuels, power, and chemicals from biomass[4]. As in a petroleum refinery, the major input is the petroleum required for processing in a multitude of products. In a biorefinery, biomass is the input needed for producing different products. "Biorefining is the sustainable processing of biomass into a spectrum of marketable products and energy"—this the most exhaustive definition provided by International Energy Agency (IEA) Bioenergy Task 42 (Cherubini, 2010).

A biorefinery, similar to what occurs in an oil refinery, should be based on feedstock upgrading processes, where raw materials are continuously upgraded and refined. That means a biorefinery should separate all the biomass feedstock components and lead, through a chain of several processes, to a high concentration of pure chemical species (e.g., ethanol) or a high concentration of molecules having similar, well-identified functions (e.g., the mixture of C alkanes in FT-fuels). It should follow the guidelines (Cherubini, 2010):

- A biorefinery should produce at least one high-value chemical/material product, besides low-grade and high-volume products (like animal feed and fertilizers), according to the specifications given above.
- A biorefinery should produce at least one energy product besides heat and electricity; the production of at least one biofuel (liquid, solid or gaseous) is then required.

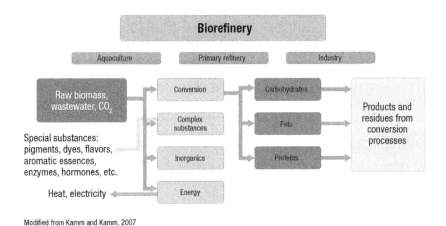

Modified from Kamm and Kamm, 2007

FIGURE 28.6

Biorefinery.

(Source: Courtesy: Bioenergy Technologies Office, DoE, 2010.)

---

[4]NREL http://www.nrel.gov/biomass/biorefinery.html (accessed 5/25/2014).

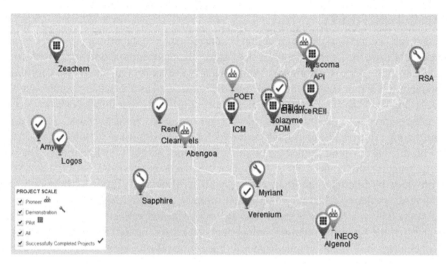

FIGURE 28.7

Integrated biorefineries funded by Bioenergy Technologies Office (BETO).

(Source: http://www.energy.gov/eere/bioenergy/integrated-biorefineries)

- Biorefinery plants should also aim at running in a sustainable way: all the energy requirements of the several biomass conversion processes should be internally supplied by the production of heat and electricity from the combustion of residues (within a properly sized set of processes/technologies).

Examples of integrated biorefineries (Figure 28.7) are:

- *POET and Abengoa*: The combined production capacity of both the facilities is estimated at more than 50 million gallons per year.
- *INEOS:* A demonstration-scale waste-to-fuel biorefinery in Vero Beach, FL, at full operational capacity is estimated to produce eight million gallons of advanced biofuels and 6 MW of renewable biomass power per year from renewable biomass, including yard, wood, and vegetative wastes.
- *Logos and EdeniQ*, CA, estimated the facility annual production capacity of 50,000 gallons of cellulosic ethanol.
- *Mascoma*: Upper Peninsula, MI, estimated 20 million gallons per year of ethanol and cogen of heat/power using biochemical-consolidated bioprocessing of woody biomass
- *Altech Envirofine*: Washington County, KY, 1 million gallons per year of ethanol based on biochemical-solid state fermentation of corn cobs.
- *Verenium Biofuels Corp*: Jennings, LA, 1.5 million gallons per year of ethanol using biochemical process of energy cane and sugar cane bagasse.
- *Amyris Biotechnologies Inc*: Emeryville, CA, 1370 gallons per year of biofuel via bioconversion of sweet sorghum.
- *Sapphire Energy Inc*: NM, 1 milllion gallons per year biofuel via conversion of algae.
- *ICM Inc.*, St. Joseph, MO, 245 K ethanol from corn fiber, switchgrass, energy sorghum.
- *Solazyme Inc*: Peoria, IL, 300 K gallon per year, biofuel from algae.

FIGURE 28.8

The Myriant biorefinery produces a valuable chemical product, succinic acid.

(Source: Myriant Corporation (reprinted with permission.)

- *RSA*: MN, 13 million gallons per year of ethanol via forest resources.
- *Algenol Biofuels Inc*: 100 K per year ethanol from algae.
- *Myriant Corp:* Lake Providence, LA. Myriant has developed a process to produce bio-succinic acid, and the facility is estimated to produce 30 million pounds of bio-succinic acid per year (Figure 28.8).

The key challenges for integrated biorefineries as identified by DoE (2013) are:

- Financial investment and the technical risk in the deployment and validation of novel and innovative technologies, particularly for pilot, demonstration, and pioneer-scale projects.
- Market and economic viability to create a product mix that is matched to market demand and can compete with fossil fuels.
- Feedstock diversity.
- Permitting.
- Sustainability modeled via economic, environmental, and social impacts monitored on life-cycle basis.
- Consistent research, development, and demonstration investments.

CO-LOCATING BIOREFINERIES

"Co-location of biodiesel and ethanol production at ethanol plants benefiting from sharing of existing infrastructure to share process essentials, an in-house feedstock such as distillers corn oil (DCO) and use of ethanol rather than methanol for biodiesel reactions—a glaring example is, the co-located 100MMgy ethanol plant processing 36 million bushels a year in Lena, Illinois by Adkins Energy LLC, with estimated capital costs $28 million for fractionation, 10 million for solvent extraction, and 15 million for a 10 MMgy biodiesel facility, totaling $53 million, and a same sized plant just selling DCO would generate $37 million in earnings before interest, taxes, depreciation, and amortization (EBITDA), compared to a plant with fractionation plus solvent extraction and biodiesel production would have $65 million before EBITDA with estimate of $28 million annual improvement" (Kotrba, 2014).

CONCLUSION

As stated in the beginning of this chapter, US "biodiesel" and "renewable diesel" production jumped to a record production of 1.8 billion gallons in 2013, and a possibility of huge profit margin by co-location of biorefineries (for instance, Lena-based plant) is indicating multifold higher chances of sustainable biofuel production in the future. Research, development, and demonstration of cost-effective technologies would be essential, especially for the integrated biorefineries that can take advantage of waste streams as throughput feedstocks for cost offsets for the production of different byproducts. Toward commercialization of biofuels for reducing fossil fuel usage, both options would have to be aggressively pursued viz integrating the biofuels with existing petroleum-based refineries via implementation of efficient processing pathways and development of new, integrated biorefineries, preferably co-located plants.

REFERENCES

Francesco Cherubini, 2010. The biorefinery concept: using biomass instead of oil for producing energy and chemicals. Energy Conversion and Management 51, 1412–1421.

Ciolkosz, Daniel, 2009. What's So Different about Biodiesel Fuel? Renewable and alternative energy factsheet. The Pennsylvania State University.

CONCAWE, 2009. Guidelines for Handling and Blending FAME. Report No. 9/09. CONCAWE, Brussels.

DoE, May 2013. Integrated Biorefienries. DOE/EE-0912.

EcoSeed, Sept. 23, 2009. Kinder Morgan Delivers Biodiesel through Oregon Pipeline. EcoSeed Category: Biodiesel. www.ecoseed.org.

Elliott, D.C., 2007. Historical developments in hydroprocessing bio-oils. Energy and Fuels 21 (3), 1792–1815.

Elliott, D.C., Hart, T.R., Neuenschwander, G.G., Rotness Jr., L.J., Roesijadi, G., Zacher, A.H., Magnuson, J.K., 2014. "Hydrothermal Processing of Macroalgal Feedstocks in Continuous-Flow Reactors." ACS Sustainable Chemistry & Engineering 2 (2), 207–215.

Draft Technical Report EPA, Environmental Protection Agency, 2002. A Comprehensive Analysis of Biodiesel Impacts on Exhaust Emissions, EPA420-P-02–001. www.epa.gov/OMS/models/biodsl.htm.

Froment, Marie, 2010. Jet Fuel Contamination with FAME (Fatty Acid Methyl Ester) World Jet Fuel Supply. FAST (Flight Airworthiness Support Technology). AIRBUS. S.A.S. publication, 8–13.

Hill, J., Nelson, E., Tilman, D., Polasky, S., Tiffany, D., 2006. Environmental, economic and energetic costs and ben-efits of biodiesel and ethanol blends. Proceedings of the National Academic Sciences 103, 11206–11210.

Holmgren, J., Gosling, C., Marinangeli, R., Marker, T., UOP LLC, Des Plaines, Illinois, Faraci, G., Perego, C., Sept. 2007. Eni S.p.A. Refining and Marketing Division, Novara, Italy. New developments in renewable fuels offer more choices. Hydrocarbon Processing, 67–71.

Huo, H., Wang, M., Bloyd, C., Putsche, V., 2008. Life-cycle Assessment of Energy and Greenhouse Gas Effects of Soybean-derived Biodiesel and Renewable Fuels. ANL/ESD/08–2. Argonne National Laboratory, Illinois.

Kotrba Ron, May/June 2014. Time has come today. Biodiesel Magazine, 26–31.

Lakeman, Michael, Michael Epstein (GE), Nicolas Jeuland (Alfa-Bird/IFPEN), Stephen Kramer (Pratt & Whitney), Kristin Lewis (Volpe), Laurie Starck (Alfa-Bird/IFPEN). "Alternative fuels specification and testing". In: Research and Development Team White Paper Series: Specifications and Testing, Commercial Aviation Alternative Fuels Initiative (CAAFI). March 2013, Pages 1–7

NBB, National Biodiesel Board's Biodiesel fact sheet, April 17, 2013. Biodiesel: Advanced Biofuel – Here, Now.

NBB, April 07, 2014. NBB to Defend Advanced Biofuel Standard in RFS Case. NBB Press Releases.

NREL, Nov. 2006. Biodiesel and Other Renewable Diesel Fuels. NREL/FS-510–40419.

NREL, 2007. National Bioenergy center. http://www.nrel.gov/docs/fy07osti/40419.pdf

NREL, January 2009. Biodiesel Handling and Use Guide. NREL/TP-540-43672, fourth ed. Revised.

ORNL, Sept. 2010. Fungible and Compatible Biofuels: Literature Search, Summary, and Recommendations. ORNL/TM-2010/120.

Kinder Morgan completes Gulf Coast ethanol terminal. Pipeline & Gas Journal 238 (4), April 2011. www.pipelineandgasjournal.com.

Sheehan, J., Camobreco, V., Duffield, J., Graboski, M., Shapouri, H., 1998. An Overview of Biodiesel and Petroleum Diesel Life Cycles. NREL/TP-580-24772. National Renewable Energy Laboratory, Golden, CO.

BIOFUEL CONVERSION PATHWAYS SERVICE LEARNING PROJECTS AND CASE STUDIES

29

Anju Dahiya[1,2]

[1] *University of Vermont, USA;* [2] *GSR Solutions, USA*

As per the World Bioenergy Association's global bioenergy statistics report (WBA, 2014), in 2012, the global bioethanol production reached 83.1 billion L, biodiesel to 22.5 billion L, and vegetable oil to 169 million tons. According to the Industrial Specialist of Center for Industrial Research and Service at Iowa State University, Rudy Pruszko (see Chapter 20), "Biodiesel production is a deceptively complex process, rather than a simple process at first sight, especially if you want to make quality biodiesel and a product that meets ASTM D 6751[1] specifications and not harm a diesel engine." A former chemistry professor at the University of Vermont, Dr Scott Gordon and his team (see Chapter 21) prepared a protocol for students to follow for the lab-based, hands-on experience. The Biomass to Biofuels program[2] has been following this protocol for the lab-based demo of biodiesel production (transesterification) process using waste vegetable oil as feedstock for biodiesel.

Irrespective of the educational background of a student, hands-on experience in a lab turns out to be an important session for a bioenergy course involving liquid biofuels as a component and that helps build a sound background for a service learning project (Figures 29.1 and 29.2). For instance, in the 2013 version of the Biomass to Biofuels course, one of the students pursuing a doctoral degree in mathematics commented, *"Though I enjoyed the lecture (on biodiesel business), my favorite part of the class was definitely the lab. I appreciated the story of the business (presented in lecture), especially the importance of the small details (loading dock) that ended up making a big difference. But mathematicians don't get to play with chemicals in the lab very often... had a blast!"* He also described his detailed experience from the lab as follows:

In the lab I made biodiesel from both virgin and used veggie oil. First I'll describe the process for the virgin oil, and then I'll mention the slight differences for the used oil. For each case it was nice to have the KOH already prepared for us (thanks to the biodiesel instructor).

Virgin oil: The KOH was used as the base in a biphasic reaction with the oil (transesterification), 3 parts virgin oil to 1 part KOH. Because the reaction is biphasic, the reactants needed to be stirred constantly and vigorously for about 20 min. The products of the reaction are methanol, biodiesel and glycerol (listed from least to most dense). The differences in density are important, allowing the

---

[1]ASTM stands for American Society for Testing and Materials
[2]http://go.uvm.edu/7y1rr.

Bioenergy. http://dx.doi.org/10.1016/B978-0-12-407909-0.00029-8

FIGURE 29.1

Biodiesel preparation from waste vegetable oil in the lab.

Source: Courtesy: A. Dahiya

separation of the products just by letting them sit in a test tube for a while (about 10–15 min in the lab, though longer would allow better separation). Then the biodiesel could be "cleaned", which is just removing more of the glycerol that hadn't quite separated, as far as I can tell. The cleaning was done by pouring the top two layers (leaving the glycerol layer behind) into a separating vessel (I forget the name of that thing), adding water (about 1:3 water to mixture), and shaking or swirling a bit (remembering to vent every few shakes). After letting the mixture settle again, the bottom layer can be dumped from the vessel, leaving a purer biodiesel. There is a so-called mixed "rag" layer between the two. If you throw all of that away, your result will be purer, but you'll be throwing away some of the product. How much to keep would depend on whether you care more about purity or production efficiency.

Used oil (from radio bean): *Another student and I tried to repeat the process using dirty oil. The only things done differently were: (1) we filtered the dirty oil (using a coffee filter) before*

FIGURE 29.2

Hands on with biodiesel preparation from waste vegetable oil in the lab.

Source: Courtesy: A. Dahiya

transesterification; and (2) we ran the reaction at a higher ratio of KOH to oil. Specifically, we tried 2:3 and 4:3 (instead of the 1:3 ratio for the virgin oil). The 4:3 reaction seemed to work, whereas the 2:3 didn't. The fact that more base is required for the reaction to work properly affects the amount (and hence cost) of inputs you'd have to consider for an operation based on converting used veggie oil.

This small experience greatly pays off when students actually visit a biodiesel facility. The same batch of students, including the mathematic doctoral student, went on a field trip to a local farm producing biodiesel from oilseed crops. Among this group one student reflected in the forum, *"This was a really great trip and really built upon our (instructor's) talk and our experience of making a small amount of biodiesel from vegetable oil with him in the lab."* The doctoral student of mathematics also chimed in, *"One of my favorite field trips as well..."* He undertook a service learning study on modeling of anaerobic digestion that included the chemistry of the process (see part IV) (Figure 29.3).

Many different biofuel conversion pathways have been explored by the Biomass to Biofuels students over the past many years. A few instances are presented below.

In the 2012 version of the Biomass to Biofuels course, three students, including Richard Barwin, an animal science high school teacher, and environmental studies students William Riggs and John O'Shea, partnered with Missisquoi Valley Union High School (MVUHS) for their service learning project: *An Educational Experience Converting Waste Cooking Oil to Biodiesel* to convert school fleet of diesel powered tractors to biodiesel. MVUHS has been planning to convert its fleet of grounds

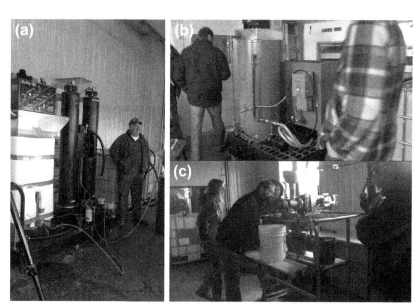

FIGURE 29.3

Field trip to oilseed crops biodiesel-producing farm. (A) The biodiesel processer at borderview farm; (B) Related equipment; (C) Oil press

Source: Photos courtesy: A. Dahiya

maintenance diesel-powered tractors from petroleum diesel to biodiesel. The Biomass to Biofuels student team started their project with these objectives: define the chemical process required to convert waste cooking oil into usable biodiesel; develop a standardized local procedure to convert waste cooking oil into biodiesel; design a machine that will implement the procedure; develop a method to evaluate the quality of the biodiesel produced; and develop a plan to utilize the waste product of biodiesel (glycerol) in a sustainable manner. See the complete report from this project as presented in this chapter by Rich Barwin.

Another innovative biofuel conversion process was recently explored by a couple of students from the 2014 batch: Tom Joslin, an environmental engineer from the Vermont Department of Environmental Conservation, worked with an environmental studies undergraduate, Vanessia Lam, on the *Use of Magnetite for Enhanced Harvesting of Wastewater Biogas Feedstock*. They partnered with a municipal wastewater treatment plant producing biogas. Their interesting study is presented in this chapter. Also in the 2014 batch, a master's student in engineering, Richard Smith III, undertook the feasibility of the *Fungal Breakdown of Lignocellulosic Biomass,* and based on interviews with three established companies as case studies for analysis, he developed a concept pilot system and explored its commercialization. See his case study in this chapter.

REFERENCE

WBA GBS report, 2014. World Bioenergy Association (WBA) Global Bioenergy Statistics. www.worldbioenergy.org.

BIODIESEL PROJECT: AN EDUCATIONAL EXPERIENCE CONVERTING WASTE COOKING OIL TO BIODIESEL

29A

Richard O. Barwin, William R. Riggs, John C. O'Shea

2012 Biomass to Biofuels Course Students, University of Vermont, Burlington, VT, USA

SUMMARY

It is the intention of the leadership of our community partner, Missisquoi Valley Union High School (MVUHS), to convert its fleet of grounds maintenance diesel-powered tractors from petroleum diesel to biodiesel. The primary goal of the program is to teach students in science, technology, engineering, and math (STEM) competencies and renewable energy. Furthermore, the school intends to produce, on campus, biodiesel produced from locally procured waste cooking oil. The full spectrum of planning, development, implementation, and quality control will serve as an experiential learning platform for traditional science and agriculture science instruction. In addition to the instructional and experiential learning benefits, the school intends to save 30% on diesel fuel costs per year.

PROJECT OBJECTIVES

1. Define the chemical process required to convert waste cooking oil into usable biodiesel
2. Develop a standardized local procedure to convert waste cooking oil into biodiesel
3. Design a machine that will implement the procedure
4. Develop a method to evaluate the quality of the biodiesel produced
5. Develop a plan to utilize the waste product of biodiesel (glycerol) in a sustainable manner

BACKGROUND

Petroleum diesel is a hydrocarbon extracted from crude oil. It is commonly used as an internal combustion engine fuel, particularly in agricultural and industrial applications. Diesel fuel typically is composed of hydrocarbon chains between 8 and 21 carbons in length (Figure 29A.1).

The two primary disadvantages of petrodiesel are that is releases large amounts of sulfur into the atmosphere during combustion, and that it is derived from petroleum oil, a nonrenewable energy source. The released sulfur is a contributing factor to acid rain. Technological and regulatory conditions exist to reduce the amount of sulfur in petrodiesel. However, the sulfur content of petrodiesel is related to the lubricity of the fuel, an important factor in engine wear and efficiency.

Biodiesel is a fuel derived from plant oil. Plant oils are composed primarily of triglycerides. Triglycerides are molecules composed of three long chain fatty acids attached to a glycerol backbone (Figure 29A.2).

Typical diesel chemical composition
cetane, or n-hexadecane is typical of diesel fuel - C$_{16}$H$_{34}$

FIGURE 29A.1

Diesel fuel chemical composition.

Source: http://www.firmgreen.com/fuel/fuel_facts.htm

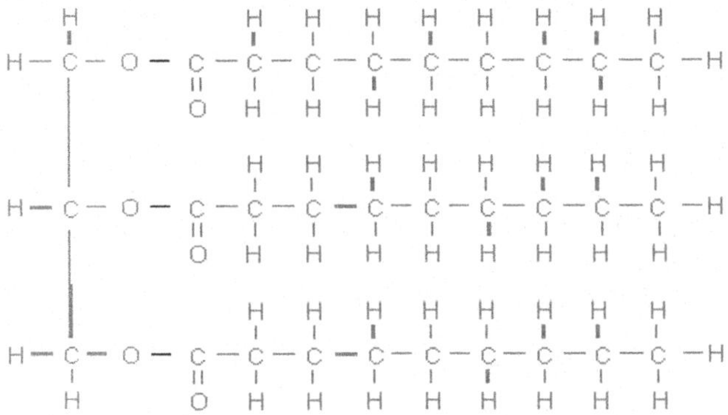

FIGURE 29A.2

Triglyceride chemical composition.

Source: http://biology.unm.edu/ccouncil/
Biology_124/Summaries/Macromol.html

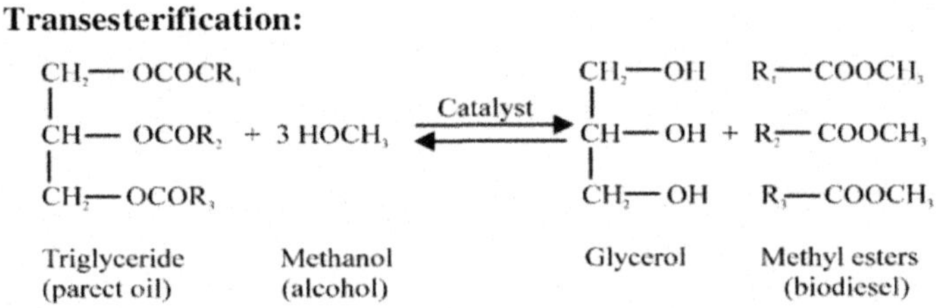

FIGURE 29A.3

Transesterification

Source: http://econuz.com/page/7/

Transesterification:

CH$_2$—OCOCR$_1$			CH$_2$—OH	R$_1$—COOCH$_3$
CH—OCOR$_2$	+ 3 HOCH$_3$	Catalyst	CH—OH	+ R$_2$—COOCH$_3$
CH$_2$—OCOR$_3$			CH$_2$—OH	R$_3$—COOCH$_3$
Triglyceride (parcet oil)	Methanol (alcohol)		Glycerol	Methyl esters (biodiesel)

To convert plant oil to biodiesel, it must undergo a chemical process called transesterification. During transesterification, triglycerides are mixed with methanol in the presence of the catalyst potassium hydroxide (Figure 29A.3). The reaction separates the long chain fatty acids from the glycerol molecule, leaving a water soluble glycerol fraction and a water insoluble biodiesel fraction (methyl esters).

The result of this reaction is a separated mix of biodiesel and glycerol. The water soluble glycerol is on the bottom, the biodiesel is on the top (4) (Figure 29A.4).

FIGURE 29A.4

Biodiesel and glycerol layers.

COMMUNITY PARTNER

Missisquoi Valley Union Middle and High School is an 800 student, 8–12th grade comprehensive secondary school located in the northern Vermont town of Swanton, close to the Canadian border. The local economy is influenced by agriculture, primarily medium- to large-sized dairy farms. MVUHS has a track record of implementing progressive educational and operational improvements. Though not a Vocational Technical Education Center, MVUHS currently has a comprehensive Diversified Agriculture program that includes a barn with livestock, a green house, a maple sugar operation, an aquaculture room, a mechanics building, and a 6 acre land lab. MVUHS embraced renewable, sustainable energy early, and converted their facility oil boiler to a wood chip boiler, not only reducing our greenhouse gas emissions, but realizing a substantial energy cost savings in the process.

METHODS USED AND/OR EXPERIMENTS (LISTED BY OBJECTIVE)
DEFINE THE CHEMICAL PROCESS REQUIRED TO CONVERT WASTE COOKING OIL INTO USABLE BIODIESEL

The chemical process to convert waste cooking oil to biodiesel is well-documented and explained in section 3. Because we are using waste cooking oil, we must determine the pH of the oil. The pH of the oil will determine the amount of catalyst (KOH) needed in the system. The volumetric ratio of oil to methanol is 5:1.

DEVELOP A STANDARDIZED LOCAL PROCEDURE TO CONVERT WASTE COOKING OIL INTO BIODIESEL

Establishing a local procedure will ensure reproducibility and consistency. The following is a concept map describing our process (Figure 29A.5).

FIGURE 29A.5

MVUHS concept map for biodiesel production. MVUHS, Missisquoi Valley Union High School.

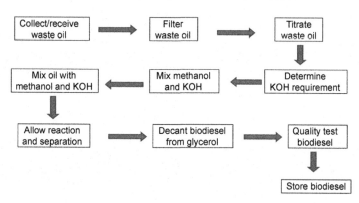

MVUHS concept map for boidiesel production

DESIGN A MACHINE THAT WILL IMPLEMENT THE PROCEDURE

Our team is currently collaborating on the design of the processor. We are currently purchasing components such as beakers, buckets, barrels, spigots, agitators, and tubing to build a process confirming prototype. We have been given space in the wood chip boiler room to use as a test and manufacture site.

DEVELOP A METHOD TO EVALUATE THE QUALITY OF THE BIODIESEL PRODUCED

We will experiment with a variety of options for determining the quality of our biodiesel.

1. Specific gravity—the specific gravity of our biodiesel should be 0.860 and 0.900.
2. PH—the pH of the unwashed biodiesel should be approximately 9.0.
3. Clarity—there should be two, and only two, distinct layers in the reaction vessel.

DEVELOP A PLAN TO UTILIZE THE WASTE PRODUCT OF BIODIESEL (GLYCEROL) IN A SUSTAINABLE MANNER

It is our long term goal to convert the waste glycerol into a locally produced soap. In order to use the glycerol for soap, we must remove excess methanol. Methanol has a boiling point of 148 °F and can be steamed off the glycerol. Eventually, we may be able to capture the methanol by condensation and reuse it in the transesterification process. However, the toxicity and flammability of methanol may argue against this step.

RESULTS/EXPECTED OUTCOME

The primary purpose of this project is education for our students. The chemistry, design, engineering, and community partnership will be driven substantially by our students. We will measure our success by the value of the education, not the return on dollar invested. We accept that this may become a multiple-semester project, and we may not reach the goal of replacing our entire diesel requirement for several years. However, we expect students from a variety of classes over several years to gain invaluable experiential learning from their involvement.

FUTURE DIRECTIONS

The possibilities for this program are limited only by our imagination and industry. The potential exists to partner with a local farm (Dr Heather Darby and farmer Roger Raineville engaged in a large-scale transesterification process from oilseed crops) to expose the school students to the full spectrum of biodiesel production, including planting, harvesting, pressing, tranesterification, washing, and the use of glycerol in soap.

BENEFITS TO COMMUNITY PARTNER

MVUHS is committed to teaching and practicing sustainable, local, socially responsible operations in all aspects of operation. Producing biodiesel is an important component of our goals. The project is still in the early stages. This plan is a framework to follow as we move toward our goal.

USE OF MAGNETITE FOR ENHANCED HARVESTING OF WASTEWATER BIOGAS FEEDSTOCK

29B

Thomas G. Joslin, Vanessia B. Lam

2014 Bioenergy—Biomass to Biofuels Course, University of Vermont, Burlington, VT, USA

SUMMARY

In March 2014 bench testing at the laboratory of the Airport Parkway Municipal Wastewater Treatment plant in South Burlington, Vermont, we have confirmed that magnetite, a fully oxidized iron oxide powder, may be used to improve settling and capture of primary sludge biosolids in municipal wastewater treatment. Preliminary calculations based on Airport Parkway plant operating data indicate that use of magnetite, for increased harvesting of primary sludge biosolids from wastewater treatment, has the potential to significantly increase biogas production in anaerobic digestion, and increase electric power generation at Airport Parkway, while also reducing electrical energy consumption for aeration of the secondary treatment process.

Project objectives: The purpose of this study is to confirm whether or not magnetite powder, a fully oxidized form of iron oxide, may be used as a ballasting agent to increase removal and harvesting of primary (gravity settled) biosolids in wastewater treatment plants, for use as feedstock to municipal biogas energy recovery projects at wastewater plants. Also, this study seeks to determine whether increased removal of primary biosolids, and subsequent reduced removal of secondary treatment process biosolids, would yield a net increase in the total methane-forming potential of all biosolids collected for a biogas project at a wastewater plant.

BACKGROUND

During the previous decade, Cambridge Water Technology, of Cambridge, Massachusetts, developed and commercialized a wastewater treatment technology, called CoMag™, that uses magnetite powder, *a fully oxidized form of* iron oxide, as a high density and chemically inert ballasting agent to improve solids settling and treatment performance. This technology was primarily driven by increasingly strict requirements for removal of total phosphorus from the treated effluent of municipal wastewater treatment plants, particularly relatively large plants located on relatively small streams in eastern Massachusetts. The first pilot study of the CoMag™ technology was conducted at the municipal wastewater plant in Concord, Massachusetts.

The CoMag™ technology[1] uses magnetite in combination with a coagulant chemical that forms chemical precipitate floc, plus polymer for enhanced floc formation, plus pH adjustment chemical as

---

[1]http://www.water.siemens.com/en/products/separation_clarification/ballasted-clarifiers/Pages/default.aspx.

needed. Magnetite and the other chemicals are added and mixed with wastewater, in mixing and flocculation tanks, and the resulting mixture of biosolids and chemical floc is allowed to settle in a clarifier tank. Some of the resulting sludge is pumped from the bottom of the clarifier back to the mixing and flocculation tankage, to further thicken the incoming wastewater and improve the solids settling. Some of the solids are pumped from the bottom of the clarifier to a magnetic drum separator. Permanent magnets in the drum separator remove more than 99% of the magnetite for return to the mixing and flocculation tankage. A small amount of makeup magnetite needs to be added to the process, to replace a small amount of magnetite that is lost to wasted sludge and effluent wastewater. CoMag™ is typically a tertiary wastewater process, treating the effluent downstream from a secondary (biological) wastewater treatment process. There is also a companion BioMag™ technology, also developed by Cambridge Water Technology, that adds magnetite ahead of suspended growth biological treatment processes. The magnetite is removed from secondary clarifier tanks located immediately downstream from suspended growth biological treatment tanks.

This study seeks to confirm that CoMag™ is technically feasible for enhanced settling of primary biosolids. Because primary biosolids are not oxidized in a biological (secondary) treatment process, they have a higher energy content and methane-forming potential for biogas than secondary biosolids. A process that increases harvesting of primary biosolids and reduces formation of secondary biosolids from the metabolism of soluble organic content leaving primary treatment could potentially increase biogas production at a wastewater plant. In addition, the resulting reduced-power requirement for secondary biosolids aeration could further improve the overall energy balance of a wastewater plant.

COMMUNITY PARTNERS

The City of South Burlington, Vermont, Water Quality Department, owns and operates two municipal wastewater treatment plants, at Bartlett Bay Road and Airport Parkway. Staff at the Airport Parkway plant participated in this study, and we appreciate their assistance. Steve Crosby is the superintendent of the Water Quality Department. Robert Baillargeon is the chief operator of the Airport Parkway plant. Jennifer Garrison is the laboratory technician at the Airport Parkway plant.

PLAN OF WORK

This study was originally inspired by the Village of Waterbury, Vermont, tertiary wastewater CoMag™ project, now under construction for phosphorus removal. Waterbury operates a secondary aerated lagoon plant.

It had been initially planned to test magnetite settling of lagoon bottom sludge at both Waterbury and Swanton, Vermont, to determine the potential for enhanced harvesting of algae for biofuel production. The first laboratory day was scheduled at Swanton on March 10, 2014. On that date, lagoon bottom sludge was not available for testing, because of continuing ice cover on the Swanton lagoons. It would not have been safe to collect sludge samples at Swanton on that date. Instead, we tested lagoon effluent, collected by plant chief operator Jim Irish. We used various magnetite doses, in a jar test gang mixer with six electrically operated paddle mixers and six 1-L beakers. We found that use of magnetite alone, without a coagulant chemical such as alum, aluminum sulfate, has virtually no effect on settling of lagoon effluent solids. Given the continuing unusually cold March weather, we abandoned plans to test at the Swanton and Waterbury lagoon plants. We contacted the staff at the Airport Parkway non-lagoon advanced wastewater treatment plant in South Burlington, not far from the UVM[2] campus. Airport Parkway also operates an advanced two-stage thermophilic-mesophilic anaerobic digester complex with cogeneration. We hoped to demonstrate that magnetite could be used to increase capture of relatively energy-rich primary biosolids, as feedstock for anaerobic digestion biogas production and electric power generation.

On March 24, at Airport Parkway, we repeated the Swanton test, but with screened and degritted raw sewage, or influent, instead of lagoon effluent. We found that use of magnetite alone, without a coagulant chemical such as alum, aluminum sulfate, has virtually no effect on settling of influent solids. On March 31, our first successful lab day, at Airport Parkway, we initially tested three sets of beakers: one set of three beakers for influent settling only, one set of two beakers for simulated cosettling of secondary treatment waste activated sludge (WAS) with influent, and one single beaker for WAS only (Figure 29B.1).

In the influent set, there was one control beaker with no magnetite or alum, one beaker with a minimum practical dose of alum only, and another beaker with a minimum practical dose of alum plus 1:1 magnetite:total suspended solids (TSS) by mass. In the cosettling series, there was one beaker with a minimum practical dose of alum only, and another beaker with a minimum practical dose of alum plus 1:1 magnetite:TSS by mass. No alum or magnetite was added to the WAS beaker. Initially the 1:1 magnetite:TSS dose for March 31 was to be based on an assumed average influent TSS concentration of 250 mg/L, as on March 24, but as local air temperatures finally rose during the last week of March,

FIGURE 29B.1

Lab experience. Clockwise, starting upper left corner: (a) Lagoon effluent sampling location, Swanton, VT (b) Swanton lab. (c) magnetite bag. (d) influent sampling at South Burlington, VT, Airport Parkway (e) septage unloading at Airport Parkway (f) Tom Joslin, PE, at Swanton lab.

Source: Photo collage by Vanessia Barros Lam.

[2]UVM University of Vermont

and the snow cover began to melt, flows in the Airport Parkway sewer system began to increase from clear inflow sources such as home sump pumps. We lowered our assumed average influent TSS concentration to 225 mg/L, on advice from Robert Baillargeon. Weighing of magnetite was simplified, compared to March 24, because there were fewer beakers containing magnetite, and we went directly to glass as the material of the container for weighing.

We used one drop of the same alum solution as used in the plant, per liter beaker, as a minimum practical alum dose. That ratio was about 25% higher than the full plant alum dose. One drop of alum solution is at least 0.05 mL alum per liter beaker. The estimated full-scale alum dose was based on 110 gallons alum per day, per 2.75 million gallons influent per day, based on figures from Robert Baillargeon. That estimated full-scale dose is about 0.04 mL per liter. On March 24, before the snowmelt, Robert gave us figures of 40 gallons alum per day and 1.8 million gallons influent per day. On March 31, Robert said that he had to increase the alum feed rate dramatically because the incoming cold meltwater had depressed the rate of biological phosphorus removal in the selector zones at the Airport Parkway plant.

For the two beakers in the cosettling series, influent and WAS were mixed at a volumetric ratio based on full plant scale volumes of 45,000 gallons per day WAS and 2.75 million gallons per day influent, per Robert Baillargeon. The 1:1 magnetite:TSS cosettling dose was based on that WAS: influent volumetric ratio, as well as the assumed influent TSS concentration of 225 mg/L; a WAS total solids (TS) (not TSS), concentration of 6700 mg/L, or 0.67% TS WAS, as measured in a moisture analyzer by Jennifer Garrison of Airport Parkway; and an assumed TSS:TS ratio of 0.95, or 95%. TS includes dissolved solids as well as suspended solids.

RESULTS/EXPECTED OUTCOME

In both the influent series and the cosettling series, where magnetite was used, it quickly settled to the bottom of the beaker. There was also a thin layer of biosolids at the bottom of each beaker. No distinct interface, or boundary, appeared in any of the beakers. The bulk of the liquid, above the bottom solids layer, remained somewhat cloudy, but to varying degrees. In the influent series, the beaker with alum only was clearer than the control beaker, and the beaker with both magnetite and alum was clearer yet. In the cosettling series, the beaker with both magnetite and alum was clearer than the beaker with alum only.

At our request, Jennifer Garrison collected and tested relatively clear samples from the tops of the influent and cosettling beakers containing both magnetite and alum. She tested both "supernatant" samples for TSS. Results were 22 mg/L TSS for the settled influent sample and 42 mg/L TSS for the cosettling sample. Jennifer reported those results by email on April 1. Next, we compared settling of WAS only, with no additives, to WAS with 1:1 magnetite: TSS and no additional alum. We say "no additional" because the WAS does contain chemical precipitates formed by addition of alum upstream from the secondary clarifiers from which the WAS is collected.

During the first 5 min of settling, the WAS with magnetite settled about 3.5 times faster than the WAS without magnetite. A distinct interface appeared in both beakers. Interface elevations were recorded at 5-minute intervals. Rates of settling declined with time in both beakers, but the settling rate declined faster in the beaker with magnetite. After 10 min, the settling rate of WAS with magnetite was less than the settling rate of WAS without magnetite. After 50 min, the interface level without magnetite was 410 mL (out of the original 1000 mL), and the level with magnetite was 305 mL. Because of schedule constraints, the WAS settling comparison had to be cut off at 50 min, but it

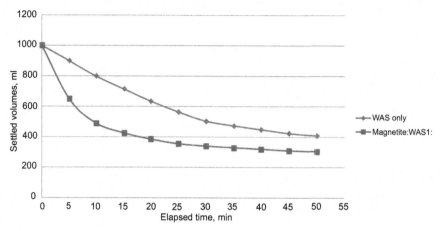

FIGURE 29B.2

City of South Burlington, Vermont, Airport Parkway wastewater treatment plan. Effect of magnetite on settling of waste activated sludge (WAS).

appeared that after one hour, both samples would have converged toward a total solids concentration of about 2%, from the original 0.67% Figure 29B.2. The effect of magnetite on WAS settling is illustrated in the graph below:

From required monthly operations reports from Airport Parkway to the Vermont Department of Environmental Conservation, for the three-month period September 2013 through November 2013, estimated median removals of primary sludge biosolids and WAS biosolids were 2520 pounds per day and 2860 pounds per day, respectively. The median primary effluent TSS concentration for that three-month period is about 190 mg/L, from primary clarifiers where WAS is cosettled with influent. By contrast, lab testing by Jennifer Garrison of Airport Parkway showed a TSS concentration of only 42 mg/L, from a simulation of primary clarifier cosettling with 1:1 magnetite:TSS and a minimum practical alum dose (one drop per liter) about 25% greater than full plant alum dose. A drop of 150 mg/L TSS during that three-month period would yield an additional 2325 pounds per day primary sludge biosolids. It is assumed that all WAS is pumped to the primary clarifiers for cosettling (rather than to mechanical WAS thickening) and that all WAS is captured by the primary clarifiers. Under the assumptions above, TSS removal across the primary clarifiers would increase from about 45% to about 90% (both percentages rounded to nearest 5%). Primary sludge biosolids removal would increase by about 92%.

Multiple wastewater reference works state that typical minimum removals of TSS and BOD_5, or five-day Biochemical Oxygen Demand (BOD), a measure of organic waste strength across primary clarifiers (and without cosettling) are about 50% and 25%, respectively. BOD_5 removal across primary clarifiers is smaller than TSS removal because much of the incoming BOD_5 is in soluble form and therefore cannot be removed by solids settling. Soluble BOD_5 must be removed by uptake and metabolism by microorganisms in secondary wastewater treatment. If it is assumed that additional BOD_5 removal attributable to magnetite treatment would be one-half that of the additional TSS removal attributable to magnetite treatment, then an additional 1162 pounds per day BOD_5 would be removed by magnetite treatment. Calculated BOD_5 removal across the primary clarifiers would increase from about 25% to about 40%. BOD_5 loading to the secondary treatment process and the associated blower energy requirement for delivering oxygen to the secondary process would drop by about 33%. Growth of WAS would drop by a comparable percentage.

The methane yield from anaerobic digestion of primary sludge is much greater than that of secondary sludge. One source (1)[3] states findings of methane yields of 470 and 179 mL methane for primary and secondary sludge, respectively, at gas standard temperature and pressure per gram sludge volatile solids. By that source, the methane yield of primary sludge is 2.62 times greater than that of secondary sludge. By all assumptions above, magnetite treatment to improve capture of primary sludge biosolids at Airport Parkway could potentially increase methane production and power generation by about 50%, and could reduce secondary treatment power cost by about 30% under optimum conditions. The total amount of sludge biosolids and total volatile solids mass fed to the anaerobic digester complex would increase by about 25%, but with the primary/secondary sludge ratio shifting toward primary, there would be much greater destruction of volatile solids associated with methane formation from the primary biosolids.

FUTURE DIRECTIONS

Preliminary calculations based on Airport Parkway plant operating data indicate that use of magnetite for increased harvesting of primary sludge biosolids from wastewater treatment has the potential to significantly increase biogas production in anaerobic digestion and increase electric power generation at Airport Parkway, while also reducing electrical energy consumption for aeration of the secondary treatment process.

In February 2014, Evoqua Water Technologies successfully demonstrated an advanced primary application of CoMag™ in plant scale testing at Terre Haute, Indiana. Harvesting of primary sludge biosolids is an area of growing interest in the wastewater and bioenergy industries: **Evoqua Water Technologies**, owner of the CoMag™ and BioMag™ technologies that use magnetite for wastewater treatment improvement, has also recently introduced its Captivator™ technology for enhanced primary biosolids harvesting. **ClearCove Systems**, of Rochester, New York, has received funding to demonstrate its own technology to increase primary biosolids capture from the New York State Energy Research and Development Authority. **Blue Water Technologies**, of Hayden, Idaho, is also marketing an advanced primary treatment technology, Eco MAT® Rotating Belt Filter. And there may be other examples.

BENEFITS TO COMMUNITY PARTNERS

There are no immediate benefits to the community partners as a direct result of this study, but it is hoped that further studies, with more time and resources available, could yield benefits to the community partners and the wastewater treatment industry in general.

[3]Kabouris et al., "*The Ultimate Anaerobic Biodegradability of Municipal Sludge and FOG*", Proceedings of the Water Environment Federation, WEFTEC 2007, Session 81 through Session 90, pp. 6776–6792(17).

SUBCHAPTER

FUNGAL BREAKDOWN OF LIGNOCELLULOSIC BIOMASS IN VERMONT

29C

Richard P. Smith

2014 Bioenergy—Biomass to Biofuels Student, University of Vermont, Burlington, VT, USA

PROJECT OBJECTIVES

The main objective of this project is to explore the commercial use of fungi to produce renewable energy from lignocellulosic biomass. A successful fungal pretreatment system requires the study of locally available fungal species and their compatibility with various available biomass sources. Based on the analysis of three case studies, a concept pilot system is developed in this report and its commercialization is explored.

INTRODUCTION

Lignocellulosic biomass is the most abundantly available raw material on the earth for the production of biofuels, mainly bioethanol (Carroll and Somerville, 2009). It is predominately made up of three different polymers: two consisting of sugar monomers (cellulose and hemicellulose) and one which tightly binds the sugars (lignin). It is the breaking of these lignin bonds to access the sugars that aids in the production of biofuel (Ravikumar et al., 2013).

BIOMASS PRETREATMENT

The conversion of lignocellulosic biomass to ethanol is currently being hindered by the difficulty to break down lignin that houses the sugars necessary for ethanol production in the form of cellulose and hemicellulose. Pretreatment refers to the process in which lignocellulose is converted to its basis, a form which increases cellulose hydrolysis efficiency. Pretreatment methods can be physical, chemical, and biological (Zheng et al., 2009). Biological pretreatment, the main focus of this report, utilizes wood degrading microorganisms, such as white rot fungi, to modify the structure of lignocellulose so that it is more susceptible to enzyme digestion.

FUNGAL PRETREATMENT

The use of white rot fungi in the biological pretreatment process is mainly to break down the lignin and hemicellulose shield that prevents access to cellulose in lignocellulosic biomass. These fungi produce many types of ligninolytic enzymes, which aid in the decomposition of these shields (Zheng et al., 2009). The fungi also possess the ability to conduct nonenzymatic reactions with lignin, which results in a more complete fracturing of lignin barriers. Therefore, solid state cultivation (SSC) of white rot fungi on lignocellulosic biomass is more effective as it is more cost-effective and more straightforward when compared with commercial ligninolytic enzyme preparation (Tian et al., 2012). The diversity of lignocellulosic biomass is matched in full by the diversity of white rot fungi. There are multiple fungi suitable for various stages of the pretreatment process of every type of biomass. Characterizing various biomass based on the optimal fungal species required for lignin breakdown and controlling the various conditions for this process to occur would lead to successful commercialization.

PILOT SYSTEM

The theorized pilot system is a small-scale pretreatment unit that can live on farms, waste management sites, or close to regional biorefineries. The main concept would be to utilize the time it takes for biomass to build up to the amount required before transport to a biorefinery. As the biomass collects, it could go through a controlled fungal pretreatment process while it waits to be sent to the refinery.

PLAN OF WORK

In order to develop a better understanding of what it takes to go from a pilot design system to commercial scale, a cost analysis was conducted. Gathering information from the three case study companies allowed for a deeper understanding of the aspects of stepping up to a commercial scaled operation. Extracting exact costs and processing specific details from these companies was not possible due to confidentiality, but the connections made through this process shed light onto what is happening currently on the commercial scale and what advancements these companies have made to be where they are.

CASE STUDIES

In order to develop a greater understanding of what is required to operate as a commercial biofuel company, three case studies were performed. The case studies included researching and connecting with three different companies focused on biofuel production. The insight gathered from these companies was used to develop a commercial-scale concept of fungal pretreatment of lignocellulosic biomass, as well as to build a cost analysis table.

   A. Mascoma: Mascoma Corporation is a U.S. biofuel company in New Hampshire that produces cellulosic ethanol made from wood and switchgrass. They use a proprietary consolidated bio-processing, or CBP, technology platform to develop genetically modified yeasts and other microorganisms to reduce costs and improve yields in the production of renewable fuels and chemicals. Mascoma's proprietary microorganisms, and the methodology for producing them, allows for a quick biomass conversion process and elimination of the need for high-cost enzymes in the process. They plan to expand the application of their CBP technology to develop advanced biorefineries that can produce multiple high-value end products, such as advanced fuels and chemicals, from many different feedstocks (Mascoma Corporation, 2014).

   B. Old Town Fuel and Fiber: Old Town Fuel and Fiber (OTFF) is based in Maine. Their focus is transforming old pulp mills into biorefineries. Pulp mills typically include an infrastructure with an efficient energy platform, water supply, waste water treatment, and skilled workforce. An onsite or adjacent colocated industrial sugar processing plan could effectively leverage an underused, idled, or noncompetitive pulp mill. In Old Town, this idea has gone through pilot and demonstration stages and is moving toward commercialization. There is a closed loop focus as most of steam and electricity needs are provided by the biomass itself and some hydropower (Zheng et al., 2009).

C. Amyris: Amyris is an integrated renewable products company based out of Emeryvill, CA. Amyris uses its industrial synthetic biology platform to convert plant sugars into a variety of hydrocarbon molecules, flexible building blocks which can be used in a wide range of products. Amyris is commercializing these products both as renewable ingredients in cosmetics, flavors, fragrances, polymers, lubricants, and consumer products, as well as renewable diesel and jet fuels. Their Synthetic Biology platform is based on genetic engineering and screening technologies that allow for modification of the way microorganisms process sugar. By controlling their metabolic pathways, microbes (primarily yeast) have been designed and used as living factories in fermentation processes to convert plant-sourced sugars into target molecules. They maintain a constant feedback loop between the laboratory, where strains are initially created and improved, and their two pilot plants, where those strains are exposed to conditions that simulate an industrial production environment (Gustafson and Bura, 2009).

COMMERCIAL SCALE

Based on the cost analysis and other insight gained through the case studies, it seems that in order to commercialize the proposed pilot system, there would need to be about 7 to 10 individual systems set up in order to facilitate one biorefinery. It is safe to assume Vermont produces statewide about 1000 metric tons of lignocellulosic biomass a day, which has the potential to produce 100,000 gallons of biofuel per day. A small-scale biorefinery requires around 100 tons/day to produce cost-effective biofuel. Therefore, it would be reasonable for Vermont to be serviced by 2 to 3 strategically located biorefineries. A fleet (7 to 10) of pretreatment systems, with the capacity of 10 tons/day, would be established in close proximity to each biorefinery across the state to collect and process lignocellulosic biomass (Gustafson and Bura, 2009).

FUTURE WORK WITH COMMUNITY PARTNER

The next step required to move forward with implementing fungal pretreatment of lignocellulosic biomass would be to develop an understanding of fungal and feedstock relationships. In order to optimize the pretreatment process, the most effective fungal strains need to be selected based on the type of feedstock (Tian et al., 2012). Further advancement would involve selecting certain fungal strains for certain steps of the pretreatment process for certain feedstocks. The results would be a roster of microorganisms assembled for maximum pretreatment efficiency. Focusing first on the locally and abundantly available waste biomass and their corresponding fungal friends, one could conduct laboratory experiments to determine optimal pretreatment ratios and conditions. A sophisticated, detailed control system could be developed based on the results of these experiments. The control systems would allow for efficient operation of the pilot system, thus creating a tested methodology of fungal pretreatment.

REFERENCES

Carroll, A., Somerville, C., January 2009. Cellulosic biofuels. Annu. Rev. Plant Biol. 60, 16582.

Gustafson, R., Bura, R., Cooper, J., McMohan, R., Schmitt, E., Vajzovic, A., 2009. Converting Washington Lignocellulosic Rich Urban Waste to Ethanol. Ecology Publication. Number 09-07-060. Washington State University Extension.

Mascoma Corporation, 2014. Mascoma Corporation.

Ravikumar, R., Ranganathan, B.V., Chathoth, K.N., Gobikrishnan, S., February 2013. Innovative and intensied technology for the biological pre-treatment of agro waste for ethanol production. Korean J. Chem. Eng. 30 (5), 10511057.

Tian, X-f, Fang, Z., Guo, Fe, 2012. Impact and prospective of fungal pre-treatment of lignocellulosic biomass for enzymatic hydrolysis. Biofuels, Bioprod. Biorefin. 6 (3), 335350.

Zheng, Yi, Pan, Z., Zhang, R., 2009. Overview of Biomass Pretreatment for Cellulosic Ethanol Production, vol. 2 (3):5168.

BIOFUELS ECONOMICS, SUSTAINABILITY, ENVIRONMENTAL AND POLICY

ANJU DAHIYA[1,2]

[1] *University of Vermont, USA,* [2] *GSR Solutions, USA*

As Prof. Bob Parsons states in his chapter (*Economics of Ethanol and Biodiesel*) in this book, "the growing interest in biofuels is driven by political, economic, environmental, and ethical forces. Characteristically, there are both proponents and opponents on each of the driving forces. For one, economics, it would seem that the numbers should do the talking and any analysis should be fairly direct and give a story," the driving areas of biofuels are economics, sustainability, environmental policy, and biofuels-related entrepreneurship opportunities.

Part VI covers a wide range of related topics including economics, sustainability, and environmental policy-related issues through nine chapters.

FIGURE 1

Sustainable and economical bioenergy.

Picture courtesy: A. Dahiya.

Chapter 30 (Biofuel Economics and Policy: The Renewable Fuel Standard the Blend Wall, and Future Uncertainties) begins with describing how biofuel markets are in a state of flux and moves on to describe the major biofuel policy today in the United States, which is the Renewable Fuel Standard (RFS). The detailed description of RFS is provided, followed by the blend wall (the physical limit on the blending of ethanol), and alternatives to the current situation.

Chapter 31 (Economics of Ethanol and Biodiesel) begins with how the current mandate in the United States to mix ethanol with gasoline has impacted agricultural production and goes through the history of economics, facts, and figures leading to the economics of biofuels, farm-scale production, and it concludes with the question, will biodiesel and ethanol survive on their own? That is a question that can only be answered in time.

Chapter 32 (Fuel Quality Policy) includes the factsheet by the National Biodiesel Board (NBB) that describes the American Society of Testing and Materials (ASTM); Government Adoption of ASTM D6751; ASTM Standards for Biodiesel Blends; BQ-9000 Certification; and government enforcement.

Chapter 33 (Renewable Oil-Heat) describes how the oil-heat industry is in the middle of a dramatic transformation as the fuel that heats more than 6 million homes in the northeastern United States is rapidly losing both volume and market share. The rise and fall of the oil-heat market; a cleaner and greener fuel; other renewable opportunities, and presents a case study: Bourne's energy; it explains how the subsequent economic consequences for oil-heat retailers has driven the industry to reshape its image by fundamentally changing the composition of its core product, and the oil-heat retailers are delivering biodiesel blended fuel oil and advocating for fuel quality standards that assures that consumers are heating with a low sulfur renewable fuel.

Chapter 34 (What's so different about Biodiesel Fuel?) describes how the biodiesel fuel has recently received much interest as an alternative fuel source; however, many people remain uncertain about whether biodiesel is a reliable, safe fuel to use for diesel engines. This chapter outlines the major differences between biodiesel and petroleum diesel, including information about biodiesel additives and blends.

Chapter 35 (Biodiesel Emissions and Health Effects testing) includes two of the NBB factsheets. The first factsheet summarizes the average biodiesel emissions (B100 & B20) compared to conventional diesel according to the U.S. Environmental Protection Agency (EPA), and presents ozone (smog) forming potential; sulfur emissions; criteria pollutants; carbon monoxide; particulate matter; hydrocarbons; nitrogen oxides; and the health risks associated with petroleum diesel. The second factsheet, presents the results from the first and second tiers of health effect testing of biodiesel and related significant benefits.

Chapter 36 (Biodiesel Sustainability Factsheet) includes the NBB factsheet that describes biodiesel-related sustainability principles; sustainability principles; water conservation; land conservation; food supply security; fuel diversity; and cleaner air and health effects.

Chapter 37 (Entrepreneurial Opportunities in Bioenergy) describes what is a bioenergy entrepreneur? And then describes bioenergy entrepreneurship motivations; the energy situation now and in the future; what drives the market? opportunities in bioenergy: the biofuel value chain; opportunities in small-scale bioenergy; opportunities in large-scale bioenergy; opportunities at the periphery; examples of entrepreneurism from the ethanol boom; challenges; and entrepreneurs in bioenergy.

Chapter 38 (Integrated Agroecological Technology Networks for Food, Bioenergy, and Biomaterial Production) introduces the application of ecological technologies to agriculture and natural resources management; agriculture and bioenergy. It then describes the role of industrial ecology in Combined Food-Energy Agro-Ecosystems (CFEA), the design and evaluation, measuring CFEA performance in a competitive world, participatory action research for CFEA primary data collection, analysis of ecotechnologies, key aspects of regionally appropriate ecotechnologies (material flows, energy flows, ultimate and proximate means, labor issues) and concludes with a CFEA case study on agro-eco park concept.

BIOFUEL ECONOMICS AND POLICY: THE RENEWABLE FUEL STANDARD, THE BLEND WALL, AND FUTURE UNCERTAINTIES

30

Wallace E. Tyner

Department of Agricultural Economics, Purdue University, West Lafayette, IN, USA

Biofuel markets are in a state of flux. The major biofuel policy today in the United States is the Renewable Fuel Standard (RFS). The RFS stipulates a minimum quantity of four categories of biofuels out to 2022 (Figure 30.1). The RFS is enforced via Renewable Fuel Identification Numbers (RINs). There is an RIN attached to each batch of ethanol produced or imported. At the end of the year, all obligated parties (refiners and gasoline importers) must turn in to the Environmental Protection Agency (EPA) RINs equal to their total blending obligation for the year. Corn ethanol RIN prices have surged from $0.03 in late 2012 to $1.35 on July 17, 2013, and back down to $0.24 in October 2013. In this chapter, we will explain what is behind the evolution in the corn ethanol RIN market. Prior to

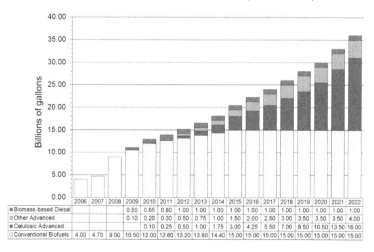

Renewable fuel standard (2007-2022)

	2006	2007	2008	2009	2010	2011	2012	2013	2014	2015	2016	2017	2018	2019	2020	2021	2022
■ Biomass-based Diesel				0.50	0.65	0.80	1.00	1.00	1.00	1.00	1.00	1.00	1.00	1.00	1.00	1.00	1.00
▨ Other Advanced				0.10	0.20	0.30	0.50	0.75	1.00	1.50	2.00	2.50	3.00	3.50	3.50	3.50	4.00
■ Celulosic Advanced					0.10	0.25	0.50	1.00	1.75	3.00	4.25	5.50	7.00	8.50	10.50	13.50	16.00
☐ Conventional Biofuels	4.00	4.70	9.00	10.50	12.00	12.60	13.20	13.80	14.40	15.00	15.00	15.00	15.00	15.00	15.00	15.00	15.00

FIGURE 30.1

US renewable fuel standard.

Bioenergy. http://dx.doi.org/10.1016/B978-0-12-407909-0.00030-4

2013, the RIN price was normally less than 5 cents/gal. Essentially, this meant there was no problem meeting the blending obligation. In other words, the RFS was not really a binding constraint.

The U.S. ethanol industry faces another obstacle known as the blend wall. When the Energy Independence and Security Act (U.S. Congress, 2007) was passed in 2007, the United States was consuming about 141 billion gallons (BG)/year of gasoline type fuel. In 2013, that level was about 133 BG. Gasoline consumption declined because of the recession and also because of the increased fuel economy in the U.S. automobile fleet. In the United States, most ethanol is blended with gasoline at a 10% level. If all gasoline were blended at 10%, the maximum ethanol that could be marketed would be 13.3 BG (0.1×133). The RFS level for conventional biofuel for 2013 is 13.8 BG, more than the blend limit. With the surge in RIN prices in 2013, the market began to perceive that the blend wall is binding, meaning it would be difficult to meet the RFS blending obligation. Furthermore, the RFS level rises in 2014 and 2015 reaching 15 BG, clearly far beyond the blend wall.

In addition, the RFS is under increasing attack from various parties. Bills have been introduced to eliminate or substantially modify the RFS. Thus, it is prudent to evaluate and understand the implications of possible changes that might occur. In this chapter, we will first describe the RFS, its history, how it works, and current implementation issues. Then, we will explore the possible implications of a set of possible changes to the RFS. Finally, we will review the status of cellulosic biofuels and the key uncertainties associated with cellulosic biofuels development.

THE RENEWABLE FUEL STANDARD

The RFS was initially created in the Energy Policy Act of 2005 (U.S. Congress, 2005). However, in short order, it was amended in the Energy Independence Act (EISA) of 2007 (U.S. Congress, 2007). The RFS created in 2007 is now sometimes referred to as RFS2, but we will use the term RFS in this chapter. The EISA RFS contained four categories of biofuels, but has a nesting structure that makes it somewhat difficult to understand. The general flow of the nesting structure is shown in Figure 30.2.

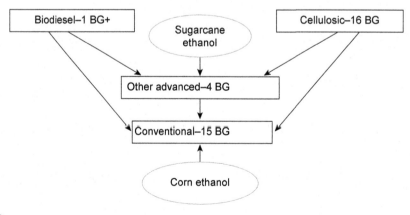

FIGURE 30.2

Nested renewable fuel standard (RFS) structure.

The overall level of required biofuels in 2022 is 36 BG ethanol equivalent. However, with the nesting structure, it is possible to meet the different components of the RFS in many ways. It is important to note that the EPA has interpreted the RFS volume levels in Figure 30.1 as ethanol equivalent. For example, biodiesel has 1.5 times the energy of ethanol, so 1.28 BG of biodiesel counts as 1.92 BG toward the overall RFS mandate. Below is a short definition of each of the four categories that also explains the nesting of the RFS components (Tyner, 2013).[1]

- *Biodiesel*: The original maximum mandate for biodiesel was 1 BG, but EPA has now increased that level to 1.28 BG. The biodiesel category must reduce greenhouse gas (GHG) emissions by at least 50%, compared with the fossil fuel option as defined by the legislation and EPA. It can be transportation fuel, transportation fuel additive, heating oil, or jet fuel. It can be ester based diesel (e.g., from soybean oil), or nonester renewable diesel (e.g., from cellulosic feedstocks). Biodiesel (as defined here) is required for the biodiesel part of the RFS. However, biodiesel can also be used to meet the other advance category or the conventional biofuel category (e.g., corn ethanol).
- *Cellulosic advanced biofuel*: Only biofuels produced from cellulosic feedstocks such as corn stover, miscanthus, switchgrass, forest residues, or short-rotation woody crops can count in this category. Cellulosic biofuels must be shown to reduce GHG emissions by 60%. By 2022, 16 BG ethanol equivalent of cellulosic biofuels are required. Since that is ethanol equivalent, if the biofuel were biodiesel, the volume would be 10.67 BG. Although no other type of biofuel can meet this category, cellulosic biofuels could in principle meet the entire 36 BG ethanol equivalent RFS, if at least 1.28 BG were renewable diesel. However, progress in developing the cellulosic biofuel industry has been slow, and EPA has been forced to waive most of the RFS each year because the product does not exist. For example, in 2013, the RFS called for 1 BG of cellulosic biofuels, but that was reduced to 14 million gal—the amount EPA expected to be available in 2013 (Agency USEP, 2013). So far, each year that EPA has reduced the cellulosic mandate to near zero, it has not reduced the overall renewable fuel mandate. For 2013, for example, even though the cellulosic category has been reduced from 1 BG to 14 million gal, the overall mandate remains at 16.55 BG. Thus, the shortfall in cellulosic biofuels must be made up by extra biodiesel or noncellulosic advanced biofuels.

One other important point in the cellulosic category is that in any year that EPA waives any part of the cellulosic RFS, blenders have an option to buy their way out of blending instead of actually blending (U.S. Congress, 2007; Tyner, 2010). To buy out of blending, obligated parties must purchase a credit from EPA plus purchase an advanced biofuel RIN. The price for the credit in 2013 is $0.42, and the October 2013 price for an advanced biofuel RIN was $0.31. Thus, the total cost of buying out of the RFS obligation was $0.73. At that time, wholesale gasoline was about $2.70, so the maximum one would pay for cellulosic biofuel is $3.43/gal gasoline equivalent. In other words, a blender would not choose to purchase cellulosic biofuel if its cost exceeded the wholesale price of gasoline plus the cost of buying out of the blending obligation. At present, there is no cellulosic biofuel available for that price or even close to that price. The consequence of this "off ramp" is that the cellulosic part of the RFS is not really a binding mandate.

---

[1]This material is taken from Tyner (2013).

- *Other advanced*: This category can be a wide range of biofuels that reduce GHG emissions at least 50%. Sugarcane ethanol that meets the GHG reduction standards qualifies, biodiesel qualifies, and cellulosic biofuels can be used. Recently, EPA approved sorghum ethanol produced under certain conditions. Corn ethanol cannot be used for this category.
- *Conventional biofuels*: This category is the only one that permits corn-based ethanol. It requires a reduction in GHG emissions of at least 20%. However, ethanol plants that were in operation or under construction as of December 2007 are grandfathered. The RFS level is 13.8 BG in 2013, reaches 15 BG in 2015, and remains at that level. In addition to corn ethanol, any of the other biofuel categories also can be used to meet the conventional biofuels category. In fact, technically, there is no mandate for corn ethanol. For example, for 2013, there is an overall mandate of 16.55 BG, of which 2.75 must be some form of advanced biofuel (1.0 cellulosic, 0.75 other advanced, and 1.0 biodiesel from Figure 30.1). The difference between the overall mandate of 16.55 and the sum of the advanced biofuels, 2.75 BG, is the amount that can be filled with corn ethanol, 13.8 BG.

The RFS is enforced by creating blending obligations for each type of biofuel. The blending obligations are based on market share for the type of fuel. For example, if you are a fuel seller, and you have 10% of the gasoline market, for 2013 with a 13.8 BG total obligation for corn ethanol, you would be required to blend 1.38 BG. To satisfy this blending obligation, you would need to supply to EPA at the end of the year certificate RINs demonstrating that you have blended 1.38 BG of corn ethanol. Each category of biofuels has a separate RIN, and there are blending obligations for each category for each obligated party. Obligated parties essentially are gasoline and diesel refiners for the domestic market and gasoline and diesel importers. Table 30.1 provides the October 24, 2013 price for each category of RIN. RINs can be bought and sold in an open market. Most RINs are actually turned into EPA at the end of the year by the party that blended the fuel. Thus, for most renewable fuel, the RINs are just the process of meeting the blending obligation. RINs are traded by those who expect to blend more than or less than their blending obligation. In general, if the RIN price is near zero, that is an indication that the RFS is not really binding. A higher RIN price suggests that the RFS, perhaps in combination with the blend wall, is driving behavior in the market place. Historically, corn ethanol RINs were usually near zero, but biodiesel and other advance were much higher. Now, corn ethanol RINs are priced near biodiesel and other advanced RINs because all three can be used to meet the blending obligation.

One other characteristic of the RINs market that bears mentioning is that RINs for up to 20% of the blending obligation can be carried forward to the next year and used later. In practice, what this means is that any carried forward RINs are used in the subsequent year, and RINs for that year replace the

Table 30.1 Renewable Fuel Identification Number (RIN) Prices for October 24, 2013	
RIN Category	**Price ($/gal)**
D6: ethanol	0.24
D4: biodiesel	0.40
D5: advanced	0.31

Table 30.2 Carry-Forward Renewable Fuel Identification Numbers (RINs) by Category

Category	Million gal of RINs
D4: biodiesel	301
D5: advanced	154
D6: renewable	2053
Total	2508

Source: EPA Moderated Transaction System, accessed April 12, 2013.

RINs that were used to be carried forward to the next year. In other words, even though the regulations states that the RINs must be used in the next year, in fact they can be continuously rolled forward. Paulson has estimated the 2013 carry forward RINs in the system, and they are shown in Table 30.2 (Paulson, 2013).

THE BLEND WALL

The blend wall is about 13.3 BG/year of E10. There is a small amount of ethanol blended as E85, and a tiny amount blended as E15, but they are really too small to matter for present purposes. As mentioned earlier, the 2013 RFS blending requirement for corn ethanol was 13.8 BG, and it grows to 15 BG by 2015. Thus, the physical limit on blending is less than the RFS, which makes the blend wall a real constraint. With the growth in the RFS, it may be impossible to meet the RFS requirements because of the blend wall. Figure 30.3 shows the evolution of the price of corn ethanol RINs from October 2012 to October 2013. Prices shot higher in July and then came back down. As is clear from Figure 30.3, the corn ethanol RIN price was around three cents for much of 4th quarter 2012, and grew to around five cents by the end of the year. In January, the RIN price started to surge topping $1/gal before stabilizing in the $0.75 to $0.95 range, and then moving up again in July. This surge in corn ethanol RIN price is because of the binding blend wall. With RIN prices near zero for 2012 and much of previous years, the RFS clearly was not binding. Absent the blend wall, we could have expected to see low corn-ethanol

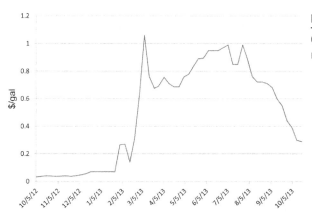

FIGURE 30.3

Corn ethanol renewable fuel identification number (RIN) prices.

RIN prices continue, as we have about 15 BG of corn ethanol production capacity, more than enough to meet the RFS. However, all the ethanol that can be produced cannot be blended because of the binding blend wall. In 2011, the United States exported 1.2 BG of ethanol, which helped provide a market for the additional ethanol that could be produced. In essence, the corn ethanol RIN price increases to close to the biodiesel and other advanced values because those fuels also can be used for conventional biofuel.

Thus far, as indicated earlier, EPA has maintained the total RFS unchanged even when it waives most of the cellulosic RFS component. If the blend wall is at 13.3 BG, and the total RFS is 16.55, the difference between the 13.3 BG of corn ethanol plus 1.28 BG of biodiesel—a total of 14.58 BG—will have to be made up from additional biodiesel and sugarcane ethanol imports plus use of carry forward RINs. We will provide a more detailed analysis below, but the point is that the blend wall is truly a binding constraint. Given that some of the RFS–blend wall gap will have to be made up from advance biofuel and biodiesel RINs, it is natural that corn ethanol RINs would move up to approximate the value of those RINs, and that is exactly what has happened. Also, it is made worse because every gallon of sugarcane ethanol that is imported to satisfy the other advanced category also falls within the blend wall and effectively reduces gallon for gallon the corn ethanol that can be blended.

The RIN prices began declining in August 2013 because EPA announced that they would be more flexible in establishing blending requirements in 2014 and beyond (U.S. Environmental Protection Agency, 2013a,c). On August 6, the EPA issued their final ruling for the 2013 RFS. In addition to providing all the final RFS levels for 2013, it also provided an indication that for 2014 and perhaps beyond, EPA intends to accept the reality that it will be difficult to meet the RFS levels that increase each year through 2022:

> Given these challenges, EPA anticipates that in the 2014 proposed rule, we will propose adjustments to the 2014 volume requirements, including to both the advanced biofuel and total renewable fuel categories. We expect that in preparing the 2014 proposed rule, we will estimate the available supply of cellulosic and advanced biofuel, assess the E10 blend wall and current infrastructure and market-based limitations to the consumption of ethanol in gasoline-ethanol blends above E10, and then propose to establish volume requirements that are reasonably attainable in light of these considerations and others as appropriate.
>
> **(U.S. Environmental Protection Agency, 2013a).**

Essentially what this implies is that EPA intends to reduce the overall RFS and the advanced RFS by the amount of reduction in the cellulosic component. So far, EPA has been forced to reduce the cellulosic mandate every year because the biofuel was not available. However, in reducing the cellulosic mandate, they have maintained the overall mandate at its original level. What this means is that the shortfall in cellulosic biofuels must be made up from other advance biofuels such as sugarcane ethanol or biodiesel. However, today with the ethanol blend wall, biodiesel RINs are needed to meet the shortfall in ethanol, so it became pretty clear that moving forward, the RFS was not workable. The above paragraph also signals that EPA will take into account the ethanol blend wall in setting the RFS levels for 2014 and beyond. The RFS problem is illustrated in Table 30.3. The table provides numbers for 2014–16. The first row provides the overall RFS, which in the past EPA has maintained even when it reduced the cellulosic mandate. Rows 2 and 3 provide the conventional total and the level of the blend wall (13.3 BG). The next row provides my assumption on the amount of carry-forward RINs that

Table 30.3 Situation without Any Change in Environmental Protection Agency (EPA) Practice (billion gal)

Year	2014	2015	2016
RFS total	18.15	20.5	22.25
Corn total	14.4	15	15
Blend wall	13.3	13.3	13.3
RIN carry forward balance	1.7	0	0
RIN CF used	1.7	0	0
Sugarcane	0.75	1.00	1.25
Corn ethanol prod (gal)	12.55	12.30	12.05
Biodiesel RINs	3.15	4.50	6.00
Biodiesel production (gal)	2.10	3.00	4.00
Advanced RFS	3.75	5.50	7.25
Advanced supplied	3.90	5.50	7.25
Total RINs	18.15	17.8	19.3

will be available for 2014. If prior practice were maintained, all the carry-forward RINs would be needed in 2014, and none would be available for 2015 and beyond. The sugarcane row provides optimistic assumptions on sugarcane ethanol imports, and these levels are much higher than recent history. The corn ethanol row is the blend wall minus sugarcane ethanol imports. The next two rows provide biodiesel RINs and assumed biodiesel production. The assumed biodiesel production is much higher than the current level. Biodiesel has 1.5 times the energy content of ethanol, so each gallon gets 1.5 RINs. Next are the advanced RFS mandate and the advanced supplied (sugarcane ethanol plus biodiesel RINs). The final row provides the total RINs generated under these assumptions. For 2014, with pretty extreme assumptions on sugarcane ethanol imports plus biodiesel production and using all carryforward RINs, it is barely possible to meet the RFS. In 2015 and 2016, we cannot even come close to meeting the RFS (row 1). Table 30.3 illustrates the problem and the reasons EPA was compelled to act. Since the 2013 final RFS ruling quoted above, EPA has actually announced their proposed RFS levels for 2014 (U.S. Environmental Protection Agency, 2013b,d). As expected, they proposed reducing the overall mandate by the reduction in the advanced biofuels level due to the lack of cellulosic biofuels. Moreover, they also proposed reducing the effective mandate for corn ethanol from the 14.4 BG in the legislation to 13.01 BG. They propose this reduction because of the blend wall. We will come back to this proposal, but first we will review other alternatives that have been proposed.

ALTERNATIVES TO THE CURRENT SITUATION

In this section, we evaluate the likely impacts of eight alternatives to the current RFS. We are not restricting ourselves to alternatives that have been proposed, but are taking a broad look at many possibilities.

The impacts for each of the eight alternatives may be summarized as follows:

1. *Eliminate the RFS*: Eliminating the RFS would kill the biodiesel and cellulosic biofuels industries immediately. This is because both biodiesel and cellulosic biofuels cost more than their fossil fuel equivalents, so they would not be used in a pure market environment. Sugarcane ethanol imports

also likely would be eliminated or substantially reduced. In the near term, corn ethanol likely would retain market up to the blend wall, as it is less expensive than gasoline and provides added octane and oxygen. In October 2013, corn ethanol was around 75 cents/gal cheaper than gasoline. Over the longer term, it is possible the petroleum industry would develop other octane enhancers to use in lieu of ethanol.

2. *Elimination of the cellulosic biofuel off ramp*: Eliminating the off ramp would be a positive step in providing a stronger mandate for cellulosic biofuels, However, it might not be sufficient to get the industry moving quickly. Off-take agreements are what is needed to get plants built.

3. *Reduction in the overall RFS any time the cellulosic mandate is partially waived*: this option would go a long way toward having a viable RFS given the blend wall problem, especially if the overall and advanced mandates were both lowered by the amount of the cellulose waiver. It would require some increased biodiesel production, but probably manageable. This is the option it appears EPA may take.

4. *Eliminating the other advanced category and expanding biodiesel*: Under this option, the other advanced category would be eliminated entirely, and biodiesel would grow 300 million gal/year for at least the next 2 years. The overall RFS would also be reduced by the sum of the previous other advanced and cellulosic categories (assuming cellulose continued to be waived). This option also would help to solve the blend wall problem, but it requires a fairly large increase in biodiesel production. It would also cause political problems with Brazil, the major supplier of sugarcane ethanol.

5. *Reduction of the overall RFS to accommodate the blend wall*: This option also accommodates the blend wall (by definition). However, it is difficult to determine exactly where the blend wall is or will be given dynamic market adjustments and RIN pricing. EPA likely would want to provide an incentive for companies to expand sales of E85. It is more attractive if combined with waiving the overall mandate when there is a cellulose waiver. It is also not clear what legal authority EPA has to reduce the conventional biofuels category.

6. *Irwin/Good proposal to freeze RFS at 2013 levels*: Scott Irwin and Darrel Good have proposed freezing the RFS for 2014 and 2015 at 2013 levels (Irwin and Good, 2013). The levels for the overall, renewable (corn ethanol), and advanced biofuels in 2013 are 16.55, 13.8, and 2.75 BG, respectively. This option gets us through 2015, but is not a long term solution.

7. *EPA approval of E15 for all vehicles*: This option if implemented at fueling stations around the country would solve the problem. However, it seems unlikely that E15 would be rapidly implemented around the country.

8. *Greater market penetration for E85*: There is some potential for expanding the size of the E85 market even with current infrastructure (E85 pumps and flex fuel vehicles). The theoretical potential market is large enough to solve the blend wall problem, but we do not know how large the market could be in reality because we do not know how many RINs might have value for E85 blending.

We have not evaluated the options with an aim of suggesting one is better or worse than another. Our purpose is not to make recommendations, but to provide an indication of what might happen under a wide range of possible changes. It is clear that the RFS is important to the biofuels sector. Because of the blend wall and perhaps other issues, the RFS has now come under increased attack. Thus, it is important for the key players to understand what might happen under a wide range of possible changes to the RFS.

What are the implications of the specific EPA proposal described above? For corn ethanol, the proposal essentially ratifies the status quo and would not lead to any growth in corn ethanol. It would remove any incentive to expand E85, as all of the 13 BG level can be met with E10. It also would not lead to any expansion of biodiesel. Cellulosic biofuel would be mandated at the level EPA believes will be produced in 2014. Certainly, the EPA proposal is a pull back from their previous posture. As of this writing, it is not clear if it will become the final set of levels or not.

CELLULOSIC BIOFUELS

Cellulosic biofuel technology development has not progressed as fast as had been hoped. In fact, the 2014 proposed RFS level for 2014 is 17 million gal instead of the original level of 1.75 BG. What is holding cellulosic biofuels backs? There are five key uncertainties impeding the development of cellulosics (Tyner, 2010):

1. Feedstock availability and cost
2. Conversion efficiency and cost
3. Future oil price
4. Environmental impacts
5. Government policy

We will discuss each of these in turn.

FEEDSTOCK AVAILABILITY AND COST

Early in the investigation of cellulosic feedstocks, it was commonly assumed that they could be available in large quantities for around $30/ton. However, recent evidence suggests that feedstock costs will be at least three times that level and sometimes higher (National Research Council, 2011). Corn stover today is estimated to cost about $90 per dry ton, and switchgrass between $98 and $133 per dry ton, depending on production conditions. With a conversion yield of 70 gal of ethanol per dry ton of feedstock, the feedstock cost alone would be $1.43/gal of ethanol.

The good news on feedstock is that all the major studies conclude that sufficient feedstock would be available to meet the RFS mandated levels of cellulosic biofuel. So the question is not availability, but cost.

CONVERSION EFFICIENCY AND COST

Much of the early research and development on cellulosic biofuel was on biochemical conversion processes that result in the production of ethanol. These processes require separation of lignin in the plant material from cellulosic and hemi-cellulose, which can be fermented to ethanol. However, it has proven to be quite difficult to do this separation cost effectively, and the costs of the enzymes to do the fermentation also remain high. In addition, the blend wall discussed above applies to cellulosic ethanol just as to corn ethanol. Given the infrastructure in the United States, there is no way to use cellulosic ethanol even if it could be produced economically.

Today, there is more attention on thermochemical conversion processes, which can produce drop-in hydrocarbons: biogasoline, green diesel, or jet fuel. The process getting the most attention is fast pyrolysis (Brown and Brown, 2013). In this process, the biomass is heated rapidly in the absence of oxygen, and a bio-oil is produced. This bio-oil can be further processed to yield hydrocarbon substitutes. The problem is one of cost. By our estimates, it would take at least crude oil prices of $110/barrel over the plant lifetime to make such a process economic. We are close to that level, but because of the uncertainties involved, it is very difficult to get private sector investment.

FUTURE CRUDE OIL PRICES

The U.S. Department of Energy predicts future oil prices in its Annual Energy Outlook (U.S. Department of Energy, 2013). They present three cases: a reference case, a low price case, and a high price case. For the 2013 outlook, the reference case price in 2035 is $145 per barrel (bbl.), but the low and high cases are $73 and $213, respectively. Thus, there is as high degree of uncertainty in future oil prices. Given that it takes at least $110/bbl. for cellulosics to be economic, even the reference case is uncertain in the early years of any potential investment. With the surge in shale oil development, many believe it is more likely that oil prices will not increase substantially in the medium term.

ENVIRONMENTAL ISSUES

In general, most studies conclude that environmental impacts of cellulosic biofuels would be more positive than corn ethanol. Estimates of GHG reduction are much higher for cellulosics. Dedicated crops like switchgrass and miscanthus have lower chemical runoff and less soil erosion than conventional crops. Dedicated energy crops also provide good wildlife habitat. The main environmental concern is potential loss of biodiversity. It is likely that the area surrounding a cellulosic conversion facility reliant on dedicated crops would see thousands of acres of those crops in a small area.

GOVERNMENT POLICY

Perhaps one of the key uncertainties is government policy. Significant opposition to the RFS has emerged from the oil industry, livestock production groups, and grocery manufacturers. They all see the RFS as increasing their costs. It is impossible to say if those groups will carry the day or if the renewable fuel lobby will continue to successfully defend the RFS.

If the RFS is eliminated, as indicated above, cellulosic biofuels would not be developed. It is very difficult to persuade private investors to put money into a plant if the only guarantee of a market is the government, when government policy is so uncertain.

The bottom line, whether it is corn ethanol, biodiesel, or cellulosic biofuels, is that the future is highly uncertain. Cellulosic biofuels are close to being economic, but they need the RFS to be developed.

REFERENCES

Agency USEP, 2013. Regulation of fuels and fuel additives: 2013 renewable fuel standards. Federal Register 78 (26), 9282–9308.

Brown, T.R., Brown, R.C., 2013. Techno-economics of advanced biofuels pathways. RSC Advances 3 (17), 5758–5764.

Irwin, S., Good, D., April 10, 2013. Freeze it–a proposal for implementing RFS2 through 2015. FarmdocDaily. http://farmdocdaily.illinois.edu/2013/04/freeze-it-proposal-implementing-RFS2.html.

National Research Council, 2011. Renewable Fuel Standard: Potential Economic Effects of U.S. Biofuel Policy.

Paulson, N., April 12, 2013. An update on the 2012 RIN carryover controversy. FarmdocDaily. http://farmdocdaily.illinois.edu/2013/04/update-2012-rin-carryover-controversy.html.

Tyner, W.E., 2010. Cellulosic biofuels market uncertainties and government policy. Biofuels 1 (3), 389–391.

Tyner, W.E., 4th quarter 2013. The renewable fuel standard–where do we go from here? Choices 28 (4), 1–5.

U.S. Congress, 2005. Energy policy act of 2005. Public Law, 109–158.

U.S. Congress, 2007. Energy independence and security act of 2007. In: H.R. 6, 110 Congress, 1st Session.

U.S. Department of Energy, 2013. Annual Energy Outlook.

U.S. Environmental Protection Agency, (August 6, 2013a. 40 CFR Part 80, Regulation of Fuels and Fuel Additives: 2013 Renewable Fuel Standards.

U.S. Environmental Protection Agency, November 2013b. 2014 Standards for the Renewable Fuel Standard Program. 40 CFR Part 80.

U.S. Environmental Protection Agency, August 6, 2013c. EPA Finalizes 2013 Renewable Fuel Standards.

U.S. Environmental Protection Agency, November 15, 2013d. EPA Proposes 2014 Renewable Fuel Standards.

ECONOMICS OF ETHANOL AND BIODIESEL

31

Bob Parsons

Department of Community Development and Applied Economics,
University of Vermont, Burlington, USA

The growing interest in biofuels is driven by political, economic, environmental, and ethical forces. Characteristically, there are both proponents and opponents on each of the driving forces. For one, economics, it would seem that the numbers should do the talking, and any analysis should be fairly direct and give a story. However, as in many of the arguments about biofuels, there is more than meets the eye because there are far-reaching implication that many do not realize, and any economic analysis greatly depends on the assumptions that mold the analysis. There are economic associations with quantities and prices of various inputs, substitutes, complements, production scale, and consumption of direct and byproducts that place different but important perspectives on the use of biofuels. For example, the use of corn derived ethanol in the United States since 2005 has grown to the point that by law, it now constitutes approximately 10% of fuel that powers our automobiles. Ethanol's impact can be seen beyond our fuel purchases, dramatically impacting the acreage and price of corn, other grains, oil crops, and other fiber, feed, and foodstuff crops resulting in financial impacts on our foods and fibers for everyone. Is this impact just because of the growth of ethanol? Is it a coincidence? Or finally, is it direct impact? Look at the evidence and decide for yourself.

AGRICULTURAL ECONOMY

The current mandate in the United States to mix ethanol with gasoline has impacted agricultural production unlike any other event, except for the grain sale to the Soviet Union in the 1970s, and the USDA Payment in Kind (PIK) program in 1983. If one remembers their history, the United States sold large amounts of grain to the Soviet Union from 1971 to 1973, resulting in a dramatic shift in farm prosperity, significant increases in consumer prices, and a 180° shift in agricultural policy from restricted production to full-scale production; farming fencerow to fencerow was termed at the time (Luttrell, 1973). What followed was arguably the most prosperous peacetime period in U.S. agriculture until that time. In 1983, to counter large surpluses that were attributed to the 1970s shift in agricultural policy, the United States instituted the PIK program that removed 20 million acres, or 25% of the U.S. total corn crop, from corn production (USDA-ERS, 1983). Comparing the ethanol mandate to these events certainly does indicate that the impact is a major event, resulting in repercussions comparable to some of the most major events in U.S. agricultural history. All three of these events have changed our acreage and prices of all major crops, caused an increase in land prices, increased the income and buying power of farmers, led to unprecedented sales of farm equipment, and led to an increase in consumer food prices. The reality for our complex food system is that altering the price and market structure for one commodity has extensive impacts that one generally fails to initially perceive. Unfortunately, most economic analyses are done on a single sector with assumptions that other sectors remain the same. This is not true in reality, and as we found out with ethanol, impacts were felt way beyond the ethanol facility.

The legislation requiring the United States Environmental Protection Agency (EPA) to mandate that ethanol derived from corn in gasoline has had far reaching effects on our complex food system. Ethanol has been around and used as a fuel for a long time. Even Henry Ford's Model T could run on ethanol. From 1986 to 2001, corn used for ethanol had increased from 290 million bushels to 707 million bushels, which in 2001 was only 7.5% of the U.S. corn crop (Figure 31.1) (National Corn Growers).[1] Interest and policy were just picking up speed, which saw 2119 million bushel (20% of U.S. production) delivered to ethanol plants by 2006, and then the growth really took off with 3049 million bushel (23.4% of U.S. production) delivered in 2007 and peaking at 5000 million bushel in 2010 and 2011 (40% of U.S. production). In 2012, corn for ethanol dropped for the first time in the twenty-first century, to 4500 million bushels, due higher corn prices and a drought that reduced corn supplies, but the ethanol use was still 42% of the corn crop. The United States was in the ethanol business, going from a minor role in 2001 to major use in 2011. The growth was such that it seemed everyone was planning to build ethanol plants. There was even a statement this author heard at a workshop that if all the ethanol plants that were on the drawing board at one time for Iowa had actually been built, Iowa would have become a net-corn importer. That's a lot of corn and a lot of ethanol.

There are various arguments supporting the promotion of corn-based ethanol. From a strategic perspective, growing our own fuel would make the United States less dependent on foreign oil sources. Politically, it made sense to promote growing our own fuel, something that was arguably sustainable,

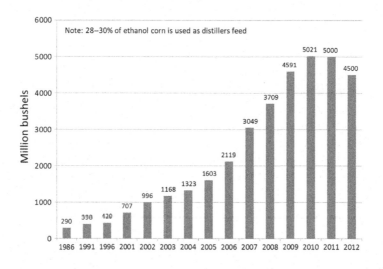

FIGURE 31.1

Corn used for ethanol production (million bushels).

(Source: http://www.ncga.com/upload/files/documents/pdf/WOC%202013.pdf.)

[1]A bushel of corn weighs 56 pounds at 15% moisture.

as we were capable of growing our "fuel" corn annually. Economically, it made sense in that although corn was our most widely planted and most valuable crop, it was not highly profitable, there were large surpluses, was subsidized. Anything that could increase its demand would be helpful to the farmers and popular with the government and taxpayer if it reduced expenditures on crop subsidies. From 1997 to 2005, average corn prices in the United States ranged from $2.43 to $I.82, with six of those years under $2.06 (USDA Quick Stats). Direct payments and other corn support programs were costing the government billions. Why not use some of our corn for ethanol to replace foreign oil and help the farmer? So, with a focus on ethanol, demand for ethanol increased. Corn prices also increased, jumping to $3.04 in 2006, to $4.20 in 2007, and peaking at 6.67 per bushel in 2012 (Figure 31.2) (USDA Quickstats).

U.S. farmers did not need much incentive to respond to the rise in prices. If corn goes up in price, farmers tend to plant more corn if they see it as being profitable. It was, and they did. Corn acreage went from 78.3 million acres in 2006 to 93.5 million acres in 2007, which was the most corn acreage since 1944 (Figure 31.3). By 2012, corn acreage peaked at 97.2 million acres, the highest acreage since 1937 (USDA Quickstats). The impact to corn country was tremendous. More acres and a higher price

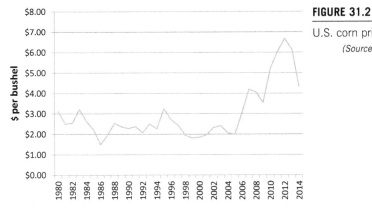

FIGURE 31.2

U.S. corn prices, 1980–2014 ($/bushel).

(Source: USDA (http://www.nass.usda.gov/Quick_Stats/.))

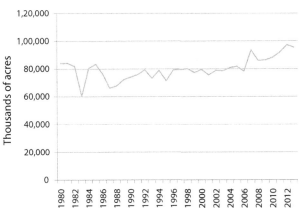

FIGURE 31.3

U.S. corn acres, 1980–2013 (Thousand acres).

(Source: USDA (http://www.nass.usda.gov/Quick_Stats/.))

FIGURE 31.4

Value of corn production ($Billions).

(Source: USDA (http://www.nass.usda.gov/Quick_Stats/.))

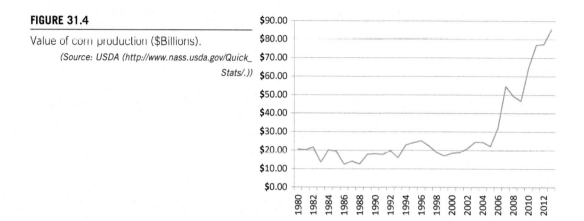

means more money. The value of corn production at farm level was $22.2 billion in 2005, a value that had been fairly consistent since 1989 (Figure 31.4). However, value of production jumped to $54.6 billion in 2007, and to $76.5 billion in 2011 (USDA Quickstats). It does not take much logic to realize that ethanol was very popular in corn country. Ethanol was not the only driver of corn prices, as the period from 2006 to 2013 also saw an increase in the demand for corn from the international market. The United States experienced an international financial crisis that also impacted prices, but what the producer saw was ethanol as the driver of corn prices, driving profitability and income to levels not experienced since the Russian grain sale.

There was no doubt that ethanol benefitted the corn farmer's bottom line. The impacts of the increased farm income had direct and indirect impacts on the local, regional, and national economy. From a generally positive perspective, when farmers have more income, they spend that income. Farm equipment like tractors, planters, and combines were in high demand as equipment dealers experienced booming sales. Other farm goods were also in higher demand, leading to greater prosperity for farm suppliers. There was a lot of money to spread around, and it was flowing. Considering that in 2008, the nation entered a financial crisis, the response in corn country was "what recession?" Times were good for the corn farmer, for their suppliers, and their community. When corn acreage increases, it also drives the price of corn inputs. Land prices, land rent, corn seed, fertilizer, and herbicide prices also increased the production cost for farmers, but not enough to dampen the growth in corn acreage.

As corn acreage increased, one has to remember that we do not make more land. Therefore, more of one crop means less of other crops. So as farmers were planning more corn, land was diverted from production of other crops, such as soybeans in the Midwest. Farmers in the south reduced cotton acreage and planted more corn. On the western edge of the Corn Belt, land was taken out of wheat and planted to corn. Across the United States, land that may have been idle or used for hay or pasture was put into corn. Some land under expiring contracts in the Conservation Reserve Program was diverted to corn. From 2006 to 2007, soybean acreage dropped 10.8 million acres, cotton acreage dropped by 4.5 million acres, and even rice acreage dropped by 65,000 acres, with most of this diverted land going to corn (USDA Quickstats).

One only has to think about supply and demand to get a full picture of the farm commodity situation. Ethanol demand went up, corn price went up, corn acreage went up, and acreage of other crops went down. Now logic also tells us that less acreage of other crops means less of those crops, and that will drive up the prices of the other crops. That is what happened: The higher price for corn resulted in significantly higher prices for soybeans, wheat, rice, and even cotton. Although that is good for the producers of those crops, the higher prices filter through the food system and resulted in higher prices for the foods consumed by the consumer. The higher price you see for oatmeal can arguably be attributed to the rise in corn prices due to ethanol. Think about that as you put ethanol in your car at your next fill up.

There remains another down side to the ethanol debate: Before the rise of ethanol, the biggest user of corn in the United States was the livestock industry. Corn is the primary feed ingredient for beef, hogs, chickens, and dairy cows. In 2012, these animals consumed 40% of the U.S. corn crop (National Corn Growers Association). Historically, there has always been a major debacle about the focus of U.S. agricultural policy. Ask any livestock producer about what is good policy, and the answer is low grain prices. Ask any grain producer about what is good policy, and the answer is high grain prices. Now how do your remedy this obvious conflict? Students of U.S. agricultural policy realize that it is no accident that we ended up with a system of direct farm payments to grain producers. The direct payments subsidized grain production, which promoted higher-than-normal grain production that kept prices lower than would have occurred without the subsidies. This policy benefited livestock producers who purchased extra grain for feeding livestock. The policy framework was a transfer of wealth from taxpayers (not all individuals pay income tax) to all consumers. How did this work? Taxpayer's money went to grain farmers to subsidize grain production, keeping grain prices lower than would have occurred otherwise. The lower price of grain encouraged additional meat production, creating more meat than otherwise would have been produced at a lower price for consumers. Everyone wins except for the taxpayer, but the taxpayer seldom knew the direct amount of their taxes going to agriculture. Ethanol policy upset this balance, causing corn-based livestock feed prices to triple from 2005 to 2012. Needless to say, the livestock farmers are not happy, and have had to make changes in their production that resulted in smaller herds, feeding less grain, and now resulting in higher meat prices at the meat counter. Ethanol is a prime factor driving prices as the consumer picks up their steaks, pork chops, chicken, or milk.

The repercussions of higher grain prices do not stop with the livestock sector. When corn becomes expensive, you look for alternatives. Barley, wheat, and other small grains are substituted for corn. This increased demand helps put pressure on their prices. Therefore, there is not only less acreage for the substitute crops, there is also additional demand, driving prices upward. There is also a complexity with soybeans. Soybeans represents the primary source of protein in human and livestock food. For livestock producers, not only did their energy rations (starch from corn and other grains) go up in price, so did their protein costs as farmers moved 10 million acres from soybeans to corn (Figure 31.5), resulting in soybeans going up from $6.43 per bushel in 2006 to $10.10 in 2007, and peaking in 2012 at $14.04 in 2013 (Figure 31.6) (USDA Quickstats). With soybeans and corn both rising in prices, it was not a good suggestion to ask for support for ethanol at a livestock producers meeting.

Ethanol does provide one upside for livestock producers: Ethanol is distilled from corn, leaving behind a fermented byproduct, dried distillers grain (DDG), which has significant nutrient value when fed to livestock. For every bushel of corn processed into ethanol, you can produce approximately 2.7 gal of ethanol and 17.4 lbs of DDG (about 31% by weight) (National Corn Growers Association). With more corn used for ethanol, there was a greater supply of DDG available for animal feed. So for 2012, while 5000 million bushels were delivered to ethanol plants, 31%, or an equivalent of

FIGURE 31.5

Corn—soybean acreage (Thousand acres).

(Source: USDA (http://www.nass.usda.gov/Quick_
Stats/.))

FIGURE 31.6

Soybean prices, 1980–2014 ($/bushel).

(Source: USDA (http://www.nass.usda.gov/Quick_
Stats/.))

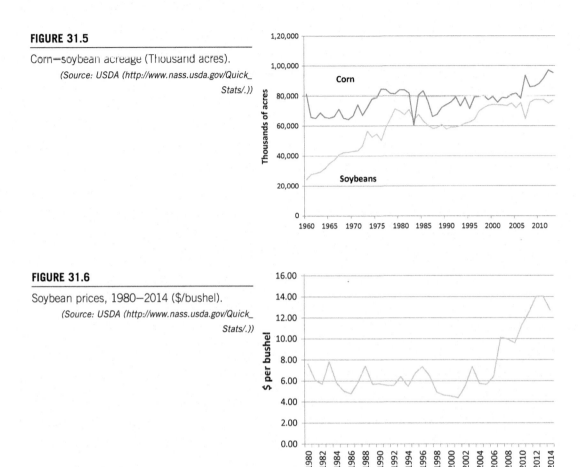

1447 million bushels, was put back into animal feed. When the ethanol total is looked at from this perspective, only 27% of the nation's corn supply is used directly for ethanol, and more than half of our corn is still used for animal feed. The value of the DDG is important in the economics of ethanol, as seen below (Figure 31.7, Figure 31.8).

The other major economic impact of the ethanol is what has happened to land prices and to whom the increased profit is flowing. There is an ironic aspect of economics that profits flow to the owners of capital, in the case of corn farming, the owners of land. In the U.S. Midwest, it is very common for most grain farmers to rent about 40–60% of their land, and to own the balance. When farming experiences above-normal profits, like during the Russian grain sale or ethanol boom, farmers try to grow more of the profitable crops. They will divert some of their own acreage to corn, and will try to rent additional acreage. They will also try to buy more acreage; however, in any given year, limited land acreage becomes available for sale.

The land rental market is similar to other markets; it is driven by supply and demand. As Will Rogers once expressed, "…invest in land because they aren't making any more of it." There is a limited

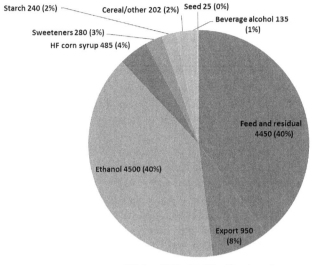

Starch 240 (2%)
Cereal/other 202 (2%)   Seed 25 (0%)
Beverage alcohol 135 (1%)
Sweeteners 280 (3%)
HF corn syrup 485 (4%)
Feed and residual 4450 (40%)
Ethanol 4500 (40%)
Export 950 (8%)

US Corn 2012 = 10,780 million bushels

FIGURE 31.7

U.S. corn Original Destination, 2012.

(Source: http://www.ncga.com/upload/files/documents/pdf/WOC%202013.pdf.)

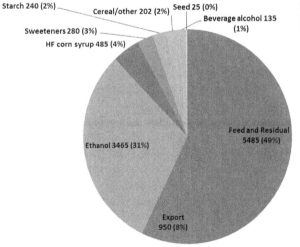

Starch 240 (2%)
Cereal/other 202 (2%)   Seed 25 (0%)
Beverage alcohol 135 (1%)
Sweeteners 280 (3%)
HF corn syrup 485 (4%)
Feed and Residual 5485 (49%)
Ethanol 3465 (31%)
Export 950 (8%)

US Corn 2012 = 10,780 million bushels

FIGURE 31.8

U.S. corn by end use, 2012.

(Source: http://www.ncga.com/upload/files/documents/pdf/WOC%202013.pdf.)

supply of land, so when demand increases, with nearly all farmers looking for more land to farm, the farmland owners benefit. When land becomes available, farmers will compete against each other, driving up the rental cost of the land. Farm land in the Midwest that was renting for $200–$250 per acre in 2006 suddenly became more valuable, with some farmers bidding rents over $400 per acre, and by some reports in Iowa and Illinois, to $500 per acres (Stebbins, 2013). How does this affect the farmer and the landlord? For the farmer who owns some or all of their land, the increased profit flows to the owner/farmer. However, for renters, the farmer now is paying far more per acre for land, thus limiting their profitability. For example, an acre of farmland that just increased from $200 to $400 per acre in rent has increased the cost of production by $1–$1.50 per bushel (depending on the yield) just to pay the rent. The farmer needs to have a higher price to pay the rent. In this case, the increased profit from farming did not accrue to the farmer, but rather to the landlord. Correspondingly, land sale prices jumped, with many parcels more than doubling from 2006 to 2012. Land prices seem to have leveled off and have declined a bit in 2013, but the overall result is that finding land to farm, either rent or own, costs much more than before ethanol.

The rise of ethanol has also indirectly jolted other products. Corn is not a low-input crop: it requires seed, fertilizer, herbicides, and equipment for planting, harvesting, and transport. Just think of the increase in corn acreage of nearly 15 million acres. These new corn acres require seed corn that the seed corn sector was not expecting. It required more nitrogen, phosphorus, and potassium fertilizer, putting a squeeze on available supplies and prices. Nitrogen, the most crucial nutrient for corn, was in far greater demand than before. In addition, natural gas, which is used to produce much of the nitrogen fertilizer, jumped in price due to rising energy costs in 2007, driving up the price of nitrogen fertilizer even further. World phosphorus and potassium supplies were in much tighter supply and facing increasing international demand, resulting in 60–120% higher prices for fertilizer (USDA-ERSb). Farmers had to pay more for diesel fuel for tillage, planting, and harvesting due to rising energy prices. Therefore, while profitability was booming due to the rising corn prices, farmers in a short time were paying higher prices for land rent, fertilizer, and fuel, and nearly doubling the cost of production for a bushel of corn from 2006 to 2012. So not only were farmers receiving higher prices for corn due to ethanol and other demands, they had to be receiving the higher prices to cover the higher cost of production.

Biofuels, as used here meaning biodiesel, have not had near the impact on the food system as has ethanol. For ethanol, 40% of 90 million acres of corn equates to 36 million acres of corn used for ethanol. For biofuels from soybeans, we are seeing a growing production, but still far less significance in quantity of fuel and acreage dedicated biofuel production. Since 2008, the United States has seen biofuel production increase from 4.5 million acres (6% of production) to 11.7 million acres (15% of U.S. production) (Agricultural Resource Marketing Center). Soybeans thus has not seen quite the growth and full scale impact as we have seen with corn. There are several key differences between the crops: Soybeans are a legume that add nitrogen back to the soil. On the production side, they requires far fewer inputs, especially no nitrogen, which it can produce itself. Soybeans are one of the biggest suppliers of protein, both for animals and humans, which make them quite valuable. Soybeans, unlike corn, have not historically been a major U.S. crop, with most of their growth coming in the last 40 years. Soybean acreage nearly doubled from 24 million acres in 1960–1973, and nearly tripled by 1979 to more than 70 million acres (Figure 31.5). Acreage varied throughout the 1980s and 1990s, but since 1997, it has always exceeded 70 million acres. The exception to this was 2007, when we experienced the major surge in corn acreage (USDA Quick Stats). Farmers use soybeans as a rotational crop with corn, relying on their nitrogen-fixing traits to reduce nitrogen needed for corn. While requiring fewer inputs, soybeans yield about 30% of the volume corn yields per acre, but generally command two to three times higher price per bushel. Acreage shifts occur as farmers determine whether they believe corn or soybeans will generate more net revenue per acre. For economic and biological diversification reasons, farmers often rotate corn and soybeans.

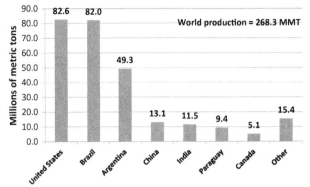

FIGURE 31.9

World Soybean production, 2012–2013 (MMT).

(Source: http://www.soystats.com/2012/Default-frames.htm.)

Soybeans represent a bit more complicated product for conversion to biofuel than does corn. Soybean is very valuable for a food source around the world, so international prices can make export far more attractive than conversion to biofuel. The United States is not the world's biggest producer of soybeans (Figure 31.9). Brazil produces almost the same volume as the United States with Argentina, producing about 60% of U.S. production. Between the two, the South American agricultural giants produce 50% of the world's soybeans. China is a distant fourth in production and imports 63% of the world's traded soybeans to be used as supplemental protein, primarily in foods with some for animal use (United States Soybean Board).

Soybeans have reacted in price similar to corn. From 1987 to 2006, soybeans ranged from $4.38 per bushel in 2001, to $7.42 in 1988. Again, the same arguments can be made for supporting biodiesel as ethanol. Prices were very consistent with a low level of profitability and held up by direct farm payments and price subsidies. Therefore, as with corn, it did not seem to be an outrageous idea to promote biofuels to help increase farm profitability. Again, as with corn, the forecasts failed to predict how the farmers and markets would react to the change in market conditions. In 2007, with a decrease of 10 million acres, soybean prices averaged $10.10, and peaked at $14.04 in 2013 (Figure 31.6). Spot prices for soybeans on some months hit $15 per bushel (USDA Quickstats). In addition to biofuels and reaction to the corn markets, economic growth around the world improved living standards, with growing demand for protein that was provided by soybeans. Therefore, due to focus on biofuels, reacting to the ethanol-induced crop changes, and increased world demand for protein, the soybean world changed overnight.

From a farm perspective, the increased price for soybeans in 2007 lead to acreage bouncing back to 2006 levels in 2008, and remained in the 74–76 million acre range through 2013. The increased price, with minimal rising input prices as compared to corn, kept soybean popular with producers. Another factor that keeps soybeans fairly steady acreage is that soybeans have few competitors that can provide as much protein. In addition, the legume provides a great choice for rotating with corn, so as corn acreage increased, there was more need for crop rotation with soybeans. Soybeans were also faced with the same dilemma as corn production with respect to land prices. Land rents and values went up no matter what you raised, being driven by the potential profits derived from either corn or soybeans.

In the big picture, it is easy to see that there have been definitive changes in our commodity production because of the rise of ethanol and arguably biofuels. We have seen a shift of millions of acres to produce more corn. We have seen soybeans rebound. Prices of both corn and soybean have hit sustained levels of high prices never seen before. Although crop farmers like these prices, livestock farmers and consumers have come to pay higher prices for feedstuffs and food, respectively, indicating unexpected economic outcomes that definitely raise questions about the political outcome of ethanol and biofuel.

Although ethanol and biofuels have definitely have affected prices and acreage of their primary inputs, there remains a question as to how economically viable are these fuel alternatives. Ethanol has seen a proliferation of processing plants across the United States, with most located in the Corn Belt. Some have been built in odd places, but that is more due to the sourcing of the funding. Some areas, for example, had access to industrial development funding that was used to build ethanol plant at locations that defied rational reasoning. If one was to consider important aspects of ethanol production, two considerations would be access to corn and access of delivering ethanol to the gas system. Therefore, one would expect ethanol plants to be built in corn producing regions, but due to availability of specific industrial development funding, some were located in strange places.

The economics of ethanol depends on a number of factors: Market conditions have to exist, such as the demand and willingness to pay for the product and supply, with the willingness to provide the product at the prevailing price. The supply and demand are also influenced by the role of alternatives and governing rules. For ethanol, the U.S. government set up a market situation that practically guaranteed economic viability. First, to create a market, the government passed laws that mandated that retail gasoline contain ethanol. Without the mandate, oil companies had little incentive to blend ethanol with gasoline. Demand was created. Second, the government laws considered the impact of substitutes. The biggest producer of ethanol in the world is Brazil, which derives its ethanol from sugar cane, a much more efficient source of ethanol. Brazil potentially could export ethanol to the United States at prices lower than for what ethanol could be produced in the United States. Therefore, to encourage ethanol production in the United States, import barriers were established to prevent the importation of Brazilian ethanol. Third, production of ethanol in the United States would initially be expensive, requiring a major investment in infrastructure. To encourage the investment of plants and assuring profitable operation, subsidies were enacted to underwrite production costs, reducing the risk to plant operators, and encouraging the construction of ethanol plants. With this legislation, the U.S. government created demand for ethanol, restricted competition, and undercut the cost of production to encourage the construction of plants in the United States to supply the domestic market. This type of legal arrangement would be envied by any business sector.

The cost of manufacturing ethanol is not a set cost, due to the differences in cost of investing in plant, equipment, plant capacity, and developing an experienced manufacturing and marketing procedure. Examination of an ongoing analysis by the Don Hofstrand, Iowa State University, provides a reliable view of a representative ethanol processing facility. In this case, the plant has an annual production capacity of 100 million gal with an investment of $211 million. By their estimate, the fixed cost of ethanol production would be $0.60 per bushel of corn processed. Variable costs include enzymes, yeast, other fermenting ingredients, labor, maintenance, and miscellaneous expenses that are estimated at $1.03 per bushel. Natural gas used to heat the fermenting grain is estimated at $0.93 per bushel. Obviously, the largest cost involved in ethanol production is the cost of corn grain.

The production process per bushel of corn (56 lbs.) produces 2.8 gal of ethanol and 16 lbs of DDG. There are marketing cost of $0.05 per gal of ethanol and $4 per ton of DDG that comes to $0.17 per bushel. In Table 31.1 below, one can observe the possible outcomes under several scenarios. The first

Table 31.1 Net Returns of Ethanol Under Various Scenarios

	Market	Breakeven Ethanol Price	Breakeven Corn Price	Worse Case	High DDG	Low DDG
Price of corn/bushel	$4.85	$4.85	$5.50	$5.50	$4.85	$4.85
Price of DDG/ton	$225	$225	$225	$225	$250	$180
Price of ethanol/gal	$2.30	$2.06	$2.30	$2.06	$2.06	$2.06
Gallon of ethanol/bu	2.80	2.80	2.80	2.80	2.80	2.80
Pounds DDG/bushel	16.0	16.0	16.0	16.0	16.0	16.0
Fixed cost/bushel	$0.60	$0.60	$0.60	$0.60	$0.60	$0.60
Variable costs/bu	$1.03	$1.03	$1.03	$1.03	$1.03	$1.03
Energy/bu	$0.93	$0.93	$0.93	$0.93	$0.93	$0.93
Marketing costs/bu	$0.17	$0.17	$0.17	$0.17	$0.17	$0.17
Revenue ethanol/bu	$6.43	$5.77	$6.43	$5.77	$5.77	$5.77
Revenue DDG/bu	$1.80	$1.80	$1.80	$1.80	$2.00	$1.44
Total revenue	$8.23	$7.57	$8.23	$7.57	$7.77	$7.21
Cost per bushel	$7.58	$7.58	$8.23	$8.23	$7.58	$7.58
Net revenue per bu	$0.65	−$0.01	$0.00	−$0.66	$0.19	−$0.37
Revenue ethanol/gal	$2.30	$2.06	$2.30	$2.06	$2.06	$2.06
Revenue DDG/gal	$0.64	$0.64	$0.64	$0.64	$0.71	$0.51
Total revenue/gal	$2.94	$2.70	$2.94	$2.70	$2.77	$2.57
Cost/gal	$2.71	$2.71	$2.94	$2.94	$2.71	$2.71
Net revenue per gal	$0.23	$0.00	$0.00	−$0.24	$0.07	−$0.13

DDG, dried distillers grain.
Source of base data: Hofstrand (https://www.extension.iastate.edu/agdm/articles/hof/HofJan08.html).

column reflects current prices for ethanol, corn, and DDG. All the fixed and variable costs from above are included in the table. At a price of corn at $4.85 per bushel, and ethanol at $2.30 per gal, the ethanol producer will clear $0.65 per bushel of corn, or $0.23 per gal of ethanol. The process is showing a profit at these prices. In Columns 2 and 3, we look at the sensitivity of the prices for ethanol and corn. From Column 2, we see that if the price of ethanol drops to $2.06, the ethanol production breaks even. In column 3, we see that at a corn price of $5.50 per bushel and ethanol at $2.30 per gal, the process breaks even. If we get reduced ethanol prices and increased corn prices, the bottom line ends up losing money, at $0.66 per bushel and $0.24 per gal.

The returns on ethanol depends greatly on several factors: Obviously, the price of corn, the primary input, is of chief importance. From 2012 to 2014, we have seen corn costing more than $7.50 per bushel and as little as $4.10 per bushel. The other key price is ethanol, which at the $2.30 mark is nearly $0.40 more than late January 2014. The price of DDG also fluctuates depending on supply and demand and exports. At a current corn price of $4.85 per bushel, and DDG up to $250, the ethanol production can show a return of $0.07 per gal (column 5). When the price of DDG drops to $180 per ton, the returns to ethanol drop to a loss of $0.13 per gal. When excluding a variation in efficiencies between plants of the same capacity and design, Table 31.1 provides a view into how the returns can vary with small movements in the prices of corn, ethanol, and DDG. Please note that although this is

Table 31.2 U.S. Ethanol Plants and Production Capacity

Year	Operating Plants	Capacity (Million Gallons)	Future Capacity (Million Gallons)
2005	82	3643.7	754
2006	95	4336.4	1746
2007	111	5533.4	6189
2008	139	7888.4	5536
2009	179	10,343.4	1450
2010	200	11,877.4	1432
2011	204	13,507.9	560
2012	209	13,859.4	487
2013	211	12,836.9	158

Source: State of Nebraska (http://www.neo.ne.gov/statshtml/122_200501.htm).

for a representative plant, local prices, efficiency, and marketing costs will influence returns. Marketing will include efforts by operators to lock in favorable prices for inputs and outputs. Corn prices can also vary according to local supply and demand, with prices yielding both a positive and negative basis depending on the time of the year.[2]

Ethanol production capacity has also leveled off at nearly 13 billion gal. Since 2010, capacity has leveled off, with only 11 new processing plants added. From 2005 to 2013, the United States went from 82 to 211 ethanol plants, with capacity increasing from 3.6 billion to 12.8 billion gal. One can conclude that the ethanol sector has matured, with the nation's production capacity planning to add only 158 million gal in 2013. In 2007, in contrast, 6.2 billion gal of production capacity was under construction (State of Nebraska) (Table 31.2).

Estimates conducted by the University of Illinois shows highly variable returns in the ethanol sector. Returns in 2007 and 2011 were excellent, while they were negative in 2009, 2010, and 2012 (Figure 31.10) (Farm Doc Daily). By any standard, an industry that shows negative returns three out of 5 years would be considered to be very risky and would have a challenge attracting new investors. However, one positive aspect for long-term returns is that the EPA mandates are still in effect and prevent the demand for ethanol from dropping.

ECONOMICS OF BIOFUELS

Biodiesel use has also grown in the past, but not nearly the same rate as has ethanol. Biodiesel is made from plant oils, particularly from soybeans, which account for 60% of all biodiesel currently being produced. Other sources are used vegetable oil, canola oil, and sunflower oil. Primarily, the biodiesel sector has revolved around soybeans, a feed crop with a major oil that is relatively easy to convert into biodiesel; however, it faces tight competition from other uses, such as a protein source for food and feedstuffs.

[2]Basis is the difference between the local price and clearing prices on the Chicago Mercantile Exchange (CME). With a local price of $4.60 and a CME price of $5, the basis would be −$0.40.

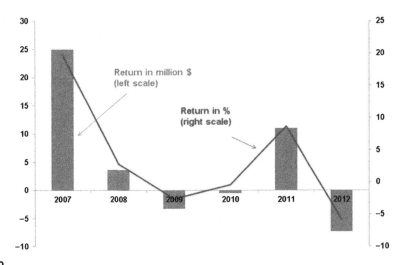

FIGURE 31.10

Return to investors in a representative Iowa ethanol plant, 2007–2012 (For interpretation of the references to color in this figure legend, the reader is referred to the online version of this book.)

(Source: http://farmdocdaily.illinois.edu/2013/06/updated-profitability-ethanol-production.html.)

Since 2008, soybean use for biodiesel has increased with 15.2% of the United States. Soybean acreage in 2013–2014 was projected to be used for biofuel (Table 31.3). This is significantly lower in percentage and acreage than the 40% the U.S. corn crop delivered to ethanol plants. Yet, biofuel use has grown from 6% in 2008–2009, primarily due to the biodiesel excise tax credit that provides a $1 tax credit per gal of biodiesel fuel. In addition, there is a federal mandate requiring the blend of specific amounts of biodiesel fuels. The $1 tax credit was put in place in 2010 and expired in December 2013. The quantity mandate remains in place. One can easily see the impact of the biofuel tax credit in

Table 31.3 Soy Oil Biodiesel Production in the United States 2008–2014

Year	Total Yield (Bill Bushels)	% of Acres for Biofuel	Million Bushels for Biofuel	Million Acres for Biofuel	Millions of Gallons of Biofuel
2005–2006					112.0
2008–2009	2.967	6.0%	179	4.51	262.5
2009–2010	3.359	4.5%	150	3.42	218.2
2010–2011	3.329	7.1%	237	5.45	355.5
2011–2012	3.094	13.6%	420	10.02	632.5
2012–2013	3.014	14.4%	435	10.99	655.8
2013–2014 (Projected)	3.238	15.2%	491	11.69	733.1

Source: Agricultural Resource Marketing Center (http://www.extension.iastate.edu/agdm/crops/outlook/biodieselbalancesheet.pdf).

2010–2011, when soybean acreage for biofuel production nearly doubled, as did the gallons of biofuel from soybeans (Agricultural Resource Marketing Center).

The economics of biodiesel has many of the same factors as ethanol. Soybeans has an alternative use, and thus is competitive in the marketplace as a protein source. Exports have risen in recent years due to an increased demand for protein from developing countries, particularly China and the Pacific Rim. For the analysis below, we are comparing the value of soy oil as extracted from processing. Soy oil has a distinct market value that is determined on the Chicago Mercantile Exchange. The prices for the two primary products of soybeans, soy oil and soy meal, are directly correlated and derived from the price of soybeans.

For this analysis, we again will use the assumptions of a University of Illinois study. This is based on an industrial-size plant with the capacity of to produce 30 million gal of biodiesel per year from soybeans. Construction cost is estimated at $47 million, or $1.57 per gal of capacity. The plant is financed with a combination of fifty–fifty debt-equity, with the debt financed at 8.25% for 10 years. The plant converts 7.55 lbs of soybean oil to produce one gal of biodiesel and 0.9 lbs of glycerin (per gal of biodiesel). Production per gal of biodiesel requires seven cubic feet of natural gas and 0.71 lbs of methanol. Variable costs are $0.25, whereas total fixed costs are $0.26 per gal of biodiesel. The above specifications are for a representative plant with recognition of variation of capacity, and efficiency between plants. Soybeans are used as the biodiesel example as it they represent more than 60% of the production of biodiesel in the United States. There are also some factors involved in analyzing bio-diesel. There was a Blender Tax Credit that expired at the end of 2013. Another factor is the biodiesel mandate that was expanded by EPA from 1.0 to 1.28 billion gal in 2013. As with ethanol, the actions by government create demand for a product, thus providing incentive for firms to invest in plant and equipment, and to develop technology to increase production efficiency.

The analysis below uses Hofstrand(b) as a guide to determine profitability. For the initial analysis, we use the average soy oil and biodiesel prices for 2014. Local prices for soy oil and biodiesel will vary depending on the time of year and local supply and demand conditions. Currently soybean prices are extremely high due to world demand. With this aspect, the average soy oil price for 2014 is $0.45 per pound and biodiesel averaged $4.62 per gal. Please note that corresponding diesel fuel price was much cheaper, in the $3.80–$4.00 range. At 2013 average prices, the biodiesel process net revenue was $0.53 per gal. For the process to break even economically, the price of soy oil has to rise to $0.52 per pound or the price of diesel fuel to drop to $4.09 per gal (Table 31.4).

The primary ingredient is soy oil, so if we look at a range of soy oil prices for 2013, we can see that an increase to $0.496 estimates a net revenue of just $0.07, whereas a decrease in soy oil price to of $0.378 can yield a net revenue of $0.32 per gal of biodiesel. If we look at when the biodiesel price was highest, at $5.03, net returns were estimated at $0.87 per gal of biodiesel.

The previous budget exercise makes biodiesel processing a potentially profitable venture. What is not shown is the variation in net returns over the year. During 2013, net returns varied from a loss of $0.08 to a positive return of $1.02 per gal (Table 31.5). Generally, prices of soy oil moved with the price of biodiesel (correlation = 0.43). These returns show a good profit from the biodiesel production; however, only 1 year of the previous 4 (2009–2012) produced a positive average return. From these slim returns, one can easily see why the mandate creating demand and the $1 tax credit was instrumental in nearly doubling biodiesel production, as it took much of the risk out of the equation, assuring more consistent returns for investors. Please note that while this is for a representative plant, local prices, efficiency, and marketing costs will influence returns. Marketing will include efforts by operators to lock in favorable prices for inputs and outputs.

Table 31.4 Net Returns of Soy Based Biodiesel Under Various Scenarios from 2014 Prices[a]

	Average 2014 Prices	Breakeven Soy Oil	Breakeven Biodiesel	Highest Soy Oil	Lowest Soy Oil	Highest Biodiesel
Price of soy oil/lb	$0.450	$0.520	$0.450	$0.496	$0.378	$0.460
Pounds soy oil/gal biodiesel	7.55	7.55	7.55	7.55	7.55	7.55
Price of biodiesel/gal	$4.62	$4.62	$4.09	$4.49	$3.90	$5.03
Value of Glycerine/gal	$0.03	$0.03	$0.03	$0.03	$0.03	$0.03
Fixed cost/gal	$0.26	$0.26	$0.26	$0.26	$0.26	$0.26
Variable costs/gal	$0.25	$0.25	$0.25	$0.25	$0.25	$0.25
Net gas/gal	$0.04	$0.04	$0.04	$0.04	$0.04	$0.04
Methanol cost/gal	$0.17	$0.17	$0.17	$0.16	$0.20	$0.17
Revenue biodiesel/gal	$4.62	$4.62	$4.09	$4.49	$3.90	$5.03
Total revenue/gal	$4.65	$4.65	$4.12	$4.52	$3.93	$5.06
Cost per biodiesel/gal	$4.12	$4.65	$4.12	$4.45	$3.61	$4.19
Net revenue/gal	$0.53	$0.00	$0.00	$0.07	$0.32	$0.87

[a]Source: Don Hofstrand (http://www.extension.iastate.edu/agdm/energy/html/d1-15.html).

Table 31.5 Variation in Prices and Costs for 2013

Month	Soy Oil $/lb	Total Costs $/gal	Biodiesel $gal	t Revenue $/gal
Jan	0.487	4.39	4.28	0.08
Feb	0.496	4.45	4.49	0.07
Mar	0.490	4.42	4.63	0.23
Apr	0.491	4.43	4.75	0.35
May	0.495	4.46	4.87	0.43
June	0.484	4.38	4.94	0.59
July	0.457	4.17	5.03	0.89
Aug	0.425	3.93	4.90	1.00
Sept	0.419	3.88	4.88	1.02
Oct	0.400	3.75	4.64	0.92
Nov	0.393	3.71	4.12	0.43
Dec	0.378	3.61	3.90	0.32
Average	0.451	4.13	4.62	0.51

Source: Don Hofstrand(b) (http://www.extension.iastate.edu/agdm/energy/html/d1-15.html).

FARM-SCALE PRODUCTION

Some promoters of biodiesel support the idea of a self-supporting, sustainable farm, raising crops that can be processed into fuel for the tractors that the farm uses produce the crops. This idea is politically popular and technically feasible, but not so practical when considering the economic realities (Stebbins-Wheelock et al., 2012). One obvious challenge is the question of scale. How can any farm-size biodiesel operation ever compete with the efficiencies of a 30 million gal plant? First off, the farms that have tried to do the whole operation with soybeans or other oil crops start with growing the plant and processing a step earlier, meaning they have to get the oil from the soybean. To do this they need a press to extract the oil. All presses are not created equal; therefore the cost of the press is proportional to the completeness of the oil removal process. A cheaper press can extract most of the oil but leave considerable amounts of fat in the meal. What does this mean for the process? An on-farm processor has to deal with the soy meal. Soy meal, at current prices of $476 per ton, is a valuable by product if it meets industry standards. If not, the meal is not as valuable as industry soy meal, selling at a significant discount and altering the economics of the entire farm scale project. The problem lies with the fat content that is important in balancing rations for dairy cows, beef, pigs, and chickens. The problem is not just with the level of fat in the meal, but also due to the variability of fat in the oil meal. Commercial-grade soy oil is guaranteed consistency, something nearly impossible to achieve with most farm-scale oil presses.

There are other costs involved in farm-scale production, including buildings, processing equipment, screening, and achieving consistent yield. The primary aspect is to consider what cost is necessary to produce a quantity that may only be, at most, several 1000 gal of biodiesel fuel per year. The farmer has to remember that the raw material, soybeans, have a ready market value that sets the initial cost of the biodiesel. The farmer may have an advantage in that they may consider their labor to be free, but labor is a very small part of the biodiesel process. Ultimately, it makes little economic sense for a single individual to invest in biodiesel processing equipment to make their own fuel oil. It may meet personal goals, but on a strict economic analysis, making one's own fuel does not appear to be a paying proposition.

An alternative to farm scale might be the creation of a cooperative type of arrangement between a group that contribute resources and labor. In this situation, individuals contribute to the overhead cost of processing equipment and share its use. Greater use reduces the cost per gal of fuel for everyone. Idle facilities only cost money, whereas greater use promotes efficiency. However, the cooperative set up would likely not reduce costs to the point where it becomes economically feasible to process one's own biodiesel. This observation is in a strict economic sense. It does not take into consideration the satisfaction of being energy independent or self-reliant.

When one looks at the economics of ethanol and biodiesel, the figures show tight margins and variable returns to investment. Therefore, the subsidy of tax credits was considered to be essential to develop the ethanol and biodiesel sectors. In theory, we expect a new economic venture to be in the growth stage and unable to meet profitable status against established business sectors. Therefore, we justify government involvement through legislation, market mandates, trade barriers, and tax policy to provide incentives for actions that some argue are in the public good. The government action allows the desired business to compete in the market place, develop the infrastructure, technology, efficiency, and market know how to get established and compete. These are the arguments for the government actions that supported the ethanol and biodiesel movements. As mentioned above, tax subsidies were given both to ethanol and biodiesel production to encourage production. Government assured markets through public mandates to produce ethanol and biodiesel fuels. Government required their use in the fuel system, both ethanol and biodiesel. As a further step, government created legal barriers to prevent imports of competitive products. One may argue that the choice was not a good one; however, it does provide a good example of how government actions can be employed to develop products that are desired by government.

Will biodiesel and ethanol survive on their own? That is a question that can only be answered in time. As the above analysis shows, without government action, the economic atmosphere was not present to develop the ethanol or biodiesel sectors. If the sectors can refine their technologies, become more efficient, and develop market savvy, they likely have to change. But to compete in a world market, making use of major food and feedstuffs that have expanding demand at the world's population increases will be challenging. Will they be able to compete economically, or will they be a bygone era where the idea was tried but failed to develop its own economic relevancy? Stay tuned.

REFERENCES

Agricultural Resource Marketing Center. Soybean Balance Sheet by Iowa State Extension. Available at: https://www.extension.iastate.edu/agdm/crops/outlook/soybeanbalancesheet.pdf.

Farm Doc Daily, June 14, 2013. An Updated Look at the Profitability of Ethanol Production. University of Illinois-Urbana-Champaign, Department of Agricultural and Consumer Economics. Available at: http://farmdocdaily.illinois.edu/2013/06/updated-profitability-ethanol-production.html.

Hofstrand, Don, Ag Decision Maker: Tracking Biodiesel Profitability, Iowa State University Extension and Outreach. Available at: http://www.extension.iastate.edu/agdm/energy/html/d1-15.html.

Hofstrand, Don(b), Ag Decision Maker: Tracking Ethanol Profitability, Iowa State University Extension and Outreach. Available at: https://www.extension.iastate.edu/agdm/articles/hof/HofJan08.html.

Luttrell, Clifton B., October 1973. The Russian Wheat Deal—Hindsight vs. Foresight. Reprint no. 81. Federal Reserve Bank of St. Louis. http://research.stlouisfed.org/publications/review/73/10/Russian_Oct1973.pdf.

National Corn Growers Association, 2013. World of Corn: Unlimited Possibilities. Available at: http://www.ncga.com/upload/files/documents/pdf/WOC%202013.pdf.

State of Nebraska. Ethanol Production Capacity by Plant. Available at: http://www.neo.ne.gov/statshtml/122_200501.htm.

Stebbins, Christine, December 24, 2013. High Cash Rents to Squeeze U.S. Midwest Grain Farmers in 2014. Available at: http://www.reuters.com/article/2013/12/24/usa-farm-rents-idUSL2N0K30LY20131224.

Stebbins-Wheelock, Emily, Parsons, Robert, Wang, Qingbin, Darby, Heather, Grubinger, Vern, December, 2012. Technical feasibility of small-scale oilseed and on-farm biodiesel production: a Vermont case study. Art #6RIB8 Journal of Extension 50 (6). http://www.joe.org/joe/2012december/pdf/JOE_v50_6rb8.pdf.

United States Soybean Board, Soystats 2103 Guide: A Reference Guide to Important Soybean Facts and Figures. Available at: http://soystats.com/.

United States Department of Agriculture, National Agricultural Statistics Service. Available at: http://www.nass.usda.gov/Quick_Stats/.

United States Department of Agriculture. Economic Research Service, April 1983. An Initial Assessment of the Payment in Kind Program.

United States Department of Agriculture. Economic Research Service, Fertilizer Use and Price. Available at: http://www.ers.usda.gov/data-products/fertilizer-use-and-price.aspx#.U1UQHFepQu8.

FUEL QUALITY POLICY

32

National Biodiesel Board, USA

Purpose: The American Society of Testing and Materials (ASTM) is the recognized standard-setting body for fuels and additives in the United States. ASTM has adopted a specification for pure biodiesel (ASTM D 6751) and a specification for biodiesel blends containing between 6% and 20% biodiesel (ASTM D7467). ASTM also recognizes that blends as high as 5% biodiesel are allowable in ASTM D975 for diesel fuel and ASTM D396 for heating oil providing the biodiesel meets ASTM D6751. When biodiesel that meets its specification is properly blended into diesel fuel which meets its specification, and is handled according to proper fuel management techniques, the resulting fuel is a high-quality, premium diesel fuel which has been shown to perform well in virtually any unmodified diesel engine. However, use of any fuel that does not meet its quality specifications could cause performance problems or equipment damage, and this includes biodiesel. The National Biodiesel Board (NBB) believes strongly that rigorous adherence to D6751 is important in order to protect consumers from unknowingly purchasing substandard fuel, in order to maintain the integrity of the nation's fuel supply, and in order to protect the reputation of biodiesel as a high-quality, high-performance fuel. Sale of off-spec fuel is usually a violation of federal and state law. Several federal and state government agencies are responsible for the regulation and enforcement of fuel quality in the United States. The NBB is a nonprofit trade association and does not have authority to regulate or enforce fuel quality. However, this "Fuel Quality Policy" outlines the measures that the NBB will take to enhance overall fuel quality in the industry.

Government adoption of ASTM D6751: The NBB is a strong proponent of ASTM D6751. The NBB expects each of its producer members to be dedicated to the consistent production of ASTM D6751-compliant B100. The NBB will continue to urge the adoption of ASTM D6751 by every appropriate level of federal, state, and local government as a legal requirement for the manufacture and sale of biodiesel. As to member and nonmember producers: the NBB will, upon receiving a complaint regarding fuel quality, offer the complainant all available information regarding fuel quality enforcement efforts by any applicable state or federal government agency.

ASTM standards for biodiesel blends: ASTM has adopted ASTM D7467 for biodiesel blends containing between 6% and 20% biodiesel. ASTM also recognizes that blends as high as 5% biodiesel are allowable in ASTM D975 for diesel fuel and ASTM D396 for heating oil providing the biodiesel meets ASTM D6751. The NBB is fully supportive of these specifications acknowledging that biodiesel must meet specification ASTM D6751 prior to blending activities.

BQ-9000 certification: The NBB created the National Biodiesel Accreditation Commission in 2000 and charged it with developing a certification program for quality biodiesel producers and

Bioenergy. http://dx.doi.org/10.1016/B978-0-12-407909-0.00032-8

marketers. The resulting certification program is BQ-9000. There are two certifications: Certified Marketer and Accredited Producer. In either case, the certified party must possess a quality manual, a quality control system, and employ best practices as required to assure the delivery of a quality product. NBB encourages all biodiesel producers and marketers to achieve and maintain approval under the BQ-9000 program.

Government enforcement: NBB will encourage active enforcement of D6751 by the Internal Revenue Service, the Environmental Protection Agency, the US Department of Agriculture, and individual state weights and measures bureaus. In addition, NBB will actively investigate which of these agencies might adopt the most effective proactive, ongoing testing program for D6751 enforcement and support that effort. Many of these agencies have indicated that they will also respond appropriately to potential violations upon credible complaint.

ACKNOWLEDGMENTS

This fact sheet is published with permission from National Biodiesel Board http://www.biodiesel.org/. The editor is grateful to Ray Albrecht, P.E. (technical representative, Northeast US Region) and Jessica Robinson (director of communications) for their time and effort in making this fact sheet available for this book.

RENEWABLE OILHEAT

33

Matt Cota

Vermont Fuel Dealers Association, Montpelier, Vermont, USA

THE RISE AND FALL OF THE OILHEAT MARKET

The oilheat industry is unique to the Northeast. Almost all of the 8 billion gal of oilheat sold every year in the United States are consumed in New York, New Jersey, Pennsylvania, and the six New England states. New England and New York are different from the rest of the country when it comes to home heating fuel consumption. While the more populated areas near Boston and the New York metropolitan area are converting to piped natural gas, this is far less likely in rural areas. The cost of installing an underground gas pipeline can exceed $2 million/mile. This investment would result in a negative rate of return for gas utilities in sparsely populated areas of Northern New York and New England. It is for this reason that deliverable liquid fuels will be part of the energy infrastructure in New England for the foreseeable future. However, the declines in market share in urban population centers have propelled the industry to explore ways of transforming their core product into a renewable source of energy.

Unlike the energy utility model, most oilheat marketers are third-or fourth-generation family businesses. The deliverable home energy industry dates back to the early 1900s with the transportation of coal and ice. These businesses would use one truck for both products, moving ice in the summer and coal in the winter. This is how most of the rural Northeast kept their food cold and their homes warm in the first half of the twentieth century (Figure 33.1). The mass production of the refrigerator by General Electric in the 1940s eventually eliminated the ice home delivery business. However, a new more profitable business model emerged as consumer's converted their heating systems from coal to #2 fuel oil. More commonly known as *oilheat*, the use of liquid home heating fuel began with the invention of the oil burner in the 1920s. It soon became a dominant fuel source throughout the cold-weather states in the Northeast.

There are several reasons for oilheat's ascendency, which peaked in the 1970s. Coal has to be shoveled into a steam boiler while an oilheat system eliminates any manual labor. The large and dirty coal bin was no longer needed, so homeowners could reclaim this space in their basement. Oilheat provides a more even heat with fewer drafts than coal, which means fewer health risks. Oil burners are also much cleaner. As anyone who has lived in a coal-heated house knows, soot can seep through the building, ruining clothes and furniture. In addition, coal ash has to be hauled away and disposed of. Oilheat also allows a homeowner to control the temperature by simply touching the thermostat in the living room. Unlike a coal or wood furnace, the family using oilheat could leave the house for weeks and the home would stay warm.

However, oilheat has lost significant market share over the past four decades. While still in a majority of homes in Maine, New Hampshire, and Vermont, the popularity of oilheat has plunged in

Bioenergy. http://dx.doi.org/10.1016/B978-0-12-407909-0.00033-X

FIGURE 33.1

Oil delivery truck, 1940s, in Rockingham, Vermont.

(Source: Courtesy of Cota and Cota, Inc.)

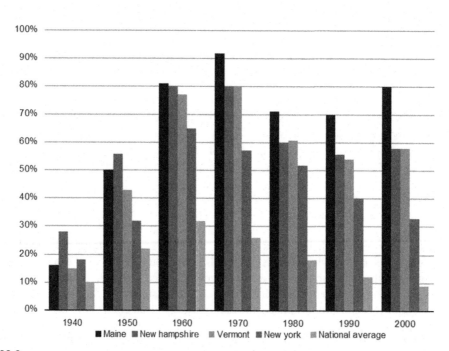

FIGURE 33.2

Growth and decline in the percentage of homes heated with oil.

nearly every other state as more homes convert to natural gas or propane. Since 1960, the oilheat industry has lost half of its customers to competing fuels (Figure 33.2). Faced with a loss of market share and declining volumes due to energy conservation, the oilheat retailers of today are faced with a similar dilemma as their grandfathers. In the 1940s, coal dealers dumped the black rocks and convinced their customers to switch to oilheat. Now oilheat dealers are on the precipice of a similar challenge.

A CLEANER AND GREENER FUEL

On September 15th, 2009, nearly 100 representatives of the oilheat industry gathered in Baltimore for a national policy summit. Attendees included nearly every national, regional, and state oilheat association, as well as industry leaders from the wholesale, retail, and home heating equipment manufacturing sectors. The purpose of the policy summit was to discuss and vote upon the steps necessary to secure the future of the industry. The majority of those who attended approved a resolution that called for the transition to an ultra-low sulfur heating oil blended with diverse-feedstock biofuels. The resolution stated that by July 2010, all heating oil will be mixed with a biofuel component, and that by July 2011, the petroleum base stock we now call oilheat will be transitioned to ultra-low sulfur diesel fuel.

The decision didn't occur overnight. The industry benefited from over a decade of research that showed it could be done. In 2001, Brookhaven Laboratories conducted a comprehensive feasibility study using biodiesel as a replacement for traditional heating oil (Krishna, 2001). The scientists at Brookhaven found that blends of biodiesel and heating oil could be used with few or no modifications to the equipment. Other research focused on the environmental benefits related to the reduction in smoke and in nitrogen oxides (NOx) by blending biodiesel with ultra-low sulfur fuel (NESCAUM, 2005). This gave oilheat dealers the confidence to start marketing biodiesel blends to their customers. A registered trademark, BioHeat®, and an advertising campaign soon followed.

The timeline established at the national policy summit was extremely aggressive. While the goals have not yet been realized, there has been tremendous progress. One of the reasons for the delay is that fuel quality standards for heating oil are not regulated by the federal government, but by individual states. The result has been a patchwork regulatory approach. Massachusetts, Rhode Island, Vermont, Connecticut, and New York all have laws on the books that set either a timeline or a trigger to require blends of biodiesel with heating oil. While it is left to the states to establish mandates, thanks to federal tax policy which encourages upstream blending of biodiesel, most oilheat is currently blended with up to 5% biodiesel.

In order for the oilheat industry to continue this transition to renewable fuels, there need to be a few transformational technologies. The first is with the heating equipment. Current equipment can handle up to B-20 blends, however it are only rated up to B-5 by Underwriters Laboratories. Once heating systems are built to burn blend levels higher than B-20, the amount of biofuel blended into oilheat will increase if the arbitrage between #2 distillate and B-100 makes it economical to do so.

The key for oilheat marketers in the northeast is to blend biofuel that is locally produced. There are three strategies for creating this fuel: through oilseed crops, waste oil, and algae. Biodiesel derived from all three sources has been successfully blended and used in heating systems. The potential market for producers of biodiesel in just the oilheat market is astounding. The U.S. oilheat market is currently 6 billion gal. In the next decade it is feasible that 1.2 billion gal, a B-20 blend, could come from biofuel. All info is current

OTHER RENEWABLE OPPORTUNITIES

Heating fuel dealers continue to experiment with a "back to the future" business model by distributing solid fuels, much like their grandfathers once did. However, instead of coal, dealers are examining the economics of delivering wood pellets. In rural forested areas of northern New England, the concept of delivering wood to the customer door to door has merit. Advances in solid fuel heating equipment have helped as customers adopt central heat systems and "pellet hoppers" that are reminiscent of the coal bins still found in the basements in some old Victorian houses.

However, the challenge is the distribution of the pellets. A 10-ton delivery truck can carry up to 2800 gal of heating oil under the weight restrictions of most roads. This amount of liquid heating fuel represents approximately 380 million BTUs. The same delivery truck can only carry 20,000 pounds of wood pellets, or the equivalent of 160 million BTUs. In addition, a pellet delivery truck can't rapidly transfer the pellets into the hopper, or the fuel will break up into "fines" and lose its BTU value. In several bulk delivery distribution experiments, the driver averaged about 200 pounds of wood pellets per minute, or 1.6 million BTUs. Liquid fuel can be delivered much faster. An average heating oil delivery truck can pump 50 gal/min, the equivalent of 7 million BTUs.

The implications are clear for those in the deliverable fuel industry. If all of the heating oil homes changed over to wood pellets, the fuel dealer would need twice as many trucks and drivers. And each truck would take four times as long to deliver the same amount of BTUs, requiring even more trucks and more drivers. When you consider that a 10-ton delivery truck gets about 7 miles/gal and travels 13,000 miles/year, transportation costs associated with pellet delivery are far greater than with oilheat. This could be ameliorated through an increase in the retail cost of pellets; however, that would make them less financially attractive to the customer.

This is why more and more heating oil dealers are embracing a transition to BioHeat® (Figure 33.3). Renewable liquid fuel is more appealing because the infrastructure is currently in place. The trucks, storage, tanks, and heating equipment do not need to be converted for consumers to benefit from a locally produced, environmentally superior heating fuel. This new renewable fuel represents the best future for the deliverable home energy industry and their customers.

CASE STUDY: BOURNE'S ENERGY

In the spring of 2010, Peter Bourne looked at the future of his company and knew something had to change. Over the past decade, improvements in thermal efficiency and heating equipment has resulted in a loss of volume as per home sales continued to decline. When his father Bob Bourne first started selling heating oil in Morrisville back in 1947, the average home used over 1400 gal (Figure 33.4). Today most homes in Vermont consume less than half that amount. Bourne's Energy wasn't the only company facing this challenge. Nearly a dozen other heating fuel providers in and around Lamoille County were fighting for the same customers. Bourne knew that in order for the company to make it to the next generation, he had to differentiate his business from the competition.

Recognizing an environmentally conscious customer base, Bourne made the decision to stop selling oilheat and start selling BioHeat®, a blend of biodiesel and heating oil. BioHeat® can be purchased pre-blended and trucked in from wholesale terminals in Burlington or Albany. The other option is to truck both products (B-100 and heating oil) to Morrisville, where it can be blended onsite.

FIGURE 33.3

Marketing of biodiesel blended heating oil with the BioHeat® brand. Courtesy of national biodiesel board.

FIGURE 33.4

While now selling renewable oilheat, Bourne's Energy began as a traditional oil service company. Courtesy of Bourne's energy.

Bourne decided to blend the product himself for both technical and economic reasons. Purchasing B-100 and mixing it with standard heating oil allows him to determine the blend levels. This is important when the weather turns extremely cold and higher blends can "cloud up" the fuel, inhibiting combustion. It also allows Bourne to secure the better pricing on behalf of his customers and blend at higher levels depending on the customer's preference.

Bourne needed to build a facility that could store the deliveries of biodiesel and injection-blending equipment. Thanks in part to a grant from the Vermont Sustainable Jobs Fund, Bourne installed a 10,000 gal fuel tank of B100, or 100 percent recycled cooking oil (Figure 33.5). The fuel is produced by White Mountain Biodiesel in North Haverhill, N.H. White Mountain purchases used cooking oil from restaurants throughout New England for $1 per gallon and turns it into biodiesel. They then sell it to Bourne's Energy and other heating fuel dealers at a price that is comparable to standard heating oil.

Bourne's storage tanks were upgraded and a blending facility was added to accommodate the biofuel. Customers and delivery truck drivers use a touch-screen computer to dial in their desired grade. Bourne's Energy is now selling blends of B-5 up to B-99 for home heating and off-road generators. All biofuel blends burn at least as efficiently as straight diesel and home heating oil products.

Despite the extra equipment needed by Bourne to maintain and blend the biofuel, the prices are competitive. The fuel also burns cleaner, and significantly reduces air pollution emissions of sulfur

FIGURE 33.5

Bourne's Energy at the grand opening of their biodiesel blending facility in Morrisville, VT, on October 2012. Courtesy: Matt Cota, VFDA.

dioxide, particulate matter (soot), and carbon dioxide, according to the U.S. Environmental Protection Agency (Biodiesel Impacts on Exhaust Emissions, October 2002). Local biodiesel producers are beginning to recognize the potential of the heating oil market. There are nearly a dozen Vermont farms growing oilseed crops for fuel. According to the Vermont Sustainable Jobs Fund (Vermont Sustainable Jobs Fund, January 17, 2014), the production on the farm has jumped from 271,000 gal/year in 2010 to over 604,000 in 2011. Marketing biodiesel blended heating has also been successful for Bourne's Energy. The company has opened up five new retail locations outside of Lamoille County and has become one of the 10 largest independently owned heating fuel providers in the state.

REFERENCES

A Comprehensive Analysis of Biodiesel Impacts on Exhaust Emissions Assessment and Standards Division, October 2002. Office of Transportation and Air Quality, U.S. Environmental Protection Agency. EPA420-P-02-001.

Krishna, C.R., 2001. Biodiesel Blends in Space Heating Equipment. Prepared for: National Renewable Energy Laboratory Program Manager. Brookhaven National Laboratory, Upton, NY. Retrieved fromhttp://homepower.com/files/webextras/Biodiesel_Space_Heating.pdf.

Northeast States for Coordinated Air Use Management (NESCAUM), 2005. Low Sulfur Heating Oil in the Northeast States: An Overview of Benefits, Costs and Implementation Issues. Boston, MA. Retrieved from: http://vtbio.org/ss_files/NESCAUM%20REPORT.pdf.

Vermont Sustainable Jobs Fund, Oilseeds and Biodiesel Project–Program Outcomes. http://www.vsjf.org/project-details/11/vermont-grass-energy-partnership (last accessed 17.01.14).

WHAT'S SO DIFFERENT ABOUT BIODIESEL FUEL?

34

Daniel Ciolkosz

Penn State Biomass Energy Center and Department of Agricultural and Biological Engineering, The Pennsylvania State University, USA

INTRODUCTION

Biodiesel is a liquid fuel that is created by chemically processing vegetable oil and altering its properties to make it perform more like petroleum diesel fuel. It was first evaluated seriously in the late 1970s, but it was not widely adopted at that time.

The topic of biodiesel fuel has recently received a great deal of interest, and large- and small-scale manufacturers have started production at locations throughout the state. However, many people are still uncertain about whether biodiesel is a reliable, safe fuel to use for diesel engines.

This chapter explains the major differences between biodiesel and petroleum diesel (also called petrodiesel), including information about biodiesel additives and blends. The companion fact sheet in this series, *Using Biodiesel Fuel in Your Engine*,[1] explains the performance you can expect when running an engine on biodiesel.

PROPERTIES OF BIODIESEL VERSUS PETROLEUM DIESEL

The sizes of the molecules in biodiesel and petroleum diesel are about the same, but they differ in chemical structure. Biodiesel molecules consist almost entirely of chemicals called fatty acid methyl esters (FAMEs), which contain unsaturated "olefin" components. On the other hand, low-sulfur petroleum diesel consists of approximately 95% saturated hydrocarbons and 5% aromatic compounds.[2] The differences in chemical composition and structure between petroleum diesel and biodiesel result in several notable variations in the physical properties of the two fuels. The seven most significant differences are as follows:

1. Biodiesel has higher lubricity (it is more "slippery") than petroleum diesel. This is a good thing because it can be expected to reduce engine wear.
2. Biodiesel contains practically no sulfur. This is also a good thing because it can be expected to result in reduced pollution from engines using biodiesel.

---

[1]http://pubs.cas.psu.edu/FreePubs/pdfs/uc204.pdf.

[2]If the biodiesel is made using ethanol rather than methanol, then the resulting molecules are fatty acid ethyl esters.

Bioenergy. http://dx.doi.org/10.1016/B978-0-12-407909-0.00034-1

3. Biodiesel has a higher oxygen content (usually 10–12%) than petroleum diesel. This should result in lower pollution emissions. However, relative to petroleum diesel, it causes slightly reduced peak engine power (~4%).
4. Biodiesel tends to thicken and "gel up" at low temperatures more readily than petroleum diesel. Some types of oil are more of a problem than others. This is a concern, especially for the cold winters that are typical to Pennsylvania.
5. Biodiesel is more likely to oxidize (react with oxygen) to form a semisolid gel-like mass. This is a concern, especially for extended fuel storage and when using engines that are only operated occasionally (such as standby power generators). A good method for storage is to use a dry, semi-sealed, cool, light-tight container.
6. Biodiesel is more chemically active as a solvent than petroleum diesel. As a result, it can be more aggressive to some materials that are normally considered safe for diesel fuel.
7. Biodiesel is much less toxic than petroleum diesel. This can be a real benefit for spill cleanups.

The quality of petroleum diesel fuel tends to be more uniform and reliable, especially when compared with small-scale production of biodiesel, in which quality control may or may not have been good. Petroleum diesel can vary in quality from plant to plant or from region to region, but the variations are typically much smaller. Poor-quality biodiesel fuel can lead to many problems in engine performance, and care should be taken to ensure that your fuel is of good quality (see the *Renewable and Alternative Energy Fact Sheet: Using Biodiesel Fuel in Your Engine*). Biodiesel that conforms to ASTM Standard D6751 should be of a consistent, high quality.

In all fairness, we should mention that petroleum diesel has also demonstrated problems with oxidative stability and low-temperature performance, although biodiesel, at present, seems to be more susceptible.

DOES THE TYPE OF VEGETABLE OIL USED MATTER?

A common question regarding biodiesel that comes up is, "Which oil crop results in the best biodiesel?" There are definite differences from crop to crop, but it is not a straightforward matter to choose a "best" one, especially when the cost of growing or buying oil can also vary quite a bit from crop to crop. Different vegetable oils have higher or lower concentrations of different chemical components (fatty acids, for the most part), which affects their performance when they are made into biodiesel. In addition, the chemical structure of the alcohol that is reacted with the oil to create biodiesel can also affect the properties of the fuel. In general, the chemical properties that matter the most are the length of the biodiesel molecule, the amount of "branching" in the chain, and the degree of "saturation" of the molecule. As shown in Table 34.1, these properties have positive and negative effects on biodiesel; therefore, it is not really possible to choose a "perfect" oil for biodiesel. As if this was not complicated enough, we need to also remember that cold-starting properties might be vital during winter in cold climates but unimportant in summertime or in warm parts of the world. On top of all that, it is possible to buy additives that improve some of the less-than-ideal properties of biodiesel. In general, longer molecules with more branching are beneficial to the performance of biodiesel but are seldom present in FAMEs. High unsaturation (high iodine number) leads to poor oxidative stability and is undesirable in biodiesel. Of the many types of fatty acid found in vegetable oils, oleic acid is probably best, whereas linoleic is less desirable and linolenic acid is most undesirable. With all of this

Table 34.1 General Comparison of Different Oil Chemical Properties Related to Their Use as Biodiesel

Property	Positive Effects	Negative Effects
Length of molecule	Increases the cetane number, heat of combustion; decreases NO_x emissions	Increases viscosity
Amount of branching	Decreases the gel point	Decreases the cetane number
Saturation	Decreases NO_x emissions, improves oxidative stability, reduces deposition	Increases melting point and viscosity; reduces lubricity[a]

NO_x, oxides of nitrogen.
[a]Technically, the reduction in lubricity is due to the removal of polar compounds containing sulfur that are natural additives by hydrogenation and the formation of saturated compounds.

in mind, it appears that canola oil, with its high proportion of long, unsaturated fats (lots of oleic acid), may be slightly better for biodiesel fuel quality than some of the other oilseed crops, although this has not been conclusively confirmed with careful testing. Tropical oils such as palm oil, with their high proportion of saturated fats, tend to have significant problems with cold-weather performance because they tend to solidify more readily than many other oils.

MAKING BIODIESEL BETTER WITH ADDITIVES

Some of the properties of biodiesel fuel are not ideal from an engine performance point of view. Thankfully, additives can be used to counteract these problems and improve the overall quality of the fuel.

- Cold-flow improvers: These additives improve the cold weather performance of biodiesel by limiting its ability to gel. They tend to only improve the operating range by approximately 5 degrees.
- Fuel stabilizers: These additives act as "antioxidants" to reduce the possibility of oxidation degradation of the fuel.
- Antimicrobial additives: It is possible for microbes to grow in biodiesel, resulting in clogged lines and fouled equipment. Antimicrobial additives prevent this by killing off any existing microbes and preventing them from returning.
- Detergent additives: These help reduce the formation of deposits on engine parts by forming a protective layer on the parts and dissolving existing deposits from the surfaces within the engine.
- Corrosion inhibitors: These also protect the engine by forming a protective layer on the components, thus preventing corrosive chemicals from reaching the surface.

A wide array of additives is available on the market today, and they can be purchased at an automotive shop or on the Internet. Often, a single product can be purchased that combines many or all of the above additives. The actual composition of these additives is usually a closely guarded trade secret, and not all additives perform the same. Users should keep track of how well a specific additive is working for them and take care to follow the manufacturer's recommendations for the concentration and proper use of the additive. Keep in mind that there are many "snake oil" salesmen in the market today. Only deal with reputable companies and suppliers that are approved by your engine's manufacturer.

WHAT ABOUT BLENDS?

Biodiesel fuel blends very easily with petroleum diesel. These blends are described by their percentage of biodiesel (e.g., "B20" has 20% biodiesel, 80% petroleum diesel). In general, the properties of a blend will lie somewhere between the properties of the biodiesel and the petroleum diesel. Blends are sometimes used to improve the lubricity of petroleum diesel or reduce its sulfur content. Probably the most useful reason for a biodiesel producer to blend would be to improve cold-operating characteristics during the winter. A mix of 70% biodiesel and 30% petroleum diesel has been reported to be effective for mild winter conditions. Kerosene, also known as #1 diesel fuel, is blended with standard (#2) petroleum diesel during winter months (usually ~40% kerosene, 60% #2 diesel) to improve its cold-weather performance. This approach is probably the easiest way to make biodiesel usable during harsh midwinter conditions in Pennsylvania. However, keep in mind that only low-sulfur kerosene that is approved as an engine fuel should be used.

SUMMARY

Biodiesel and petroleum diesel are very similar fuels, but they are not identical. However, the differences are remarkably small when we consider the radically different procedure for making biodiesel as compared to petroleum diesel. Many additives are available that can modify the properties of biodiesel fuel, and biodiesel can be easily blended with petroleum diesel fuel if desired. For additional information, please refer to the following Penn State Cooperative Extension fact sheets and reports:

- *Biodiesel: A Renewable, Domestic Energy Resource*
- *Renewable and Alternative Energy Fact Sheet: Using Biodiesel Fuel in Your Engine*
- *Making Your Own Biodiesel: Brief Procedures and Safety Precautions*
- *Biodiesel Safety and Best Management Practices for Small-Scale Noncommercial Production*

ACKNOWLEDGMENTS

This chapter was originally published as a fact sheet of the same name by The Pennsylvania State University College of Agricultural Science (Pub Code UC205). The author acknowledges the review of this fact sheet by Joseph Perez, Dennis Buffington, and Glen Cauffman of Penn State.

FURTHER READINGS

Agarwal, A.K., Bijwe, J., Das, L., 2003. Wear assessment in a biodiesel-fueled compression ignition engine. Journal of Engineering for Gas Turbines and Power 125, 820–826.

Bhale, P., Deshpande, N., Thombre, S., 2008. Improving the low temperature properties of biodiesel fuel. Renewable Energy 1–7.

Bruwer, J.J., van D Boshoff, B., Hugo, F., du Pleiss, L.M., Fuls, J., Hawkins, C., van der Walt, A., Wenglebrecht, A., June 11, 1980. Sunflower seed oil as an extender for diesel fuel in agricultural tractors. In: Paper Presented at the 1980 Symposium of the South African Institute of Agricultural Engineers.

Cambray, G., December 2007. Helping biodiesel become unstuck. Science in Africa. http://www.scienceinafrica.com/old/index.php?q=2007/december/biodiesel.htm.

Cetinkaya, M., Ulusoy, Y., Tekin, Y., Karaosmanoglu, F., 2005. Engine and winter road test performances of used cooking oil originated biodiesel. Energy Conversion and Management 46, 1279–1291.

Fernando, S., Karra, P., Hernandez, R., Jha, S.K., 2007. Effect of incompletely converted soybean oil on biodiesel quality. Energy 32, 844–851.

Flitney, R., 2007. Which elastomer seal materials are suitable for use in biofuels? Sealing Technology 9, 8–11.

Graboski, M., McCormick, R., 1998. Combustion of fat and vegetable oil derived fuels in diesel engines. Progress in Energy Combustion Science 24, 125–164.

Hancsok, J., Bubalik, M., Beck, A., Baladincz, J., 2008. Development of multifunctional additives based on vegetable oils for high-quality diesel and biodiesel. Chemical Engineering Research and Design 86, 793–799.

Knothe, G., 2005. Dependence of biodiesel fuel properties on the structure of fatty acid alkyl esters. Fuel Processing Technology 86, 1059–1070.

Lapuerta, M., Armas, O., Rodrıguez-Fernandez, J., 2008. Effect of biodiesel fuels on diesel engine emissions. Progress in Energy and Combustion Science 34, 198–223.

Ryan, T., Dodge, L., Callahan, T., 1984. The effects of vegetable oil properties on injection and combustion in two different diesel engines. Journal of the American Oil Chemists' Society 61, 1610–1619.

Sharma, Y., Singh, B., Upadhyay, S., 2008. Advancements in development and characterization of biodiesel: a review. Fuel 87, 2355–2373.

Zheng, M., Mulenga, M., Reader, G., Wang, M., Ting, D., Tjong, J., 2008. Biodiesel engine performance and emissions in low temperature combustion. Fuel 87, 714–722.

BIODIESEL EMISSIONS AND HEALTH EFFECTS TESTING

35

National Biodiesel Board, USA

BIODIESEL EMMISSIONS

Biodiesel is the first and only alternative fuel to have a complete evaluation of emission results and potential health effects submitted to the US Environmental Protection Agency (EPA) under the Clean Air Act Section 211(b). These programs include the most stringent emissions testing protocols ever required by EPA for certification of fuels or fuel additives. The data gathered complete the most thorough inventory of the environmental and human health effects attributes that current technology will allow.

EPA has surveyed the large body of biodiesel emissions studies and averaged the Health Effects testing results with other major studies. The results are seen in the Table 35.1. To view EPA's report titled "A Comprehensive Analysis of Biodiesel Impacts on Exhaust Emissions" visit: www.epa.gov/otaq/models/analysis/biodsl/p02001.pdf.

The ozone (smog)-forming potential of biodiesel hydrocarbons is less than diesel fuel. The ozone-forming potential of the speciated hydrocarbon emissions is 50% less than that measured for diesel fuel.

Sulfur emissions are essentially eliminated with pure biodiesel. The exhaust emissions of sulfur oxides and sulfates (major components of acid rain) from biodiesel are essentially eliminated compared to diesel.

Criteria pollutants are reduced with biodiesel use. Tests show the use of biodiesel in diesel engines results in substantial reductions of unburned hydrocarbons, carbon monoxide, and particulate matter. Emissions of nitrogen oxides stay the same or are slightly increased.

Carbon monoxide—The exhaust emissions of carbon monoxide (a poisonous gas) from biodiesel are on average 48% lower than carbon monoxide emissions from diesel.

Particulate matter—Breathing particulate has been shown to be a human health hazard. The exhaust emissions of particulate matter from biodiesel are about 47% lower than overall particulate matter emissions from diesel.

Bioenergy. http://dx.doi.org/10.1016/B978-0-12-407909-0.00035-3

Table 35.1 Average Biodiesel Emissions Compared to Conventional Diesel, according to US Environmental Protection Agency

Emission Type	B100	B20
Regulated		
Total unburned hydrocarbons	−67%	−20%
Carbon monoxide	−48%	−12%
Particulate matter	−47%	−12%
NOx	+10%	+2% to 2%
Nonregulated		
Sulfates	−100%	−20%[a]
PAH (polycyclic aromatic hydrocarbons)[b]	−80%	−13%
nPAH (nitrated PAHs)[b]	−90%	−50%[c]
Ozone potential of speciated HC	−50%	−10%

[a]Estimated from B100 result.
[b]Average reduction across all compounds measured.
[c]2-Nitroflourine results were within test method variability.

Hydrocarbons—The exhaust emissions of total hydrocarbons (a contributing factor in the localized formation of smog and ozone) are on average 67% lower for biodiesel than diesel fuel.

Nitrogen oxides—NOx emissions from biodiesel increase or decrease depending on the engine family and testing procedures. NOx emissions (a contributing factor in the localized formation of smog and ozone) from pure (100%) biodiesel increase on average by 10%.

However, biodiesel's lack of sulfur allows the use of NOx control technologies that cannot be used with conventional diesel.

Additionally, some companies have successfully developed additives to reduce NOx emissions in biodiesel blends.

Biodiesel reduces the health risks associated with petroleum diesel.

Biodiesel emissions show decreased levels of polycyclic aromatic hydrocarbons (PAH) and nitrated polycyclic aromatic hydrocarbons (nPAH), which have been identified as potential cancer causing compounds. In Health Effects testing, PAH compounds were reduced by 75–85%, with the exception of benzo(a)anthracene, which was reduced by roughly 50%. Targeted nPAH compounds were also reduced dramatically with biodiesel, with 2-nitrofluorene and 1-nitropyrene reduced by 90%, and the rest of the nPAH compounds reduced to only trace levels.

HEALTH EFFECTS TESTING
HISTORY

In June 2000, representatives of the US Congress announced that biodiesel had become the first and only alternative fuel to have successfully completed the Tier I and Tier II Health Effects testing requirements of the Clean Air Act Amendments of 1990. The biodiesel industry invested more than two million dollars and four years into the health effects testing program with the goal of setting biodiesel apart from other alternative fuels and increasing consumer confidence in biodiesel.

TESTING

The first tier of health effects testing was conducted by Southwest Research Institute and involved a detailed analysis of biodiesel emissions. Tier II was conducted by Lovelace Respiratory Research Institute, where a ninety-day subchronic inhalation study of biodiesel exhaust with specific health assessments was completed.

RESULTS

Results of the health effects testing concluded that biodiesel is nontoxic and biodegradable, posing no threat to human health. Also among the findings of biodiesel emissions compared to petroleum diesel emissions in this testing are the following:

- The ozone (smog)-forming potential of hydrocarbon exhaust emissions from biodiesel is 50% less.
- The exhaust emissions of *carbon monoxide* (a poisonous gas and a contributing factor in the localized formation of smog and ozone) from biodiesel are 50% lower.
- The exhaust emissions of *particulate matter* (recognized as a contributing factor in respiratory disease) from biodiesel are 30% lower.
- The exhaust emissions of *sulfur oxides and sulfates* (major components of acid rain) from biodiesel are completely eliminated.
- The exhaust emissions of *hydrocarbons* (a contributing factor in the localized formation of smog and ozone) are 95% lower.
- The exhaust emissions of *aromatic compounds* known as Polycyclic Aromatic Hydrocarbons (PAH) and Nitrated Polycyclic Aromatic Hydrocarbons (nPAH) compounds (suspected of causing cancer) are substantially reduced for biodiesel compared to diesel. Most PAH compounds were reduced by 75–85%. All nPAH compounds were reduced by at least 90%.

SIGNIFICANCE

The health effects testing results provide conclusive scientific evidence using the most sophisticated technology available to validate the existing body of testing data. The comprehensive body of biodiesel data serves to demonstrate the significant benefits of biodiesel to the environment and to public health. This will lead to increased consumer confidence and increased use of biodiesel. Since the majority of biodiesel is made from soybean oil, a promising new market is materializing for soybeans.

ACKNOWLEDGMENTS

The editor is grateful to Ray Albrecht, P.E. (technical representative, Northeast US Region) and Jessica Robinson (director of communications) for their time and effort in making the two fact sheets available for this book with permission from National Biodiesel Board http://www.biodiesel.org/.

BIODIESEL SUSTAINABILITY FACT SHEET

36

National Biodiesel Board, USA

SUSTAINABILITY PRINCIPLES

- The National Biodiesel Board (NBB) has made sustainability in climate change mitigation, human rights, food security, and respect for all natural resources a top priority. That's why the industry has adopted and follows guiding principles that demonstrate our commitment to a full spectrum of sustainability tenets.
- Biodiesel producers are already providing a very sustainable fuel, and these principles are another way we're ensuring that, as our industry grows, it continues to improve quality of life, safeguard the environment, and strengthen economies.
- Biodiesel improves air quality, it's renewable, and it's creating green-collar jobs in our communities. The NBB is committed to keeping biodiesel on the cutting-edge of sustainability.

ENERGY BALANCE

- Biodiesel has a very high "energy balance." Newly published research from the University of Idaho and U.S. Department of Agriculture show that for every unit of fossil energy needed to produce biodiesel, the return is 5.54 units of energy. Biodiesel made from soybean oil has a high energy balance because the main energy source used to grow soybeans is solar.
- The "energy balance" takes into account the planting, harvesting, fuel production and fuel transportation to the end user. As a result of modern farming techniques and energy efficiencies, biodiesel's energy balance continues to improve.
- In contrast, regular fossil fuel biodiesel has a negative energy balance.

Bioenergy. http://dx.doi.org/10.1016/B978-0-12-407909-0.00036-5

WATER CONSERVATION

- Crops are not irrigated or planted solely to produce biodiesel. Conversion of these co-products and byproducts uses very little water—the entire US biodiesel industry used less processing water in 2008 than it took to irrigate two golf courses in the Sunbelt.
- A 1998 jointly produced U.S. Departments of Agriculture and Energy "cradle to grave" analysis of biodiesel's production found it reduces wastewater by 79% and reduces hazardous waste production by 96% compared to petroleum diesel.

LAND CONSERVATION

- The USDA reports that US acreage for crop production has not increased since 1959.
- Major land use changes in the United States that would endanger environmentally sensitive lands are not expected due to biofuels. In fact, there are very solid federal and state laws in place to help ensure these lands remain undisturbed.
- Crop production in the United States is trending significantly toward utilizing more conservation practices, and advances in agriculture are leading to higher yields and lower inputs with the same acreage.
- The United Nations Food and Agriculture Organization has calculated that, of the land that could be used for agriculture today, only 3.7 billion acres of the 10.4 billion acres are used, and of that, only 1% of that area is used for biofuels, which includes ethanol.

FOOD SUPPLY SECURITY

- Biodiesel is not made by grinding up soybeans into fuel. Soybeans have two components—oil and protein meal. The meal represents the majority of the soybean, and is used in food and livestock feed. Biodiesel uses only the oil portion of the soybean.
- By creating a new market for the soybean oil that is co-produced, the overall value of the bean increases, and the meal portion becomes more cost competitive for protein markets. This has a net positive impact on the food supply.
- Biodiesel does not impact food prices like the big food companies would have you believe. For example, in the last quarter of 2008, biodiesel production was near an industry high at more than 60 million gallons per month. Yet in that same time frame, soybean commodities were selling at near record lows. And if that's not enough, even as commodity prices fell, food prices barely budged.
- Biodiesel produced from America's soybeans only uses approximately 3 percent of the nation's soybean harvest annually.
- Biodiesel uses only the oil portion of the soybean, leaving all of the protein available to nourish livestock and humans.
- In 2008, biodiesel produced from soybeans co-produced enough soybean meal for the equivalent of 115 billion rations of protein for the hungry in developing countries.

DIVERSITY

- Biodiesel is the most diverse fuel on the planet. It is made from regionally available, renewable resources that are abundant in the United States, including soybean oil, other plant oils, recycled restaurant grease, and beef tallow and other fats.
- The increased demand for biodiesel is stimulating research and investment in developing new materials to make biodiesel, such as algae, camelina, jatropha, other arid land crops, and waste materials like trap grease. The result is that we will see additional feedstock volumes coming from fallow or low-production lands and utilizing innovative technologies.
- According to a study by the National Renewable Energy Laboratory in Golden, Colorado, domestic feedstocks for biodiesel totaled 1.6 billion gallons (including greases, animal fats, and vegetable oils). NREL anticipates the natural growth and expansion of existing feedstocks (soy, canola, and sunflowers) will expand feedstock supplies for an additional 1.8 billion gallons by 2016.

CLEANER AIR AND HEALTH EFFECTS

- DOE and USDA say biodiesel reduces life cycle carbon dioxide, a greenhouse gas, by 78 percent.[1] Biodiesel also significantly reduces EPA-regulated emissions with direct impact to human health.
- Biodiesel is the only alternative fuel to voluntarily complete EPA Tier I and Tier II testing to quantify emission characteristics and health effects.
- Breathing particulate has been shown to be a human health hazard. The exhaust emissions of particulate matter from biodiesel are about 47 percent lower than overall particulate matter emissions from diesel.
- Biodiesel emissions show dramatically decreased levels of polycyclic aromatic hydrocarbons (75–85 percent) and nitrated polycyclic aromatic hydrocarbons (90 percent to trace levels), which have been identified as potential cancer causing compounds.
- As a result of the health benefits of biodiesel, some chapters of the American Lung Association have pledged their support for use of the alternative fuel.

ACKNOWLEDGMENTS

This fact sheet is published with permission from National Biodiesel Board http://www.biodiesel.org/. The editor is grateful to Ray Albrecht, P.E. (technical representative, Northeast US Region) and Jessica Robinson (director of communications) for their time and effort in making this fact sheet available for this book.

---

[1] In May of 1998, the US Department of Energy (DOE) and US Department of Agriculture (USDA) published the results of the Biodiesel Life Cycle Inventory Study. This 3.5-year study followed US Environmental Protection Agency (EPA) and private industry approved protocols for conducting this type of research.

ENTREPRENEURIAL OPPORTUNITIES IN BIOENERGY 37

F. John Hay

Department of Biosystems Engineering, University of Nebraska Lincoln Extension, Lincoln, NE, USA

BIOENERGY ENTREPRENEURISM

Entrepreneurism is important because entrepreneurs drive innovation and technological change, and thus generate economic growth (Schumpeter, 1934). Bioenergy entrepreneurs drive growth of the bioenergy sector. This chapter discusses bioenergy entrepreneurism, including market drivers, entrepreneurial motivations, and opportunities for bioenergy entrepreneurs. Market drivers such as policy, energy use, and the environment are moving the energy sector toward more renewables. Entrepreneurs are motivated by economics, the environment, social factors, and a pioneering spirit. By exploring the market drivers and entrepreneurial motivations, one can more clearly visualize and seek out opportunities to participate in the bioenergy sector.

Bioenergy entrepreneurs participate in the market of converting biomass into energy. The term *entrepreneur* takes on many meanings to many people. The Merriam-Webster definition of entrepreneur is one who organizes, manages, and assumes the risks of a business or enterprise. For the purposes of this chapter, we will use the definition from Shane and Venkataraman (2000), who described entrepreneurship as the process by which "opportunities to create future goods and services are discovered, evaluated, and exploited."

Although not all entrepreneurs are as successful as the top few, many share the same traits: self-starter, dedicated, risk taker, boundless energy, and vision (University of Maryland Extension). Yet, even with these traits, success is not guaranteed. The bioenergy world is full of successes, failures, and everything in between.

The difference between a success and failure in modern bioenergy can come down to timing. The modern corn ethanol industry boomed starting in 2005, with biofuel and environmental policy favoring the blending of ethanol into fuel. Ethanol facilities popped up all over, with some able to pay off debt in as little as 1–2 years, making enormous profits. The facilities that came online toward the end of the decade struggled with high corn prices and low margins (Hofstrand, 2008). Timing, either by luck or vision, is a key.

THE ENERGY SITUATION NOW AND IN THE FUTURE

The demand for energy will continue as the human population rises and developing countries gain in wealth. The world's population is increasing and developing countries are rapidly increasing the affluence of their people. The Energy Information Administration (EIA) estimates that world energy consumption will rise from 523 quadrillion Btu in 2010 to over 819 quadrillion Btu by the year 2040

Bioenergy. http://dx.doi.org/10.1016/B978-0-12-407909-0.00037-7

(EIA International Energy Outlook, 2013). Current sources of energy worldwide are as follows: 33% from petroleum, 28% from coal, 22% from natural gas, 11% from renewables, and 6% from nuclear (Figure 37.1). Electricity is predominately made from coal, natural gas, nuclear, hydroelectric, wind, and solar. Transportation fuel is predominately from petroleum. Transportation fuel's heavy reliance on one feedstock means that price spikes in the oil market heavily influence the price of fuel. The petroleum industry has a 100+ year head start on infrastructure, and a decade of high prices has given them huge sums of capitol to invest in new drilling, exploration, and extraction technologies. Conversely, the bioenergy industry has had to start from near scratch in scale-up development of technology to bring biofuels and bioenergy to large-scale production while relying on the oil industry to buy their product. The mandates coupled with the chance for profit have motivated some oil companies to invest in renewable energy. Predictions of energy use out to the year 2040 show increasing use of renewables yet steady use of fossil fuels (Figure 37.2) (EIA International Energy Outlook, 2013). The bioenergy entrepreneurs of tomorrow will face similar challenges when competing against oil, coal, natural gas, and nuclear to produce low-cost energy.

BIOENERGY ENTREPRENEURSHIP MOTIVATIONS

For all bioenergy entrepreneurial projects, the primary movers or key people behind the project may be motivated by different goals. Lin (2008) identified three major development motivations for entrepreneurs: environmental, social welfare, and entrepreneurial. A project with environmental motivations will use bioenergy to solve an environmental problem or issue. An example of environmental motivation is a swine operation whose desire to expand is stalled by concerns of increased odor. The addition of an anaerobic digester that creates biogas solves the air quality odor problem while at the same time creating energy, which the farm uses to create electricity and run farm equipment (biogas is scrubbed to remove impurities, then compressed and used as compressed natural gas). Social welfare

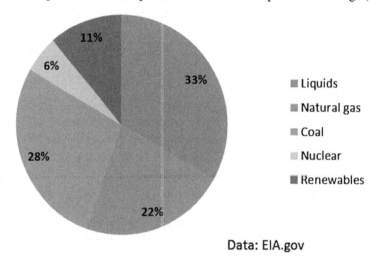

Data: EIA.gov

FIGURE 37.1

World energy consumption by fuel in 2012 (quadrillion Btu).

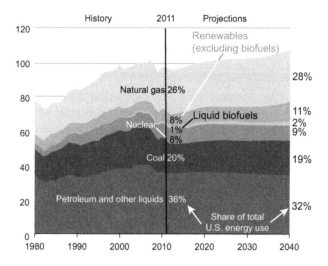

FIGURE 37.2

Primary energy use by fuel, 1980–2040 (quadrillion Btu).

(Source: EIA, 2013)

motivation stems from a goal to improve the lives of a community or group. Improving lives can have many meanings and examples, such as providing energy to a community at a lower cost than conventional generation, providing jobs to members of the community, or generating increased tax revenue for the community. The corn ethanol boom offers many examples of social welfare motivation where community leaders worked to finance, build, or attract ethanol facilities to their community. Once built, the ethanol facility provided high-wage jobs, large sums of tax revenue, and increased corn prices.

Finally, entrepreneurial motivations are shown when prime movers take risks with innovative technologies in the absence of a mature market. These entrepreneurs are motivated to be a leading enterprise in a new bioenergy market. An example of this type of entrepreneurial motivation can be seen by the first-generation cellulosic facilities struggling to begin production. Another example of entrepreneurial motivation is the many small companies working with algae to perfect a technology to grow and harvest algae for bioenergy.

Beyond the motivations identified by Lin, Shane argued that entrepreneurism involves human agency and a creative process where "by rearranging resources in a new way, entrepreneurs engage in creative activity" (Shane et al., 2003). This could be described as a fourth motivation of "pioneering spirit"—or, as described by Shane, "human motivation plays a critical role in the entrepreneurial process." Bioenergy entrepreneurs will be aided by this pioneering spirit as they work to bring alternative energy sources to market in competition with fossil fuels. An entrepreneur must have the motivation to react to the market conditions. These market drivers are discussed further in the next section.

MARKET DRIVERS

Bioenergy has always had a piece of the energy market. Burning wood for heat and steam is one example of bioenergy use, which has remained a part of the energy mix in areas where wood is prevalent or other fuels are higher cost. Liquid fuels from biomass, such as ethanol, had a share of the market prior to the 1920s when prohibition ended the industry. The modern fuel ethanol industry started as a way to use corn locally and gain more value.

The corn ethanol industry's growth has been clearly tied to subsidies, mandates, and environmental policy from state and federal governments (Figure 37.3). Starting with a $0.40 per gallon tax exemption subsidy in 1978, the industry increased production slowly until 2005, when the U.S. Congress passed the Energy Policy Act of 2005 (part of the act is referred to as the Renewable Fuel Standard, or RFS), which required that 7.5 billion (10^9) gallons of renewable fuel be blended with gasoline by 2012. This, combined with increasing environmental issues related to the fuel additive methyl tert-butyl ether (MTBE), caused the fuel industry to switch from MTBE to ethanol as an oxygenate and octane booster. The RFS was soon followed by the Energy Policy Act of 2007 (containing the RFS-2), which required the blending of 36 billion gallons of renewable fuel into gasoline and diesel by 2022. The RFS-2 went beyond the mandate of quantity and designated the quality of renewable fuels, classifying fuels by their greenhouse gas reduction compared to gasoline. Categories were created for each type of biofuel. A limit was set for how much of each category was required to be blended. The limit for starch based biofuel was set at 15 billion gallons. The goal of the RFS-2 was to stimulate demand for each different category of biofuel, thus building up the industry. Both RFS and RFS-2 have been crucial in establishing market certainty and thus stimulated investment in renewable fuel industries. Ethanol as a fuel has proven to be a popular additive to gasoline due to its low cost and high octane, providing the fuel industry an inexpensive octane booster. This fact has helped spur demand for ethanol at a rate faster than the RFS mandated and maintain demand past the expiration of the federal ethanol excise tax credit (ethanol subsidy) at the end of 2011. Figure 37.3 shows the progression of corn-based ethanol production in the United States, demonstrating how federal policy and environmental regulations combined to drive the demand for ethanol.

The future of RFS-2 is in the hands of the U.S. Environmental Protection Agency (EPA). In 2014, the EPA issued a decision to reduce the required blending levels of biofuels, citing lower-than-projected gasoline consumption. This reduction will add uncertainty to the future of the RFS-2 and

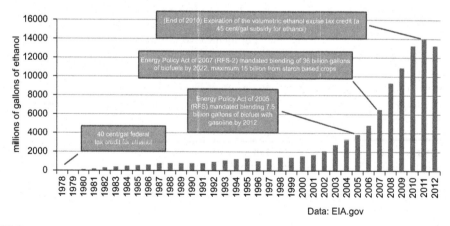

Data: EIA.gov

FIGURE 37.3

U.S. Fuel Ethanol Production 1978–2012.

the growth of advanced biofuels to meet mandated blending levels. As of 2014, the policy has stimulated the demand and production of biodiesel as an advanced biofuel used for diesel vehicles and added to home heating oil. Yet, it has not stimulated the production of cellulosic biofuels at the rate initially predicted.

Stimulation of electrical forms of renewable energy has followed a similar model of regulation and subsidy. Renewable portfolio standards (RPS) refer to policies that require a percentage of electricity be created using renewable sources. Such policies have been adopted by numerous states and territories (RPS states/territories include WA, OR, CA, MT, NV, AZ, NM, TX, HI, KS, MN, WI, IA, IL, MO, MI, OH, NY, PA, ME, NH, MA, RI, CT, NJ, MD, DE, NC, DC, and Puerto Rico; renewable portfolio goal states include UT, ND, SD, OK, IN, WV, VA, and VT; Database of State Incentives for Renewables and Efficiency, dsireusa.org). Where adopted, the RPS have been strong drivers in much of wind turbine and solar development. Some states include a solar portfolio standard, where their RPS requires a percentage of solar. In these states, large megawatt-scale solar photovoltaic arrays have been installed. These arrays would not be possible without the RPS.

The uncertain future for mandates and tax credit subsidies is a major question for investors and entrepreneurs alike. The question can be asked: Should a business venture be started to support an industry with an uncertain market? There is both great potential and great risk for such endeavors (E2, 2012) Exploration of opportunities for entrepreneurial activity starts with study of the supply chain and determinants of supply and demand for the bioenergy industry (Table 37.1).

OPPORTUNITIES IN BIOENERGY: BIOFUEL VALUE CHAIN

Feedstocks may include any number of "bio" options, including dedicated energy crops, agricultural crop residues, waste streams such as municipal solid waste or manure, and plant oils (virgin or postconsumer). Each feedstock has a cost of production, harvest, storage, and transport. Studying the steps in production of feedstocks will expose opportunities for entrepreneurism.

The biofuel value chain can be described as follows:

Feedstock breeding and genetics → Feedstock production → Feedstock logistics → Biomass processing → Conversion processes → By-product/coproduct utilization → Marketing distribution.

Technology encompassing the genetics of biomass crops and conversion biotechnology are both areas where innovative people and small companies can make important discoveries related to better ways to convert biomass to fuels. Enzymes, catalysts, yeasts, and plant breeding are all areas where biotechnology can impact biofuels production. Engineering related to equipment development and refinement, both at the biorefinery and throughout the supply chain, are needed to solve and improve the system. Engineers and tinkerers alike have developed innovative solutions to equipment problems. These solutions can be marketed.

The distribution of fuel has historically been done through major companies and the gasoline and diesel infrastructure. Increasingly, biofuel companies are playing a part in the distribution systems by becoming blenders and marketing through special pumps that add biofuel choices for the consumer.

By-products and coproducts are extremely important to the bottom line of first-generation ethanol plants. Similarly, the ability to market by-products and coproducts will lead to improved economics of bioenergy facilities. The corn ethanol industry's major coproducts are distiller grains and corn gluten feeds. Additionally, carbon dioxide and postfermentation corn oil are by-products marketed by the

Table 37.1 Determinants of Bioenergy Supply and Demand

Determinants of Demand	Determinants of Supply

1. Tastes or preferences
How people feel about their energy: Do they want bioenergy? Are they willing to pay more for it?
	1. Costs of production
	  a. Change in input prices
	  b. Change in technology increases in quality, efficiency, etc.
	  c. Organization changes leading to increase/decrease in efficiency
	  d. Government policy, including taxes and subsidies

2. Amount and price of related goods *Substitutes*: The higher the price for substitute goods, the higher the demand for this good. If the price of oil rises, oil companies will maximize blending of lower-cost biofuels. *Complements*: As the price of the complement rises, demand for complement falls and so too will demand for the good in question. If gasoline price rises, the demand for cars falls and with it demand for gasoline.	**2. Profitability of alternative goods and supply** For a sugar cane facility in Brazil, if greater profit can be made from sugar than ethanol, the supply of ethanol will decrease and sugar will increase.
3. Income As people's income rises, demand for goods and services rise. As incomes rise, the number of cars and miles driven rise.	**3. Nature, random shocks** These include weather, earthquakes, wars, and industrial disputes.
4. Expectations of future price changes If people expect prices to change in the near future, they will alter their buying habits accordingly. Expectations for higher oil prices could lead to purchases of more efficient cars or flex fuel cars.	**4. Expectations for future prices** If the price of a good is expected to rise, the supplier may hoard stock (reducing current supply) in order to benefit later.
5. Population The size and makeup of the population affects demand. If the population is growing and becoming more affluent, the demand for energy will increase.	**5. Profitability of goods in joint supply** If the supply of ethanol increases, there will be a corresponding supply of distillers grains as one of the coproducts of the other.

corn ethanol industry. It is common to have a separate company partner to capture and sell the by-product. In some cases, companies have formed to add value to the by-product, including pelleting distillers grains for sale as a more dense feed, carbon dioxide capture and use in the bottling industry, and corn oil conversion to biodiesel. By-products are worth a serious look in terms of opportunities for entrepreneurs to add value to a low-value product.

EXAMPLES OF ENTREPRENEURS IN THE BIOFUEL VALUE CHAIN

A short list of real-world examples of entrepreneurs who have discovered, evaluated, and exploited opportunities in the existing biofuel value chain is provided here.

Feedstock breeding and genetics:
- Corn breeders may focus on high amylase corn hybrids that, when used for ethanol production, reduce the cost of production and save on enzyme costs.
- Algae have great potential for bioenergy, so multiple entrepreneurial companies have started algae breeding and production. Early markets have been for specialty chemicals.

Feedstock Production:
- The contract growing of high amylase corn is one example of where farmers can participate today in the bioenergy industry.

Feedstock Logistics:
- Two Nebraska farmers commercialized their invention of cob harvesting attachments for combines and grain carts, which can be used in the harvest and delivery of cob and husk biomass to advanced biofuel facilities.

Biomass Processing:
- An old feed mill in Nebraska has been converted into a wood pellet mill for the production of biomass pellets for a diverse market.
- A group of Missouri farmers started the Show Me Energy Cooperative as an outlet for their switchgrass and other biomass. The cooperative has developed proprietary technologies for blending and pelleting biomass into fuel and feed pellets.

Conversion Processes:
- New gasification technologies are being commercialized by former students stemming from research done at Iowa State University's Bioeconomy Institute.

By-product/coproduct Utilization:
- Private companies have contracted with ethanol facilities to harvest carbon dioxide for use in beverages, dry ice, and other markets.
- Corn oil was underutilized in dry mill ethanol production prior to 2010. Today, many dry mill facilities spin off postfermentation oil and sell it to companies who use it for feed or biofuel.
- The commercialization of biochar (a by-product of the thermochemical conversion of biomass) was started by some eager entrepreneurs. This market may grow fast with second-generation biofuels.

Marketing and Distribution:
- Ethanol has been used as a marketing tool in many midwestern fueling stations. The same is true for biodiesel, where some consumers will select to pay more for homegrown fuels.
- An entrepreneurial startup in Nebraska developed, built, and markets ethanol-powered irrigation engines.

OPPORTUNITIES IN SMALL-SCALE BIOENERGY

Bioenergy has a long history at small-scale and niche markets, with entrepreneurs making their own biodiesel conversion systems, capturing methane from manure and other waste, or using waste wood and other materials for heat.

BIODIESEL

Biodiesel is a good example of small scale bioenergy because of its relatively simple conversion from vegetable oil to methyl ester (biodiesel). A simple chemical reaction with the correct ingredients will yield biodiesel. Farmers have led this trend for homegrown biodiesel by growing their own oil seeds, then processing on-farm to yield biodiesel. The biodiesel produced can be used in tractors and other diesel operations. A major challenge to small-scale production is fuel quality. It is challenging for small-scale producers to attain the fuel quality to meet American Society for Testing and Materials (ASTM) standards. Without meeting ASTM standards, the market for sale of fuel is limited. Some small-scale biodiesel facilities that are still in operation today have grown enough to justify the standards testing or operate on a cooperative model, where oil seeds or oil is delivered by members and fuel is used by members. Other sources of oil for biodiesel could be waste cooking grease, animal fats, and vegetable oils such as canola, camelina, rapeseed, mustard, soybean, sunflower, corn, algae, etc. Each oil source has different characteristics and the processing will slightly different for each based on these characteristics. Biodiesel entrepreneurs who are skilled enough to deal with feedstock diversity may be able to exploit the lowest-cost feedstocks.

METHANE

This section considers biogas methane from anaerobic digestion; other possible sources of biogas methane, such as land fill gas are not covered here. Biogas methane is generated when high-carbon materials are sealed in an anaerobic environment where microorganisms eat the organic materials and give off biogas (predominately methane). Waste streams are common feedstocks for biogas methane production, including manure, food processing wastes, human sewage, and other organic waste. The methane can be used in internal combustion engines to generate electricity, burned for heat in a furnace or boiler, cleaned and added to the natural gas pipeline, or flared (burned in an open flame). Manure makes a good feedstock for a methane digester, and it is common for an anaerobic digester to have manure as its primary feedstock. Subsequently, dairy farms, municipal waste water systems, and swine operations are common locations for such systems. Once installed, the systems produce methane continuously. The methane must be used or flared to prevent the pressure from building inside the system.

- The flaring of methane can be an environmental benefit, taking a one carbon methane molecule and turning it into a carbon dioxide molecule with 24 times less greenhouse gas impact than the original methane molecule.
  - Methane capture and flaring may have value if there is a value for reduction in carbon dioxide equivalent greenhouse gases, such as renewable energy credits or carbon trading.
- Anaerobic digestion also reduces odors and has been used successfully as a way to reduce the odors of a farm or processing facility, allowing the facility to remain in a populated area or expand the operation with fewer objections from neighbors.
  - Farmers can be bioenergy entrepreneurs by using methane technologies to maintain and expand operation with fewer odors. The energy produced may be only a secondary product instead of the primary.

BIOMASS AND WOOD

Wood heating at the residential scale has been common throughout human civilization. Today in the United States, wood heating in homes is most common in heavily timbered parts of the country. In some cases, the technology has changed little. In most cases, individual home owners are cutting and processing their own wood while others may purchase wood from another individual or small business.

Pellet stoves that use wood or other biomass pellets have become more common and have more controls than wood stoves. Wood pellets are uniform, dense, and flow easily for mass transport and packaging. Because wood is renewable and has less greenhouse gas impact than coal or other fossil fuels, wood pellets have also become quite popular in Europe. Markets have developed to ship pellets for commercial and other heating uses.

- Wood harvest
  - Slash, thinning, trimmings, lumber mill waste, factory waste, old pallets, shelterbelt renovation, and dead tree removal, are all streams of wood other than commercial timber harvest.
- Wood processing: conversion of wood to wood pellets or other higher value product
  - The international market for wood pellets has increased greatly in recent years.

COOPERATIVES

Cooperatives are made up of a group of people acting together to meet the common needs of its members. Members share in the ownership and governance of the cooperative. Bioenergy cooperatives have formed to supply bioenergy to members or use feedstocks from members.

Many ventures start as cooperatives to take advantage of shared resources. Some are formed because a venture is too large for one person or family to undertake on their own. The formation of a cooperative can help overcome the initial scale. Others are formed to provide a good or service that is higher quality than individuals could provide on their own. The organization will differ depending on the goals of the cooperative, yet generally a group of like-minded individuals buy in with their money, labor, or other investment to build a business that will help all members. Recalling the developmental motivations, a cooperative's motivation could be environmental*, social welfare**, or entrepreneurial***.

- Examples of bioenergy cooperatives
  - Small-scale biodiesel* (oilseeds delivered, biodiesel is the end product, used by members = environmental motivation)
  - Biomass pellet mill*,*** (members bought shares to pay for pellet mill, members deliver biomass, they get paid for their biomass = environmental and entrepreneurial)
  - Methane digester *,*** (neighboring farms built community digester, all deliver manure to the digester, they share in profits from energy produced, and all benefit from odor reduction = environmental and entrepreneurial)

OPPORTUNITIES IN LARGE-SCALE BIOENERGY

Large-scale bioenergy refers to enterprises with capacities of tens of millions of gallons per year or millions of tons of biomass material throughput. Examples include a multimillion gallon per year biorefinery or megawatt-scale biomass to heat and power facility. Such facilities require large financial backing. Entrepreneurship can participate at this scale, but many times we think of entrepreneurs as small startup companies. For the purposes of this chapter, we have considered how a small company can participate in small- and large-scale bioenergy. At all scales, bioenergy entrepreneurs consider the supply chain and how problems can be solved in new and unique ways. Bioenergy facilities are constantly reliant on technology providers for specialized skills and equipment. These technology providers cover a wide range of disciplines, including heavy equipment, computers, chemistry, analytical equipment, and waste management.

OPPORTUNITIES AT THE PERIPHERY

Many opportunities exist beyond the strict boundaries of the biofuel or bioenergy industries. Some entrepreneurs may be content to work the periphery as a way to participate. Such periphery opportunities may not directly support the biofuel and bioenergy industries but do take advantage of their existence. Businesses that start up to support the personal needs of construction workers during the intense building phases are examples of this peripheral activity. These businesses also rely on the community for patronage after the boom of construction. Such businesses get a leg up with an initial year of excellent business to help retire debt early.

Peripheral businesses started as part of bioenergy booms include the following:

- Restaurants
- Service stations
- Clothing stores
- Banks
- Motels
- Apartment complexes

EXAMPLES OF ENTREPRENEURISM FROM THE ETHANOL BOOM

During the ethanol boom from 2001 to 2007, hundreds of entrepreneurs exploited opportunities related to the ethanol industry. This section explores the diverse ways that entrepreneurs benefited from the expansion of one industry. Between 2001 and 2007, hundreds of ethanol plants were constructed across the Corn Belt (Figure 37.4). Nearly all were built in rural communities near the corn resources and animal agriculture that uses the coproduct feeds. Each facility cost $50–200 million to build and took 1–2 years to construct. Most facilities were constructed by large bioenergy companies, although a few were built with local investments or by cooperatives of farmers and community members. The motivation for these facilities with local investments was mostly social welfare. An ethanol facility located in a community will provide jobs and an increase in corn price locally. The long-term direct jobs at the ethanol facility are 25–50 full-time jobs. These communities also experienced indirect job growth related to the facility, with new businesses supporting the ethanol industry.

In 2008, the University of Nebraska–Lincoln Extension convened a study and interviewed the prime movers and community leaders from five rural Nebraska communities to determine the impact the ethanol facility had on each community and its people. A goal of the study was to use the information to help other rural communities prepare for future bioenergy industry expansion. The results of the study clearly showed how local entrepreneurs stepped in by starting needed businesses or growing existing businesses (Hay et al., 2008).

During the construction phase of each facility, nearly 200 workers must live, eat, and play in the rural community. Local entrepreneurs jumped at the chance to start and expand businesses. The initial

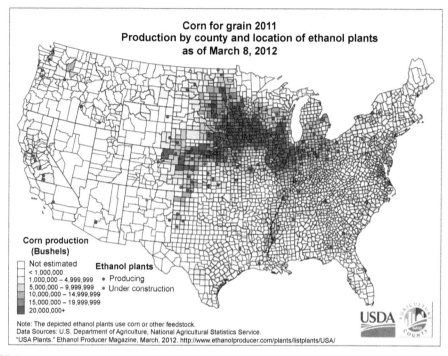

Corn for grain 2011
Production by county and location of ethanol plants
as of March 8, 2012

Corn production
(Bushels)

☐ Not estimated	**Ethanol plants**
< 1,000,000	
1,000,000 – 4,999,999	• Producing
5,000,000 – 9,999,999	• Under construction
10,000,000 – 14,999,999	
15,000,000 – 19,999,999	
20,000,000+	

Note: The depicted ethanol plants use corn or other feedstock.
Data Sources: U.S. Department of Agriculture, National Agricultural Statistics Service.
"USA Plants." Ethanol Producer Magazine, March, 2012. http://www.ethanolproducer.com/plants/listplants/USA/

USDA

FIGURE 37.4

Corn ethanol facilities increased from just a handful in the 1980s to over 200 in 2012. To reach 16 billion gallons of cellulosic biofuel, more than 300 cellulosic facilities (50 million gallons/year) would need to be built.

year of excellent business gave them a jumpstart to return their investment. A good example is the addition of a fast-food restaurant in Albion, Nebraska. There, a local entrepreneur built and opened a regional chain restaurant. The Albion community has 1623 residents and did not have a fast food restaurant prior to construction. The influx of construction workers who heavily patronize restaurants created the perfect time to build. Similarly, the need for housing stimulated the building and remodeling of motels, houses, apartments, and even camping pads. All of the housing choices filled quickly during construction and have added to the quality housing choices available in the community after construction.

After the construction phase, ethanol facilities begin operation. The most common business startup and expansion during the operation phase was trucking. Each ethanol plant needs a steady stream of corn and few have grain delivery by train due to their proximity to the corn production. For example, a 100-million gallon per year facility will require 100 semitruck loads of corn per day to remain in continuous operation. Additionally, the same facility will need 30–90 trucks per day to move distiller grains out to local feedlots. The major increase in trucking stimulated locals to start trucking businesses or expand. For example, in Trenton, NE, a local trucking firm with two trucks expanded to 28 trucks and 22 employees; it now spends more than $100,000 per month in fuel at the local fueling station.

Entrepreneurial opportunities exist in bioenergy—not only in the direct production of the bio-energy, but also in the support of the bioenergy industry and its employees. At times, such opportunities fall in your lap, but in most cases the entrepreneur must have the pioneering spirit to exploit the opportunity.

New businesses started in rural Nebraska during the corn ethanol boom included the following:
* A motel in Osmond, NE
* A fast-food restaurant in Albion, NE
* A drive-through coffee shop in Albion, NE
* Trucking businesses

Expanded businesses started in five towns in rural Nebraska during the corn ethanol boom, including the following:
* Banks with extended hours
* Remodeled motel
* Remodeled rental houses
* Trucking businesses in all locations

CHALLENGES

In many cases, the energy from biomass sources has not become price competitive with the status quo. This is a great challenge for bioenergy entrepreneurs who wish to sell energy into the marketplace in direct competition with fossil fuels. For example, as of January 2014, the gasoline national average of $3.27 per gallon was equal to $28 per million Btu. Ethanol in January 2014 was $2.61 per gallon or $34 per million Btu. The same is true for coal and natural gas compared to renewable electricity (EIA.gov). Changes may occur when environmental regulation pressures the fossil fuel industry. One main reason for the higher cost of renewable fuels and energy is the fact that feedstocks are not low cost. An example of a low-cost biomass feedstock is trees. Even if the trees are free (federal thinning for wildfire prevention is an example), the cost of harvest, chipping, and transportation was 2–4 times higher per million Btu than the price of Wyoming coal in January 2014 (Han, 2007). The basic scenario of expensive feedstocks will likely continue with increasing competition.

Markets for biofuels are immature and uncertain, which will challenge entrepreneurs in their access to financial capital and a stable market for their products. Government mandates are in place for biofuels, yet discussions of changes and repeals creep in each year, leading to uncertainty for investors. This immaturity and uncertainty is a major—if not the major—issue facing the bioenergy industry.

The traits of an entrepreneur—self-starter, dedicated, risk taker, boundless energy, and visionary—all come into play for bioenergy entrepreneurs. These traits are critical to the success of the next generation of bioenergy development. An entrepreneur must have the vision to prepare for what is coming, the strength to risk it in this uncertain market, and the energy to charge ahead.

REFERENCES

Database of State Incentives for Renewables and Efficiency, U.S. Department of Energy. http://dsireusa.org.
Environmental Entrepreneurs (E2), Solecki, M., Dougherty, A., Epstein, B., 2012. Advanced Biofuel Market Report. http://www.e2.org/ext/doc/E2AdvancedBiofuelMarketReport2012.pdf.
U.S. Energy Information Administration, 2013. Annual Energy Outlook. http://www.eia.gov/forecasts/aeo/pdf/0383(2013).pdf.

EIA.gov, Energy Information Administration, Energy production and Consumption. Data online at http://eia.gov.

Han, Han-Sup, 2007. Powerpoint Presentation Getting Biomass to Market: Harvesting Methods and Costs. Humboldt State University. http://ucanr.edu/sites/WoodyBiomass/files/78827.pdf.

Hay, F.J., Corr, A., Teel, D., 2008. Impact of Ethanol Production on a Community or Region. Unpublished. University of Nebraska–Lincoln Extension.

Hofstrand, D., Corn-Ethanol Profitability, AgMRC Renewable Energy Newsletter November/December 2008. http://www.agmrc.org/renewable_energy/ethanol/corn-ethanol-profitability/.

Lin, N., 2008. Bioenergy Entrepreneurship in Rural China [IIIEE thesis]. Lund University, Sweden. Merriam Webster, http://m-w.com.

Schumpeter, J.A., 1934. The Theory of Economic Development. Harvard University Press, Cambridge, MA.

Shane, S., Venkataraman, S., 2000. The promise of entrepreneurship as a field of research. Academy of Management Review 25 (1), 217–226.

Shane, S., Locke, E.A., Collins, C.J., 2003. Entrepreneurial motivation. Human Resources Management Review 13, 257–279. http://faculty.utep.edu/Portals/167/52%20Entrepreneurial%20Motivation.pdf.

University of Maryland Extension, Traits of Entrepreneurs. Online at http://extension.umd.edu/learn/traits-entrepreneur.

INTEGRATED AGROECOLOGICAL TECHNOLOGY NETWORKS FOR FOOD, BIOENERGY, AND BIOMATERIAL PRODUCTION

38

Samuel Gorton[1], Jason McCune-Sanders[2], Anju Dahiya[3,4]

[1] *Gund Institute for Ecological Economics, Rubenstein School of Environment & Natural Resources, University of Vermont, USA;* [2] *Department of Engineering, University of Vermont, USA;* [3] *University of Vermont, USA;* [4] *GSR Solutions LLC, USA*

INTRODUCTION
APPLICATION OF ECOLOGICAL TECHNOLOGIES TO AGRICULTURE AND NATURAL RESOURCES MANAGEMENT

Ecological technologies could be best defined as technologies that transform natural resources into anthropocentric goods and services, while returning beneficial materials or energy to the ecosystem and minimizing pollution. Ecological technologies may provide for the needs of the growing human population with positive societal and ecological impacts relative to current practices, namely restoring land and reducing pollution.

There are three primary types of agroecological eco-technologies:

1. *Natural resource management eco-technologies* such as grazing, tillage, and harvesting systems.
2. *Natural resource conversion eco-technologies* including post-harvest, energy generation, and material transformation processes.
3. *Structural eco-technologies* like living and farming structures and equipment for above technologies.

Despite the best intentions, ecological technologies must be evaluated to determine their overall environmental and economic impacts. Environmental and economic performance assessment of eco-technologies may be accomplished by continuous monitoring of material and energy flows as well as ecosystem health. In order for eco-technologies to be economically viable and ecologically beneficial, data feedback loops may be developed to relate agroecosystem mass and energy flows to agroecosystem health.

This chapter focuses on existing and emerging eco-technologies suitable for a given region, and provides an introduction to eco-technologies and how to model and measure their performance. The state of Vermont is the geographic region used as a case study in this chapter. The remainder of Section Introduction introduces the entwined importance of agriculture and energy to society, describes the concept of sustainable intensification and introduces the ecological design concept of combined food-energy agroecosystems (CFEA). Section The role of industrial ecology in CFEA design and evaluation further describes the CFEA concept. Then Section Measuring CFEA performance in a competitive

Bioenergy. http://dx.doi.org/10.1016/B978-0-12-407909-0.00038-9

world reviews why CFEA systems and the eco-technologies applied in such systems should be rigorously analyzed for effectiveness, namely the competition for resources by humans and the ecological base on which CFEA systems inherently stand. Section PAR for CFEA primary data collection outlines one important method for this analysis, participatory action research (PAR). Last, Sections Analysis of eco-technologies for Vermont-based CFEA and CFEA case study: the Burlington Area Agro-Eco Park concept provide examples of applying these techniques to examine eco-technologies, including the role of CFEA and related eco-technologies in renewable energy production via a case study of the Burlington Area Agro-Eco Park. It is hoped that the reader is encouraged to apply their insight to the development and evaluation of eco-technologies.

AGRICULTURE AND BIOENERGY

Energy, particularly its application to agriculture, is a key factor in the rise of modern human civilizations (Smil, 1994). Farms function by transforming solar energy, soil, and atmospheric nutrients into desirable plant biomass. Agriculture also requires physical work, mainly preparing and maintaining the soil, harvesting and processing crops. This work requires the input of additional, nonsolar energy for constructing and operating farm tools. Historically, this energy input has been derived from humans, draft animals, biomass, wind, water, fossil fuel combustion and electricity from coal and nuclear power.

Pre-industrial civilizations employed human, animal, biomass, wind and hydropower in agro-ecosystems. The equipment used to transform these energy resources into usable work was often inefficient; the industrial revolution and associated scientific advancements improved the efficiency of these traditional tools (Smil, 1994). But, the industrial revolution also introduced new sources of energy, namely fossil fuels, to agroecosystems. This has led to massive increases in the labor productivity of agroecosystems, but a general decline in the ratio of food calories produced to fuel calories expended for production and distribution of those food calories.

Agricultural practices are a key leverage point in addressing global issues of ecological degradation caused by modern industrial practices. While prior to the advent of advanced energy technologies, farms were substantially material and energy self-sufficient (Smil, 1994); today's industrialized food system is heavily dependent upon external inputs with significant economic and ecological impacts on and beyond the farm (Pimentel and Pimentel, 1996; Conway and Pretty, 2009). In response to the specialized nature of industrialized agriculture, the ecological design concept of CFEA seeks to re-integrate food, feed, fiber, and energy production activities in the working landscape (Porter et al., 2009). This re-integration may be accomplished by the application of agroecological technologies, discussed further in Section Analysis of eco-technologies for Vermont-based CFEA.

Over the next 50 years and beyond, agriculture—including the management of forests and fisheries—faces extraordinary demand pressure due to four primary factors: (1) human population growth, (2) increasing affluence and awareness of food security issues, (3) degradation of productive natural ecosystems and agroecosystems (e.g., fisheries depletion, soil erosion, crop failure, etc.) due to human mismanagement and natural disaster, and (4) perceived ecological and human benefits of biomass based fuels and products (e.g., lower greenhouse gas (GHG) emissions) compared with fossil fuels (Godfray et al., 2010). To address these pressures, Godfray et al. and others (Royal Society, 2009) describe the practice of "sustainable intensification." Sustainable intensification refers to the production of more food or other agricultural products from the same area of land while reducing the requisite environmental impacts (Godfray et al., 2010).

THE ROLE OF INDUSTRIAL ECOLOGY IN CFEA DESIGN AND EVALUATION

One form of sustainable intensification is the concept of CFEA, in which multiple products (including food, feed, and energy crops) are produced from an integrated agro-industrial system. One example described by Porter et al. is an experimental farm managed by the University of Copenhagen, Denmark. While operating under CFEA production since 1995, this CFEA produced substantially more food (wheat and barley), pasture fodder (clover and grass), biomass energy crops (fast-growing willow hedgerows) and ecosystem services (pollinator habitat, GHG regulation, etc.) with fewer nonrenewable inputs than an average conventional agroecosystem (Porter et al., 2009).

Such sustainable intensification concepts represent the synergy of agroecology and industrial ecology (IE) principles. The field and practice of agroecology focuses on the application of ecological principles to study and manage the interactions between agroecosystem components, such as such as plants, animals, soils, climate, and humans for the purposes of producing food, fiber, energy, and other agricultural products (Wezel et al., 2009). Thus, while motivated toward improving the ecological and socioeconomic impacts of agriculture, agroecology is principally concerned with farm-based activities and cycles in the greater agro-industrial system.

In order to fully analyze impacts, agroecological researchers, farmers, and service providers must adopt IE principles and tools. IE is based on the principle that—particularly in today's globalized economy—every industrial activity, including agriculture, is linked to thousands of other transactions and activities and to their associated environmental impacts (Graedel and Allenby, 2003). Furthermore, like the agroecologist who seeks to foster beneficial interactions between agroecosystems components, the industrial ecologist is concerned with the material, energy, and other exchanges between firms in the economy.

While industrial ecologists have largely focused their efforts on manufacturing sectors such as personal electrics, consumer goods, and automobile production activities, agricultural system analysts have employed IE simulation modeling tools to calculate energy return on energy investment (EROI) (Cleveland, 1995; Pimentel and Pimentel, 1996; Pimentel, 2009) or perform a life-cycle assessment (LCA) (Andersen, 2002; Pizzigallo et al., 2008; Peters et al., 2010) for various agro-industrial systems. While these assessment tools only partially represent impacts of a given industrial system, the data upon which IE assessments are based may be used to generate an array of metrics detailing the ecological and economic performance of agroecological practices and technologies. Such data and associated metrics can provide important feedback to decision-makers, including agricultural entrepreneurs, researchers, policy-makers and technical consultants working to develop and improve appropriate eco-technologies.

MEASURING CFEA PERFORMANCE IN A COMPETITIVE WORLD

The CFEA concept seeks to balance the material needs of society for food and energy goods with the conservation of the natural resource base from which these basic goods are derived. With the growing numbers and affluence of the global population, the productivity of CFEA systems is a key factor in their overall socio-environmental implications. How to balance these needs and continuously improve research methods is critical and requires methods of continuous evaluation and reassessment.

Biomass is currently the only alternative energy resource that may provide a direct replacement for every major form of conventional energy—electricity, heating fuel, cooking fuel, liquid transport fuel, and gaseous fuel (Green and Byrne, 2004). Demirbas described 20 technologies for rural-based energy production from biomass (Demirbas and Demirbas, 2007). Biomass may even replace energy-intensive construction materials and chemicals. However, biomass and the resources required to produce it have competing values as energy, food, and ecosystem services. Crop residues, in particular, may be used for energy or animal fodder (Green and Byrne, 2004; Devendra and Leng, 2011; Devadas, 2001). Thus, examining the material and energy flows and transformations in rural agricultural communities must be central to research methods for farm-based energy systems.

Whether a technology is appropriate in a given context is determined primarily by the social and political climate and the resources available. Resources include environmental (the physical landscape, climate, biology, and nutrients) and human-provided (money, labor availability and costs, machinery, fuel, and electricity) inputs. Unequal geographic distribution and lopsided political power have created huge disparities in agricultural energy sources, practices, and technologies. In rural areas of Nepal and India, for example, nearly 90% of household energy needs are met by fuel wood or other traditional biomass resources (Malholtra et al., 2004; Salerno et al., 2010; Devadas, 2001). As recently as the late 1990s, Devadas found that in agricultural districts of Tamil Nadu, India, the primary energy sources for field traction, transport, and irrigation are draft animals and humans (Devadas, 2001). Additionally, these farmers rely on energy-intensive inorganic fertilizers to boost crop productivity.

Thus, a technology appropriate for the rural areas described above might differ dramatically from one appropriate for the developed world where, relatively speaking, labor is expensive, energy is cheap, and motorized transportation is widely accessible. Analytical methods that employ PAR build on dialogue with local people and stakeholders to measure the performance and appropriateness of technologies.

PAR FOR CFEA PRIMARY DATA COLLECTION

PAR refers to a family of research methods, which incorporates dialog with lay participants for education and empowerment (Chambers, 1994). In Nepal and India, PAR has been applied to rural farm and energy systems. As interest in farm-derived biomass energy sources for mitigating GHG emissions grows (Green and Byrne, 2004), PAR approaches may be useful for generating primary data to evaluate the agroecological and socio-economic trade-offs of alternative energy systems.

Incorporating stakeholder interviews, site visits, and surveys with quantitative data collection and mathematical models, some researchers have employed PAR methods for evaluating farm-based biomass production (Vayssieres et al., 2007; Devadas, 2001; Devendra and Leng, 2011) as well as farm and household energy consumption (Devadas, 2001; Malholtra et al., 2004; Salerno et al., 2010). These researchers, while not explicitly stating the agroecological nature of their work, have intently applied "whole systems" methods that view human systems and actions as part of complex interactions with their biophysical environment. The products of their research provided actionable information, based on farmer input, for farm-based biomass and energy management.

Household energy consumption in rural areas has been a major focus of international development research. Traditional cooking practices generate indoor air pollution and GHG emissions while contributing to deforestation. To study these impacts in Nepal's Sagarmatha National Park and Buffer

Zone (SNPBZ), Salerno applied a participatory modeling approach (Salerno et al., 2004). Structured interviews were used to collect energy demand and fuel source data. Devadas employed a similar approach for collecting household- and farm-level data of important agroecosystem material and energy flows (Devadas, 2001). In the SNPBZ study, a qualitative model of management levers, performance indicators, system concepts, and connections between them was developed with input from SNPBZ citizens. This qualitative model was then translated into a spreadsheet-based mathematical model for studying interactions between energy management systems, indoor air quality, and forestry practices. While the model provided an adaptive management tool for improving the economic and ecological vitality of the SNPBZ, it was unclear if and how it accounted for performance substitutability and cultural acceptability of proposed energy alternatives for cooking food and heating water.

Devadas's participatory dynamic modeling research focused primarily on modeling existing biomass transformation methods (Devadas, 2001). A linear programming model was developed using energy, food, fodder, and fertilizer values of biomass resources. The researcher sought to incorporate the following six categories of rural energy research into the model: (1) household energy consumption, (2) energy consumption in agriculture, (3) energy interactions in the rural systems, (4) assessing economic feasibility of technologies in the rural systems, (5) impact of technology on rural systems, and (6) rural energy planning at the micro level. Special emphasis was placed on collecting primary data from farms and villagers. Major results of the model are a series of tables describing the financial, material, and energy flows of an actual Indian farming community. However, in addition to tabulation, this data might best be represented to the broader community in a series of descriptive block flow diagrams.

ANALYSIS OF ECO-TECHNOLOGIES FOR VERMONT-BASED CFEA

As a first step in applying IE and PAR methods to Vermont-based CFEA evaluation, potentially appropriate eco-technologies must be uncovered and described in context. Aside from hosting the authors' academic institution, Vermont was chosen as the geographical focus of this chapter due to the state's demonstrated progress—backed by social pressure and political will—toward concrete goals of renewable energy generation and energy self-sufficiency. Specifically, Vermont has committed to the 25%-by-'25 challenge for the state to generate 25% of its energy from in-state renewable sources such as biomass, solar, and wind (Spring Hill Solutions, 2008). While this commitment is not legally binding, Vermonters have recently lobbied their legislature to toughen these standards and targets. Based on the social and political climate, it is appropriate to consider eco-technologies for achieving goals of renewable energy generation and energy self-sufficiency. A similar climate is evident in Vermont's food system, with consumers and stakeholders interested in ecological, economic, and other benefits of increased local food consumption (Schattman, 2009; Kahler et al., 2011).

With the social and political climate ripe for appropriate eco-technologies to address Vermont's food and energy self-sufficiency goals, what resources are available for the state to actually fuel and feed itself through domestic production activities? Vermont has 21% of its land area in agricultural production, with some counties approaching 50% (Kahler et al., 2011). Additionally, over 70% of the state's land acreage is in forest cover (Spring Hill Solutions, 2008). Based on these facts, the state's 25%-by-'25 plan projects that by 2025, Vermont could obtain over 20% of its energy from energy crops, farm residues, and woody biomass, but achieving these goals, the researchers admit, is based on untested or economically infeasible technologies such as cellulosic ethanol and algae-based biodiesel (Spring Hill Solutions, 2008). There are also nonrenewable energy costs associated with producing biofuel and bioenergy products, such as those associated with cultivating, harvesting, processing, and transporting biomass feedstocks (Mears, 2007).

Applying an 8:1 energy output to input ratio as found by a recent energy analysis of Vermont-based organic biodiesel production (Garza, 2011) to all biomass-based energy technologies for a first approximation, producing 20% of Vermont's renewable energy from biomass would require that 2.5% of the state's total energy demand come from nonrenewable energy sources for domestic production of energy. A similar calculation, with appropriate assumptions, could approximate the energy input required to produce various food, feed, fiber, and forest products. And, of course, these production activities often have nonenergy implications as well, including impacts such as soil erosion and compaction, nutrient runoff and leaching, or biodiversity loss, and the general health of humans, livestock, soil and other agroecosystem components.

Finally, there is ongoing debate about the ecological benefits of biomass-based eco-technologies, not to mention skepticism and uncertainty regarding their economic viability. Recently, researchers in Massachusetts found that forest-based biomass electricity may be more carbon-intensive (CO_2 equivalent emissions per unit of energy generated) than coal-fired power plants (Manomet Center, 2010). Methods that evaluate a wide range of ecological and economic factors will support research and development efforts to improve eco-technologies. Such methods may be based upon the inherent material and energy flows of eco-technologies.

KEY ASPECTS OF REGIONALLY APPROPRIATE ECO-TECHNOLOGIES

While any given technology might have several variations in terms of its specific economic and ecological performance[1]—the basic resource flows, both inputs and products—are similar. Four key aspects of eco-technologies are:

1. *Material flows* include biomass crop inputs; food, fuel, fiber, and nutrient products and associated by-products; fertilizer and pesticide applications; and bio-catalysts (enzymes, minerals, seed cultures). Specific emphasis is placed on agricultural nutrient (nitrogen-N, phosphorus-P, potassium-K, carbon-C), food nutrient (calories, fat, protein, carbohydrates, micronutrients) and potentially toxic flows.
2. *Energy flows* are heat and power (electrical or mechanical, including traction) demand or generation. Particular attention is paid to the ultimate fuel source and form of both fuel and energy flow.
3. *Ultimate and proximate means* incorporate core upstream infrastructure for raw material extraction and manufacturing, including associated co-product (i.e., other industrial inputs, wastewaters, and toxic discharges) management systems.
4. *Labor issues* relate to the duration, repetition, and intensity of work; general occupational and public safety, such as exposure to hazardous chemicals and interaction with heavy equipment and livestock; potential income levels (and thus labor costs) and training requirements of the eco-technology workforce.

---

[1]Different variations of a given eco-technology might be designed by competing firms, as is the case with commercial-scale processing facilities designed by various engineering firms.

According to these aspects, three appropriate eco-technologies for Vermont are examined below. A more comprehensive list of available appropriate eco-technologies for Vermont agroecosystems is provided in Table 38.1.

Management intensive grazing

Management intensive grazing (MIG) is a holistically integrated land management and animal husbandry practice that seeks to maximize pasture forage quality and productivity through the judicious application of livestock grazing pressure and disturbance. Both pasture and crop land are ultimate means required for MIG production systems in Vermont since pasture-based production is virtually impossible in cold and wet periods, and the animals' feed must be supplemented with machinery-harvested crops. Pasture-based production of animal products also requires livestock fencing and watering infrastructure.

The MIG eco-technology may be applied to generate any range of animal products, including meat, milk, fiber, and eggs. Pasture-based food products have been shown to contain higher levels of essential fatty acids and amino acids than those derived from confined animals. While the animals are on pasture, the farmer may spare the expense of harvesting, storing, and distributing feed to them, but must instead have pasture management skills and time to arrange and repair fencing and watering systems. Livestock waste management costs in MIG systems are likely to be substantially lower than those for confinement operations. However, a Vermont-based MIG farmer must still maintain farm structures and equipment to harvest, store, and manage feed for wet and wintry conditions when the animals must be off pasture.

A key aspect of the MIG system is that significant primary inputs are sunlight and fresh water, which are directly utilized in the production system. Such natural inputs are common for *natural resource management eco-technologies* like MIG and offer the opportunity for nature to perform functions, such as harvesting and converting solar energy into live animal biomass and fresh fertilizer and ultimately new soil organic matter that conventional systems, based on nonrenewable inputs, simply cannot replicate. As such, MIG is incorporated into agroecosystems as an eco-technology to build, restore, and maintain soils in a highly productive, low external input system.

A major challenge for the MIG eco-technology is the lower yield per acre obtained by raising livestock on natural, unrefined, fresh feed as opposed to the totally mixed rations (TMR) of fresh and processed feeds common in conventional livestock production systems. There are certainly trade-offs between conventional and MIG systems in the pasture area, cropland area, and energy resources required for meat, milk, and fiber production over an animal's lifetime. Such details are context specific and thus require simulation modeling and accounting to quantify these trade-offs. Generally speaking, confinement systems are more energy intensive in terms of on-farm inputs of fuel and electricity due to a higher reliance on cropland and farm structures relative to MIG systems.

The metabolic activity of animals generates substantial quantities of heat and gaseous emissions. There is on-going debate regarding the life-cycle GHG (CO_2 equivalent) emissions of pasture- and confinement-based livestock production (Peters et al., 2010). Interestingly, some agricultural innovators have developed farm structures, which integrate livestock and plant production. In such bioshelters, small livestock, such as laying hens and rabbits, generate CO_2 and heat supplementary flows for enhanced, year-round solar-based greenhouse production (Edey, 1998). Finally, livestock production generates substantial quantities of biomass residues including manure, bedding, processing wastewater, and inedible or undesirable animal flesh and bones which eventually, through further energy-dependent processing and management, could be returned to crop and pasture lands as fertilizer.

Table 38.1 Nonexhaustive List of Possible Appropriate Eco-Technologies for Vermont Agroecosystems

Eco-Technology	Technology Type	Key Proximate and Ultimate Means	Key Inputs	Key Outputs	Key Co-products	Vermont Example
Anaerobic digestion	NR conversion	Confined animals; transportation networks; heavy equipment; crop land; fresh water; petrochemical processing and other manufacturing sectors	Manure and bedding; food, feed and other organic residues; heat and power; heavy equipment and fuels; farm structures	Renewable natural gas[a], heat and power[a]; fertilizer[b]; bedding or peat moss replacement[a]	Milk, meat	Foster Brothers' Farm
Biodiesel processing	NR conversion	Transportation networks; heavy equipment; crop land; fresh water; petrochemical processing and other manufacturing sectors	Vegetable oil; methanol; sodium hydroxide; fresh water; heat and power	Biodiesel; glycerol[a]; wastewater[b]	Seed meal; personal and resource mobility; transportation greenhouse gases (GHGs)	State Line Biofuels
Alcohol processing	NR conversion	Transportation networks; heavy equipment; crop land; fresh water; petrochemical processing and other manufacturing sectors	Sugars and starches; fresh water; enzymes and cultures; heat and power	Ethanol; spent solids[a]; wastewater[b]	Equipment could be used for food, fuel and medicinal fermentation and distillation; personal and resource mobility; transportation GHGs	State Line Biofuels
Dairy processing	NR conversion	Transportation networks; heavy equipment; crop land; fresh water; petrochemical processing and other manufacturing sectors	Milk; fresh water; enzymes and cultures; heat and power; sterilization chemicals	Cheese; whey[a]; wastewater[b]	Skim milk or cream products; consumer organic residues	Cabot Creamery Cooperative
Woody biomass gasification and combustion	NR conversion	Forest land; harvest equipment; petrochemical processing and other manufacturing sectors	Woody biomass; heat and power; harvest equipment and fuels	Renewable synthesis gas[a], heat and power[a]; wood ash; air emissions[b]	Forest products; forest life-cycle ES	McNeil Generating Station
Woody biomass pyrolysis	NR conversion	Forest land; harvest equipment; petrochemical processing and other manufacturing sectors	Woody biomass; heat and power; harvest equipment and fuels	Renewable synthesis gas[a], heat and power[a]; biochar[a]; bio-oil[a]; air emissions[b]	Forest products; forest life-cycle ES	Green Fire Char

Integrated aquaculture	NR management	Fish hatcheries; crop land and fisheries; fresh water; petrochemical processing and other manufacturing sectors	Fish feed; heat and power; seeds; aquaponic media; (sun)light	Fish meat[a], edible and inedible plants; fertilizer[b] (fish sludge)	Inedible or undesirable fish biomass; consumer organic residues	Carbon Harvest Energy (R, D & C)
Integrated bioshelters	Structural	Seed crop land; glazing material; petrochemical processing and other manufacturing sectors	Heat and power; seeds; (sun)light; fresh water	Edible and inedible plants	Biomass residues for compost; consumer organic residues	Wild Branch Medicinals
Management intensive grazing	NR management	Partial animal confinement; fresh water; crop land; young stock; transportation networks; petrochemical processing and other manufacturing sectors	(Sun)light; fencing; farm structures; fresh water; harvested feed	Meat[a]/milk[a]/eggs[a]; animal heat and air emissions	Inedible or undesirable carcass biomass; consumer organic residues; young stock	Carbon Farmers of America
Composting	NR conversion	Partial animal confinement; organic residue management systems; transportation networks; petrochemical processing and other manufacturing sectors	Heavy equipment and fuels; organic residues; woody biomass and other amendments	Compost; heat[a]; air emissions	Food, feed and other compost-fertilized products	Diamond Hill Custom Heifers
Commercial kitchen or food micro-processor	NR conversion	Crop land; packaging; fresh water; transportation networks; petrochemical processing and other manufacturing sectors	Raw and pre-processed foods; fresh water; heat and power; heavy equipment and fuels	Processed foods; organic residues; wastewater[b]	Consumer organic residues; other consumer residues	Vermont Food Venture Center
Nutrient recovery	NR conversion	Confined animals; transportation networks; heavy equipment; fresh water; crop land; petrochemical processing and other manufacturing sectors	Slurry manure or anaerobic digestate; power; heavy equipment and fuels	High N fertilizer[b]; high P fertilizer[b]; bedding or peat moss replacement	Edible and inedible plants	Vermont Organics Reclamation

[a] May require additional equipment to generate a usable or marketable product.
[b] Possible toxicity concern for humans and wildlife.

Anaerobic digestion

Anaerobic digestion (AD) is a process through which organic materials are converted into biogas fuel and stable fertilizer in an oxygen (O_2)-free bio-reactor. Feedstock inputs available in Vermont include animal manures, primarily from dairy cows, as well as organic residues from food scrap collection and dairy processing. Such inputs are ultimately dependent upon heavy machinery and the associated fuels and infrastructure.

The biogas product is composed of methane (60–70% by volume), carbon dioxide (30–40%), with trace amounts of water vapor, hydrogen sulfide, ammonia, and other volatile compounds. These trace compounds may necessitate some form of gas treatment prior to efficient utilization for energy or other applications. High-quality (low contaminant) biogas may be used as a substitute for natural gas or propane gas fuels for heating water, cooking, or processing food, feed, and fuel products. On-farm usage of biogas requires a storage and distribution system along with minor pressurization (electro-mechanical energy) to transmit the fuel from the bio-reactor to the point of use. In order to maximize energy and resource efficiency, off-farm transport of biogas must incorporate CO_2 removal and compression to high pressures. No such renewable natural gas (RNG) systems exist in Vermont, and few are operating in all of North America.

A more common application for biogas is in farm-based combined heat and power (CHP) generation. These renewable energy generation systems typically operate between 20% and 30% efficiency from biogas to electricity, with excess heat captured to heat the bio-reactor and other farm heat sinks (Goodrich, 2005). The anaerobically-treated slurry by-product of AD technology may be further processed into liquid fertilizer and solid bedding or potting material via mechanical screw separation and dewatering. The value of such by-products is critical to the financial viability of AD systems (Goodrich, 2005). And, given the substantial resources invested into the animal confinement, bio-reactor, and CHP structures, as well as into maintaining and operating heavy equipment for cropping, incorporating the value-added solid–liquid separation process is advantageous.

Integrated bioshelters

Integrated bioshelters (IB) are greenhouse structures designed to optimize light capture, thermal regulation, and gas cycles for plant (vegetable, ornamental, perennial, tree/shrub, etc.) production. Ultimately, IB structures are dependent upon the manufacture of advanced glazing and other unique structural materials. Key inputs into IB systems include sunlight, seeds, fresh water, and heat and power. Like both MIG and AD, a large volume of these inputs may be generated locally, minimizing input transportation costs. However, this cost minimization must be balanced against the resource costs of constructing and operating the bioshelter system.

Year-round greenhouse production in Vermont is limited by extremely cold temperatures and low levels of light exposure in the winter months. In the wintertime, most locally available vegetables are root vegetables, cold-stored from the fall harvest; leafy green vegetables are shipped in from warmer climates with accompanying freshness and energy impacts. With insulation and supplementary heating, production of certain low-light crops such as vegetable shoots and spinach could be contained in IB systems designed for year-round production. There is also an opportunity to heat and power IB systems with renewable energy, such as that from anaerobic digester technologies. Such eco-technology integrations are revealed through the analysis summarized in Table 38.1, but require additional study in order to develop appropriate solutions to Vermont's food and energy self-sufficiency objectives.

This section analyzed eco-technologies that may be appropriate for achieving Vermont's food and energy self-sufficiency goals based on the state's forest and agriculture resources. All of the eco-technologies analyzed are currently in development or operation within Vermont, but are ultimately based on significant nonrenewable inputs and means while generating potentially detrimental ecological impacts. To achieve competing, yet interconnected, goals of human well-being and ecological health, this analysis will support simulation modeling research efforts to evaluate agro-industrial eco-technologies.

CFEA CASE STUDY: THE BURLINGTON AREA AGRO-ECO PARK CONCEPT

A toolkit of eco-technologies is constantly being added to and improved upon by innovative ecological designers in the state of Vermont and beyond. Vermont is well-suited for the application of MIG, AD, and IB systems as well as composting, gasification, and pyrolysis processes in CFEA. These eco-technologies may be integrated directly with sustainable agriculture and forestry operations to form low-input, rich-output enterprises.

The Burlington, Vermont area possesses the natural resources required to support a grass-based dairy-centric Agro-Eco Park. There is sufficient demand in the area for sustainably produced food and an emerging market for products generated from low-input, off-the-grid CFEA. The Agro-Eco Park proposed herein describes a network of ecological unit operations that internally cycle mass and energy to generate exports of fresh and preserved foods, fresh and cultured dairy products, and garden-scale quantities of fertilizers. In order for the proposed park to become a reality, a dynamic model, derived from chemical engineering practices and principles, must be developed and applied to iterative testing of various design scenarios. An early stage conceptual block flow diagram of ecological technologies and associated material and energy flows in presented in Figure 38.1.

LAND RESOURCES

An agro-eco park is essentially a modern farm operating virtually independently from conventional energy sources, namely fossil fuels. Such a park consumes minimal external inputs, instead providing the large majority of raw materials from within its own boundaries. While operating in this way, the park must be economically viable, providing its laborers with a livable wage, food, and shelter. In order to achieve such self-reliance, the farm will require substantial land resources for the *natural resource management eco-technologies* to generate sufficient food, forage, and energy crops. Additional land is required for *natural resource conversion eco-technologies* to process biomass (food, feed, or fiber) materials into saleable products, while collecting and transforming the remaining biomass residues into usable forms. The basic land resources defined for a Burlington Area Agro-Eco Park are: grazing paddocks for livestock; grass and small grain fields for harvesting wintertime livestock feed; forested lands for building materials, wood chips, and sawdust; and flat, contiguous lands for structures and raised bed gardens.

DAIRY-CENTRIC PROCESS

Given these land resources, the processes that operate within the park may be defined to achieve economic viability. First, since the park has sufficient land resources to support a livestock herd, a

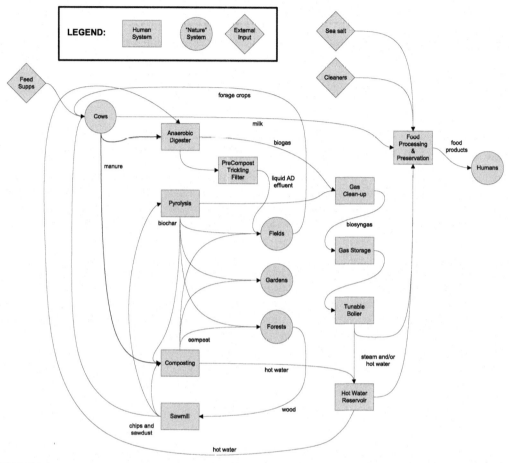

FIGURE 38.1

A block flow diagram of the Burlington Area Agro-Eco-Park.

dairying operation was chosen as a primary activity. With a dairy operation as its centerpiece, the park produces valuable, energy-laden materials—milk, meat, and manure. Milk may be processed directly into bottles for sale as well as converted into butter, yogurt, cheese, sour cream, and buttermilk. Meats may be processed into conventional cuts or by smoking and curing into various charcuteries with farm-produced herbs. Manure may be passed through an anaerobic digester to sequester carbon emissions in the form of biogas for energy production. Manure may also be mixed with woody biomass wastes from an in-park sawmill, food, and forage wastes to produce compost.

COMPOSTING

Compost production is vital to low-input farming, providing stable nutrients and erosion protection for crop fields, gardens, and even forests. Chickens may forage the top of active compost piles, producing eggs and meats. The composting process may also provide thermal energy for food processing (and backup heat for greenhouse production) and finished compost has potential for gas and liquid biofiltration. In the proposed park system, finished compost mixed with alkaline wood ash or biochar may be used to remove ammonia and hydrogen sulfide from biogas and syngas. This high-sulfur compost might have an application in berry production or other sulfur-demanding agricultural processes. Also, AD effluent may be passed through a matrix of finished compost and wood chips/sawdust to uptake nitrogen and phosphorus and filter out particulates. The matrix would then be mixed into new compost batches. The liquid portion of the AD effluent may be spread on crop fields while some might be diverted for aquaponic operations for algae production. A vermicompost operation might provide additional feedstocks for aquatic life and vermicompost for garden and crop production.

GARDENS, FORESTS, AND AQUAPONICS

Garden, forest, and aquaculture management may provide sufficient value-added production for the park, thus enhancing its economic viability. Outdoor, raised-bed gardens may produce an abundance of fresh vegetables for immediate sales in summer and storage/preservation (powered by waste heat, see below) for winter sales. Nurseries based on IB systems may produce saplings to replenish forest harvests and seed orchard operations. These systems would be fueled primarily by compost (which could also be sold to local gardeners). Solar aquaponics systems, designed into living quarters, could provide heat and grow medicinal and culinary herbs as well as fish. The main inputs would be liquid AD effluent and worms from vermicompost.

WASTE HEAT UTILIZATION

Waste heat could be produced consistently from active compost piles and supplemented with the biogas or syngas fuel from AD and thermochemical reactors, as well as solar thermal generation. Some of this heat will be required to heat the digester, but there may be significant waste heat remaining. The heat may be stored in hot water reservoirs to be used in food processing and preservation. The main food processing needs will be for control of temperature-sensitive processes like yogurt, cheese, and sour cream production. An important balance may need to be struck between using gaseous fuels for heating water or directly for heating or cooking food products. For food preservation, waste heat from compost can be upgraded to steam with an on-demand steam generator fueled by the biogas or syngas. Steam can be used in an autoclave to sterilize returnable glass jars for storing dairy products and canned fruit and vegetable products. These processes can provide significant value-added production with off-the-grid, renewable energy management practices.

ADDITIONAL ECO-TECHNOLOGIES FOR THE AGRO-ECO PARK

A thermochemical reactor for gasification and pyrolysis requires woody biomass feedstocks. These would be provided by a managed forest landscape and a sawmill operation. Grass crops may also be added to the thermochemical process. Woody biomass consumption may be partitioned between composting and pyrolysis processes. Mechanical energy is required for the sawmill. There will be a substantial need for transportation within the park, which might be provided by a team of workhorses who forage with the livestock. Baseload electrical energy might be provided by a mixture of wind and solar power, coupled with energy storage infrastructure. There will also be significant human waste produced by park laborers. This waste may be composted separately for bioenergy crop production. An interesting opportunity for the human waste stream is a constructed wetland of cattails that may be harvested for alcohol distillation, providing liquid fuels for harvesting equipment. Finally, flue gases from stationary combustion processes (steam generation) might feed algae for aquaponics operations.

MODELING OF UNIT OPERATIONS FOR ECONOMIC PROJECTIONS

The proposed Agro-Eco Park design is a highly interconnected network of agricultural, manufacturing, and ecological processes. Due to the integrated nature of the park, a dynamic model of mass and energy flows is required to assess the economic viability of the venture. One way to approach park modeling is to consider the main processes (fields, gardens, forests, AD, composting, pyrolysis, sawmill, etc.) as unit operations. Each of these unit operations may be modeled based on their inherent inputs and outputs. With a model of each unit operation, connections may be made so that the output of one unit operation generates the required input of another, thus generating a second layer of modeling.

This modular approach to process modeling is applied in computer-based process simulation modeling software, used by chemical engineers to model complex chemical processes including petrochemical refineries. Such simulation modeling programs provide models for reactors, heat exchangers, distillation towers, boilers, compressors, vapor–liquid separators, and other equipment typically found in chemical plants. Networks of these unit operations may be modeled in a graphical user interface (GUI) to generate a process flow diagram for a chemical plant. Materials flowing between the unit operations are defined along with equations of state to model their behavior at different temperatures and pressures. Utility (heat, cooling, electricity) consumption or generation models may also be incorporated into such models.

In the case of agro-eco parks, a simulation modeling software program would be valuable for modeling material properties and how they are affected by the various unit operations laid out for the park. The ultimate goal for the park is to generate products from the agroecosystem that can either be used internally or sold to customers. Thus, the simulation modeling program must be able to define and calculate material and energy properties so that unit operations may be connected, and flow rates of system outputs may be estimated.

A software program must also be capable of accounting for seasonal variations in biomass production in order to match with baseline demands for animal feed, human food, and process energy. For instance, a substantial amount of energy would be expended during harvest times. This energy may be derived from human labor, but might also include tractors and implements. Most of the soil cultivation would take place in the gardens as the dairy would be grass-based. The model would also need to include flywheel operations developed by farmers and engineers to handle seasonal variations in biomass productivity. Canning operations would coincide with year-end gluts of produce. Thus, a substantial amount of thermal energy may be required to prepare the foods for canning. The model may be used to simulate means of matching such demands with energy production and storage unit operations. Ultimately, there would be periods of baseload and peak economic activity, and that cycling would need to be matched with the mass and energy resources and storage capabilities of the system.

CASE STUDY CONCLUSION

There are several ancient and modernized eco-technologies available to construct a hypothetical Agro-Eco Park for the Burlington area. Defining available land-based resources is the basis for park design. The park proposed in this section contains field, forest, pasture, and water resources to support dairy and light sawmilling operations. These are central components of the system, providing basic raw material outputs that propagate through highly integrated processes, generating large volumes of required dairy and sawmilling inputs. Integrated processes include woody biomass pyrolysis and gasification, AD, composting, solar thermal and electricity generation, windmills, dairy processing, meat processing, food preservation and canning operations, gardens and nurseries, aquaponics, constructed wetlands, and various food, fuel, and energy storage systems. Manure, milk, woody biomass, compost, liquid manure, medicinal crops, gaseous fuels, and flue gases are the primary internally cycled materials. Exported materials include raw and cultured dairy products (cheese, yogurt, sour cream, milk, butter, and buttermilk), fresh and smoked meats (fish and beef), fresh and preserved fruits and vegetables, compost, saplings, and possibly electrical or thermal energy. Vital external inputs are rock powders, livestock feed supplements and medications, sea salt, lubricants, and food-grade cleaning chemicals. In order to determine the economic viability of the integrated system, a dynamic modeling approach is required to account for seasonal variations in mass and energy production. A computer-based modular modeling program based on chemical engineering fundamentals and unit operations may provide the necessary interface to generate data and design scenarios for CFEA performance evaluation.

CONCLUSION

This chapter has elucidated the potential application of IE principles and evaluation techniques as well as PAR methods in CFEA system design, implementation, and management. It is important to recognize the relationships between the three types of agroecological eco-technologies presented in Section Introduction. *Natural resource management eco-technologies* are the systems, which generate the raw materials for *natural resource conversion eco-technologies*. Generally speaking, *natural resource conversion eco-technologies* are processes to transform biomass and other natural resources into valuable products. *Structural eco-technologies* are the equipment and structures utilized in natural resource management systems and natural resource conversion processes.

By defining potential CFEA-appropriate eco-technologies in this way, agro-industrial ecologists may uncover existing and possible synergies based on inherent material and energy flow relationships. Agro-industrial ecologists are advised to incorporate PAR methods into their efforts at uncovering agro-industrial synergy. In the authors' experience, farmers and agricultural professionals provide critical and valuable feedback on CFEA system concepts. Furthermore, these important stakeholders are vital nodes in the data feedback loops necessary to evaluate current and emerging agroecological eco-technologies against economic and environmental metrics.

Despite their developers' best intentions for ecological impact minimization, eco-technologies must be measured against their conventional alternatives based on the efficiency (numerator/denominator) with which they consume land, water, energy, money, and other ultimate and proximate means (denominator) for productive ends (numerator). These fundamental mathematical metrics provide crucial—yet, individually incomplete and collectively only partial—information for evaluating eco-technology solutions to a given region's energy and food self-sufficiency goals. Basically, this chapter represents a step toward "uncovering" the agro-industrial symbiosis opportunities in Vermont's food, fiber, forest production, and renewable energy generation sectors. From an IE perspective, such analyses are a key first step in the sustainable development process (Chertow, 2007).

The preceding analysis did not quantify the costs and benefits of agroecological eco-technologies relative to their conventional alternatives; instead the relevant resource-related concerns and generalized trends were mapped based on the authors' own experiences in the field of sustainable agriculture development. In future analyses, it will be critical to establish baseline or reference values for key metrics. This framing would emphasize that progress toward sustainability is continuous and avoid directly pitting producer against producer.

Of critical importance to renewable resource development is the fact that all Vermont-appropriate eco-technologies are reliant upon nonrenewable power grids and power plants as well as petrochemical processing and other manufacturing sectors. Certainly, in the case of a broad array of specialized industrial products such as lubricants, structural reinforcement, and specialized engines and bioreactors, Vermont must rely on trading domestic surpluses of its own specialized products. In the ideal case, Vermont's trade partners would be progressing toward energy and material self-sufficiency simultaneously, based on their own internal resources and trade networks. Achieving this in practice would likely require substantial social, political, and economic change informed by well-designed simulation models.

The preceding analysis provides a foundation upon which to model trade-offs across ecological and economic dimensions using process-based computer simulation (Urban et al., 2010). As an example, AD, Vermont's principle farm-based energy generating eco-technology, has key nonrenewable inputs and potential ecological impacts associated with confinement-based animal husbandry and crop production systems. A Vermont-specific baseline simulation model of these and other resource flows resulting from AD eco-technology will contribute substantially to ensuring AD's appropriate application within the state and its unique array of CFEA contexts. Such quantitative modeling research will be performed after further iteration of the methods presented in this chapter.

ACKNOWLEDGMENTS

This chapter is made possible by contributions of resources and funding from the Gund Institute for Ecological Economics and the Rubenstein School of Environment and Natural Resources at the University of Vermont, as well as the National Science Foundation's Graduate Research Fellowship Program.

REFERENCES

Andersen, O., 2002. Transport of fish from Norway: energy analysis using industrial ecology as the framework. Journal of Cleaner Production 10 (6), 581–588.

Chambers, R., 1994. The origins and practice of participatory rural appraisal. World Development 22 (7), 953–969.

Chertow, M.R., 2007. 'Uncovering' industrial symbiosis. Journal of Industrial Ecology 11 (1), 11–30.

Cleveland, C.J., 1995. Resource degradation, technical change, and the productivity of energy use in US Agriculture. Ecological Economics 13 (3), 185–201.

Conway, G.R., Pretty, J.N., 2009. Unwelcome Harvest: Agriculture and Pollution. Taylor & Francis.

Demirbas, A.H., Demirbas, I., 2007. Importance of rural bioenergy for developing countries. Energy Conversion and Management 48, 2386–2398.

Devadas, V., 2001. Planning for rural energy system: Parts I, II and III. Renewable and Sustainable Energy Reviews 5 (3), 203–297.

Devendra, C., Leng, R.A., 2011. Feed resources for animals in Asia: issues, strategies for use, intensification and integration for increased productivity. Asian-Australasian Journal of Animal Sciences 24 (3), 303–321.

Edey, A., 1998. Solviva. Trailblazer Press, Vineyard Haven, MA.

Garza, E., 2011. Reacting to the Peak: Multiple Criteria Analysis and Energy Return on Energy Invested in Energy Decision Making. Natural Resources. University of Vermont, Burlington, VT.

Green, C., Byrne, K.A., 2004. Biomass: impact on carbon cycle and greenhouse gas emissions, pp. 223–236. In: Cutler, J.C. (Ed.), Encyclopedia of Energy. Elsevier, New York City, NY, USA.

Godfray, H.C.J., Beddington, J.R., Crute, I.R., Haddad, L., Lawrence, D., Muir, J.F., Pretty, J., Robinson, S., Thomas, S.M., Toulmin, C., 2010. Food security: the challenge of feeding 9 billion people. Science 327 (5967), 812–818.

Goodrich, P.R., 2005. Anaerobic Digester Systems for Mid-Sized Dairy Farms. The Minnesota Project.

Graedel, T.E., Allenby, B.R., 2003. Industrial Ecology. Pearson Education, Upper Saddle River, NJ.

Kahler, E., Perkins, K., Sawyer, S., Pipino, H., St Onge, J., 2011. Farm to Plate Strategic Plan Executive Summary. V. S. J. Fund, Montpelier, VT.

Malhotra, P., Neudoerffer, R.C., Dutta, S., 2004. A participatory process for designing cooking energy programmes with women. Biomass and Bioenergy 26 (2), 147–169.

Manomet Center for Conservation Sciences, 2010. Massachusetts Biomass Sustainability and Carbon Policy Study: Report to the Commonwealth of Massachusetts Department of Energy Resources. T. Walker, Brunswick, Maine. Natural Capital Initiative Report NCI-2010-03.

Mears, D.R., 2007. Energy use in production of food, feed and fiber. In: Fleisher, D.H., Ting, K.C., Rodriguez, L.F. (Eds.), System Analysis and Modeling in Food and Agriculture. UNESCO/EoLSS, Oxford, UK. Encyclopedia of Life Support Systems.

Peters, G.M., Rowley, H.V., Wiedemann, S., Tucker, R., Short, M.D., Schulz, M., 2010. Red meat production in Australia: life cycle assessment and comparison with overseas studies. Environmental Science and Technology 44 (4), 1327–1332.

Pimentel, D., 2009. Energy inputs in food crop production in developing and developed nations. Energies 2 (1), 1–24.

Pimentel, D., Pimentel, M., 1996. Food, Energy, and Society. University Press of Colorado, Niwot, CO.

Pizzigallo, A.C.I., Granai, C., Borsa, S., 2008. The joint use of LCA and energy evaluation for the analysis of two Italian wine farms. Journal of Environmental Management 86 (2), 396–406.

Porter, J., Costanza, R., Sandhu, H., Sigsgaard, L., Wratten, S., 2009. The value of producing food, energy, and ecosystem services within an agro-ecosystem. Ambio 38 (4), 186–193.

Royal Society of London, 2009. Reaping the Benefits: Science and the Sustainable Intensification of Global Agriculture. Royal Society, London.

Salerno, F., Viviano, G., Thakuri, S., Flury, B., Maskey, R.K., Khanal, S.N., Bhuju, D., Carrer, M., Bhochhibhoya, S., Melis, T., Giannino, F., Staiano, A., Carteni, F., Mazzoleni, S., Cogo, A., Sapkota, A., Shrestha, S., Pandey, R.K., Manfredi, E.C., 2010. Energy, forest, and indoor air pollution models for Sagarmatha National Park and buffer zone, Nepal implementation of a participatory modeling framework. Mountain Research and Development 30 (2), 113–126.

Schattman, R., 2009. Sustainability Indicators in the Vermont-Regional Food System. Natural Resources (MS). University of Vermont, Burlington, VT.

Smil, V., 1994. Energy in World History. Westview Press, Boulder, CO.

Spring Hill Solutions, 2008. Vermont 25 × '25 Initiative: Preliminary Findings and Goals. Vermont Department of Public Service.

Urban, R.A., Bakshi, B.R., Grubb, G.F., Baral, A., Mitsch, W.J., 2010. Towards sustainability of engineered processes: designing self-reliant networks of technological-ecological systems. Computers and Chemical Engineering 34 (9), 1413–1420.

Vayssieres, J., Lecomte, P., Guerrin, F., Nidumolu, U.B., 2007. Modelling farmers' action: decision rules capture methodology and formalisation structure: a case of biomass flow operations in dairy farms of a tropical island. Animal 1 (5), 716–733.

Wezel, A., Bellon, S., Dore, T., Francis, C., Vallod, D., David, C., 2009. Agroecology as a science, a movement and a practice. A review. Agronomy for Sustainable Development 29 (4), 503–515.

QUIZZES

Quizzes and Self-Test Questions

INTRODUCTION TO BIOENERGY

(**Note to reader taking the self-test:** Refer to the chapter titled, "Introduction to Bioenergy" in this book authored by Carol Williams et al., and the BioEN1 program described in the acknowledgements below)

SECTION 1 QUESTIONS

1. Why is bioenergy considered a renewable form of energy?
   a. Feedstocks for bioenergy production are derived from materials formed thousands of years ago through geological processes
   b. Feedstocks for bioenergy production are derived from materials formed from recently living biological material, which can be harvested and replanted
   c. The energy contained in the biomass is solar energy captured through natural processes of photosynthesis
   d. Both a and c
   e. Both b and c
2. What are the main sources of biomass for bioenergy?
   a. Aquatic invertebrates, fungi, and bacteria
   b. Corn grain, corn stover, and switchgrass
   c. Agriculture, forests, and waste
3. What are the three types of conversion technology?
   a. Thermal, chemical, and biochemical
   b. Combustion, pyrolysis, and gasification
   c. Chemical agents, enzymes, and microorganisms
4. True or false: Most of all current ethanol production in the United States uses advanced processes to derive sugars from cellulosic biomass crops to ferment into ethanol.
   a. True
   b. False
5. What are the drivers of bioenergy development?
   a. Policy initiatives that promote research and development
   b. Concern about United States' energy independence
   c. Environmental and climate concerns associated with climate change due partially to use of nonrenewable energy resources
   d. Potential for new means of economic development
   e. All of the above
6. What role does policy have in bioenergy development?
   a. Providing incentives for farmers to grow bioenergy crops, in the absence of a mature and reliable market
   b. Providing incentives to industry to develop commercial-scale bioenergy facilities, in the absence of tested technologies
   c. Providing assistance in developing infrastructure needed to support a nascent industry

    d. Providing funding for research and development of potential technologies, in that commercial enterprises cannot commit extensive financial resources to technologies and products that may not be commercially viable for many years

    e. All of the above

ANSWERS TO SECTION 1 QUESTIONS

1. Answer: e. Bioenergy feedstocks are recently living biological materials; the energy for their growth comes from the sun, and if they are harvested at sustainable rates, bioenergy feedstocks are renewable.

2. Answer: c. Biomass is derived from multiple sources; most of these come from agriculture forests and waste. Biomass can also come from algaculture, that is, the production of microalgae. Option b lists the three main sources of biomass from agriculture alone, but currently most feedstock material used for energy production is woody biomass.

3. Answer: a. Thermal conversion technologies use heat to convert biomass into other forms. Chemical conversion technologies involve the use of chemical agents to convert biomass into liquid fuels. Biochemical conversion technologies use enzymes of bacteria or other microorganisms to break down biomass.

4. Answer: False. Nearly all US ethanol production utilizes corn grain in fermentation processes (first-generation biofuels, which are derived from sugars or starches). Most research is aimed at identifying cost-effective and sustainable means to convert cellulosic plant material into biofuels.

5. Answer: e. All mentioned factors contribute toward efforts to develop bioenergy resources and technologies.

6. Answer: e. Policy initiatives can provide incentives to entrepreneurs that spark development of a new industry.

SECTION 2 QUESTIONS

  1. What is a first-generation biofuel?

    a. Ethanol derived from cellulosic materials, such as switchgrass or *Miscanthus*

    b. Ethanol derived from the fermentation of sugars, derived from crops such as wheat, corn, or sugarcane

    c. Pretreatment of cellulosic biomass to break apart and release sugars for fermentation to ethanol

  2. What are second-generation biofuels?

    a. Ethanol derived from cellulosic materials, such as switchgrass or *Miscanthus*

    b. Ethanol derived from the fermentation of sugar and starch crops, such as wheat, corn, or sugarcane

    c. Pretreatment of cellulosic biomass to break apart and release sugars for fermentation to ethanol

  3. True or false: The technical parameters for biopower (electricity generated from combustion of biomass) are well developed.

    a. True

    b. False

4. How are liquid and gas biofuels produced?
   a. Fermentation (to produce bioethanol)
   b. Gasification (to produce syngas)
   c. Pyrolysis (to produce syngas and char)
   d. Torrefaction (to produce biocoal)
   e. Transesterification (to produce biodiesel)
   f. Anaerobic digestion (to produce pipeline-quality gas)
   g. All of the above

5. Why is it important to develop combined heat and power (CHP) units?
   a. Federal policy mandates it
   b. Large-scale electricity generation facilities have invested heavily in it
   c. CHP is many times more efficient than combustion for electricity generation alone

6. What are potential reasons that an electric utility might opt to burn woody biomass in a power plant?
   a. To meet renewable energy standards
   b. To reduce amount of materials that would have to be transported to the boiler
   c. To cut back on pollution from coal burning
   d. To improve boiler efficiency
   e. a and c
   f. b and d

7. What are the main sources of woody biomass?
   a. Forests
   b. Agriculture
   c. Sawmill waste
   d. a and b
   e. a and c

8. What are the main types of agriculture-based energy crops?
   a. Corn, switchgrass, and *Miscanthus*
   b. Soybean, sunflower, and canola
   c. Willow and poplar
   d. Perennial lignocellulosic crops, lignocellulosic residues, sugar and starch crops, and oil-producing crops
   e. Corn stover, rice chaff, oat husks

9. What are examples of waste-based feedstocks?
   a. Sawdust
   b. Landfill gases
   c. Construction wastes
   d. Compostable materials
   e. Manure
   f. All of the above

10. True or false: Bioenergy by-products are insignificant in determining the profitability of a bioenergy enterprise.
   a. True
   b. False

ANSWERS TO SECTION 2 QUESTIONS

1. Answer: b. Nearly all current ethanol production in the United States uses corn grain for ethanol. In Brazil, sugarcane is the primary feedstock. In other tropical countries, oil seeds are used for the manufacture of biodiesel. These are all first-generation biofuels.
2. Answer: a. Second-generation biofuels are derived from cellulosic plant materials. Much of the current research and development work is aimed at second-generation biofuels.
3. Answer: False. There are technical challenges in burning biomass, including feedstock quality, boiler chemistry, ash deposition, and ash disposal.
4. Answer: g. All of the above.
5. Answer: c. CHP is the simultaneous production of electricity and heat from a single fuel source. Overall energy system efficiency is relatively high in CHP facilities because they make use of waste heat that would otherwise be lost to the environment.
6. Answer: e. Burning woody biomass can help a utility to meet renewable energy standards and cuts back on pollution from coal burning. However, it would increase the amount of materials that would have to be transported, and would decrease boiler efficiency slightly.
7. Answer: d. Forests are the primary source of woody biomass. Woody material is also sourced from agriculture, as fast-growing short-rotation woody crops are an agricultural feedstock. Sawmill waste is not a significant source of woody biomass, and is typically treated as a waste-based feedstock.
8. Answer: d. These four categories represent types of agriculture-based energy crops, whereas the other choices are plants that fall within a single category.
9. Answer: f. The use and conversion of waste materials into bioenergy is referred to as waste to energy, and diverts material from landfills.
10. Answer: False. Bioenergy by-products have economic use, and can contribute significantly toward the profitability of a bioenergy enterprise.

SECTION 3 QUESTIONS

1. What is life cycle assessment when applied to biofuels?
   a. An evaluation of the energy balance in production and utilization of the biofuel
   b. An accounting of all greenhouse gas (GHG) emissions generated in the production of the biofuel, including both agricultural and industrial processes
   c. An accounting of human health impacts of production and utilization of the biofuel
   d. All of the above
2. What are the food vs fuel concerns with respect to bioenergy?
   a. Diversion of food crops for use in bioenergy
   b. Diversion of land that could be utilized for food crops toward cultivation of bioenergy crops
   c. Concern that small-scale farmers in developing countries will produce and sell bioenergy commodity crops to the detriment of utilizing land, labor, and capital for the production of food crops
   d. Concern that policy initiatives will incentivize research and development of energy crops to the deficit of research and development initiatives that increase the productivity (and other attributes) of food crops
   e. All of the above

3. True or false: Bioenergy has the potential to be carbon neutral.
   a. True
   b. False
4. Why consider using marginal lands for bioenergy crops?
   a. Cultivation of agricultural row crops on marginal lands can lead to high rates of soil erosion; cultivation of perennial grass crops on these lands would minimize environmental degradation
   b. Cultivation of perennial crops on marginal lands would not compete with land for food crops
   c. Conversion of steep or wet land currently in food crop production to less intensive bioenergy crops has the potential to generate more ecosystem services
   d. All of the above
5. What is the relationship between ecosystem service trade-offs and bioenergy crops?
   a. Bioenergy crop production will reduce the ability of natural ecosystems to regulate natural processes
   b. Use of land to grow bioenergy crops will reduce the net productive output of that land and the ecosystem associated with it
   c. Bioenergy crop production will impede regulating and supportive functions of ecosystems
   d. Use of land for one function may alter the ecosystem such that it detracts from other functions or services provided by that ecosystem

ANSWERS TO SECTION 3 QUESTIONS

1. Answer: d. Life cycle assessment is an approach to evaluate the direct and indirect effects associated with a product or activity on human health and the environment in a cradle-to-grave analysis of benefits and costs.
2. Answer: e. The spike in food prices seen in 2007–2008, and again in 2011, were and are due to many interrelated factors, including biofuel policies that put pressure on land use and crop production systems. Nevertheless, the "food vs fuel" concerns that have been voiced, and continue to be voiced, must be addressed.
3. Answer: True. Bioenergy has the potential to be carbon neutral by balancing the amount of carbon released in the production and use of bioenergy products with an equivalent amount put into and stored in soils, plant and animal tissues, or other material such as the ocean floor. However, there is no assurance that bioenergy will be carbon neutral, as much depends on agricultural and industrial production practices.
4. Answer: d. However, some researchers caution that conversion of marginal lands to more intensively managed bioenergy cropping systems could lead to permanent land degradation and/or increases in net GHGs.
5. Answer: d. Options a, b, and c are possible outcomes, but not necessary outcomes. Option d is a necessary outcome, and true whether the land is utilized for production of food crops or energy crops.

SECTION 4 QUESTIONS

1. What factors affect the land that is available for biomass production?
   a. Policy incentives
   b. Availability of markets for biomass

    c. Opportunity costs—other uses that the farmer might have for the land
    d. Land owner experience and knowledge of production practices
    e. All of the above

2. Farmers utilizing crop residue as a bioenergy crop might do the following to assure a sustainable production system:
    a. Remove as much residue as the market can handle
    b. Consider how residue removal will impact the land based on the geophysical characteristics of the specific location
    c. Always apply extra N fertilizer to subsequent crops to offset the impact of residue removal
    d. Utilize scientific models (such as RUSLE2[1]) to estimate the likely impact of residue removal for specific conditions
    e. Both a and c
    f. Both b and d

3. True or false: Biotechnology is a tool being used to advance the development of bioenergy crops and a bioenergy industry?
    a. True
    b. False

4. Which of the following physical infrastructure elements need to be in place for expansion and growth of the bioenergy industry?
    a. Commercially viable conversion facilities
    b. Farm and forest harvest equipment
    c. Transportation networks
    d. Means to store quantities of biomass
    e. Preprocessing facilities
    f. All of the above

5. What is meant by "intangible infrastructure"?
    a. Software programs for trading feedstock in the futures market
    b. Enzymatic processes for cellulosic feedstock conversion
    c. Public works projects, such as road construction, that support the bioenergy industry
    d. Public policy, regulatory structures, quality standards, and marketing institutions

6. How can storage cost of biomass materials be reduced?
    a. Allow the materials to decompose prior to transporting them to the storage facility
    b. Pretreat with chemicals to convert materials into a smaller form
    c. Compress biomass materials into pellets or bricks to increase bulk density (densification)
    d. Locate all biomass combustion or conversion facilities in areas where land values are low

7. True or false: A "fuelshed" is the area of feedstock supply surrounding a bioenergy conversion facility?
    a. True
    b. False

8. Why is it more costly to convert cellulosic material into ethanol compared to use of grain for ethanol?

---

[1]RUSLE2 is an advanced, user-friendly software model that predicts long-term, average-annual erosion by water. It runs under Windows, and can be used for a broad range of farming, conservation, mining, construction, and forestry sites. http://bioengr.ag.utk.edu/rusle2/

a. On a per weight basis, cellulosic feedstock materials take up greater volume of storage and transportation space than grains

b. Pretreatment of cellulosic biomass is required to break apart hemicellulose into sugars for fermentation to ethanol

c. With thermochemical technologies, additional processing of product is necessary in order to achieve a drop-in transportation fuel

d. All of the above

ANSWERS TO SECTION 4 QUESTIONS

1. Answer: e. Producers make decisions about whether or not to grow bioenergy crops based on complex decision making taking into account an array of factors.
2. Answer: f. Residue removal has been associated with increased soil erosion, soil compaction, loss of soil carbon, and consequent negative impact on soil quality and associated waterways. Farmers should exercise caution, using sustainable harvest guidelines, in the removal of residue from farmland.
3. Answer: True. Biotechnology is currently being aimed both at addressing feedstock limitations (e.g., higher yield, faster growing crops) and at speeding up components of process technologies (e.g., better enzymes, more efficient organisms for fermentation of ethanol).
4. Answer: f. All of the above elements will probably need to be in place for the bioenergy industry to succeed.
5. Answer: d. Public policy, in particular, can drive the development of a new industry. Regulatory structures, quality standards, and marketing institutions are all necessary to meet mandated renewable energy goals and to develop a mature bioenergy industry.
6. Answer: c. Densification reduces storage and transportation costs. Biomass materials must be stored to reduce incidence of decomposition. Pretreatment is used to break about cellulose so that it can be converted to sugars or starches and then ethanol (not as a means to reduce bulk). Biomass combustion facilities should be colocated, to the extent possible, in proximity to users. Ethanol conversion facilities are typically located in regions where biomass feedstocks are plentiful in order to reduce transportation costs.
7. Answer: True. The cost of transportation of feedstocks (particularly cellulosic materials) can contribute significantly to the costs of cellulosic ethanol production. To limit costs, most conversion facilities will locate where required feedstock is abundant.
8. Answer: d. The technologies for converting cellulosic materials into ethanol are being developed, but as yet are costly. As such, industry and government incentive programs are working to move the technology from pilot to commercial scales.

ACKNOWLEDGMENTS

This material (sections 1-4) is based upon online curriculum for Based on BioEN1, Bioenergy Crop Production & Harvesting Bioenergy & Sustainability Course Series, developed with support from the National Institute of Food and Agriculture, US Department of Agriculture, under Agreement No. WISN-2007-03790. BioEN1 was led by Carol Williams (Agricultural Ecosystems Research Group, University of Wisconsin–Madison). Online: http://blogs.extension.org/bioen1/

WOOD ENERGY

1. Harvesting trees can contribute to sustainable forest management.
   a. True
   b. False
2. What is the closest number of 16-inch-diameter trees that would be required to heat an average-efficiency, modest-sized, single-family-sized house with a wood stove in Vermont-like weather? (Hint: the cords of wood needed per winter, and the cord of wood provided by a tree of 16 inch diameter)
   a. 1
   b. 3
   c. 12
   d. 32
3. Forests that are harvested contain woods roads. Water bars are road structures that protect the forest by:
   a. Moving surface water off the road surface
   b. Allowing surface water to gradually infiltrate the road surface
   c. Providing native wildlife with a source of drinking water and social interaction
   d. Concentrating water flow into a road channel
   e. Keeping heavy logging machinery out of sensitive areas
4. Whole tree harvesting is currently the fastest, most efficient means to convert standing trees into chips for biomass
   a. True
   b. False
5. The term "coppicing" means
   a. Measuring the addition of ethanol to gasoline in "cups"
   b. The attempt of oil investors to prevent entrance of biofuels into the market
   c. A traditional method of woodland management
   d. Forecasting the future through simulation modeling

ANSWERS TO WOOD ENERGY SECTION

1. a. True
2. c. 12
3. a. Moving surface water off of the road surface
4. a. True
5. c. A traditional method of woodland management

QUIZZES AND SELF-TEST QUESTIONS

(**Note to reader taking the self-test:** Refer to the chapter titled, "Bioenergy Crops" in this book authored by Dennis Pennington, and the BioEN2 program described in acknowledgments below)

SECTION 1

1. What are some questions producers should ask before planting a biofuel crop?
   a. Is there a reliable market?
   b. What additional equipment or labor is needed?
   c. What is the cost of production?
   d. Will the yield and price be comparable to my current crop?
   e. All of the above
2. From Table 7.2 in chapter 7 (see note above) in the Cost–Benefit Analysis section, what is the most profitable (net return/A) biofuel crop priced at $60/ton?
   a. Corn + stover
   b. Switchgrass
   c. Native prairie
   d. *Miscanthus* spp. (cheap rhizomes)
3. Biomass resources in the United States vary by state and even within a state. Look at Figure 7.1 in Chapter 7, in the Analyzing Potential Markets section to identify the 1000 tons/year of biomass in your area (NREL—National Renewable Energy Lab map—see note above):
   _____ 1000 tons/year of biomass
4. What are the likely markets for biomass?
   a. Gasoline or diesel
   b. Chemical or enzymatic
   c. Methanol or butanol
   d. Thermal or pyrolysis
5. When determining the value of biomass delivered, weight in tons and moisture content is extremely important. The recommended biomass moisture content percent should be:
   a. <20%
   b. >20%
   c. >30%
   d. <30%
   e. a and d
6. Under the Biomass Crop Assistance Program (BCAP), what is the maximum CHST matching payment ($/dry ton) to producers delivering eligible biomass material? (CHST: Collection, Harvest, Storage and Transportation)
   a. $30
   b. $45
   c. $50
   d. $65
7. In the BCAP, what is the biomass crop establishment and annual payments contract length?
   a. 2 years (5 years for woody crops)
   b. 3 years (10 years for woody crops)
   c. 5 years (15 years for woody crops)
   d. 10 years (20 years for woody crops)
8. Ethanol coproduct streams include:

   a. Distillers wet grains
   b. Condensed solubles (syrup)
   c. Distillers dried grain with solubles
   d. All of the above
9. As a fuel for heat and power applications, rank the lowest to the highest greenhouse gas (GHG) emissions for the following fuels:
   a. Coal
   b. Natural gas
   c. Corn stover
10. Which biomass feedstocks below are NOT eligible for BCAP?
   a. Switchgrass
   b. *Miscanthus* spp.
   c. Algae
   d. None of the above

ANSWERS

1. Answer: e. All of the above. All are very good questions to ask.
2. Answer: d. *Miscanthus* spp. (cheap rhizomes) $166, (corn + stover, $126; switchgrass, $34; and native prairie, $25).
3. Answer: Varies with area of the United States.
4. Answer: a and d. From the Analyzing Potential Markets section.
5. Answer: a. Biomass should be less than 20% moisture to prevent mold and ensure longer storage life.
6. Answer: b. Qualifying producers will be eligible for up to $ 45/dry ton of biomass delivered to an approved biomass conversion facility.
7. Answer: c. Reimbursement for some establishment costs and annual contract payments for land use can be up to 5 years for crops and 15 years for woody crops.
8. Answer: d. All of the above.
9. Answer: Corn stover as a first-generation fuel source has the lowest GHG emissions. Rank, highest to lowest: coal, 3; natural gas, 2; corn stover, 1.
10. Answer: d. All of the crops listed above are eligible for BCAP.

SECTION 2

(Note that you will need to refer to the crop fact sheets to answer several of these questions—see note above).
1. _____ is essential for stand establishment for switchgrass and other perennial biomass grasses.
   a. Fertilizer
   b. Weed control
   c. Soybean as previous crop
   d. Properly tilled seedbed
2. *Miscanthus* is planted using which of the below methods? (Check all that apply.)

    a. Vegetative propagation with rhizomes
    b. Seeded with a drill
    c. Transplanted
    d. a and c above
3. Sorghum and tropical maize grow very large when planted in the midwest because:
    a. The region has highly fertile soils
    b. More sunlight means increased photosynthesis
    c. Day length does not initiate reproductive growth
    d. Cooler nights slow respiration, causing more growth
4. Woody biomass crops are harvested every _____ years
    a. 3
    b. 5
    c. 2
    d. 7
5. Corn grain yields a national average 152 bushels per acre. How much aboveground corn stover would this produce?
    a. 6.0 tons
    b. 2.5 tons
    c. 3.4 tons
    d. 4.2 tons
6. Which sugar crops can be grown in the northern United States?
    a. Sweet millet
    b. Sugar beets
    c. Sweet sorghum
    d. Sugarcane
    e. a and c
    f. b and c
7. How much biodiesel can be made from 1 gallon of canola oil?
    a. 1.4 gallons
    b. 2.3 gallons
    c. 0.75 gallons
    d. 1.0 gallons
8. Most US biodiesel is made from what oilseed crop?
    a. Soybean
    b. Canola
    c. Sunflower
    d. Corn
9. Rank these perennial biomass crops in order of highest to lowest yield potential in tons per acre per year.
    _____ Soybean
    _____ Canola
    _____ Sunflower
    _____ Corn

ANSWERS

1. Answer: b. Weed control during establishment is critical to have a good stand. No fertilizer is recommended during the establishment year. Previous crop is not a major issue. Switchgrass can be planted no till or in tilled seedbed.
2. Answer: d. *Miscanthus* does not have viable seeds.
3. Answer: c. The length of the day/night cycle does not initiate reproductive growth, causing plants to grow vegetatively for a long period of time.
4. Answer: a. Woody biomass crops like willow and hybrid poplar are harvested every 3–4 years.
5. Answer: d. Corn stover yield corresponds closely to corn grain yield (nearly 1:1). So 152 bushels at 56 lbs per bushel produces 4.2 tons of grain and about 4.2 tons of stover.
6. Answer: f. Sugar beet and sweet sorghum can be grown in temperate regions. Sugarcane is a subtropical to tropical crop and there is no such thing as sweet millet.
7. Answer d. 1 gallon of biodiesel can be made from 1 gallon of vegetable oil. 1 gallon of oil plus 10% methanol or ethanol and a catalyst yields 1 gallon of biodiesel and 10% glycerin.
8. Answer: a. Soybean is by far the most used oil for biodiesel. Its large-scale production and crushing facilities make it the oil available in the greatest quantities.
9. Answer: 3,4,2,1. *Miscanthus* has the highest potential with upward of 15 or more dry tons per acre. Corn is second with close to 14 tons per acre from a 250 bu/acre yield. Switchgrass can yield 5–7 tons per acre and hybrid poplar yields 4–6 tons per acre per year.

SECTION 3

1. What ecosystem services are provided by perennial biomass crops?
   a. Erosion control
   b. Reduced pesticide use
   c. Reduced fertilizer use
   d. All of the above
2. The idea of fossil energy ratio is:
   a. How much fossil fuel is used to produce a biofuel crop
   b. Fossil energy used to produce a crop/the energy in the biomass crop
   c. Diesel fuel used in farming/ethanol produced by the biomass
   d. Fossil energy used to produce a crop/the bioenergy derived from the crop
3. The renewable fuel standard calls for _____ gallons of biofuels by the year 2020.
   a. 20 billion gallons
   b. 36 billion gallons
   c. 50 billion gallons
   d. 30 billion gallons

ANSWERS

1. Answer: d. All of the above are ecosystem services provided by perennial biomass crops.
2. Answer: d. The fossil energy ratio for cellulosic ethanol would be the fossil energy used to produce the biomass/energy in the ethanol produced by the biomass. By some calculations

ethanol made from switchgrass has a fossil energy ratio of 5:1, which is very good compared to the estimate for corn ethanol, which is approximately 1.6:1.

3. Answer: b. 36 billion gallons, as follows: 15 from starch-based ethanol, 5 billion from advanced biofuels, and 16 billion from biomass.

ACKNOWLEDGMENTS

This material (sections 1-3 above) is based upon online curriculum for Based on BioEN2, Bioenergy Crop Production & Harvesting Bioenergy & Sustainability Course Series, developed with support from the National Institute of Food and Agriculture, US Department of Agriculture, under Agreement No. WISN-2007-03790. BioEN2 was led by Dennis Pennington (Bioenergy Educator, Michigan State University Extension). Online: http://blogs.extension.org/bioen2/

GASIFICATION

1. Heat required for the gasification reaction $H_2O + C$ is generated from the process or reaction that releases energy called:
   a. Endothermic reaction
   b. Exothermic reaction
   c. Nothing—it is not needed
2. Pyrolysis of carbon-rich material involves it being heated in the presence of:
   a. Chemically inert atmospheres (no oxygen)
   b. Oxygen
   c. Air
3. *Syngas*, or synthesis gas, is a fuel gas mixture consisting primarily of:
   a. Hydrogen, carbon monoxide, and very often some carbon dioxide
   b. Oxygen, nitrogen, and some carbon dioxide
   c. Both of the above
4. The reaction of carbon (C) with water can only take place at high temperatures ($>700\,°C$).
   a. True
   b. False
5. Gas combustion produces minimal or no "ash," reducing concerns about fouling heat transfer surfaces and contamination of materials being heated by direct contact with the products of combustion
   a. True
   b. False
6. Typical medium-BTU[2] gas mixtures are mixtures of:
   a. CO, H_2, and N_2 (carbon monoxide, hydrogen, and nitrogen)
   b. Mainly methane (CH_4) with small amounts of H_2
   c. CO and H_2
7. In a moving-bed gasifier:
   a. The gasifier moves

[2]BTU: British Thermal Unit

b. The bed moves

c. Both move

8. The basic technology for gasification of carbon-rich materials was developed in:
    a. The early nineteenth century
    b. World War II
    c. World War I

9. Low-BTU gas mixtures are always used at the same location as the gasifier because:
    a. They are difficult to store
    b. They are cheap
    c. The cost of compressing the mixture into a long-distant pipeline is too expensive

10. Great care has to taken in handling carbon monoxide because:
    a. It is expensive
    b. It is a colorless, odorless, poisonous gas
    c. It has a bad odor

ANSWERS

1. b. Exothermic reaction
2. a. Chemically inert atmospheres (no oxygen)
3. a. Hydrogen, carbon monoxide, and very often some carbon dioxide
4. a. True
5. a. True
6. c. CO and H_2
7. b. The bed moves
8. a. The early nineteenth century
9. c. The cost of compressing the mixture into a long-distant pipeline is too expensive
10. b. It is a colorless, odorless, poisonous gas

BIOGAS, ANAEROBIC DIGESTION

(**Note to the reader taking the self-test:** Refer to the chapter titled, "Bioenergy and Anaerobic Digestion" in this book authored by M. Charles Gould, and the ANDIG program described in acknowledgments below)

SECTION 1 QUESTIONS

1. What are examples of complex organic matter?
    a. Carbohydrates, proteins, and fats
    b. Sugars and amino acids
    c. Fatty acids
    d. Methane

2. Anaerobic digestion will not occur if this major element is present?
    a. Carbon dioxide
    b. Oxygen
    c. Calcium chloride

    d. Ammonium sulfate

3. Biogas has a typical carbon dioxide content of what percent?

    a. Over 60%

    b. 50–60%

    c. 40–50%

    d. 30–40%

4. What are the steps of anaerobic digestion (in order)?

    a. Methanogenesis, hydrolysis, acidogenesis, acetogenesis

    b. Hydrolysis, acidogenesis, acetogenesis, methanogenesis

    c. Acidogenesis, acetogenesis, methanogenesis, hydrolysis

    d. Acetogenesis, methanogenesis, hydrolysis, acidogenesis

5. What is the typical methane composition of biogas?

    a. Less than 35%

    b. 35–45%

    c. 45–65%

    d. More than 80%

SECTION 1 ANSWERS

1. Answer: a. Carbohydrates, proteins and fats

2. Answer: b. Oxygen

3. Answer: c. 40–50%

4. Answer: d. Acetogenesis, methanogenesis, hydrolysis, acidogenesis

5. Answer: c. 45–65%

SECTION 2 QUESTIONS

1. A digester goes "sour" when fermenting bacteria are producing more organic acids than methanogens can convert into methane, causing the pH of the slurry to drop. When the pH drops below 6.8, methanogens die off and methane production ceases.

    a. True

    b. False

2. Anaerobic conditions occur:

    a. When oxygen is present

    b. When specific fermenting bacteria are present in the slurry

    c. In the absence of oxygen

    d. When the slurry is at pH 4.0 or higher

3. Fermenting bacteria and methanogens do not need to work together to produce biogas.

    a. True

    b. False

4. Methanogens need organic acids produced by fermenting bacteria to generate methane.

    a. True

    b. False

5. The primary conversion route of complex organic compounds to methane formation is: complex organics → acidic acid › methane.
   a. True
   b. False
6. A feedstock is any organic material with biogas production potential.
   a. True
   b. False
7. Anaerobic digestion increases odors and pathogen levels.
   a. True
   b. False
8. Volatile solids are the portion of the total solids (i.e., organic matter) that can be converted to gas.
   a. True
   b. False
9. Once a digester is "sour," the best option is to:
   a. Inoculate the slurry with enzymes to jump start biogas production
   b. Empty the digester and start over with fresh feedstocks and new digestate
   c. stop feeding the digester and stir the slurry for a day
   d. Infuse biogas into the slurry to feed the methanogens and restart biogas production
10. Stabilization is:
   a. A biological process that, over time, makes the slurry less odorous and reduces pathogens that pose a health risk
   b. Shoring up the walls of a digester so that they do not fall over
   c. A necessary step to maximize the production of biogas
   d. Both a and c

SECTION 2: ANSWERS

1. Answer: a. True
2. Answer: c. In the absence of oxygen
3. Answer: b. False
4. Answer: a. True
5. Answer: a. True
6. Answer: a. True
7. Answer: b. False
8. Answer: a. True
9. Answer: b: Empty the digester and start over with fresh feedstocks and new digestate.
10. Answer: d: Both a and c

SECTION 3: QUESTIONS

1. What is the typical methane composition of biogas?
   a. Less than 35%
   b. 35–45%

 c. 45–65%

 d. More than 80%

2. How long does it normally take to achieve steady-state gas production from an anaerobic digester?

 a. Less than 4 weeks

 b. 4–8 weeks

 c. 2–3 months

 d. At least 6 months

3. Purging an anaerobic digester with an inert, nonoxygen gas will reduce the amount of oxygen that goes into solution and gives the anaerobic microorganism an opportunity to reproduce faster.

 a. True

 b. False

4. How do you start biogas production in a digester?

 a. Fill the digester half full of seed sludge and meter in the feedstock until the digester is at capacity

 b. Fill the digester with manure until it is at capacity

 c. Fill the digester with water, bring the water to operational temperature, and then slowly add feedstock

 d. Inject a blend of 75% food waste, 25% manure, and 15 pounds of yeast into the digester

5. What is the advantage of adding "seed sludge or effluent" for digester start-up?

 a. There is no advantage

 b. Biogas production in the winter

 c. Aids in the breakdown of cellulosic materials

 d. Seed sludge or effluent will contain all the fermenting bacteria to ensure a quicker start-up for biogas production

6. Starting a digester during the winter is easy to do because methanogens are cold-tolerant organisms and will produce copious amounts of biogas when temperatures are below freezing.

 a. True

 b. False

7. The purpose of operation procedures is to enable a digester to operate on a steady-state basis.

 a. True

 b. False

8. Operational procedures are not designed to prevent possible problems and improve feedstock digestion.

 a. True

 b. False

9. What are the reasons why digesters fail?

 a. Bad design and/or installation

 b. Selection of poor equipment and materials

 c. Poor farm management

 d. Introducing products into the digester that are toxic to anaerobic microorganisms

 e. All of the above

10. Why have failure rates been reduced in recent years?
    a. Improved system design
    b. Better construction practices
    c. Increased number of qualified companies to develop anaerobic digestion projects
    d. All of the above
11. An engineering review should be conducted to expose any technology failures and validate the design plan with the specific materials for the digester under the specific farm conditions.
    a. True
    b. False
12. Farmers who are considering installing a digester on their farm should give due diligence in selecting a technology provider to make sure the entity is reputable and uses proven digester technology.
    a. True
    b. False
13. What toxic products can kill anaerobic microorganisms if they are introduced into the digester?
    a. Rumensin and products like it
    b. Pesticides
    c. Cattle foot bath
    d. All of the above
14. Asphyxiation, explosion, burns, electrical shock, and falls are human health risks associated with working around an anaerobic digester and associated equipment.
    a. True
    b. False
15. _____ is an inflammable, colorless, highly poisonous gas.
    a. Sodium chloride
    b. Copper sulfate
    c. Hydrogen sulfide
    d. Carbon dioxide
16. Hydrogen sulfide inhibits _____ transport in the blood, changing the red blood pigment to a brown or olive color.
    a. Calcium
    b. Oxygen
    c. Sodium
    d. Iron
17. Hydrogen sulfide has the characteristic odor of _____.
    a. Vinegar
    b. Sour milk
    c. Rose petals
    d. Rotten eggs
18. Exposure to low concentrations of hydrogen sulfide causes:
    a. Irritation of the mucous membranes (including the eyes)
    b. Nausea, vomiting, and difficulty in breathing
    c. Cyanosis (discoloration of the skin)
    d. Delirium and cramps, then respiratory paralysis and cardiac arrest

e. All of the above

19. Improper training on confined space entry is a leading cause of accidents related to anaerobic digestion systems.
    a. True
    b. False

20. When entering a confined area, it is not necessary to have another person observing outside the confined area.
    a. True
    b. False

21. _____, the major component of biogas, is odorless, colorless, and difficult to detect. It is highly explosive if allowed to come into contact with atmospheric air. It is lighter than air and can collect at the top of confined areas.
    a. Hydrogen sulfide
    b. Hydrogen peroxide
    c. Methane
    d. Sulfur dioxide

22. Clear warnings must be placed in the respective parts of the facility. Examples include:
    a. Fences around pits
    b. Signs
    c. Both a and b
    d. Equipment operation training

23. The self-assessment tool provides guidance for process and job evaluation with suggestions based on typical potential hazards for farm digester systems and their associated preventive measures.
    a. True
    b. False

SECTION 3 ANSWERS

1. Answer: a. Carbohydrates, proteins, and fats
2. Answer: b. Sugars and amino acids
3. Answer: c. 40–50%
4. Answer: d. Acetogenesis, methanogenesis, hydrolysis, acidogenesis
5. Answer: c. 45%–65%
6. Answer: b. 4–8 weeks
7. Answer: a. True
8. Answer: c. Fill the digester with water, bring the water to operational temperature, and then slowly add feedstock.
9. Answer: d. Seed sludge or effluent will contain all the fermenting bacteria to ensure a quicker start-up for biogas production.
10. Answer: b. False
11. Answer: a. True
12. Answer: b. False
13. Answer: e. All of the above

14. Answer: d. All of the above
15. Answer: a. True
16. Answer: a. True
17. Answer: d. All of the above
18. Answer: a. True
19. Answer: c. Hydrogen sulfide
20. Answer: b. Oxygen
21. Answer: d. Rotten eggs
22. Answer: e. All of the above
23. Answer: a. True
24. Answer: b. False
25. Answer: c. Methane
26. Answer: c. Both a and b
27. Answer: a. True

ACKNOWLEDGMENTS

This material (sections 1-3 above) is based upon online curriculum for Based on ANDIG, Bioenergy Crop Production & Harvesting Bioenergy & Sustainability Course Series, developed with support from the National Institute of Food and Agriculture, US Department of Agriculture, under Agreement No. WISN-2007-03790. ANDIG1, ANDIG2, & ANDIG3 were led by M. Charles Gould (Agriculture and Agribusiness Institute, Michigan State University). Online: http://blogs.extension.org/andig1/modules/

Index

Note: Page numbers followed by "f" and "t" indicate figures and tables respectively.

Printed in the United States
By Bookmasters